NATO ASI Series

Advanced Science Institutes Series

A series presenting the results of activities sponsored by the NATO Science Committee, which aims at the dissemination of advanced scientific and technological knowledge, with a view to strengthening links between scientific communities.

The Series is published by an international board of publishers in conjunction with the NATO Scientific Affairs Division

A Life Sciences B Physics	Plenum Publishing Corporation London and New York
C Mathematical and Physical Sciences D Behavioural and Social Sciences E Applied Sciences	Kluwer Academic Publishers Dordrecht, Boston and London
F Computer and Systems Sciences G Ecological Sciences H Cell Biology I Global Environmental Change	Springer-Verlag Berlin Heidelberg New York London Paris Tokyo Hong Kong Barcelona Budapest

PARTNERSHIP SUB-SERIES

1. Disarmament Technologies	Kluwer Academic Publishers
2. Environment	Springer-Verlag
3. High Technology	Kluwer Academic Publishers
4. Science and Technology Policy	Kluwer Academic Publishers
5. Computer Networking	Kluwer Academic Publishers

The Partnership Sub-Series incorporates activities undertaken in collaboration with NATO's Cooperation Partners, the countries of the CIS and Central and Eastern Europe, in Priority Areas of concern to those countries.

NATO-PCO DATABASE

The electronic index to the NATO ASI Series provides full bibliographical references (with keywords and/or abstracts) to about 50000 contributions from international scientists published in all sections of the NATO ASI Series. Access to the NATO-PCO DATABASE compiled by the NATO Publication Coordination Office is possible in two ways:

- via online FILE 128 (NATO-PCO DATABASE) hosted by ESRIN, Via Galileo Galilei, I-00044 Frascati, Italy.
- via CD-ROM "NATO Science & Technology Disk" with user-friendly retrieval software in English, French and German (© WTV GmbH and DATAWARE Technologies Inc. 1992).

The CD-ROM can be ordered through any member of the Board of Publishers or through NATO-PCO, Overijse, Belgium.

Series I: Global Environmental Change, Vol. 40

Springer
Berlin
Heidelberg
New York
Barcelona
Budapest
Hong Kong
London
Milan
Paris
Santa Clara
Singapore
Tokyo

Forest Ecosystems, Forest Management and the Global Carbon Cycle

Edited by

Michael J. Apps
David T. Price
Natural Resources Canada
Canadian Forest Service, Northwest Region
5320-122 Street, Edmonton, Alberta T6H 3S5
Canada

Springer
Published in cooperation with NATO Scientific Affairs Division

Proceedings of the NATO Advanced Research Workshop "The Role of Global Forest Ecosystems and Forest Resource Management in the Global Cycle", held in Banff, Canada, September 12–16, 1994

Cataloging-in-Publication Data applied for

Die Deutsche Bibliothek - CIP-Einheitsaufnahme

Forest ecosystems, forest management, and the global carbon cycle : [proceedings of the NATO Advanced Research Workshop "The Role of Global Forest Ecosystems and Forest Resource Management in the Global Cycle", held in Banff, Canada, September 12 - 16, 1994] / ed. by Michael J. Apps ; David T. Price. Publ. in cooperation with NATO Scientific Affairs Division. - Berlin ; Heidelberg ; New York ; Barcelona ; Budapest ; Hong Kong ; London ; Milan ; Paris ; Santa Clara ; Singapore ; Tokyo : Springer, 1996

(NATO ASI series : Ser. I, Global environmental change ; Vol. 40)
ISBN 3-540-60684-X

NE: Apps, Michael J. [Hrsg.]; Advanced Research Workshop The Role of Global Forest Ecosystems and Forest Resource Management in the Global Cycle <1994, Banff, Alberta>; NATO: NATO ASI series / I

ISBN 3-540-60684-X Springer-Verlag Berlin Heidelberg New York

Printed in Germany

Typesetting: Camera ready by authors/editors
Printed on acid-free paper
SPIN: 10466232 31/3137 - 5 4 3 2 1 0

Foreword

Forests cover 28 percent of the Earth's land area and play a vital role in maintaining the health of the global environment. They influence the climate, prevent soil erosion, regulate stream flow, improve air and water quality, contribute significantly to biogeochemical cycles, and are a major and dynamic reservoir for carbon. Forests are also an important economic resource—providing homes, jobs, food and energy for hundreds of millions of people world-wide.

All countries face the challenge of determining the optimal balance between the environmental and economic roles in order to manage them sustainably. Global environmental change associated with increasing development and fossil fuel use, deforestation, and forest degradation continues to threaten the world's forests. Over the past 10 thousand years of human development, global forest area has been reduced by one third. The need to act now is urgent.

Canada, as the custodian of 10 percent of the world's forest area, is a strong promoter of international efforts aimed at sustainable management of global forests. For example, it took a lead role in 1992 at the United Nations Conference on Environment and Development, in Rio de Janeiro. That conference resulted in more than 150 countries signing the Framework Conventions on Climate Change and Biodiversity, and the Statement on Forest Principles. In 1993, Canada hosted the Montreal Seminar of Experts on Sustainable Development of Boreal and Temperate Forests, in conjunction with the Conference on Security and Cooperation in Europe. This led to significant efforts by the international community to develop environmental indicators for sustainable forest management.

The NATO Advanced Research Workshop on *The Role of Global Forest Ecosystems and Forest Resource Management in the Global Carbon Cycle*, was held September 12–16, 1994, in Banff, Canada. It is a further example of the strong support for and benefits of international co-operation in research and assessment directed at global environmental issues—in this case forest management and the global carbon cycle.

This book is a valuable outcome of that workshop. It assesses the state of knowledge concerning the role the world's forests and forest management play in the global carbon cycle and in mitigating the possible effects of climate change.

Nearly 40 leading scientists from around the world participated in its preparation, addressing a wide range of forest issues including:

- the current and future roles of forest ecosystems in global carbon management
- the development and potential directions of national and international forest management strategies, and
- the identification of critical scientific knowledge and data gaps.

The results will contribute significantly to the sustainable management of forests world-wide and to the scientific efforts of the Intergovernmental Panel on Climate Change. The Panel is the major international scientific body responsible for assessing the state of our knowledge of climate change in support of the Conference of the Parties to the Framework Convention on Climate Change.

International forums, such as the NATO Advanced Research Workshop, are critical to these ongoing discussions. They bring together and synthesise the best information available. This, in turn, helps governments in decision-making to meet both their domestic needs and international obligations.

The preservation and management of global forest resources are critical to human well-being and the health of the global environment. Forests play a critical role in the global carbon cycle. As they grow, they remove carbon dioxide from the atmosphere. Later, they release the stored carbon through decay, wildfires, harvesting and burning of logging debris. The build-up of greenhouse gases such as carbon dioxide is believed to be one of the primary causes of climate change. Global warming could significantly affect the world's forests, resulting in changes in the growth rates of trees and other vegetation, changes in forest species composition, shifts in forest boundaries, a period of increased forest fires, and changes in the patterns of insect and disease outbreaks.

If global climate change continues as projected and global warming becomes a reality, forest management strategies world-wide will have to be modified to adapt to the changing climate and to mitigate the impacts. Forests exist within the borders of a country, but their environmental role and the consequences of sustainable management or of mismanagement extend far beyond those boundaries. International forest management and policy decisions must be made quickly and should be based on the best information available if governments and scientists are to successfully deal with these vital issues.

I commend the scientists who have contributed to this book, and I look forward to even greater global co-operation as we strive for the sustainable management of our world's forests.

A. Anne McLellan
Minister of Natural Resources
Government of Canada

Preface

This book is the product of an international group of scientists with expertise in carbon cycling issues as they pertain to the world's forests and their management. The scientists were brought together at a NATO Advanced Research Workshop (ARW) held 12–18 September 1994, at the Banff Conference Centre in Banff, Alberta (in the heart of Canada's Rocky Mountains). The location, with opportunities for extended evening walks through wooded mountain trails, proved invaluable to the success of the workshop because it provided both aesthetic and recreational release from the frequently intense debate, while reminding the participants of the very forests whose future was under discussion.

The motivation for the ARW came from contemporary policy and scientific debates surrounding the observed increases in atmospheric concentrations of CO_2 and other radiatively active gases (the so-called "greenhouse gases" or GHGs)—and the possible role of terrestrial ecosystems in these changes. Underlying these debates, of course, is the potential for increased GHG concentrations to cause significant global warming.

The fact that some 150 countries had recently signed the UN Framework Convention on Climate Change (FCCC) (Anonymous 1992), with its objective of achieving:

> "*... stabilization of greenhouse gas concentrations in the atmosphere at a level that would prevent dangerous anthropogenic interference with the climate system*"

indicated that the policy community has been aware of the potential problem, although it clearly remains cautious in accepting what is often perceived to be uncertain scientific theory. This situation presents several challenges to the scientific community beyond the "ivory tower" label that is often publicly (and lamentably) applied to fundamental scientific research—for, it must be stressed, there are *fundamental scientific* questions to be resolved if the policy-driven research needs posed by the prospects of climate change are to be adequately addressed. But in addition, there are distinct applications of existing science which offer significant tests of our understanding of how the biosphere operates—and intense scientific debate often focuses on such tests. One of the 'hottest' of these debates—that of the *missing C sink*—has been unresolved for at least 20 years and has more recently directed attention to the role of global and regional forests. This issue is associated with our continuing inability to close the global carbon balance—to reconcile the known mass fluxes of carbon between the planetary surface and the atmosphere with the observed changes in atmospheric content—the law of mass conservation demands a balance. It was at the forefront of many of the workshop discussions and is considered in several of the Chapters of this book.

The missing C sink issue also has significant policy implications—to balance the global C budget, there must be a significant sink for atmospheric C *whose location and mechanism are as yet unknown.* The missing sink is thought to represent a biospheric *feedback* that presently acts to reduce the rate of atmospheric CO_2 increase and thereby slow any enhanced greenhouse effect. The absence of a clear understanding of this sink mechanism is troubling, however, because it leads to the possibility that the sink may one day vanish, or even become a net source of atmospheric CO_2, accelerating the possible onset of greenhouse warming. This

NATO ASI Series, Vol. I 40
Forest Ecosystems, Forest Management and the Global Carbon Cycle
Edited by Michael J. Apps and David T. Price

concern should be of particular importance to policy makers considering forest-related mitigation options, if global forests already play a major role in the missing sink mechanism.

The missing sink is not, however, the only global change issue with both policy-relevant implications and significant scientific challenges. How do we assess potential mitigation options? Just what is a 'good' forest carbon budget? What is the scope for mitigation options—i.e., to what degree can forests be used to reduce the increase of atmospheric CO_2? And what if global warming does occur—are our carbon budget forecasts still appropriate? These are all questions presenting major scientific challenges and were some of the underlying motivations for the Banff ARW.

Another significant motivation for the workshop was the acknowledged need to bridge the gap between the (physical and biological) science of global change as it relates to forest carbon cycles and the socio-economic dimensions of this change. There are two aspects to this linkage, of different dimensions. Firstly, one of the major causes of increased atmospheric GHG concentrations is the emission associated with deforestation—presently estimated to be as much as 20% of net annual anthropogenic CO_2 emissions to the atmosphere. Clearly, attempts to reduce such emissions *must* take into account the social and economic processes driving the underlying land-use changes. Secondly, since the signing of the UN FCCC, the enhanced use of forests and forest products as carbon reservoirs to mitigate atmospheric CO_2 increases has become an increasingly talked-about component of the policy-makers' tool-kit. Evaluating the technical feasibility of such options—the C offset potential—is a non-trivial scientific task involving development of methods for data scaling and synthesis as well as protocols for monitoring and assessment. (Another NATO workshop was held just two months prior to the Banff ARW to look at Carbon Sequestration in the Biosphere, Beran 1995). A yet bigger challenge is presented by acknowledging that the current utilisation and dependencies on existing land and forest resources present major socio-economic constraints that must be overcome, if implementation of C offsetting programs is to succeed. Must the apparently conflicting values, such as are associated with commercial forestry interests, forest ecosystem carbon storage, the needs of indigenous peoples, sustaining biodiversity, and multiple-use forest ecosystem management, be mutually exclusive? Although the specific answer will likely vary by site, forest, and local economic and social conditions, a question put to the workshop participants was whether, and how, socio-economic interests could be balanced with environmental ones.

There is little doubt that human activities are significantly changing the 'face'—the surface features—of planet Earth. Vivid satellite images have become readily available in the last decades to provide dramatic visual evidence for this. What is not so obvious is the extent to which these changes have altered ecosystem function at the global scale and, in human terms, begun to adversely affect the human habitat. What we 'see' from these satellite images are usually the most obvious and *direct* impacts of human activity—e.g., the changing pattern of land use in South America, the absence of forest cover across large parts of industrialised Europe, and extensive clearcut logging throughout much of the boreal zone. But are there also signals of other indirect changes associated with more subtle and *indirect* impacts of human activity such as changes in wildfire regimes in boreal forests, or in the productivity of temperate region forests due to increased nutrient inputs from airborne pollutants? If phenomena such as these are direct and indirect indicators of significant global change, what will be the impact on the forests, their carbon budgets, and the people who dwell within them? And, if greenhouse warming takes place, will the acceleration of changes in the global forests be an overwhelming or an insignificant problem? In the words of the UN FCCC applied to Earth's forests: is there a level which may be called "dangerous interference with the climate system"?

There are two *feedback* factors which give reason to be concerned if forest responses to global warming prove significant. Firstly there is the importance of forests for human social and economic well-being: changes in this resource would have significant impacts on the global economy and the livelihood of millions of people who derive their living from forest resources. Would even greater damage to forests result as environmental considerations take second place to the quest for maintaining, or improving, living standards? This could be called a socio-economic feedback to global change, and our ability to estimate it is poorly developed. Secondly, there is a postulated biophysical feedback role of forests in the global carbon cycle (as well as in other components of the physical climate system, particularly the hydrological cycle). That this feedback may be positive (i.e., act so as to *accelerate* the change), at least in the short term of 50 to 100 years, is a concern shared by many terrestrial ecologists. A recent book with the evocative title "Biotic feedbacks in the global climate system: will the warming feed the warming" (Woodwell and Mackenzie 1995) provides a timely general summary of these biotic feedback issues. The participants at the Banff ARW were well aware of these feedback factors and most of the Chapters of this book touch on at least one element of the problem.

The Chapters in this book were derived from presentations and discussions by the ARW participants. They were all extensively revised after the workshop, however, and in some cases, manuscripts were combined with others. All Chapters were subject to at least two external peer reviews (see List of Reviewers at the end of the volume), as well as to internal reviews by other workshop participants and the Editors.

The workshop was organised around four major themes: (1) Forest ecosystems and the global carbon cycle, (2) Sustainable forest land use: past, present and future, (3) Forest products, biomass and the carbon cycle, and (4) Human dimensions and socio-economic issues. All presentations were made in plenary however, leaving the working groups to focus on more specific aspects of their themes. Working groups reported back to the plenary midway through the workshop and again in the closing session. The relaxed and quiet nature of the surroundings provided significant additional opportunity for inter-group feedback and consultation.

The book is divided into four Parts reflecting the four dominant themes of the workshop. Each Part starts with an Editor's introduction, followed by the individual Chapters written by members of the working group and their colleagues. Each of the four Parts concludes with a summary paper that reflects the detailed discussions of the working group. Following the main parts is an Epilogue written by George M Woodwell which argues the need for global consensus on forests and human habitat, the motivations for which underline many of the concerns addressed during the ARW and throughout this book. Finally, a Glossary of terms relating to forests, carbon cycling and climate-biosphere interactions is presented at the end, together with the List of Reviewers and a list of the common acronyms occurring throughout the book. We hope that the book provides a valuable reference for those interested in understanding the role of forests in the global carbon cycle, and how human use and management of these natural resources can affect levels of CO_2 in the atmosphere. We also believe that the focus given by the workshop, its discussions and the preparation of this book has already made an important contribution to the ongoing debates within the IPCC and the various agencies seeking to implement the requirements of the UN Framework Convention on Climate Change.

ACKNOWLEDGEMENTS

We wish to thank all contributors to this volume, and the many external reviewers of these chapters, for the excellence of their work. We particularly thank the eight Co-chairs of the working groups for their leadership and the extra effort that they put into this volume.

We greatly appreciated the advice and guidance of our fellow members of the Science Steering Committee—Drs MGR Cannell, P Kauppi, AZ Shvidenko, WT Sommers, and RB Stewart—who played such an important role in structuring the NATO ARW from which this book evolved. We wish especially to recognise—and we are certain that all contributors will join us in this—Mrs Sue Nalder, for her tireless effort and attention to schedules; truly, without her persistence and cheerful attention to many details, this volume would have been neither as complete nor as timely.

The financial support of the NATO Scientific and Environmental Affairs Division (Brussels) and the additional support, in kind and in funds, of: Natural Resources Canada, Canadian Forest Service; Environment Canada, Strategic Science Branch; The Canadian Global Change Program of The Royal Society of Canada; The Institute of Terrestrial Ecology (NERC) UK; United States Department of Agriculture, Forest Service; and the International Institute for Applied Systems Analysis (IIASA) are also greatly appreciated. Apps also thanks IIASA for the hospitality and institutional support during the final preparation of this volume while he was a guest IIASA Research Scholar and both of us thank the Northern Forestry Centre (Canadian Forest Service, Edmonton) for supporting this publication by making facilities and staff freely available during its preparation.

Michael J. Apps
David T. Price

Edmonton, Alberta, July 1995.

REFERENCES

Anonymous (1992) Text of United Nations Framework Convention on Climate Change, UNEP/WMP Information Unit on Climate Change, Climate Change Secretariat, Palais des Nations, Geneva, Switzerland, 29 pp.

Woodwell GM, Mackenzie FT (eds) (1995) Biotic Feedbacks in the Global Climate System: Will the Warming Feed the Warming? Oxford University Press, New York and Oxford, 416 pp.

Beran M (ed) (1995) Carbon Sequestration in the Biosphere. NATO ASI Series, Subseries I "Global Environmental Change", Springer-Verlag, Heidelberg. (In press).

Table of contents

Workshop participants

Vladislav Alexeyev
VN Sukachev Institute for Forest Research, Siberian Branch of Russian Academy of Sciences, 660036 Krasnoyarsk, Russia, *Ph:* +7(3912)49-51-28, *Fax:* +7(3912)43-88-37, *E-mail:* alex%gis.ifor.krasnoyarsk.su@ussr.eu.net

Michael J Apps
Natural Resources Canada, Canadian Forest Service, Northwest Region, Northern Forestry Centre, 5320-122 Street, Edmonton, AB T6H 3S5, Canada, *Ph:* +1(403)435-7305, *Fax:* +1(403)435-7359, *E-mail:* mapps@nofc.forestry.ca

Richard A Birdsey
Richard A. Birdsey, USDA Forest Service, 5 Radnor Corporate Center Suite 200, Radnor, PA 19087-4585, USA, *Ph:* +1(610)-975-4092, *Fax:* +1(610)975-4095, *Email:* rbirdsey@lunar.nena.org

Sandra Brown
Department of Forestry, University of Illinois/US Environmental Protection Agency, 200 SW 35th Street, Corvallis, OR 97333, USA, *Ph:* +1(503)754-4346, *Fax:* +1(503)754-4799, *E-mail:* british@vmd.cso.uiuc.edu

Peter N Duinker
Faculty of Forestry, Lakehead University, Thunder Bay, ON P7B 5E1, Canada, *Ph:* +1(807)343-8508, *Fax:* +1(807)343-8116, *E-mail:* pduinker@thunder.lakeheadu.ca

Philip M Fearnside
Department of Ecology, National Institute for Research in the Amazon (INPA), C.P. 478, 69.011-970 Manaus, Amazonas, Brazil, *Ph:* +55(92)642-3314, *Fax:* +55(92)236-3822, *E-mail:* pmfearn@cr-am.rnp.br

Andreas Fischlin
Systems Ecology ETHZ, Institute of Terrestrial Ecology, Department of Environmental Sciences, Swiss Federal Institute of Technology Zürich, ETH-Zentrum, Grabenstrasse 3, CH-8092 Zurich, Switzerland, *Ph:* +41(1)633-6090, *Fax:* +41(1)633-1123, *E-mail:* fischlin@ito.umnw.ethz.ch

Inez Fung
School of Earth and Ocean Sciences, University of Victoria, PO Box 1700, Victoria, BC V8W 2Y2, Canada, *Ph:* +1(604)472-4005, *Fax:* +1(604)472-4030, *E-mail:* inez@garryoak.seaoar.uvic.ca

Wojciech Galinski
Forest Research Institute, Division of Ecology and Environmental Protection, 3 Bitwy Warzawskiej St., 00-973 Warsaw, Poland, *Ph:* +48(22)464660 Ext. 416, *Fax:* +48(22)224935, *E-mail:* ecol-fri@ikp.atm.com.pl

Alan Grainger
School of Geography, University of Leeds, Leeds, LS2 9JT, UK, *Ph:* +44(113)233-3300, *Fax:* +44(113)233-3308, *E-mail:* grainger@geography.leeds.ac.uk

John C Hollins
Environment Canada, Stategic Science Branch - Office of the Science Advisor, Jules Léger Building, 25 Eddy Street, 3rd Floor, Hull, QC K1A 0H3, Canada, *Ph:* +1(819)994-5167, *Fax:* +1(819)953-0550

Richard A Houghton
Woods Hole Research Center, 13 Church Street, PO Box 296, Woods Hole, MA 02543, USA, *Ph:* +1(508)540-9900, *Fax:* +1(508)540-9700, *E-mail:* rahwhrc@mcimail.com

Eric S Kasischke
Center for Earth Sciences Research, Environmental Research Institute of Michigan, PO Box 134001, Ann Arbor, MI 48113-4001, USA, *E-mail:* EricK@env.duke.edu

Pekka Kauppi
Finnish Forest Research Institute, Unioninkatu 40 A, Fin-00170 Helsinki, Finland, *Ph:* +358(0)857051, *Fax:* +358(0)625308, *E-mail:* Pekka.Kauppi@metla.fi

Seppo Kellomäki
University of Joensuu, Faculty of Forestry, PO Box 111, Fin-80101 Joensuu, Finland, *Ph:* +358(73)151-1, *Fax:* +358(73)151-3590, *E-mail:* kello@joyl.joensuu.fi

Werner A Kurz
ESSA Technologies Ltd., 3rd Floor, 1765 West 8th Ave, Vancouver, BC V6J 5C6, Canada, *Ph:* +1(604)733-2996, *Fax:* +1(604)733-4657, *E-mail:* wkurz@essa.com

J Piers Maclaren
NZ Forest Research Institute, Forest Technology Division, PO Box 465, Rangiora, New Zealand *Ph:* +64 (03)-313-8053, *Fax:* +64 (03)-313-8051, *E-mail:* maclarep@tawa.fri.cri.nz

Gregg Marland
Environmental Sciences Division, Oak Ridge National Laboratory, PO Box 2008, Oak Ridge, TN 37831-6335, USA, *Ph:* +1(615)576-0560, *Fax:* +1(615)574-2232, *E-mail:* gum@gaia.esd.ornl.gov

Pertti J Martikäinen
National Public Health Institute, Department of Environmental Microbiology, PO Box 95, SF-70701 Kuopio, Finland, *Ph:* +358(71)201-211, *Fax:* +358(71)201-265, *E-mail:* Pertti.Martikainen@ktl.fi

Robert W Matthews
Forestry Commission Research Station, Alice Holt Lodge, Wrecclesham, Farnham, GU10 4LH, UK, *Ph:* +44(1420)22255 Ext. 2279, *Fax:* +44(1420)23653, *E-mail:* yfcw08@festival.ed.ac.uk

Gert-Jan Nabuurs
Institute for Forestry and Nature Research (IBN-DLO), Bosrandweg 20, PO Box 23, NL-6700 AA Wageningen, The Netherlands, *Ph:* +31(8370)77700/77897, *Fax:* +31(8370)24988, *E-mail:* g.j.nabuurs@ibn.agro.nl

Ian A Nalder
Department of Renewable Resources, Faculty of Agriculture, Forestry and Home Economics, University of Alberta, 751 General Services Building, Edmonton, AB T6G 2H1, Canada, *Ph:* +1(403)492-1425, *Fax:* +1(403)492-4323, *E-mail:* inalder@gpu.srv.ualberta.ca

David T Price
Natural Resources Canada, Canadian Forest Service, Northwest Region, Northern Forestry Centre, 5320-122 Street, Edmonton, AB T6H 3S5, Canada, *Ph:* +1(403)435-7241, *Fax:* +1(403)435-7359, *E-mail:* dprice@nofc.forestry.ca

Edward B Rastetter
Ecosystems Center, Marine Biological Laboratory, Woods Hole, MA 02543, USA, *Ph:* +1(508)548-3705, *Fax:* +1(508)457-1548, *E-mail:* erastett@lupine.mbl.edu

Vjacheslav A Rojkov
Dokuchaev Soil Institute, Russian Academy of Agricultural Sciences, Pyzhevski Lyne, 7, 109015 Moscow, Russia, *Ph:* +7(095)231-0925, *Fax:* +7(095)231-5037, *E-mail:* rojkov@agropc.msk.su

Jayant A Sathaye
International Energy Studies Group, University of California Berkeley, Lawrence Berkeley Laboratory, 1 Cyclotron Road, Berkeley, CA 94720, USA, *Ph:* +1(510)486-6294, *Fax:* +1(510)486-6996, *E-mail:* jasldc@dante.lbl.gov

Roger A Sedjo
Resources for the Future, 1616 P Street NW, Washington, DC 20036, USA, *Ph:* +1(202)328-5065, *Fax:* +1(202)939-3460, *E-mail:* sedjo@rff.org

Anatoly Z Shvidenko
Forestry and Climate Change Project, International Institute for Applied Systems Analysis (IIASA), Schlossplatz 1, A-2361 Laxenburg, Austria, *Ph:* +43(2236)807*357, +43(2236)807*497, *Fax:* +43(2236)71313, *E-mail:* shvidenk@iiasa.ac.at

Birger Solberg
European Forest Institute, Torikatu 34, Fin-80100 Joensuu, Finland, *Ph:* +358(73)252-020, *Fax:* +358(73)124-393, *E-mail:* solberg@efesos.efi.joensuu.fi

Robert B Stewart
Natural Resources Canada, Canadian Forest Service, Science and Sustainable Development Directorate, 20th Floor, 351 St. Joseph Blvd., Hull, QC K1A 1G5, Canada, *Ph:* +1(819)997-1107, *Fax:* +1(819)994-3389, *E-mail:* rstewart@am.ncr.forestry.ca

Brian J Stocks
Natural Resources Canada, Canadian Forest Service, Ontario Region, PO Box 490, 1219 Queen Street E., Sault Ste. Marie, ON P6A 5M7, Canada, *Ph:* +1(705)949-9461 Ext. 2181, *Fax:* +1(705)759-5700, *E-mail:* stocks@fcor.glfc.forestry.ca

Valentin V Strakhov
Federal Forest Service of Russia, All-Russian Research & Information Centre for Forest Resources, ARICFR, Novocheriomushkinskaya Str., 69a, 117877 Moscow, Russia, *Ph:* +7(095)332-5338, *Fax:* +7(095)331-0533, *E-mail:* strakhov@aricfr.msk.su

Martin T Sykes
Global Systems Group, Ecology, Lund University, Ekologihuset, S-223 62 Lund, Sweden, *Ph:* +46(46)222 92 98, *Fax:* +46(46)222 37 42, *E-mail*: martin@planteco.lu.se

Alan R Townsend
Department of Organismic and Evolutionary Biology, Harvard University, 16 Divinity Avenue, Cambridge, MA 02138, USA, *Ph:* +1(617)495-8791, *E-mail:* atownsend@oeb.harvard.edu

W Jan A Volney
Natural Resources Canada, Canadian Forest Service, Northwest Region, Northern Forestry Centre, 5320-122 Street, Edmonton, AB T6H 3S5, Canada, *Ph:* +1(403)435-7329, *Fax:* +1(403)435-7359, *E-mail:* jvolney@nofc.forestry.ca

George M Woodwell
Woods Hole Research Center, 13 Church Street, PO Box 296, Woods Hole, MA 02543, USA, *Ph:* +1(508)540-9900, *Fax:* +1(508)540-9700, *E-mail:* gmwwhrc@mcimail.com

Steven C Zoltai
Natural Resources Canada, Canadian Forest Service, Northwest Region, Northern Forestry Centre, 5320-122 Street, Edmonton, AB T6H 3S5, Canada, *Ph:* +1(403)435-7304, *Fax:* +1(403)435-7359, *E-mail:* szoltai@nofc.forestry.ca

1 Introduction

Michael J Apps and David T Price

Natural Resources Canada, Canadian Forest Service, Northwest Region,
Northern Forestry Centre, 5320 - 122 Street, Edmonton, AB T6H 3S5, Canada

THE GLOBAL PERSPECTIVE

Throughout the world, forest ecosystems and their management are the focus of much human interest and concern. The traditional uses, as sources of food, shelter, construction materials and fuel, are as critical today as they ever were, while other values—such as havens for human relaxation and recreation, as wildlife habitats, and as maintainers of water supplies—have gained greater importance in many regions since the Industrial Revolution. But, as the 20th century draws to a close, additional values of global significance are increasingly being attributed to forests. One such attribute of the global forest resource is its ability to sequester atmospheric carbon. Examining the importance of this attribute and how it is influenced by human activities is, broadly speaking, the subject of this book.

Since Keeling's initial observations of atmospheric CO_2 concentrations in 1958 (Keeling 1960), it has been widely acknowledged that global anthropogenic emissions of CO_2 and other greenhouse gases (GHGs) may seriously affect the global climate system (Hansen 1988; Schneider 1989; IPCC 1990). This has in turn motivated research into the potential impacts of a changing climate on the biosphere, including possible feedbacks which could accelerate or retard the change. In particular, the role of forest ecosystems and their management in the global carbon cycle has moved to the forefront of both scientific enquiry and international negotiations within the last decade (Woodwell and Mackenzie 1994; see also Epilogue, Chapter 34).

Attempts to explain the observed increases in atmospheric concentration by balancing known emissions from fossil fuel combustion and land-use changes against the relatively well-established oceanic uptake (currently thought to be about 2 Pg C yr^{-1}), revealed a significant shortfall—a residual, unidentified uptake, or *missing carbon sink*, has persistently been required to balance the global C budget. To date, attempts to locate the missing sink in the oceans have not been very successful. On the other hand, the observed inter-hemispheric differences in atmospheric CO_2 concentration are smaller than expected given that 95% of fossil energy emissions originate from the Northern Hemisphere. This strongly suggests the presence of a significant terrestrial sink in the predominantly land-covered Northern Hemisphere.

Many mechanisms have been suggested to explain how global forests might contribute to the missing carbon sink. The observed latitudinal gradients in atmospheric CO_2 concentration led Tans et al. (1990), Enting and Mansbridge (1991) and others, to focus attention on northern forests. Taylor and Lloyd (1992), amongst others, have proposed tropical rainforests as a likely location, based on their relatively high productivity and potential response to CO_2 fertilisation. Other possible explanations have associated the sink with natural variations and cycles (e.g., Dai and Fung 1993) or with the indirect impacts of human activity on the global environment through pollution or changes in climate. These indirect impacts include changes

NATO ASI Series, Vol. I 40
Forest Ecosystems, Forest Management and the Global Carbon Cycle
Edited by Michael J. Apps and David T. Price

in boreal disturbance regimes (Kurz et al. 1995b; Chapter 14), nutrient fertilisation of temperate and boreal forests associated with air-borne pollutants (Kauppi et al. 1995; Schindler and Bayley 1993; Chapters 3 and 16), and enhanced regional productivity due to warmer average temperatures. Regardless of the mechanism, there is a widespread realisation that tropical, temperate and boreal forest ecosystems could all contribute significantly to resolving the discrepancy in the global carbon budget. Indeed, if all the recently reported sink explanations are valid, it may be necessary to abandon the search for a missing C sink and start anew with a search for the missing source!

If indeed forests do contribute significantly to the missing C sink, then what are the prospects for the future continuation of this role? Earth's human population is expected to exceed 10 000 million individuals sometime in the 21st Century. Much of this population is, and presumably will continue to be, subject to poverty, famine, disease and conflict. These human factors add enormous socio-economic obstacles to attempts to reverse global environmental degradation and to develop sustainable land management practices. Instead, the net effect of population growth is one of ever increasing pressure on many of the world's forested ecosystems. Thus there is an imminent and widespread potential for acceleration of global change, caused both directly through land-use pressures, and indirectly through changes in the global climate due to increasing emissions of CO_2 and other pollutants. Assessing the effects of these socio-economic pressures on global forests, and their biological feedbacks, is another key issue addressed by this book.

The politics of global forests and global change—and the need for scientific debate

Issues such as the above lead to a pressing policy question: How much anthropogenic perturbation can the Earth's biosphere accept before unacceptable—and possibly irreversible—change occurs? The UN FCCC casts this question in its Article 2 (statement of objectives) in the following words (Anonymous 1992)

> *"The ultimate objective of this convention...is to achieve...stabilization of greenhouse gas concentrations in the atmosphere at a level that would prevent dangerous anthropogenic interference with the climate system."*

What constitutes 'dangerous' is not made explicit in the UN FCCC document, but clearly the use of the word implies a *value judgement* as to what may be an acceptable level of impact. It also implies that there exists the *scientific* ability to predict the responses (or impacts) of the systems being affected. Article 2 goes on to stress that it is the responses of ecosystems and the human dependence on these ecosystems that are at issue:

> *"Such a level should be achieved within a time-frame sufficient to allow ecosystems to adapt naturally to climate change, to ensure that food production is not threatened, and to enable economic development to proceed in a sustainable manner."*

It is not sufficient therefore to consider the contribution of forests to the global carbon budget as it is today, nor is it enough to determine how human management actions may affect that budget in the future. Understanding the present-day contribution of forests to the global C budget, and being able to predict the possible future impacts of forest management are challenging problems, but solving them will not be enough to prevent 'dangerous interference with the climate system'. Even if we can successfully identify and implement management strategies for global forests that will allow GHG concentrations to be stabilised, there is also the even greater concern of whether we can manage these resources *sustainably*, given the pressures of an expanding human population and the possibility of significant climate change within the coming decades. Thus to serve global policy needsthe scientific community should consider not only the carbon implications of the *direct* anthropogenic changes (deliberate or accidental) being imposed on global forests, but also the *indirect* changes which may be imposed on these forests by an anthropogenically changed climate. Such indirect changes play an important part in the feedback effects associated with possible future carbon budgets.

Recent and ongoing deforestation and non-sustainable forestry operations, particularly (though not exclusively) in the tropics, have contributed to significant net additions of carbon to the atmosphere (e.g., Houghton 1994; Houghton et al. 1985, 1991; Harmon et al. 1990; see also Chapters 10, 11, 17, 27 and 28). A number of studies however have concluded that many of these regions are currently net C sinks (Dixon et al. 1993; Apps et al. 1993; Heath et al. 1993; Brown et al. 1993; see also Chapters 3, 4, 10, 12 and 17). For some regions the present sink effect is the result of significant and direct perturbations to natural forest succession brought about by past human intervention.

Almost paradoxically, it is partly because these latterly deforested or degraded regions were known to store carbon in the past, that they are also frequently viewed as potential future C sinks for mitigating the build-up of atmospheric CO_2. Much of the initiative for using forests as C sinks is perhaps politically motivated, driven by a fossil energy-dependent society anxious to do something about emissions, but not keen to reduce consumption! It is perhaps not surprising therefore, that some of the expectations for C sequestration in forests are unrealistic, or even impossible. That existing forests do have further C sequestering capacity is not doubted; but making scientifically sound estimates of this potential, and assessing their socio-economic feasibility, are more demanding requirements. These challenges are addressed in part by this book. Comparison of alternative strategies for using forests for C sequestration is an important topic that is also frequently touched upon in this book, from both scientific and socio-economic viewpoints, particularly in the second half (Chapters 5, 13, 18–30).

Scaling—from ecosystem science to global policy

The role of forests in the carbon cycle is not static—at any spatial or temporal scale. This underlies one of the major challenges to our understanding that forms an interweaving thread throughout the book—the *scaling problem.* Scaling may be described as the means by which observed data and process knowledge (commonly formulated as models) obtained at one scale of organisation may be applied correctly to another, generally larger, scale. A rigorous discussion of scaling in forest ecosystems is beyond the scope of this volume, as much has been written on this topic elsewhere (e.g., Apps 1993; Ehleringer and Field 1993; McNaughton and Jarvis 1991; King 1991; King et al. 1990).

It is pertinent, however, to consider a different type of scaling—the transfer of scientific knowledge to the policy arena. As suggested previously, and as will be reinforced in Chapters 2, 30 and 31, many of the overriding policy questions are global in scale, necessarily biased towards long-term management agendas, and likely to be aimed at entire biomes. Scientific knowledge and data on the other hand, often gleaned from laboratory or controlled environment studies, are much better developed for short-term responses at small spatial scales. The potential effects of global climate change have been described as an experiment at the planetary scale (e.g., Schneider 1989; IPCC 1990), but as Shuttleworth (1991) pointed out, there can be no control plots for experiments carried out at this scale. This causes something of a mismatch between the scientific desire to deal with real data and the policy maker's desire for informed estimates about future conditions, for which there *are no* data. While in the end there is always a value judgement to be made, scientists must attempt to separate their personal value biases from their scientific advice—or at least to make their biases explicit (e.g., Schneider 1995). One way of providing policy-relevant analysis is through a form of sensitivity analysis with models. Such analyses can provide an estimate of the uncertainty in future responses—both the policy and the scientific communities are used to dealing with uncertainties. It is partly for these reasons and because the future is unknowable, that most scientific evaluations of the possible effects of global change are cast as future *scenarios* rather than predictions. To serve policy needs effectively, the future scenarios should include a range of both possible environmental change and alternative resource management strategies (Apps 1993).

The observational record—atmospheric CO_2 and anthropogenic emissions

Over the past 150 years, since the onset of the industrial revolution, global releases of terrestrial carbon have been dominated by fossil fuel emissions. Keeling et al. (1995) have recently demonstrated a remarkable correlation between industrial emissions (fossil fuel combustion and cement production) and increases in atmospheric CO_2 concentration observed in air samples taken at Mauna Loa and elsewhere (see Figure 1b, *op. cit.*). The correlation is imperfect however, because on a decadal time scale, small discrepancies (CO_2 anomalies) exist between the observed changes in CO_2 concentrations and the expected trend resulting from industrial emissions. Keeling et al. (1989) attributed these CO_2 anomalies to climate forcing, since they exhibit strong coherence with variations in air temperature and terrestrial precipitation, particularly those following El Niño events, which occur on a 3–5 year time scale (Heimann 1995). A much greater and more sustained anomaly (i.e., actual CO_2 concentration rising faster than predicted from industrial emissions) is also apparent for the period following 1980.

It is important to note that the relationship between industrial emissions and atmospheric CO_2 concentrations shown in Figure 1b of Keeling et al. (1995) is not complete. Only a little over 50% of the emitted additional CO_2 remains in the atmosphere—the balance is actively removed and sequestered somewhere on the planet's surface. The amount of carbon being actively removed is not small—it is equivalent to storing more than one-quarter of Canada's entire forest biomass *every year.* We do know that CO_2 removal from the atmosphere, at least on the short time scales of interest here, is due to photosynthesis by terrestrial vegetation and oceanic phytoplankton. Delineating the relative contributions of these two systems is an important element of current scientific research.

Analysis of the $^{13}C:^{12}C$ ratio in the air samples allowed Keeling et al. (1995) to estimate the relative contributions of the oceans and terrestrial ecosystems to global C cycling. It seems that on a global scale, the oceans tend to be a large sink for atmospheric CO_2 (and terrestrial systems a source) during the warm phase of El Niños, and conversely during the cool phase. The larger, more sustained anomaly after 1980 is not explainable solely by El Niño events. Could this be related to the warmer conditions of the 1980s, causing greater releases of terrestrial biospheric carbon from respiration and disturbance events such as forest fires in the boreal regions? (e.g., Kurz et al. 1995a; see also Chapters 8 and 14). Keeling et al. stop short of suggesting this as an explanation, but they do make the observation that a dramatic decline in the rate of increase in average CO_2 concentration following 1990 cannot be attributed to land-use changes. Hence, the most likely explanation lies in natural responses of the terrestrial biosphere, related to slightly cooler temperatures (attributed at least partially to the 1991 eruption of Mount Pinatubo), possibly coupled with a recovery from terrestrial carbon losses undergone during the 1980s.

Forests, global carbon cycling and climate

Hence, there are two distinct aspects of the forest–atmospheric CO_2–climate interaction that are germane to this book. Firstly, human activities directly influence both the structure and function of forests on a global scale. Since structure and function largely govern the exchanges of CO_2 between forest and atmosphere, there are clearly significant direct impacts on global forest carbon dynamics—deforestation is but one example. Secondly, there is widespread belief that forests provide important biological *feedbacks* to the change (e.g., Smith et al. 1992). It is not clear however, whether the dominant feedbacks (at the regional and global scales) will be positive or negative—strong arguments have been posed in favour of both possibilities (see Woodwell and Mackenzie 1994).

The above discussion also points out several recent developments in the study of global C cycling. Firstly, it is probably misleading to think of forests as being in equilibrium, at any scale. The concept of equilibrium has the notion of stability associated with it. When perturbed, forests change and do not necessarily return to their original state, although the full ecosystem expression of the change (particularly in soils) may take a very long time to appear. Secondly, it is probably incorrect to assume that global forest carbon pools are in steady state (i.e., with net exchanges equal to zero). (By *steady state* is meant a condition where inputs are balanced by outputs. This may be achieved when steady environmental conditions prevail, but do such conditions actually exist?). Thirdly, the departures from steady state are likely to occur at all scales of organisation. Forest ecosystems respond to climatic variations and to human perturbations, but there are elements of inertia in these responses. The different time constants of the forest system components—associated with stand development, decomposition of dead material, soil carbon dynamics and disturbance episodes—are all significantly longer than the normal year-to-year variations in the exogenous environment, whether caused by climate or management interventions. These time constants result in *hysteresis* in ecosystem responses to changes in environmental conditions. More succinctly: present forest dynamics depend on their past history, both natural and anthropogenic, as well as present forcing. The remainder of the chapter will examine this topic in more detail.

FOREST CARBON DYNAMICS

Forest ecosystems can be compartmentalised into the three major carbon pools shown on the left side of Figure 1.1, namely: live biomass, detritus and soils. The exchanges of carbon among these pools and between them and the atmosphere, form the basis of the forest carbon cycle. Understanding the factors controlling these pools and the exchanges between them is the key to understanding the role of forests in C cycling at any scale. Forests utilised by humans also contribute to a fourth compartment, shown in Figure 1.1 as the products pool. When human management actions are applied to the forest, a proportion of the biomass carbon is typically transferred to the products pool, of which a smaller proportion is eventually discarded in the landfills pool. The concomitant use of fossil energy in management, harvesting and wood processing, results in the release of additional carbon from the fossil fuel pool. The C dynamics of the products, landfills and fossil fuel pools are determined largely by socio-economic factors. Hence Figure 1.1 demonstrates very broadly, the distinction between the forest ecosystem (three left-hand compartments) and the forest socio-economic system (three right-hand compartments) as the two aspects of forests affecting the global C cycle.

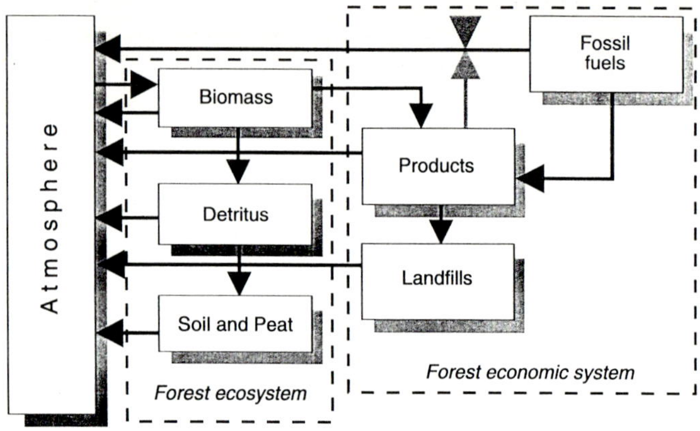

Figure 1.1. Forest system carbon (C) cycle. The main C reservoirs are in the atmosphere and forest ecosystems (left side) and in the forest sector (right side). Forest sector release of fossil fuel C, and its displacement by direct or indirect means are shown as components of the forest product pool. Releases of C to the atmosphere include all forms of decomposition, combustion and respiration in pools derived from forest vegetation biomass. Uptake of C from the atmosphere occurs only through photosynthesis. Losses of forest C to aquatic systems, and releases to the atmosphere from them, are not shown.

Carbon stores and dynamics in forest ecosystems

Temporal changes in the carbon pools of forested ecosystems are driven mainly by the dynamics of the living biomass. Accumulations of organic carbon in litter and soil reservoirs change significantly as forest stands develop or decay, and whenever disturbances occur. Biomass dynamics are clearly sensitive to environmental changes, including feedbacks resulting from changes in the forest itself. For example, canopy development as a stand ages causes significant changes in the radiation and rainfall arriving at the forest floor, which in turn affect the light, thermal and hydrologic conditions below the canopy, and hence subsequent stand development. Nutrient availability is also significantly affected by such changes in the microenvironment as the stand matures, as well as directly by exogenous inputs from fertilisation and air-borne pollutants. Keeping track of ecosystem processes, including its population dynamics is therefore a necessary part of a forest C assessment. The forest stand is the appropriate structural unit for such analyses.

Life cycle factors influencing stand dynamics

In most forest ecosystems, stand development is governed by competition among individual trees with *intrinsic* mortality resulting from several factors, including limitations on available resources (light, nutrients, water), endemic pests and disease, climatic variation and genetics. At the stand level, competition may be viewed as an *endogenous* stand development process, upon which *exogenous* disturbance events, of variable frequency and intensity, are superimposed. Often cutting indiscriminantly across existing stand boundaries, forest disturbances such as wildfire and insect outbreaks typically cause large-scale mortality. Thus disturbances can reorganise the landscape into distinct structural patterns while simultaneously changing the allocation of ecosystem resources (including C) and changing the subsequent dynamics, e.g., through changes in vegetation structure. The relative importance of intrinsic mortality and the frequency, spatial extent and intensity of such meso-scale disturbance events, largely determines whether the natural forest is composed predominantly of a mosaic of stands of more-or-less uniform age, often containing relatively few species, or of communities containing individuals of many species and ages in intimate association with one another. In the latter case, the forest succession is rarely interrupted by catastrophic disturbances (in the absence of harvesting), and replacement is typically dependent upon gaps emerging in the canopy whenever large old individuals die. Such stands are believed to be well represented by the "gap phase dynamics" paradigm (Watt 1947; Shugart 1984).

Forest carbon dynamics are therefore highly dependent upon stand development as a factor that influences both the vegetation distribution and the forest's age-class structure. Apps and Kurz (1993) and Kurz et al. (1995a, b) have demonstrated the importance of stand dynamics for carbon cycling in boreal ecosystems, where individual stands are predominantly even-aged, but the same concepts should apply to any disturbance-regulated forest ecosystem. In such an ecosystem, the disturbance regime (intensity, frequency, and spatial connectedness) leads to the formation of a heterogeneous mosaic of stands whose ages can be quantified relatively easily.

Once a stable age-class distribution has been established and if growth conditions are unchanging (i.e., such that average recruitment and growth are in balance with losses from both intrinsic mortality and disturbances), the biomass C pools will tend to a steady-state condition where the spatially averaged net biomass C accumulation is effectively zero. Under what situation, however, are unchanging growth conditions ever realised? Such conditions, at the scale of the whole forest or at the stand level, are probably an exception rather than the rule. We shall return to this point. Variations in stand productivity add an extra dimension of calculation and data complexity (requiring separate summations over different stand types) but do not alter the underlying temporal dynamics of biomass C accumulation.

When the soil and detritus pools are included, rather more complexity is introduced because of the differences in time scales for those ecosystem processes governing the C dynamics in these compartments, e.g., transfers of long-lived woody material following wildfire to the detritus and soil pools, or in the case of harvesting, to forest products. Decomposition releases from the forest floor and soil of a particular stand depend on past disturbance history and the accumulation of decomposable material as well as on the present environmental conditions and stand structure. Significant lags to the release of stored carbon through decay and decomposition are therefore introduced, leading to hysteresis effects as discussed previously. Simulation modelling is one of the few ways to examine these complex dynamics, but such modelling at the forest-level is in its infancy (e.g., Apps and Kurz 1993; Kurz et al. 1995a; Smith and Shugart 1993). Most efforts to date have been directed at a patch or gap scale and have focused on successional processes in the above-ground biomass (e.g., Smith et al. 1994).

The higher the disturbance frequency, the greater will be the proportion of younger age-classes. Therefore, if all other conditions are unchanged, a more frequently disturbed forest will contain less carbon (i.e., because younger forests contain less biomass). Note that if a higher disturbance rate is maintained (and growth conditions do not change), net forest biomass C accumulation will again tend to steady-state, once a stable age-class structure has been established. When a change in disturbance regime occurs, however, the existing age-class structure will no longer represent the steady-state condition for the new disturbance regime, so a change in the net C exchange (uptake or release) will take place as the distribution shifts towards older age classes (reduced disturbances) or younger classes (increased disturbances). Simply stated, *decreasing* the disturbance rate causes the average age to *increase* with a consequent *increase* in biomass C and *vice versa*.

A corollary to the above discussion is that where disturbances are an important factor influencing forest dynamics, spatially-averaged forest carbon pools and fluxes are susceptible to large-scale changes that extend over periods of decades to centuries. Because it takes a relatively long time for trees to grow, episodes of frequent disturbances lasting several years can cause significant perturbations in the age-class structure, and hence carbon storage, that may persist for several life spans (see Chapters 14, 16 and 21 for specific examples). Such a situation may prevail when the frequency and intensity of forest fires, for example, are related to a change in climate (e.g., Kurz et al. 1995a, b; Chapter 8). Conversely for those forests where large-scale disturbances are rare, the spatially-averaged biomass carbon storage may change relatively little over long time periods provided other environmental conditions remain constant. This stability however remains subject to changes in net primary productivity, and more specifically to environmental factors that constrain the maximum potential biomass, or *carrying capacity*, such as changes in available soil nutrients or moisture.

It would seem therefore that the combined effects of stand growth processes, competition and exogenous disturbances contribute to accumulations of forest biomass that rarely, if ever, reach the theoretical maximum biomass carbon density for the site (at the stand level). Endemic ecosystem processes, such as insect herbivory and disease, tend to reduce actual stand C density still further. The long time scales associated with soil processes cause changes in soil C density to lag behind the changes in biomass, so that total ecosystem carbon may be even further removed from the site's total carbon carrying capacity, C_{max}. At a very local scale, the biomass of individual mature stands (or of climax communities where disturbances do not impede successional development) may tend towards maximum carbon storage potential. In such stands, net primary productivity (NPP) may saturate, and hence net biomass C increment may tend to zero. Zero biomass increment, however, need not result in a balance between litterfall inputs to the forest floor and carbon losses through decomposition, so NEP may remain positive, or become negative in an overmature phase before stand replacement (following disturbance or through gap phase renewal) takes place. At the larger scale of the forest, not all stands or communities can possibly be at maturity (with respect to age and biomass density) simultaneously, hence the spatially-averaged carbon storage must also be less than the theoretical carrying capacity.

Similar arguments to the preceding can be made for uneven-aged forest stands that are more typical of temperate and tropical regions, where the role of disturbances in killing extensive areas of forests is often outweighed by intrinsic mortality affecting individual trees or small discrete patches. Small environmental changes, including those resulting from stand development such as shading and lowered water tables, tend to favour some species relative to others, leading to successional changes (see Shugart 1984; Smith et al. 1994), and related changes in both NEP and ecosystem C_{max}. Nevertheless, observations from the field and from models (*op. cit.*) suggest that given enough time, and assuming environmental conditions remain stable, the biomass pools will also tend to steady state.

Perhaps not surprisingly, if changes in environmental conditions do occur, then the steady-state forest structure (species composition and biomass density) will likely also change. Furthermore, if the change in conditions also leads to more frequent disturbance events, the time required for the new forest community to adapt and develop under a new stable environment is significantly reduced (Overpeck et al. 1990) although, as already indicated, changes in the soil components of the new ecosystems will take much longer to reach steady state. Increased mortality from any cause will result in reduced average age and reduced biomass carbon, but at the same time generate increased inputs of organic matter to the soil pools. Although there is little direct evidence for these dynamics in uneven-aged forest stands, modelling results (e.g., Townsend et al. 1992) based on the relative turnover times suggest that consequent changes in the litter layer and soil C pools will inevitably lag behind changes in vegetation structure, at least in northern latitudes. Smith and Shugart (1993) included soil pool transients under a changing climate by allowing them to relax exponentially towards assumed new equilibrium values. Such observations, and the previous discussion, suggest the following hypothesis: *forest ecosystems in reality are never at steady state or at their theoretical maximum C carrying capacity.*

Because ecosystem processes, such as decomposition and herbivory, in reality operate at all stages of stand development (in both even and uneven aged stands), it follows that the time course of actual carbon accumulation is likely to follow a trajectory below the theoretical maximum (e.g., that would obtain if disturbances were fully suppressed but all other factors were equal, i.e., $C(t) < C_{max}$ for all t. This leads to the conclusion that there is no *a priori* ecological reason why the average NEP should be zero. Hence, the notion that average C uptake in natural ecosystems is invariably at steady-state is probably misleading. We feel this is critically important because forest NEP (including, where applicable, losses from disturbances, changes in forest product pools, and associated energy utilisation) represents a direct measure of the forest system's contribution to the global C balance. Any change in large-scale average NEP would therefore imply a corresponding (opposite) change in the rate of carbon gained (or lost) by the atmosphere.

Is there however, a yet larger scale (spatial and temporal) at which the average environmental conditions (including variable disturbance regimes) have remained constant long enough for the carbon pools and fluxes to reach steady state? We believe this is very doubtful. At the regional scale and even at the global scale, the paleo-record shows that climate fluctuates not only from year to year but also on the scales of centuries (e.g., the Little Ice Age), millennia (e.g., the Hypsithermal) and even longer scales (e.g., Milankovitch Cycles). Consequently, regional and global forests are never in steady state with respect to biomass and C accumulation. Hence the question arises: to what extent do natural climatic variations account for changes in atmospheric C, and can these changes explain the present global C imbalance? This issue is addressed in Part I.

Carbon stores and dynamics in forest sector pools

Carbon storage in forest products typically represents a very small fraction of the total forest carbon stores (Vitousek 1991; see also Chapters 10, 20, 23 and 24) but can play an important role in the net forest-atmosphere exchange over the short term by delaying the return of photosynthetically fixed carbon to the atmosphere. It is important, however, to note that treating the forest product pool as a carbon sink independent of the forest ecosystems from which the fibre derives is faulty logic (see Houghton, Chapter 10); as Figure 1.1 illustrates, carbon is removed from the atmosphere only by photosynthesis and virtually all other reservoirs are temporary storage systems that only delay the eventual release of carbon they contain, through respiration, decomposition or combustion.

Hendrickson (1990) pointed out that the storage of carbon in wood products from an extensively managed forest can be compared to the storage in detritus on the forest floor of a natural one; taking the simplest view, relative carbon gains may be expected in the managed forest when the average half-life of harvested materials is greater than that of the forest floor debris. For plantation forests, a more comprehensive analysis that takes into account the relative life times of the forest biomass (determined from the rotation length), soil turnover rates and forest product half-lives has been provided in a series of theoretical analyses by Dewar (1990, 1991) and Dewar and Cannell (1992). The conclusions reached are similar—short lived products derived from, and replacing, long lived natural systems result in terrestrial carbon loss; long lived products derived from, and replacing, short lived natural systems result in terrestrial carbon gains.

There is, however, one key difference between the decomposition of forest floor detritus and that of forest products: the retention time and subsequent fate of carbon in forest products can be directly influenced by human decisions—in principle at least. Recycling (and storage in landfills) may significantly extend the effective half-life of forest product carbon, but the use of fossil fuel in the recycling process must be carefully examined. Several chapters in this book provide case studies that examine the combined carbon storage potential in managed forest ecosystems and the forest economic system. Kellomäki and Karjalainen (Chapter 5) also explicitly consider the implications of climatic change for future storage potential.

The potential importance of fossil fuel replacement by forest products is emphasised in a number of chapters, particularly in Part III of this book—see Chapter 24 for the Working Group summary. Forest products can offset the use of fossil fuel energy both by *direct* replacement (bioenergy) and *indirectly*, by substituting for materials that require substantial energy expenditures during their production. Using forest products (directly or indirectly) to displace fossil fuel is conceptually equivalent to storing those forest products as if they had the half-life of the fossil carbon they replace. The use of recycled forest products as a bioenergy source can also provide a significant cumulative carbon benefit. Some of the important factors influencing the carbon offsets possible through innovative forestry sector activities, including energy plantation management, are provided by Schlamadinger and Marland in Chapter 18.

Changes in carbon storage in the forest product sector, and the consequent impacts on the forest ecosystem from which the products are derived, are strongly dependent on social and economic factors. Part III of this book deals specifically with these issues. Here, we merely note that 'forest degradation' due to human activities—the second most important contributor to increasing atmospheric CO_2 after fossil fuel use—can be regarded as the result of an imbalance between the subsystems (forest ecosystem and economic system) indicated in Figure 1.1. Many Chapters in this book, particularly in Part IV, stress the need to identify—and achieve—a balance between the immediate material needs of society and the longer term requirement to sustain the global environment. In order to achieve sustainability, it will be necessary to first take stock of the current state of the world's forests, including deforested and degraded land. It will then be necessary to classify the degraded lands, with a view to restoring them to the status of economically *and* ecologically sustainable forests.

One possible approach is introduced by Grainger (Chapter 27) who proposes a new classification for degraded forest lands and shows how this can be used to examine carbon mitigation potentials in forests undergoing serious impacts from land-use change. Although specifically outlined only for the forests of the humid tropics, the concept, which combines elements of both *cultural biography* and *natural biogeography* may find useful applications elsewhere. Grainger suggests that physical degradation (which can be interpreted as a reduction in biomass carbon density below C_{max}), is sometimes due to human activity (or management) which may have persisted for hundreds of years—i.e., it may be strongly representative of the normal local culture. Hence, he argues, the vegetation may be *culturally optimal*, in that it meets local needs and is likely to continue doing so in the absence of change; attempts to raise productivity and C storage however, may introduce, or further exacerbate, socio-economic problems. Such cultural optimality may imply that the local C_{max} has in fact changed, so that the biomass C density maintained will be invariably less than that which would have occurred in the absence of human activity. Should such an effect be considered *non-sustainable* management? We shall return to this question in the context of sustainable forest management and the global carbon cycle.

Forest ecosystem monitoring and carbon budget assessment

For those ecosystems not dominated by natural disturbances, the detection of responses to changes in climate may be difficult, because the past record of vegetation development may be unknown and will likely be seriously confounded by human land-use practices. Yet the detection of such changes is a prerequisite to assessing changes in the carbon cycling role of such forests. It is widely believed that forest ecosystems of high diversity are probably more resilient to change, and tend to adjust by a continuum of subtle changes in species behaviour and composition. Monitoring the floristic composition, through botanical surveys and by maintaining detailed forest inventories, may be the key to detecting change. It will be seen that the topic of ecosystem monitoring and detection of change is another recurrent theme within the book.

The need for *comprehensive* carbon budget assessments receives considerable attention, particularly in Parts II and III. As we have outlined above, understanding the processes operating within forest ecosystems are critical to predicting possible effects on carbon dynamics in responses to changes in the global environment. The key processes operate at all scales of organisation, and hence C budget assessments must be *complete*, not only in considering the effects of underlying ecosystem processes, but also in accounting for the transfers among pools including the atmosphere, soils, aquatic ecosystems, and forest products as shown in Figure 1.1. The specific conclusions reached by the various authors differ largely because the system boundaries (i.e., the number of the compartments and interactions indicated in Figure 1.1) vary among studies. This variation also illustrates a point made by Matthews (Chapter 19): on the basis of an extensive literature review, he argues that the results of carbon budgeting exercises depend strongly on the scope of the analysis and the methodologies applied.

Carbon budget assessments and environmental monitoring have assumed particular importance and relevance to the debate which led to the development of international policy agreed to within the UN FCCC. First, there is a need to have some form of internationally sanctioned, and scientifically sound, carbon monitoring system to track performance of proposed carbon mitigation strategies and country-level carbon inventories as set forth in the convention. These needs are articulated by Brown et al. in the Working Group summary for Part II (Chapter 17) and will not be discussed further here. A second aspect is the larger issue dealing with sustainable development and the management of global forests, which we now address briefly.

Carbon cycles and sustainable forest management

To realise sustainable and verifiable carbon storage gains in forest systems (i.e., considering both ecological and economic components), it is obvious that forest ecosystems must be maintained *sustainably*. But, can we define what 'sustainable' means in a changing environment? What is the benchmark against which achieved performance can be gauged? Formulation of criteria for sustainable development and of indicators for monitoring system performance is outside the scope of this book (e.g., see Riley 1995), but clearly carbon stores and fluxes must be included. Changes in the global environment will not have uniform impacts on forest resources; as explained in several chapters in Part IV, heterogeneity of response is likely to lead to shifting geographic patterns in forest product supply and demand and, in turn, to spatially different and changing *cultural* impacts on global forests. Assessing the magnitude and direction of such *socio-economic feedbacks* is problematic. On the other hand, there is an immediate and ongoing need to reduce the rate of impoverishment of the global forest resource (Woodwell and Ramakrishna, Chapter 34). Stewart and Maini (Chapter 31) examine these issues and enumerate a number of social factors that must be considered. At the international level, the state of the world's forests and their possible susceptibility to global change have thus come to the forefront of the political agenda. Not only is there increasing recognition of the need to protect global forests for their role as natural sinks for atmospheric carbon—in addition deforestation and forest degradation are seen both as the result and the causes of human suffering in many parts of the world (as pointed out by several authors in Part IV). It seems clear that forests must continue to be managed for their traditional economic benefits, but sustainably. The World Commission on Environment and Development (WCED) defines *environmentally sustainable economic development* in the following way:

> *"...economic development that meets the needs of the present without compromising the ability of future generations to meet their own needs"* (WCED 1987).

Taking the view that the enhanced greenhouse effect poses a potential threat to future forests, it is clear that long term environmental sustainability must therefore include maintenance of benefits such as C storage as well as their other aesthetic and traditional attributes. Such issues are the central themes of Part IV of this book.

QUESTIONS ADDRESSED IN THIS BOOK

To summarise the foregoing discussion and introduce the remainder of the book, we present here the topics that are addressed. The book is organised in four Parts, corresponding to the main themes of the workshop. Part I addresses the global issues of forest ecosystems and the global carbon cycle:

- What are the causes of the atmosphere–biosphere carbon imbalance?
- How do nutrients impose constraints on carbon storage within forest ecosystems?
- What is the contribution of forested wetlands to the global carbon cycle?
- Can we predict forest responses to climate change and the possible impacts on carbon sequestration?
- How important are disturbances to carbon cycling, particularly in northern ecosystems?

Part II focuses on the effects of human activities on forest ecosystems, including changes in land use. The consequences for the record of carbon releases and storage during the recent past, the present, and probable trends for the future, are examined. The following topics are covered:

- What are past and current rates of carbon loss from global forest ecosystems?
- How do regional C budgets differ: can we locate the missing C sink?
- What is the role of natural disturbances, as compared to those of anthropogenic origin?
- What needs to be done to bring a halt to deforestation, particularly in the tropics?
- Can we manage forests to sequester more carbon; if so, where are the maximum potential gains?
- How do we deal with the possible effects of global change?
- The need for standardised procedures for regional C budget accounting.

In contrast to Part II, where the focus was at the global scale to compare and contrast forest C cycling within tropical, temperate, and boreal ecosystems, Part III takes a more detailed and functional perspective. The roles of forest management and the forest products industry are examined to assess the potential impacts on and possible benefits for the global C balance. The topics addressed include:

- Which forests have the greatest potential to sequester carbon?
- Which management regimes sequester the most carbon?

- How can forest products affect the local and global C balance?
- Can conflicts between C storage and other forest management objectives be resolved?
- The need for standardised methods for forest C budget accounting.
- Looking up to the global perspective.

The interaction between human socio-economic considerations and the global carbon cycle are the major focus of Part IV:

- How can economic analysis be used to evaluate and compare strategies for mitigating carbon emissions?
- What are the relative costs and benefits of the different management strategies that might be used to mitigate carbon emissions?
- What are the causes and socio-economic impacts of deforestation and land degradation?
- How can deforestation be prevented and reversed, given population pressures?
- What are the implications of adding carbon to forest management and policy agendas?
- What are the differences between developed and developing nations in using forests for mitigating carbon emissions?
- The need for a global policy on sustainable management of the world's forests.

As we enter the 21st century, the concept of the global village has become a reality. This offers both enormous benefits and enormous challenges. Optimistically, a possible benefit is the opportunity for almost immediate dissemination of new information; rapid changes in paradigms or "shared world views" can, and do, occur as new knowledge comes to light, and old ideas are superseded. We are already witnessing such a paradigm shift in the science of the Earth system. There is a rapidly growing awareness that Earth's ecosystems do not function independently of one another, but rather are intimately interconnected by the shared global environment. Thus we perceive that the world's ecosystems all play significant roles in the changing dynamics of the global carbon cycle.Direct human impacts on forests in the form of deforestation and other land-use changes may proceed tree by tree and stand by stand. The cumulative effect however, accounts for at least 30% of the annual anthropogenic perturbations to the global carbon cycle. This book focuses on the specific global influences of the world's forests on global carbon dynamics, as well as human impacts thereon. Anticipated changes in climate predicted to occur within the next century as a result of anthropogenic perturbations to the carbon cycle add an element of urgency and political motivation for the topic—but they also create a scientific challenge of enormous complexity.

This introductory chapter provides an overview of some of the recent developments in understanding, and a sense of the depth of existing scientific knowledge, while also indicating some of the significant scientific challenges still to be addressed. The science of the global environment *is* rapidly changing and evolving. The contributions of the scientists who took part in the Advanced Research Workshop in Banff, as recorded in this book, should be recognised as an important contribution to progress in this field.

ACKNOWLEDGEMENTS

The content of this chapter benefitted significantly from careful readings by IA Nalder and J van Minnen.

REFERENCES

Anonymous (1992) Text of United Nations Framework Convention on Climate Change, UNEP/WMP Information Unit on Climate Change, Climate Change Secretariat, Palais des Nations, Geneva, Switzerland, 29 pp.

Apps MJ (1993) NBIOME: A biome-level study of biospheric response and feedback to potential climate changes. *World Resour. Rev.* 5(1): 41–65.

Apps MJ, Kurz WA (1993) The role of Canadian forests in the global carbon balance. In: Carbon Balance of World's Forested Ecosystems: Towards A Global Assessment. *Publ. Acad. Finland* No. 3/1993 Helsinki, pp 14–28.

Apps MJ, Kurz WA, Luxmoore RJ, Nilsson LO, Sedjo RA, Schmidt R, Simpson LG, Vinson TS (1993) Boreal forests and tundra. *Water, Air, Soil Pollut.* 70: 39–53.

Bolin B (1986) How much CO_2 will remain in the atmosphere? The carbon cycle and projections for the future. In: Bolin B, Döös BR, Jäger J, Warrick RA (eds) The Greenhouse Effect, Climatic Change, and Ecosystems. SCOPE 29, Wiley, Chichester, UK, pp 93–156.

Brown S, Iverson LR, Prasad A, Liu D (1993) Geographical distributions of carbon in biomass and soils of tropical Asian forests. *Geocarto International* 4: 45–59.

Dai AG, Fung IY (1993) Can climate variability contribute to the "missing" CO_2 sink? *Glob. Biogeochem. Cycles* 7: 599–609.

Dewar RC (1990) A model of carbon storage in forests and forest products. *Tree Physiol.* 6: 417–428.

Dewar RC (1991) Analytical model of carbon storage in trees, soils and wood products of managed forests. *Tree Physiol.* 8: 239–258.

Dewar RC, Cannell MGR (1992) Carbon sequestration in the trees, products and soils of forest plantations: an analysis using UK examples. *Tree Physiol.* 11: 49–71.

Dixon RK, Brown S, Houghton RA, Solomon AM, Trexler MC, Wisniewski J (1994) Carbon pools and flux of global forest ecosystems. *Science* 263: 185–190.

Ehleringer J, Field CB (eds) (1993) Scaling Physiological Processes: Leaf to Globe, Academic Press, San Diego, CA, 388 pp.

Enting IG, Mansbridge JV (1991) Latitudinal distribution of sources and sinks of CO_2: results of an inversion study. *Tellus* 43B: 156–170.

Hansen J, Fung I, Lacis A, Rind D, Russell G, Lebedeff S, Reudy R, Stone P (1988) Global climate changes as forecast by GISS's three-dimensional model. *J. Geophys. Res.* 93: 9341–9364.

Harmon ME, Ferrell WK, Franklin JF (1990) Effects on carbon storage of conversion of old-growth forest to young forests. *Science* 239: 42–47.

Heath LS, Birdsey RA (1993) Carbon trends of productive temperate forests of the coterminous United States. *Water, Air, Soil Pollut.* 70: 279–293.

Heimann M (1995) Dynamics of the carbon cycle. *Nature 375: 629–630.*

Hendrickson OQ (1990) How does forestry influence atmospheric carbon? *Forestry Chron.* 66: 469–472.

Holling CS (1992) Cross-scale morphology, geometry and dynamics of ecosystems. *Ecological Monographs* 62: 447–502.

Houghton JT, Jenkins GJ, Ephraums JJ (eds) (1990) Climate Change: The IPCC Scientific Assessment. WMO/UNEP/IPCC, Cambridge University Press, Cambridge, UK, 365 pp.

Houghton RA (1994) The worldwide extent of land-use change. *BioScience* 44: 305–313.

Houghton RA, Schlesinger WH, Brown S, Richards JF (1985) Carbon dioxide exchange between the atmosphere and terrestrial ecosystems. In: Trabalka JR (ed) Atmospheric Carbon Dioxide and the Global Carbon Cycle. U.S. Dept. Energy, DOE/ER-0239, Washington, DC.

Houghton RA, Skole DL, Lefkowitz DS (1991) Changes in the landscape of Latin America between 1850 and 1980. II. A net release of CO_2 to the atmosphere. *Forest Ecol. Manage.* 38: 173–199.

Houghton RA, Unruh JD, Lefebvre PA (1993) Current land use in the tropics and its potential for sequestering carbon. *Global Biogeochemical Cycles* 7: 305–320.

IPCC Workshop statement for Carbon Balance on World's Forested Ecosystems: Towards a Global Assessment, Proc. Intergov. Panel on Climate Change Workshop, Joensuu, Finland, 11–15 May 1992, Publications of the Academy of Finland, Helsinki,

Kauppi PE, Tomppo E, Ferm A (1995) C and N storage in living trees within Finland since the 1950s. *Plant and Soil* 168–169: 633–638.

Keeling CD (1960) The concentration and isotopic abundance of carbon dioxide in the atmosphere. *Tellus* 12: 200–203.

Keeling CD, Bacastow RB, Carter AF, Piper SC, Whorf TP, Heimann M, Mook WG, Roeloffzen H (1989) A three dimensional model of atmospheric CO_2 transport based on observed winds: 1. Analysis of observational data. In: Peterson DH (ed) Aspects of Climate Variability in the Pacific and the Western Americas. *Geophys. Monogr.* 55, American Geophysical Union, Washington, DC, pp 165–236.

Keeling CD, Whorf TP, Wahlen M, van der Plicht J (1995) Interannual extremes in the rate of rise of atmospheric carbon dioxide since 1980. *Nature* 375: 666–670.

King AW (1991) Translating models across scales in the landscape. In: Turner MG, Gardner RH (eds) Quantitative Methods in Landscape Ecology, Springer-Verlag, New York, pp 479–517.

King AW, Emanuel WR, O'Neill RV (1990) Linking mechanistic models of tree physiology with models of forest dynamics: problems of temporal scale. In: Dixon RK, Meldahl RS, Ruark GA, Warren WG (eds) Process Modeling of Forest Growth Responses to Environmental Stress, Timber Press, Portland, OR, pp 241–248.

Kurz WA, Apps MJ, Beukema SJ, Lekstrum T (1995a) Twentieth century carbon budget of Canadian forests. *Tellus* 47B: 170–177.

Kurz WA, Apps MJ, Stocks BJ, Volney WJA (1995b) In: Woodwell GM, Mackenzie FT (eds) Biotic Feedbacks in the Warming of the Earth: Will the Warming Feed the Warming? Oxford University Press, New York, pp 119–133.

McNaughton KG, Jarvis PG (1991) Effects of spatial scale on stomatal control of transpiration. *Agricultural and Forest Meteorology* **54**: 279–301.

Overpeck JT, Rind D, Goldberg R (1990) Climate-induced changes in forest disturbance and vegetation. *Nature* 343: 51–53.

Riley LF (1995) Criteria and indicators of sustainable forest management in Canada. *Water, Air and Soil Pollution* 82: 67–70.

Schindler DW, Bayley SE (1993) The biosphere as an increasing sink for atmospheric C: estimates from increased nitrogen deposition. *Glob. Biogeochem. Cycles* 7: 717–733.

Schneider SH (1989) The changing climate. *Scientific American* 261: 70–79.

Schneider SH (1995) Foreword. In: Apps MJ, Price DT, Wisniewski J (eds) Boreal Forests and Global Change, Kluwer, Dordrecht, The Netherlands, pp vii–ix.

Shugart HH (1984) A Theory of Forest Dynamics: The ecological implications of forest succession models. Springer-Verlag, New York, 278 pp.

Shuttleworth WJ (1991) The Modellion Concept. *Rev. Geophys.* 29: 585–606.

Smith TM, Shugart (1993) The transient response of terrestrial carbon storage to a perturbed climate. *Nature* 361: 523–526.

Smith TM, Leemans R, Shugart HH (1994) The application of patch models of vegetation dynamics to global change issues. GCTE Workshop Summary, Kluwer, Dordrecht, The Netherlands, 22 pp.

Smith TM, Weishampel JF, Shugart HH (1992) The response of terrestrial C storage to climate change: modeling C dynamics at varying temporal and spatial scales. *Water, Air and Soil Pollution* 64: 307–326.

Tans PP, Fung IY, Takahashi T (1990) Observational constraints on the global atmospheric CO_2 budget. *Science* 247: 1431–1438.

Taylor JA, Lloyd J (1992) Sources and sinks of atmospheric CO_2. *Austral. J. Bot.* 40: 407–418.

Townsend AR, Vitousek PM, Holland EA (1992) Tropical soils could dominate the short-term carbon cycle feedbacks to increased global temperatures. *Climatic Change* 22: 293–303.

Vitousek PM (1991) Can planted forests counteract increasing atmospheric carbon dioxide? *J. Environ. Qual.* 20: 348–354.

Watt AS (1947) Pattern and process in the plant community. *J. Ecology* 35: 1–22.

WCED (World Commission on Environment and Development) (1987) Our Common Future (The Brundtland Report). Oxford University Press, Oxford, UK, 383 pp.

Woodwell GM, Mackenzie FT (1994) Biotic Feedbacks in the Warming of the Earth: Will the Warming Feed the Warming? Oxford University Press, New York, 416 pp.

Part I

Forest ecosystems and the global carbon cycle

I Forest ecosystems and the global carbon cycle

Introduction

Michael J Apps and David T Price

Transfer, storage and release of carbon (C) between the atmosphere and terrestrial and oceanic sinks typically occur over several different time scales. These processes are strongly dependent upon, and can potentially influence, the global climate. Part I is intended to provide a large-scale, long-term perspective on the role of forest ecosystems in the C cycle. The Working Group summary paper reports the group analysis of the extent to which climate variation over various time scales have influenced global terrestrial contributions to changes in the atmospheric C record. The individual articles review past changes in terrestrial C pools as determined from the observational record. They examine changes in the geologic past (e.g., due to long term shifts in climate), in the recent historical past (including the contributions of land-use change and deforestation) and contemporary 20th century trends (considering, for example, the effects of post-industrial pollution and large-scale tropical deforestation). Some Chapters also review current projections of ecosystem C dynamics into the foreseeable future.

Compared to the other Parts of the book, the individual papers in Part I cover a much broader range of topics with relatively little overlap. Discussion among members of the Working Group addressed the results of changes in global forest sector C pools (standing biomass, soils and detritus, and harvested materials) and drew upon the results reported in Parts II and III. Similarly, the comparison of regional variations in C dynamics used data and experience from across all Parts and Working Groups to better delineate the global variations. Thus differences in tropical, temperate and boreal systems were considered and compared, with particular attention paid to the processes internal to these ecosystems and to those affecting exchanges of C between forests and the atmosphere (in soils, biomass accumulation, and large-scale disturbances).

Understanding the atmosphere–terrestrial ecosystem carbon imbalance

In Chapter 2, Fung reviews the role of terrestrial ecosystems from the viewpoint of the atmosphere. The decades-long record of atmospheric CO_2 concentrations provides a direct measure of past net exchanges at the planetary surface. She emphasises that:

> *"...understanding the mechanisms responsible for CO_2 uptake is crucial for projections of future CO_2 levels in the atmosphere."*

She then discusses the uncertainties in the global C budget. Methods for estimating the magnitude of the terrestrial C sink (and the *missing sink*) are reviewed, and a mathematical framework for analysing the sink mechanisms is presented.

NATO ASI Series, Vol. I 40
Forest Ecosystems, Forest Management and the Global Carbon Cycle
Edited by Michael J. Apps and David T. Price

The term *missing sink* refers to the C flux inferred as the residual in the atmospheric mass balance; it is generally considered to be the unidentified part of the global net biotic flux after accounting for the effects of land-use modification. The inferred sink includes: the effects of natural variability in biospheric processes on decadal to centennial time scales (e.g., short-term climatic variations affecting NPP, organic matter decomposition, and disturbance regimes); direct human actions (e.g., incorrect accounting for the effects of land-use change); and alteration of biospheric functions through atmospheric changes (e.g., CO_2-fertilisation, and eutrophication caused by other air-borne pollutants). Using some examples of work reported elsewhere in this book, Fung concludes by pointing to the complexity surrounding the many interactions between human and natural perturbations of the abiotic environment and the biotic responses they evoke. Identification of the missing sink, and understanding future changes, will require comprehensive analysis of C fluxes for the entire terrestrial biosphere.

Nutritional constraints on carbon storage within forest ecosystems

Townsend and Rastetter (Chapter 3) examine how essential nutrients, particularly nitrogen (N), may affect the responses of forest ecosystems to changes in environmental conditions, both though the direct effect of nutrient deposition associated with atmospheric pollution, and through indirect effects on nutrient cycling caused by changes in the climate system and disturbance regime. Because external inputs of N are usually low, the nutritional requirements of most forests are met mainly by internal cycling, and therefore constrained. When a disturbance such as clear-cutting or wildfire occurs, however, the tight coupling between soil and vegetation pools is broken and N losses from the system may be greatly increased. Large changes in ecosystem C storage are typically accompanied by comparable changes in site nutrient capital. C:N ratios within vegetation and soils (and allocation of nutrients between these pools) vary widely among geographic regions.

Townsend and Rastetter argue that changes in ecosystem C storage must be accompanied by some change in the nutrient supply. This can be accomplished by a change in total ecosystem nutrient capital, a change in the allocation between soils and vegetation, or a change in the C:N ratios within either (or both) of these pools. The likely impacts of different environmental changes on site nutrient dynamics are reviewed, and the authors conclude that, at least for the short-term, all but disturbances should stimulate C sinks in forests. The long term responses are poorly understood, however, although the sink effect is expected to decline with time.

Townsend and Rastetter also discuss the role of phosphorus which is known to be a major limitation to C uptake in some tropical forest ecosystems. Phosphorus deposition from fertiliser inputs and industrial sources is significant and appears to be increasing, so there is potential for significant increases in C uptake in some regions. The simultaneous interactions between these nutrient inputs and global change factors are extremely important, but must be better understood before meaningful assessments of global ecosystem responses can be made.

Contribution of global wetlands to global carbon cycling

In Chapter 4, Zoltai and Martikainen review the definitions of the major wetland types and present the results of a new review and analysis of global peatlands. This analysis features a first attempt to distinguish the areal extent of forested from non-forested peatlands. Global peatlands are estimated to cover approximately 352 Mha ±5%, of which they estimate 70–88 Mha (20–25%) to be forested (i.e., with peat accumulation > 30 cm and tree canopy cover that is closed, or has the potential to become closed).

Forested peatlands store substantial amounts of C in forest biomass as well as in peat. In tropical peatlands, forest biomass is estimated to reach 500 Mg C ha^{-1} compared to only

120 Mg C ha^{-1} in northern forested peatlands. They point out, however, that the C density of forest biomass is typically much less than that of the peat, and the ratio of these densities is much higher in tropical than in high latitude forested peatlands (12–13% compared to 2–3%, respectively). The proportion of tropical peatlands that are forested is also much greater (37 Mha forested of 50 Mha total, compared to 60 Mha of 302 Mha total in high latitude ecosystems).

Although global peat accumulation is only a small fraction of global NPP (ca. 7.5%), it appears to be an important component that may be significantly affected by global change. Zoltai and Martikainen examine the influences of various factors on peat accumulation, both through new peat production and the decomposition of previously deposited peat. They examine both autogenic changes (dynamic changes within the peat ecosystem itself) and allogenic changes (induced by external factors, such as changes in climate or peatland hydrology). Temperature and water table fluctuations are shown to play decisive roles in the C dynamics of forested peatlands. A major allogenic change to forested peatlands is drainage ditching to promote aerobic conditions for better tree growth. The impacts of drainage on the C balance of peatlands depend significantly on the chosen time-horizon. Some studies report increased net C uptake for several decades following drainage, while increased biomass productivity exceeds the increased rate of peat decomposition. Other studies indicate however, that long-term changes will depend on the peatland type and its geographic location. Eventually as the forest reaches maturity, biomass C uptake slows and eventually ceases, typically causing the peatland to become a net source of CO_2.

The authors also address methane (CH_4) and nitrous oxide (N_2O) emissions from peatlands. Methane emissions vary widely, both temporally and spatially, showing both site-specific and regional variations. Currently, only crude estimates for global CH_4 emissions from peatlands are possible. The temperature dependence of CH_4 emission is also complex, making estimates of potential climate change impacts difficult. Melting of permafrost would however generally increase CH_4 emissions in affected regions. Nitrous oxide emissions from nutrient-poor northern and temperate peatlands are generally low, but evidently increase when nutrient-rich fens are drained.

Zoltai and Martikainen conclude that C and N fluxes in northern peatlands can be expected to respond to global change in many different ways. There are many interactions among climate, soil, vegetation and gas exchange in natural peatland ecosystems, which are poorly understood—as also noted by Townsend and Rastetter for forest nutrient dynamics in Chapter 3—particularly when combined with human interference. These make prediction of the full impacts of global change on global peatlands a difficult problem. The authors argue however, that the most effective C management strategy for peatland ecosystems under a changing climate will be their preservation as *functioning units*. They express concern that under a warming climate, forested peatlands would become attractive targets for conversion to agriculture or other uses; this would not only result in considerable loss of stored C, but also greatly reduce the potential for further peat formation.

Predicting forest responses to climate change and their potential to store atmospheric carbon

The potential of managed forests to sequester C is examined by Kellomäki and Karjalainen in Chapter 5, using the boreal forests of Finland as a case study. They perform simulations to investigate this potential under both current and changed climate conditions, keeping track of both forest ecosystem C and transfers of C to forest products and landfills.

In recent decades, forest increment in Finland has exceeded harvesting removals, leading to a net uptake of C in the standing crop and soils of Finnish forest ecosystems (a finding also reported by Kauppi in Chapter 16 for European forests in general). Kellomäki and

Karjalainen attribute the increased forest increment to intensive management, possibly enhanced by the fertilisation effects of increased air-borne pollutants and/or CO_2.

They also examine simulated interactions among forest composition, management and climatic conditions and conclude that growth conditions would be improved under projected climate changes, leading to increased net C sequestration by Finnish forests. CO_2 enrichment is also considered but appears to be less important than the effects of projected temperature increases.

Using simulation results, the authors conclude that unmanaged Finnish forests will sequester considerably more C in the forest ecosystem than intensively managed ones, finding that over a single rotation, the average C stored is only about one third that of the unmanaged forest ecosystem. This difference tends to disappear, the authors point out, when the C retained in wood products is included in the comparison. This difference between managed and unmanaged forest ecosystem storage is in agreement with Burschel (1993) who argued that silvicultural management does not improve the C storage of existing forests to any great extent. Kellomäki and Karjalainen therefore conclude that afforestation provides a much greater potential contribution, although availability of suitable land is a clear constraint. For this reason, they stress that a halt to global deforestation should be an obvious, immediate and direct means for mitigating atmospheric CO_2 increases (some of these issues are discussed further in Chapters 18–24).

Sykes and Prentice (Chapter 6) also examine the potential influence of climate change on future C stocks in forest ecosystems, and focus on uncertainties introduced by the methods used to estimate forest ecosystem response. They use Swedish forests to compare two common modelling approaches. The first is an equilibrium model (BIOME1.1) which uses plant functional types to extend observed global distributions of potential vegetation and climate variables to project expected distributions under a future (equilibrium) climate. The other approach is a dynamic patch-level model (FORSKA2) which uses a more mechanistic simulation of growth physiology and species competition. They caution that their analysis should not be taken as a comprehensive attempt to predict future Swedish forest C budgets—it is intended only as a comparison of the two methodologies.

They compare the forest biomass estimates under present climate and the changes that would result from a 'mean $2\times CO_2$ climate' derived from four GCM projections of local climate anomalies. CO_2-fertilisation effects were not considered. For the dynamic simulations with FORSKA2, a linear change to the $2\times CO_2$ climate conditions was assumed over a 100-year period, followed by a stabilised climate under the new conditions.

BIOME1.1 was found to simulate present-day vegetation distributions that match very well with observation, but it systematically overestimates (by factors of 1.5–4.5) actual ecosystem C storage as derived from Swedish forest inventory data. The authors point out that this finding is similar to previously reported results (Price et al. 1993) for Canadian boreal forests, when BIOME1.1 results were compared to Canadian inventory data and independent measurements of Botkin and Simpson (1990). It appears that the differences are attributable to the intrinsic differences between potential and actual vegetation stocks. In contrast, FORSKA2 produces much closer estimates of true forest C stocks (also observed in the Canadian study).

Under the simulated future climate, Sykes and Prentice report that BIOME1.1 predicts significant changes to the current vegetation distribution, driven mainly by the extreme winter warming predicted by the GCMs for high latitudes in general and Scandinavia in particular. Generally, conifer-dominant and mixed forests that are adapted to cool climates are predicted to be replaced by temperate deciduous forests. In southern Sweden, warm-temperate evergreen forests become established—far north of their present-day distributions. Overall, these changes are predicted to increase total C storage in the study area by about 8%. FORSKA2 confirms the general directions of these predicted changes, but whereas biomass (and C storage) in the north are predicted to increase, in response to extended growing seasons, in the south they are predicted to decline. The authors provide a cautionary note that this latter phenomenon may be a modelling artefact and reflect an incomplete species list in the FORSKA2 simulations.

The simulations reported in Chapters 5 and 6 provide some insight into how rapid climate change could affect C storage in northern forests, using Fennoscandian forests as case studies. Sykes and Prentice also speculate that some of the decline in biomass productivity predicted by FORSKA2 for southern Sweden could be averted by the introduction of exotic tree species, e.g., from southern Europe.

Role of disturbances in northern ecosystems

In Chapter 7, Volney examines the role of herbivorous insects in unmanaged forest ecosystems, citing studies of the dynamics of spruce and jack pine budworms, major insect pests occupying an extremely wide range of habitats across Canada. Highlighting the importance of insect disturbances to forest C stores, he suggests that insect outbreaks cause losses of forest timber significantly greater than those from forest fires and harvesting combined. He summarises the insects' life histories and their impacts on forest productivity. Climate variation is demonstrably an important factor in several possible explanations for severe outbreaks of such insects.. There do not appear to be any genetic or ecological reasons why insect success should be affected by warmer climates, and indeed climate variations during recent decades may be responsible for an increase in the observed number of outbreaks.

Volney remarks:

> *"... the overwhelming body of evidence suggests that the local climate has a profound effect on the dynamics of the insects and the consequent damage to forest stands."*

He argues that it is reasonable therefore, to expect changes in climate to exacerbate this influence, particularly when combined with anticipated effects on the frequency and intensity of wildfires (see below). Volney even speculates that the scale of C emissions resulting from increased destruction of infested forests could contribute to a positive feedback effect under a warming climate (Kurz et al. 1995).

It is widely believed that management interventions can alter dynamics of some insect pests and hence increase net forest productivity thereby reducing net C emissions, but for budworms at least, there has been little success except at a very local level. On the other hand, inadvertent changes resulting from human activities have been implicated in altered outbreak patterns for western spruce budworm over the American south-west during much of the 20th century. Volney proposes that new regional and global management practices must be implemented which avoid environmental changes that lead to more frequent and severe outbreaks. Direct control of pest insect populations may also be possible using biological agents, as has already been used with some success in several regions on other species.

Referring to the North American spruce and pine budworm populations, Volney draws the provocative conclusion that:

> *"These insects may in fact be among the agents which inadvertently but ultimately extirpate their host trees from landscapes affected by an altered climate."*

The other widely recognised cause of boreal forest disturbances is wildfire. Using the boreal forests of North America as an example, Stocks et al. (Chapter 8) point out that wildfire has been a natural force in ecosystem dynamics for millennia: boreal forests are not merely adapted to, but actually depend upon, periodic stand-replacing wildfires to maintain their existence. Hence, they argue, even if total fire exclusion were technically possible, it would not be ecologically desirable.

One of the principal objectives of their paper is to demonstrate the central role that fire plays in the ecology of boreal forests, and to relate this to ecoclimatic conditions and the level of fire protection imposed by management. Referring to highly aggregated maps of burned areas, they show that although close to 100 000 individual fires occurred in the boreal forests of Canada during the 1980s, most of the area lost was caused by only about 3% of them. The major causes of fires are human action and lightening strikes, with significant differences in location, scale and impacts. Human-caused fires typically start near population centres, and hence are amenable to rapid detection and possible suppression. On the other hand, lightening activity can occur anywhere, and strikes may ignite multiple fires which can spread quickly before they are detected, hence overwhelming suppression efforts. The remoteness of many areas greatly exacerbates the difficulties of fire control, because the probability of successful suppression is critically dependent upon rapid detection and response.

The authors review the historical efficacy of fire protection measures and conclude that where resources have been focused on limited areas of high value forest, the area burned has been significantly reduced compared to other areas where less intensive efforts have been applied. They go on to conclude however, that economic pressures will likely result in reduced investment in fire protection and a return to more natural fire regimes over larger portions of the boreal landscape.

REFERENCES

Botkin DB, Simpson LG (1990) Biomass of the North American boreal forest. A step towards accurate global measures. *Biogeochem.* 9: 161–174.

Burschel P, Kürsten E, Larson BC (1993) Die Rolle von Wald und Forstwirtschaft im Kohlenstoff-haushalt-Eine Betrachtung für die Bundesrepublik Deutschland. *Forstliche Forschungsberichte* 126/1993. München, Deutschland, 135 pp.

Kurz WA, Apps MJ, Stocks BJ, Volney WJA (1995) Global climate change: disturbance regimes and biospheric feedbacks of temperate and boreal forests. In: Woodwell GM, Mackenzie FT (eds) Biotic feedbacks in the global climate change system will the warming speed the warming? Oxford University Press, New York, pp 119–133.

Price DT, Apps MJ, Kurz WA, Prentice IC, Sykes MT (1993) Simulating the carbon budget of the Canadian Boreal forest using an integrated suite of process-based models. In: Huor-Ung C (ed) Forest Growth Models and their uses. Canadian Forest Service, Sainte-Foy, QC, pp 251–264.

2 The global carbon cycle and the atmospheric record: "The problem definition"

Inez Fung

NASA Goddard Institute for Space Studies,
2880 Broadway, New York NY 10025, USA
and
School of Earth and Ocean Sciences, University of Victoria
PO Box 1700, Victoria BC, V8W 2Y2 Canada

INTRODUCTION

The abundance of CO_2 in the atmosphere has increased from ~315 ppmv in 1958 to ~350 ppmv in the 1990s. The increase in the decade of the 1980s is ~55% of the CO_2 release from fossil fuel combustion, estimated at 5.5 Pg C yr^{-1}, or ~40% of the total anthropogenic source, from fossil fuel combustion plus land use modification (estimated at 1.6 Pg C yr^{-1}). Thus, mass balance requires the removal of ~60% of the anthropogenic CO_2 by the surface. The partitioning of the CO_2 sink between the ocean and land is a subject of intense debate and research. As the residence time of CO_2 in the terrestrial biosphere is much shorter than that in the ocean, CO_2 sequestered now in the terrestrial biosphere may be returned to the atmosphere in the next 50–100 years to accelerate greenhouse warming (see Woodwell 1995). Therefore, understanding the mechanisms responsible for the CO_2 uptake is crucial for projections of future CO_2 levels in the atmosphere.

Direct measurements of the CO_2 exchanges between the atmosphere and the underlying surface are sparse, and extrapolations of instantaneous site-specific flux measurements to global, annual mean uptakes are difficult. The atmospheric increase is therefore used to constrain interpretation of the changes in the surface fluxes.

In the following, I will re-examine the contemporary carbon (C) budget. I will take the perspective of the atmosphere, and use the terms CO_2 sources and sinks to denote fluxes into and out of the atmosphere, respectively. The corresponding sign convention would be that a positive flux increases the atmospheric abundance.

In a synthesis of the atmospheric and oceanic measurements of carbon dioxide in a global three-dimensional tracer transport model, Tans et al. (1990) found that the north–south gradient of CO_2 in the atmosphere between 1980 and 1987 implied a terrestrial C sink in the middle latitudes of the northern hemisphere that is larger than the global oceanic sink by as much as a factor of three. They stated that the mechanism of the sink was unknown. In a similar study, Keeling et al. (1989) used the atmospheric gradient of $\delta^{13}C$ in addition to CO_2 in their analysis, and inferred a terrestrial C sink that is comparable to the oceanic sink. The small number of monitoring sites with $\delta^{13}C$ measurements and the uncertainties in the fractionation coefficients of ^{13}C limit the strength of the $\delta^{13}C$ constraint on the partitioning of the C sink between land and sea (Fung 1993; Tans et al. 1993). Nevertheless, the three-dimensional modelling results all point to a terrestrial uptake of anthropogenic CO_2, with the uptake approximately equally partitioned between tropical and mid-latitude ecosystems. The

NATO ASI Series, Vol. I 40
Forest Ecosystems, Forest Management and the Global Carbon Cycle
Edited by Michael J. Apps and David T. Price

mid-latitude terrestrial sink is constrained by the latitudinal gradients of CO_2 and $\delta^{13}C$; while the tropical sink is obtained as a residual in the global $\delta^{13}C$ budget (Keeling et al. 1989).

Fung (1995) discussed the uncertainties in the estimate of terrestrial C sink. This paper reviews how the magnitude of the terrestrial C sink (the "missing" sink) is estimated and presents a framework for analysing the mechanisms responsible for the sink.

STEADINESS OF THE NATURAL STATE

Imagine a system without external (e.g., human) perturbations. The system consists of several reservoirs of C and CO_2 is exchanged among the reservoirs. Imagine also complete records of the temporal and spatial variations in reservoir sizes and the fluxes into and out of the reservoirs. Records on 100 000 year time scales would contain glacial–interglacial cycles, and other climatic cycles of millennial, centennial and shorter periods. Here I focus on the recent interglacial period, say, over the last 1000 years. The CO_2 sources and sinks at a particular location would show natural fluctuations on time scales that reflect seasonality, El Niños/Southern Oscillations, fires and fluctuations associated with other naturally recurring phenomena.

The atmosphere registers the net fluxes, ΔF, of CO_2 exchanged with the different reservoirs:

$$\Delta F = F_{in}(x,t) + F_{out}(x,t) \qquad [1]$$

where x is space, t is time, and $F_{in}(x,t)$ and $F_{out}(x,t)$ are the gross fluxes into and out of the atmosphere, respectively.

Define the averages of the fluctuations in the fluxes over some area, A, and time duration, T, thus:

$$\langle q \rangle_T = 1/(AT) \int_T \int_A q(x,t) dx\, dt$$

where q is any parameter. When T is longer than several of the longest cycles and A is a sufficiently large region (such as an ecosystem), the net flux $\langle \Delta F \rangle$ approaches zero and so a steady state obtains:

$$\langle \Delta F \rangle_T = \langle F_{in}(x,t) + F_{out}(x,t) \rangle_T \rightarrow 0 \quad \text{as } T \rightarrow \infty \qquad [2]$$

Consider the following simple example. A forest is divided into 100 stands and the fire cycle of the forest is 100 years. Let the initial biomass be denoted by $B(0)$, the biomass after the fire by $B(1) = 0$, and the biomass recovery curve by $B(t)$ at time t. Then a stand may have a fire in one year, and recovers from the fire in the following 99 years. At $(k-1)$ years after the fire, the biomass is $B(k)$, the annual biomass increment is $B(k) - B(k-1)$, and eventually $B(100) = B(0)$. At this site, the net exchange with the atmosphere is zero when integrated over the 100 year period. An alternate view is that of the entire forest for a single year. Every year, there is a fire on one of the stands. Of the remaining 99 stands, stand m is recovering from a fire $(m-1)$ years earlier, where $m = 2,3,\ldots,100$. The annual biomass increment (recovery rate) on stand m is $B(m) - B(m-1)$, and the total annual biomass increment on the entire area is $\Sigma_m[B(m) - B(m-1)] = B(100)$. Thus over the entire forest ecosystem with 100 stands, the amount of biomass lost by fire on one stand is recovered at different rates by the other 99 stands. In other words, the annual net flux is zero when integrated over the entire forest over one year.

In actuality, fires and other disturbances are not uniformly distributed in time or space: their frequencies are closely related to climate variability. Therefore, when viewed over a shorter observing period, say $T = 100$ yr or the decade of the 1980s, $\langle \Delta F \rangle_T$ may show an increasing or decreasing trend in time.

Table 2.1. Fluxes of carbon (C) viewed over the decade of the 1980s. The single arrows ↓ and ↑ denote gross fluxes out of and into the atmosphere, respectively; while the double arrow ↓↑ denotes net fluxes. See text for an explanation of the symbols.

	C FLUX (Pg C yr^{-1})					
	FOSSIL FUELS	OCEAN		LAND		$\partial C_a/\partial t$
Natural	0	$-85\downarrow+\delta_1$ $(F_{ao}\downarrow+\Phi_{bo})$	$+85\downarrow+\delta_2$ $(F_{oa}\downarrow+\Phi_{oa})$	$-120\downarrow+\delta_3$ $(F_{ab}\downarrow+\Phi_{ab})$	$+120\downarrow+\delta_4$ $(F_{ba}\downarrow+\Phi_{bo})$	
Perturbation	$5.5+\varepsilon_F\uparrow$	$\varepsilon_{od}\downarrow\uparrow$ $(D_o\downarrow\uparrow)$	$I_{oe}+\varepsilon_{oi}\downarrow\uparrow$ $(I_o\downarrow\uparrow)$	$D_{be}+\varepsilon_{bd}\downarrow\uparrow$ $(D_b\downarrow\uparrow)$	? $(I_b\downarrow\uparrow)$	
Total	$5.5+\varepsilon_F\uparrow$	$I_{oe}+(\delta_1+\delta_2+\varepsilon_{od}+\varepsilon_{oi})\downarrow\uparrow$		$D_{be}\downarrow\uparrow$ + "Missing" ↓↑		$3.2+\varepsilon_a$

Table 2.1 summarises the fluxes of C into and out of the atmosphere for the decade of the 1980s. The fluxes have been divided into natural or background fluxes and a perturbation, with the symbols F_{ij} denoting the unperturbed gross fluxes from reservoir i to reservoir j, and the subscript i or j stands for a (atmosphere), o (ocean) and b (terrestrial biosphere). Here,

$$F_{ij} = \langle F_{ij} \rangle_T = \langle F_{ij} \rangle_\infty + \delta_k \qquad [3]$$

where $\langle F_{ij} \rangle_T$ is the actual gross flux for the decade; $\langle F_{ij} \rangle_\infty$ is the long-term steady-state flux, and δ_k, for $k = 1,...4$, represents the difference between the two. In the natural unperturbed state, there is also a river flux Φ_{bo} into the oceans, which is returned from the oceans to the atmosphere, Φ_{oa}. On land, the riverine export Φ_{bo} is balanced by a flux Φ_{ab} from the atmosphere. As $\Phi_{bo} \approx \Phi_{oa} \approx \Phi_{ab} \approx 0.5$ Pg C yr^{-1}, I include their effects in the δ_k's.

For atmosphere–ocean exchange, $-\langle F_{ao} \rangle_\infty = \langle F_{oa} \rangle_\infty \approx 85$ Pg C yr^{-1} (cf. Tans et al. 1993). For atmosphere–biosphere exchange, $-\langle F_{ab} \rangle_\infty = \langle F_{ba} \rangle_\infty \approx 120$ Pg C yr^{-1}, the value of gross primary productivity for the entire globe. Note that the δ_k's do not denote uncertainties in the gross fluxes, but the departure of the actual gross flux for the 1980s from the long-term mean gross flux. Because year-to-year variability in the gross fluxes may be large for a 10 or 100 year period, neither $F_{ao}+F_{oa} = \delta_1+\delta_2$, nor $F_{ab}+F_{ba} = \delta_3+\delta_4$, is likely to equal zero.

CONTEMPORARY BUDGET

The study of the contemporary C cycle focuses on perturbations to the natural state. It is clear that each perturbation is made up of several fluxes, and could have been expressed in a manner similar to the exchanges with the natural state, as discussed above. However, I choose to separate the perturbation fluxes into two major categories: deliberate perturbations (FF, D_o

and D_b) as a result of human activities, and induced (or indirect) perturbations (I_o and I_b) in response to deliberate perturbations themselves and/or other changes (e.g., climate) resulting from the deliberate perturbations. The separation is in deference to perturbation budgets commonly used in studies of the global C cycle, and is based primarily on the methodologies used to estimate the fluxes. Unlike the natural state, the deliberate and induced perturbation fluxes are expressed as net, rather than gross, fluxes. Large uncertainties surround the estimates of the perturbation fluxes. I shall use the subscript *e* to denote estimates of the fluxes and the symbol ε to denote errors in the estimates.

Enting and Mansbridge (1991) pointed out that changes in surface CO_2 fluxes do not translate directly into changes in atmospheric CO_2 storage. This is because other C-bearing gases (e.g., CO, CH_4) are emitted from the surface and are oxidised in the atmosphere, thus contributing to an increase of CO_2 abundance. This point is reiterated by Tans et al. (1995). The fluxes in the following will be C, rather than CO_2, fluxes.

Deliberate fluxes

Deliberate perturbations include CO_2 released to the atmosphere by human action, principally fossil fuel combustion (*FF*) and land-use modification (D_b).

Estimates of "deliberate" fluxes are grounded in national statistics on, *inter alia*, energy, agriculture, economics and trade. The fossil fuel flux is a unidirectional flux, estimated to be 5.5 ±0.5 Pg C yr^{-1} for 1980–1989 (Marland 1989; Marland et al. 1994). Here the fossil fuel flux can be expressed as

$$FF = FF_e + \varepsilon_F \qquad [4]$$

where FF_e is the estimated flux (5.5 Pg C yr^{-1}) and ε_F is the error in the estimate.

The flux associated with land-use modification (D_b) is the net result of incomplete cancellation of C storage in wood products and CO_2 absorption via agriculture and forest regrowth CO_2 on the one hand; and CO_2 releases via biomass burning, decomposition of detritus, oxidation of soil C, and slow decay of the wood products on the other. Houghton's (1991, 1994) direct estimate of 1.6 Pg C yr^{-1} for the decade of the 1980s accounts for changes in the tropics. The magnitude of C flux due to land-use changes in the mid-latitudes is uncertain (e.g., Dixon et al. 1994). Instead of assigning a value to this term in Table 2.1, it is represented by the symbol

$$D_b = D_{be} + \varepsilon_{bd} \qquad [5]$$

where D_{be} is the estimated source strength, and ε_{bd} is the error in the estimated flux. If a value of 1.6 Pg C yr^{-1} is chosen for the direct flux, then

$$D_b = 1.6 + \varepsilon_{bd}$$

The degree to which human actions have affected the oceanic C cycle directly (e.g., via fisheries) and atmosphere–ocean CO_2 fluxes (D_o) has not been quantified, but is likely small:

$$D_o \approx 0 + \varepsilon_{od} \qquad [6]$$

where ε_{od} is the error in the estimate of the deliberate oceanic source.

Indirect fluxes

The altered composition in the atmosphere induces changes in the states of the atmosphere, oceans and biosphere, which, in turn, alter the gross exchanges among these reservoirs. Estimation of the indirect fluxes with the ocean are carried out using ocean models. Both models that include, and models that exclude, changes to the background oceanic C cycle have been used (e.g., Maier-Reimer and Hasselmann 1987; Sarmiento et al. 1992). Because a true steady-state background is mathematically imposed in these models, the resultant fluxes are indirect perturbation fluxes. The estimates of the ocean sink for fossil fuel CO_2 for the 1980s range from 1.47 Pg C yr^{-1} (Maier-Reimer 1993) to 2.3 Pg C yr^{-1} (Siegenthaler and Sarmiento 1993). Some of the models do not include the land-use term in the forcing, and so the terrestrial uptake required to balance the budget would be a net biotic flux, $(D_b + I_b)$, and would be smaller in magnitude than the induced terrestrial sink I_b. Like the other terms discussed above, the symbol I_o is used to represent the real (as yet unknown) oceanic uptake of anthropogenic CO_2, so that

$$I_o = I_{oe} + \varepsilon_{oi} \qquad [7]$$

If the uptake is estimated to be I_{oe} = 2 Pg C yr^{-1}, then $I_o = 2 + \varepsilon_{oi}$, where ε_{oi} is the error in the uptake estimate. In view of the wide range of the estimates, ε_{oi} is not likely to be small.

A similar analysis can be applied to the biospheric fluxes. The term terrestrial "missing" sink is sometimes used to refer to the atmosphere–biosphere fluxes induced by changing climate and atmospheric composition, i.e., the indirect flux I_b. However, I shall use the term "missing" sink to represent the flux inferred as the residual in the mass balance in the atmosphere. As will be shown below, the missing sink incorporates terms other than I_b.

Mass balance

The atmospheric accumulation of C can be written in a form similar to the other terms:

$$dC_a/dt = dC_a/dt_e + \varepsilon_a \qquad [8]$$

As the atmospheric accumulation term is a direct measurement of the change in atmospheric abundance, it comprises both changes in the natural state as well as the perturbations. Thus, in Table 2.1, it appears in the row denoted "total" (bottom line) rather than in the row denoted "perturbations". The net oceanic and biospheric fluxes, ΔF_o and ΔF_b, respectively, for the decade of the 1980s can be obtained by adding the respective natural and perturbation fluxes:

$$\begin{aligned} \Delta F_o &= \langle F_{oa}\rangle_T + \langle F_{ao}\rangle_T + D_o + I_o \\ &= (\langle F_{oa}\rangle_\infty + \delta_1) + (\langle F_{oa}\rangle_\infty + \delta_2) + \varepsilon_{od} + (I_{oe} + \varepsilon_{oi}) \\ &= I_{oe} + (\delta_1 + \delta_2 + \varepsilon_{od} + \varepsilon_{oi}) \qquad [9] \end{aligned}$$

$$\begin{aligned} \Delta F_b &= \langle F_{ba}\rangle_T + \langle F_{ab}\rangle_T + D_b + I_b \\ &= \langle F_{ba}\rangle_\infty + \delta_3 + \langle F_{ba}\rangle_\infty + \delta_4 + D_{be} + \varepsilon_{bd} + I_b \end{aligned}$$

$$= D_{be} + I_b + (\delta_3 + \delta_4 + \varepsilon_{bd}) \qquad [10]$$

where ΔF_b is the net biotic flux. The missing sink is generally defined to be the unidentified part of the net biotic flux, after the deliberate flux due to land use modification has been accounted for:

$$\text{Missing}\downarrow\uparrow = \Delta F_b - D_{be}$$

$$= I_b + (\delta_3 + \delta_4 + \varepsilon_{bd}) \qquad [11]$$

Equation [11] states that the missing sink includes not only the induced fluxes I_b, but also imbalances in the background state in the 1980s and errors in the estimate of the land-use flux.

The terms on the right hand side of Equation [11] are not known globally, and so the magnitude of the missing term cannot be evaluated directly. Instead, the magnitude of the missing sink is determined as a residual in the atmospheric mass balance:

$$dC_a/dt = FF + \Delta F_o + \Delta F_b \qquad [12]$$

or

$$\Delta F_b = dC_a/dt - FF - \Delta F_o$$

$$= dC_a/dt_e - FF_e - I_{oe} + \varepsilon_a - (\varepsilon_F + \delta_1 + \delta_2 + \varepsilon_{od} + \varepsilon_{oi}) \qquad [13]$$

so that

$$\text{Missing}\downarrow\uparrow = \Delta F_b - D_{be}$$

$$= dC_a/dt_e - FF_e - I_{oe} - D_{be} + \varepsilon_a - (\varepsilon_F + \delta_1 + \delta_2 + \varepsilon_{od} + \varepsilon_{oi}) \qquad [14]$$

For an estimated atmospheric increase dC_a/dt_e = 3.2 Pg C yr^{-1}, fossil fuel release FF_e = 5.5 Pg C yr^{-1}, land-use source of D_{be} = 1.6 Pg C yr^{-1}, and an estimated oceanic uptake of I_{oe} = -2.0 Pg C yr^{-1},

$$\text{Missing}\downarrow\uparrow = 3.2 - 5.5 + 2 - 1.6 + \varepsilon_a - (\varepsilon_F + \delta_1 + \delta_2 + \varepsilon_{od} + \varepsilon_{oi})$$

$$= -1.9 + \varepsilon_m$$

where

$$\varepsilon_m = \varepsilon_a - (\varepsilon_F + \delta_1 + \delta_2 + \varepsilon_{od} + \varepsilon_{oi})$$

Equation [14] represents an indirect estimate of the missing flux, while Equation [11] is a direct estimate of the same flux from the biosphere. From Equation [14], it is clear that the magnitude of the missing terrestrial sink includes errors in the estimates of the atmospheric trends (ε_a), anthropogenic emissions (ε_F), and assumption of a negligible human impact on the natural oceanic C cycle (ε_{od}), errors in the calculation of the oceanic sink (ε_{oi}), as well as the disequilibrium in the background oceanic exchanges (δ_1 and δ_2). Equation [11] states that both flux imbalances (δ_3+δ_4) in the natural biosphere, errors in the land-use flux estimate (δ_{bd}), and induced fluxes ($I_b\downarrow\uparrow$) due to feedbacks between the atmosphere–biosphere system contribute to the missing sink. In other words, identification of the missing sink must involve a comprehensive analysis for changes in the entire terrestrial system. Also, while the contemporary C budget places a constraint on how the biosphere responds to perturbations in climate and atmospheric composition, the constraint is only as good as our understanding of the fluctuations in the natural biosphere and our estimates of the land-use source term.

Separation of the budget in the form in Table 2.1 aids in the analysis of the budget. For example, the flux measurements of Wofsy et al. (1993) reveal a contemporary net sink of C into Harvard forest, which was recovering from the hurricane destruction of 1938. This sink, to a first approximation, reflects a non-steady natural state (δ_3). Enhancement of the regrowth due to indirect effects (if they exist) would belong to the term I_b in the perturbation budget. Other factors affecting the natural state, such as fires (Stocks et al., Chapter 8; Kurz and Apps, Chapter 14), insect infestation (Volney, Chapter 7), peat accumulation and ablation (Zoltai and Martikainen, Chapter 4) and climate variability (Dai and Fung 1993), must all be included in δ_3 and δ_4 and hence in the "missing" sink for the decade of the 1980s.

Similarly, Houghton's (1991) estimate of the land-use term (D_b) includes the effects of regrowth of forests after the abandonment of agriculture in North America and Europe, but excludes the indirect effects on the regrowth (I_b) in their budgets (likely to be small). Thus interpretation of measurements of tree growth in terms of the missing sink, such as done by Kauppi et al. (1992), must first exclude the portion of the regrowth already accounted for in D_b by Houghton (1991). This point was made by Rastetter and Houghton (1992).

FACTORS INFLUENCING ESTIMATES OF THE MISSING BIOTIC SINK

The atmosphere registers the net, rather than gross, fluxes of CO_2. Changes in the net fluxes, however, generally involve changes in several or all of the gross fluxes. For example, changes in photosynthesis are likely to result in changes in respiration. Temperature changes are likely to affect all processes at different rates. Therefore the need for a comprehensive analysis cannot be over-emphasised. Some of the mechanisms are listed in Figure 2.1. The following discusses as examples, several terms in Figure 2.1.

Land-use modification (ε_{bd})

A comprehensive inventory must include a comprehensive analysis of the fate of all the C pools, including the below ground pools, and wood products removed from the site. Any accounting must quantify the response of biomass to CO_2, N perturbations and climate variability, response of the soils to the disturbance, as well as the inventory and fate of wood products. For example, Schlesinger (1986), and Post and Mann (1990) among others, have quantified the depletion of soil C (source to the atmosphere) due to agriculture. However, changes in agricultural practices since the 1950s may result in slower depletion, and hence a net sink of atmospheric CO_2 for the decade of the 1980s compared to the 1950s.

$F_{ba}\uparrow$ +120 + δ_4	$F_{ab}\downarrow$ -120 + δ_3
Autotrophic respiration	GPP
Heterotrophic respiration	
Disturbance	Recovery
Peat ablation	Peat accumulation
...	...

$D_{be} + \varepsilon_{bd}\uparrow\downarrow$		Missing + $\varepsilon_m\uparrow\downarrow$	
$\uparrow$	$\downarrow$	$\uparrow$	$\downarrow$
Burning	Regrowth	Enhanced $F_{ba}\uparrow$	Enhanced $F_{ab}\downarrow$
Detritus decomp.	Storage as wood products	(CO_2-fertilisation	
Soil oxidation	...	N-fertilisation	
Decomp. of wood products		Climate change ....)	
...			

Figure 2.1. Biospheric processes that may contribute to the net carbon sink in the 1980s. The fluxes are in Pg C yr^{-1}. Values of the fluxes are discussed in the text.

CO_2 fertilisation and N fertilisation (I_b)

CO_2 fertilisation has been demonstrated in laboratories and in fields with unlimited nutrient supplies, and N limitation has been demonstrated for most forested ecosystems in temperate and boreal regions (Vitousek et al. 1988). However, the *in-situ* limitations on ecosystem response to increasing CO_2 levels in the atmosphere and to increasing N deposition have not been mapped and quantified globally. Suppose there is an increase in photosynthesis. It is expected that respiration and decomposition would change as a result, and that C–N interactions would be modified in the presence of CO_2 fertilisation. It is not clear if these changes would act to enhance or cancel the CO_2 sink. While the missing sink has been attributed to CO_2 fertilisation of the terrestrial biosphere (e.g., Keeling et al. 1989), Enting and Mansbridge (1986, 1991), Enting (1992) and Friedlingstein et al. (1995) have pointed out that the time history of the missing sink as obtained by deconvolution (historical analysis), is different from the time history of atmospheric CO_2, thus underlining the role of the other abiotic factors. In a calculation of the effects of nitrogen deposition alone, Townsend et al. (1995) report a net sink of 0.3–0.6 Pg C yr^{-1} in 1990. The interactions of CO_2 and N fertilisation need to be investigated (Rastetter et al. 1991, 1992).

Climate interactions (δ_3, δ_4, I_b)

The documented interannual variations of atmospheric CO_2 at Mauna Loa and other monitoring sites point to very sensitive response of ecosystems to small perturbations in temperature, precipitation and solar irradiance. Dai and Fung (1993) suggest that the year-to-year imbalance in photosynthesis and respiration due to climate variability could account for approximately half the missing sink over the past 80 years. Kurz et al. (1995) suggest that, in Canada, recovery of forests from fires may account for ~0.1 Pg C yr^{-1} (sink) in the 1980s. Zoltai and Martikainen (Chapter 4) point out that peat accumulation in the high-latitude is a CO_2 sink of 0.1–0.15 Pg C yr^{-1}. These examples are snapshots of the fluxes. With the prospect of global changes in temperature and precipitation, it is crucial to quantify the sensitivities of these C fluxes to climate, so that projections of the C balances may be made with greater confidence.

The above examples highlight the need for a comprehensive analysis of the C fluxes into and out of entire ecosystems, and the variability of these fluxes through time.

ACKNOWLEDGEMENTS

The author acknowledges discussions with Dr I Enting and comments by the reviewers. This work is supported by NASA's Office of Mission to Planet Earth, and the Carbon Dioxide Research Program of the US Department of Energy.

REFERENCES

Dai AG, Fung IY (1993) Can climate variability contribute to the "missing" CO_2 sink? *Global Biogeochemical Cycles* 7: 525–535.

Dixon RK, Brown S, Houghton RA, Solomon AM, Trexler MC, Wisniewski J (1994) Carbon pools and flux of global forest ecosystems. *Science* 263: 185–190.

Enting IG, Mansbridge JV (1986) The incompatibility of ice-core CO_2 data with reconstructions of biotic CO_2 sources. *Tellus* B 39: 318–325.

Enting IG, Mansbridge JV (1991) Latitudinal distribution of sources and sinks of CO_2: results from an inversion study. *Tellus* B 43: 156–170.

Enting IG (1992) The incompatibility of ice-core CO_2 data with reconstruction of biotic CO_2 sources II. The influence of CO_2 fertilized growth. *Tellus* B 44: 23–32.

Friedlingstein P, Fung I, Holland E, John J, Brasseur G, Erickson D, Schimel D (1995) On the contribution of the biospheric CO_2 fertilisation to the missing sink. *Global Biogeochemical Cycles*, (In press).

Fung I (1993) Models of oceanic and terrestrial sinks of anthropogenic CO_2: A review of the contemporary carbon cycle. In: Oremland RS (ed) The Biogeochemistry of Global Change: Radiative Trace Gases, Chapman and Hall, New York, USA, pp 169–189.

Fung I (1995) Perturbations to the biospheric carbon cycle: uncertainties in the estimates. In: Woodwell GM, Mackenzie FT (eds) Biotic Feedbacks in the Global Climatic System: Will the Warming Feed the Warming? Oxford University Press, Oxford, pp 367–374.

Houghton RA (1991) Tropical deforestation and atmospheric carbon dioxide. *Climatic Change* 19: 99–118.

Houghton RA (1994) The worldwide extent of land-use change. *Bioscience* 44: 305–313.

Kauppi PE, Mielikainen K, Kuusela K (1992) Biomass and carbon budget of European forests, 1971–1990. *Science* 256: 70–74.

Keeling CD, Piper SC, Heimann M (1989) A three dimensional model of atmospheric CO_2 transport based on observed winds: 4. mean annual gradients and interannual variations. In: Peterson DH (ed) Aspects of climate variability in the Pacific and the Western Americas, American Geophysical Union, Washington DC, pp 305–363.

Kurz WA, Apps MJ, Beukema SJ, Lekstrum T (1995) Twentieth century carbon budget of Canadian forests. *Tellus* B 47: 170–177.

Kurz WA and Apps MJ (1995) Retrospective assessment of carbon flows in Canadian boreal forests. (Chapter 14, this volume).

Maier-Reimer E, Hasselmann K (1987) Transport and storage of CO_2 in the ocean—an inorganic ocean-circulation cycle model. *Climate Dynamics* 2: 63–90.

Maier-Reimer E (1993) The biological pump in the greenhouse. *Global and Planetary Change* 8: 13–15.

Marland G (1989) Fossil fuel CO_2 emissions. In: CDIAC Communications, Winter 1989, Carbon Dioxide Information Analysis Center, Oak Ridge National Laboratory, Oak Ridge, Tennessee, pp 1–3.

Marland G, Andres RJ, Boden TA (1994) Global, regional, and national CO_2 emissions. In: Boden TA, Kaiser DP, Sepanski RJ, Stoss FW (eds) Trends '93: A Compendium of Data on Global Change, ORNL/CDIAC-65. Carbon Dioxide Information Analysis Center, Oak Ridge National Laboratory, Oak Ridge, Tennessee, pp 505–584.

Post WM, Mann LK (1990) Changes in soil organic carbon and nitrogen as a result of cultivation. In: Bouwman AF (ed) Soils and the Greenhouse Effect, Wiley, New York, pp 401–407.

Rastetter EB, Ryan MG, Shaver GR, Melillo JM, Nadelhofer KJ, Hobbie JE, Aber JD (1991) A general biogeochemical model describing the responses of the C and N cycles in terrestrial ecosystems to changes in CO_2, climate, and N deposition. *Tree Physiology* 9: 101–126.

Rastetter EB, McKane RB, Shaver GR, Melillo JM (1992) Changes in C storage by terrestrial ecosystems: how C–N interactions restrict responses to CO_2 and temperature. *Water, Air and Soil Pollution* 64: 327–344.

Sarmiento JL, Orr JC, Siegenthaler U (1992) A perturbation simulation of CO_2 uptake in an ocean general circulation model. *J. Geophys. Res.* 97: 3621–3645.
Schlesinger W (1986) Changes in soil carbon storage and associated properties with disturbance and recovery. In: Trabalka JR, Reichle DE (eds) The Changing Carbon Cycle: A Global Analysis, Springer-Verlag, New York, pp 194–220.
Siegenthaler U, Sarmiento JL (1993) Atmospheric carbon dioxide and the ocean. *Nature* 365: 119–125.
Stocks BJ, Lee BS, Martell DL (1995) Some potential carbon budget implications of fire management in the boreal forest. (Chapter 8, this volume).
Tans PP, Fung IY, Takahashi T (1990) Observational constraints on the global atmospheric CO_2 budget. *Science* 247: 1431–1438.
Tans PP, Berry, JA, Keeling RF (1993) Oceanic $^{13}C/^{12}C$ observations: A new window on ocean CO_2 uptake. *Global Biogeochemical Cycles* 7: 353–368.
Tans PP, Fung IY, Enting IG (1995) Storage versus flux budgets: The terrestrial uptake of CO_2 during the 1980's. In Woodwell GM, Mackenzie FT (eds) Biotic Feedbacks in the Global Climatic System: Will the Warming Feed the Warming? Oxford University Press, Oxford, pp 351–366.
Townsend AR, Rastetter EB (1995) Nutrient constraints on carbon storage in forested ecosystems. (Chapter 3, this volume).
Vitousek PM, Fahey T, Johnson D, Swift M (1988) Element interactions in forest ecosystems: succession, allometry and input–output budgets. *Biogeochemistry* 5: 7–34.
Volney WJA (1995) Global change and the management of insects in forest ecosystems. (Chapter 7, this volume).
Wofsey SC, Goulden ML, Munger JE, Fan S-M, Bakwin PS, Bassow SL, Bazzaz FA (1993) Net exchange of CO_2 in a mid-latitude forest. *Science* 260: 1314–1317.
Woodwell GM (1995) Biotic feedbacks from the warming of the Earth. In: Woodwell GM, Mackenzie FT (eds) Biotic Feedbacks in the Global Climatic System: Will the Warming Feed the Warming? Oxford University Press, Oxford, pp 3–21.
Zoltai SC, Martikainen PJ (1995) Estimated extent of forested peatlands and their role in the global carbon cycle. (Chapter 4, this volume).

3 Nutrient constraints on carbon storage in forested ecosystems

Alan R Townsend and Edward B Rastetter[1]

Cornell Center for the Environment
200 Rice Hall, Cornell University
Ithaca, New York, USA 14853

INTRODUCTION

Carbon (C) cycling in terrestrial ecosystems is inextricably linked to the cycles of essential nutrients like nitrogen (N) and phosphorus (P) (Bolin and Cook 1983; Melillo and Gosz 1983; Hobbie et al. 1984; Vitousek et al. 1988; Rastetter et al. 1991, 1992; Rastetter and Shaver 1992). Consequently, ecosystem C uptake and storage should be tightly constrained by the availability of these nutrients. In particular, both vegetation and soil micro-organisms must maintain a nutritional balance to survive. This nutritional balance sets limits on the amount of C that can be sequestered in an ecosystem with limited nutrient capital. From this perspective, changes in the amount of C sequestered in ecosystems must be accompanied by at least one of the following changes in the chemical properties of ecosystems (Rastetter et al. 1992):

1. A change in the total nutrient capital of the ecosystem.
2. A change in the carbon-to-nutrient ratio of vegetation and/or soils.
3. A change in the proportion of the total ecosystem nutrient capital that is in vegetation (with high carbon-to-nutrient ratios) versus soils (with low carbon-to-nutrient ratios).

In this paper we will examine how these three chemical properties of ecosystems constrain the responses of forests to changes in the global environment. Specifically, we will discuss responses of forests to changes in atmospheric CO_2 concentration, temperature, N deposition, and disturbance. First, however, we discuss the mechanisms underlying the constraints imposed by each of the three properties. We will concentrate on N, although most of what we will say can be generalised to other essential nutrients, and we will discuss the potential effects of increased P deposition in tropical areas.

Changes in ecosystem N capital

Most forested ecosystems have relatively tight N cycles (Bormann and Likens 1967; Cole and Rapp 1981; Johnson and Van Hook 1989). That is, the nutritional needs of forests are supplied almost entirely by internal recycling rather than from external sources. This reliance on internally recycled N arises because the rates at which N is supplied from external sources are generally low relative to demand for N by vegetation and soil micro-organisms. Thus, to maintain high rates of production and large stocks of organic matter, N has to be recycled internally and retained efficiently.

---

1 The Ecosystems Center, Marine Biological Laboratory
Woods Hole, Massachusetts, USA 02543

NATO ASI Series, Vol. I 40
Forest Ecosystems, Forest Management and the Global Carbon Cycle
Edited by Michael J. Apps and David T. Price

The major external sources of N are symbiotic and asymbiotic fixation of N_2, and deposition of both oxidised and reduced forms of N from the atmosphere. Such deposition can either be as rainfall, aerosols, or direct gaseous uptake by the vegetation (Logan et al. 1981). Changes in the N capital of an ecosystem are obviously the result of an imbalance between sources (atmospheric deposition and fixation) and losses (leaching, erosion, trace gas emissions, and deep burial). Except in forests with an abundance of N-fixing species (e.g., many tropical forests; Jordan 1985), the ecosystem has little control on the rate of N supply. Nitrogen losses, on the other hand, are strongly controlled by ecosystem processes. In particular, the strong demand for N by both vegetation and soil micro-organisms appears to preclude the build up of inorganic N that can be leached or emitted from the ecosystem. A disturbance such as clear-cutting, however, which decouples the tight linkages between soil and vegetation pools can dramatically increase N losses from the system (Vitousek et al. 1979). Unless accompanied by either an increase in C:N ratios or a movement of N from components with low C:N ratios to components with high C:N ratios, this loss of N will result in a commensurate loss of C.

Changes in vegetation and soil C:N ratios

The C:N ratios of vegetation vary widely in forests (Vitousek et al. 1988). This variance is mostly the result of changes in the abundance of woody tissues relative to foliar biomass and fine roots. However, changes in C:N ratios can also result from environmentally induced differences in the C:N ratio of individual tissues (e.g., high C:N ratios in leaves exposed to high CO_2 concentrations; low C:N ratios in high fertility sites). In general, most forests have a high potential for sequestering C by increasing the amount of woody tissues, but could lose C if disturbances reduce the average amount of wood in forest stands.

The C:N ratios of soils can also vary widely in forests (Schimel et al. 1994). These ratios will depend mostly upon the extent of decomposition of the soil organic matter. Fresh litter tends to have high C:N ratios (>40) that decline as the organic matter is processed by micro-organisms (approaching 10). In cold, water-logged soils, microbial processes proceed slowly and the organic matter can have high C:N ratios even after thousands of years. In warm soils with more optimal water content, microbial processes proceed more rapidly and organic matter is decomposed to a recalcitrant residual with very low C:N ratios within a few decades. Because of the high C:N ratios in water-logged soils, they can lose C if drained without a concomitant loss of N (i.e., by decreasing soil C:N ratios). On the other hand, most upland forests have soil C:N ratios near their lower limit. Thus, any loss of C from these soils is likely to be accompanied by a substantial increase in N mineralization, which should in turn stimulate plant growth and cause storage of C in vegetation at a higher C:N ratio.

Changes in the allocation of N between vegetation and soils

The potential for storing C by moving N from soils to vegetation is clearly related to how widely the C:N ratios of vegetation and soils differ from one another. Because of the predominance of woody tissues in forest vegetation, the C:N ratios of vegetation and soils differ by about an order of magnitude (e.g., Edwards et al. 1989, Johnson and Henderson 1989). This wide difference creates the potential for about a 10:1 trade-off between C gained in vegetation to C lost from soils for each atom of N mineralised from soils and incorporated into vegetation (Figure 3.1). However, this potential gain may be somewhat of an overestimate because a large amount of the soil organic matter is both recalcitrant and has a low C:N ratio (Trumbore 1993; Schimel et al. 1994).

The second factor determining the potential for storing C by moving N from soils to vegetation is the initial distribution of N. Soils are by far the largest reservoir of N in forests. In temperate and boreal forests, over 90% of the total N is in soil organic matter and litter (Cole and Rapp 1981, Johnson and Henderson 1989). In tropical forests this number may be as low as 80% (Jordan 1985). Because of this distribution, even minor losses of N from the soil could greatly increase C storage in vegetation if the N is made available for plant growth. This, of course, assumes that the N lost from soils is taken up and incorporated into vegetation rather than being lost from the ecosystem as discussed above.

Forest responses to environmental change

Changes in CO_2 concentration

The direct effect of elevated CO_2 concentrations is to increase the rate of C fixation by the carboxylating enzymes in leaves. The indirect effects are complex and poorly understood, but may include a downward adjustment in the photosynthetic capacity of the canopy (Bazzaz 1990). From the nutrient cycling perspective, the more important responses are likely to be an increase in leaf C:N ratios, perhaps accompanied by an increase in wood production, and an increase in the C:N ratio of litter. Decomposition of this higher C:N ratio litter will result in greater immobilisation of inorganic N by soil micro-organisms, which will in turn have two opposing effects on the N cycle. First, it will make N less available to vegetation and may

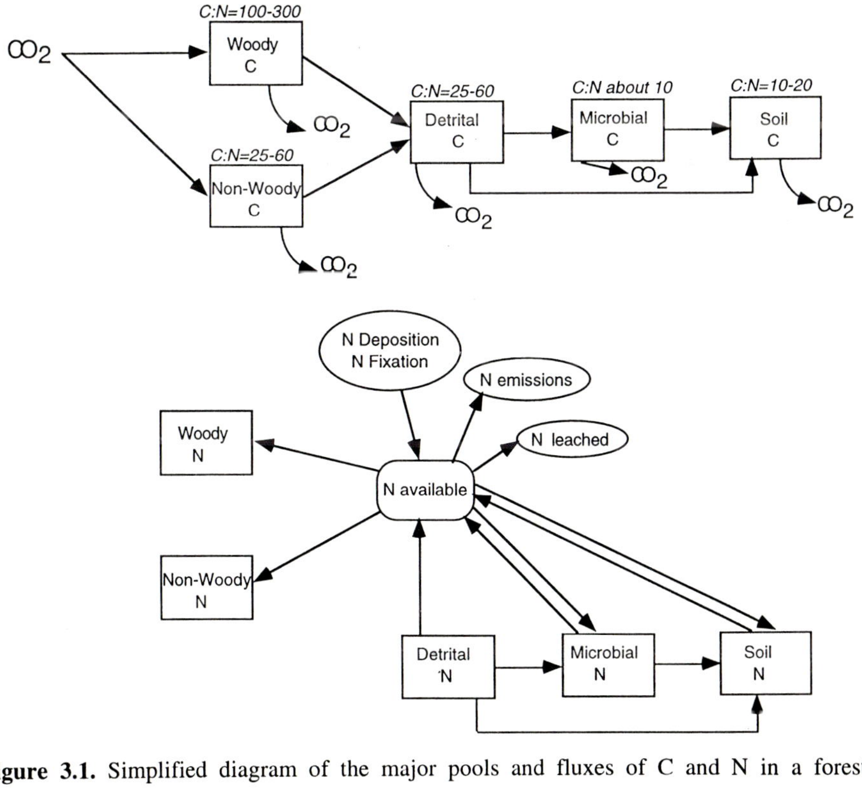

Figure 3.1. Simplified diagram of the major pools and fluxes of C and N in a forest.

limit productivity. Second, because inorganic N concentrations are reduced, less N is available to be leached or denitrified, thereby increasing the N-retention efficiency and increasing the total N capital of the ecosystem.

These responses to CO_2 enrichment will affect all three of the factors listed above. The N capital of the ecosystem will increase in the long-term (depending on the rate of external supply) because of the increased N-retention efficiency. The C:N ratios of vegetation and soils will increase because of an increase in woody tissues, and an increase in the C:N ratio of individual tissues and of litter. Both of these factors will tend to increase C sequestration. However, there should also be a net movement of N from vegetation to soils because the N uptake by vegetation will decrease as a result of the lower N availability. The net effect of all three factors will probably be dominated by the increase in C:N ratios, at least in the short term. In the long term the increase in N capital might also contribute significantly to a net increase in C sequestered.

Changes in temperature

The direct effect of an increase in temperature is to increase the rates of all metabolic processes including photosynthesis, plant respiration, and decomposition. Based on the direct effects alone, the rates of respiration and decomposition are likely to increase more than the rate of photosynthesis (Townsend et al. 1992). However, the indirect effects can again be complex and may change the long-term response. The key to these indirect effects is the increase in decomposition. Although this increase is likely to release a large amount of soil C to the atmosphere, it will also mineralise soil N, which should increase the availability of inorganic N. The increase in inorganic N will again have two opposing effects. First, there will be more N available to be leached or lost in gaseous forms. Thus, the N-retention efficiency of the ecosystem should decrease, resulting in a long-term loss of N capital and a decrease in the C-storage capacity of the ecosystem. Second, uptake of N by vegetation should be stimulated by higher N availability, resulting in a net movement of N from soils to vegetation, which will tend to increase C storage in vegetation. Increased temperature will probably decrease C:N ratios of both vegetation and soils. Vegetation C:N ratios should decline because of both the increase in N uptake and the temperature-induced increase in plant respiration. However, this decrease may be a short-term effect and the ratio may recover as the vegetation acclimates to the new conditions. Soil organic matter should become more thoroughly decomposed as a result of the increased decomposition rates and should therefore also have a lower C:N ratio.

Simulations by Rastetter et al. (1992) indicate that the movement of N from soils to vegetation could dominate the long-term response of forests that are N limited. In areas with high N deposition, the external supply of N is likely to hide any effects of N release from soils. However, the analogous mechanisms should work for other nutrients like P. Thus, even if the forest is not N limited, release of P from soils could stimulate vegetation growth and result in a net increase in C sequestered in the ecosystem.

Changes in disturbance frequency

Disturbances such as fire, pest outbreaks or clearcutting cause a rapid transfer of C from forests to the atmosphere. An increase in disturbance frequency will mean that succession is reset more often, so that forest stands will be younger on average, and therefore will have less wood. If the disturbance also causes a significant loss of nutrients from the system, the ability of the forest to return to the original C levels is constrained. There can be a large loss of N immediately after the disturbance because the uptake capacity of the vegetation can not keep up with the release in N from soils, which will be stimulated by a large pulse of fresh litter (Vitousek et al. 1979). Secondary effects of increased soil temperatures and erosion can exacerbate this N loss.

These responses to disturbance will affect all three of the factors listed above. In the likely event that disturbance causes permanent loss of nutrients, the N capital of the ecosystem will decline as the frequency of disturbance increases. There will also be a decline in the overall C:N ratio of vegetation because younger vegetation has less wood. Soil C:N ratios may increase in the short term because of the large influx of litter after a disturbance, but are likely to decline fairly rapidly thereafter as the litter decomposes. The pulse of litter will also move N from vegetation to soils. Once the vegetation begins to recover, the accumulation of plant biomass will be constrained by the rate at which this N is re-released by the soil. Thus, in the case of disturbance, all three factors act to decrease C sequestration in forests.

Finally, the probabilities of C loss following disturbance versus subsequent gain during regrowth are not necessarily equal, even if the nutrient capital remains relatively unaffected. Kurz and Apps (1993) point out that in an increased disturbance regime, it is "easier to kill than to grow"; in other words, the C loss is guaranteed, but commensurate regrowth is not.

Changes in N deposition

Emissions of reactive N to the atmosphere have increased several fold in recent decades (Galloway et al. 1994). By reactive N, we mean either N oxides ($NO_x = NO + NO_2$) or ammonium (NH_3). The primary sources of this excess N are the combustion of fossil fuels, the manufacture and use of fertilisers, livestock, and burning of biomass as a result of changing land-use practices. All NO_x and NH_3 that is emitted will be deposited back onto land or sea in a variety of different forms collectively termed NO_y and NH_x (Logan et al. 1981; Penner et al. 1991). Several authors have pointed out that deposition of this "excess" N could stimulate a significant increase in terrestrial C uptake (Peterson and Melillo 1985; Schindler and Bayley 1993; Townsend et al. 1995b). The basis of this hypothesis is twofold: N limitation of terrestrial productivity is widespread (Vitousek and Howarth 1991), and C:N ratios in terrestrial vegetation are high, so that large amounts of C may be fixed for every unit of N.

The problem with estimating the effects of N deposition is that a number of factors which will influence C uptake vary strongly in time and space. Uptake in any given year will be equal to the net primary production stimulated by the current year's deposition minus the respiration from excess C fixed in previous years. In other words, C storage in the present depends upon N inputs in the past. In addition, global N deposition is not evenly distributed across the terrestrial landscape; it varies by more than an order of magnitude among world biomes (Penner et al. 1991; Galloway et al. 1994). Different ecosystem types will also vary greatly in their potential for C sequestration. For example, woody tissue can have C:N ratios in excess of 300 and lifetimes greater than 100 years, therefore any excess N which stimulates wood production will cause a relatively large and long-term removal of C from the atmosphere. In contrast, foliar biomass in forests and grasslands in general have much lower C:N ratios and tissue lifetimes, and therefore have a limited capacity for C storage. Deposition on to cultivated areas will not contribute significantly to any terrestrial sink.

Townsend et al. (1995a) attempted to account for these spatial and temporal issues by using a simple biogeochemical model in combination with an atmospheric transport model for N and a database on biome types. Some of the results are shown in Figure 3.2, which displays the spatial heterogeneity in global C uptake for a single year (1990). Most of the sink is in three distinct mid-latitude regions: the eastern US, Europe, and eastern China. Not surprisingly, these regions also have high N deposition, but vegetation type has a profound effect on the amount

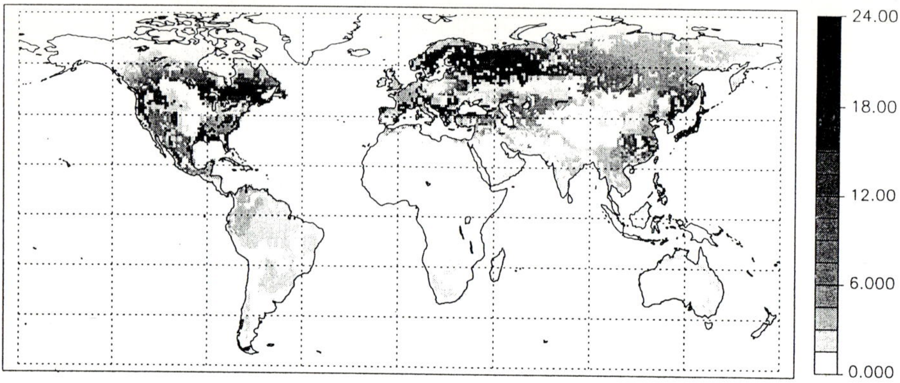

Figure 3.2. Simulated spatial distribution of the 1990 carbon sink (assimilation *minus* respiration) resulting from fossil-fuel derived NO_y deposition. Values are in g C m^{-2}. Figure taken from Townsend et al. (1995a).

of C stored per unit N. The largest sinks are in the forested regions, where significant amounts of N are incorporated into woody tissue at high C:N ratios. Much of the area experiencing high N deposition, however, is covered by grasslands or cultivated areas, where the ability of N to stimulate C sequestration is severely limited. Thus, it is both the dynamics *and* distribution of the forested biomes that control the terrestrial sink.

However, the magnitude of the increase in C capital will depend on how the added N is distributed within the ecosystem. The sink shown in Figure 3.2 is highly sensitive to the fraction of deposited N which is allocated to wood: Figure 3.3 shows the effects of varying the percent allocation to wood in forests from 10–100% of the excess fixed by the vegetation, a change which produces a five-fold difference in global C uptake. Allocation to wood is unlikely to be near the high end of this range anywhere, but even a range of 10–50%—a range typical of forested systems—results in a large range in potential C storage. An approximate global mean is extremely difficult to estimate as allocation can vary substantially with vegetation and soil type, nutrient status, and stand age (Aber and Melillo 1991). Furthermore, the distribution of N among ecosystem pools may change over time as the availability of N changes, as may the C:N ratios of the plant tissue (Boring et al. 1988; Vitousek et al. 1988).

Finally, the fraction of N deposited which is actually incorporated into vegetation will also vary among systems (e.g., Johnson 1992). Some of the N inputs will be lost via leaching or gaseous emissions, and some may be bound up in soil organic matter pools for long periods of time. In addition, N retention within a given system may change over time as N inputs change. A steady increase in N loading should cause a decrease in overall retention as the systems become "N saturated" (Aber et al. 1989), which would cause a decrease in C storage (Figure 3.3). Continued exposure to increasing N inputs may even prove harmful, resulting in ecosystem decline and an overall reduction in C stocks (Aber et al. 1989).

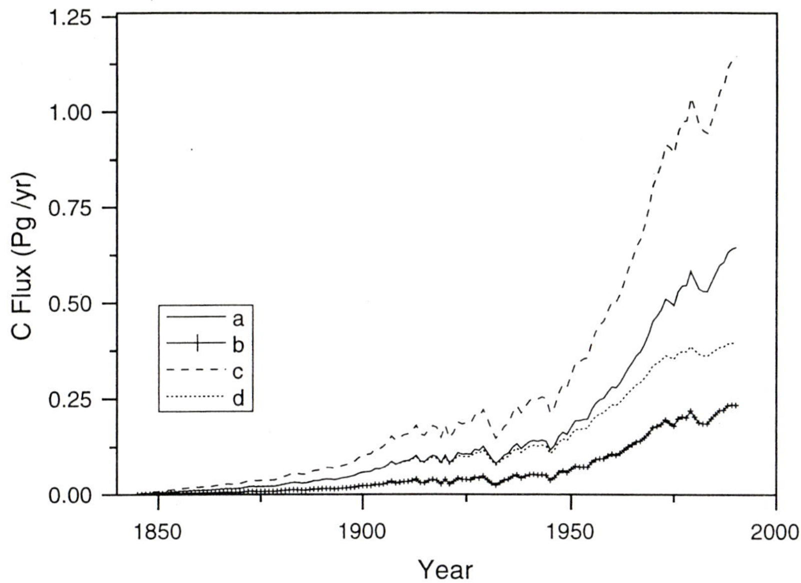

Figure 3.3. Sensitivity of the carbon (C) sink shown in Figure 3.2 to the fraction of C allocated to wood, and to the amount N deposited that is lost from the system. Figure is from Townsend et al. (1995a), and the results are those of a simulation of the effects of N deposition between 1800 and 1990. The four lines are: (a) 50% of C fixed to wood, 20% N lost from system; (b) 10% of C to wood, 20% N loss; (c) 100% of C to wood, 20% N loss; (d) 50% of C allocated to wood, N loss varies between 20% and 100% as a function of the amount of N deposited.

Overall, it is extremely difficult to estimate any current C sink due to N deposition, let alone how that sink might change in the future. Certainly forests have the greatest potential for significant longer term (centuries) storage. However, the steady global increase in available N must be causing a decrease in N limitation of terrestrial ecosystems. Many areas of chronically high deposition do appear to be losing as much N as they receive (Schulze 1989). Simply put, even if excess N does not cause ecosystem damage, the system will eventually cease to respond to additional inputs, therefore the persistence of any N-derived C sink is inherently limited.

Phosphorus and tropical carbon storage

Many areas that have not been subjected to high deposition of N will experience rapid increases over the next few decades, but these changes will be primarily tropical (Galloway et al. 1994). Here, N limitation of forested areas does not appear widespread; instead, growth is often limited by P (Vitousek 1984). Thus, increased N deposition to tropical areas is much less likely to stimulate C uptake and storage. We know, however, that approximately half of the current missing terrestrial C sink is in equatorial latitudes (Enting and Mansbridge 1991), and the mechanisms behind such uptake are not understood.

If a substantial fraction of tropical forests are limited by P, then any increase in inputs of P to these regions might stimulate C uptake. The atmosphere is rarely thought of as a significant source of P for terrestrial systems, but fertiliser use, industrial activities, and land conversion have caused substantial increases in the amounts of P-containing aerosols in the atmosphere (Graham and Duce 1979). Total P deposition on land has been estimated at about 3 Tg yr^{-1} (Graham and Duce 1979), and a recent study of P deposition on a relatively remote Mediterranean island attributed more than 60% of annual P inputs to continental fertiliser and industrial sources (Bergametti et al. 1992). This global rate of P deposition is about 5% of that for N, but C:P ratios in vegetation are, on average, about an order of magnitude greater than C:N ratios (Vitousek et al. 1988), therefore the potential exists to store extremely large amounts of C for every additional P in a P-limited system. A simple hypothetical calculation demonstrates this potential:

- P deposition near Manaus, Brazil in 1972 was 0.29 kg ha^{-1} yr^{-1} (Ungemach 1972).
- Deposition due to fertilisers and industrial sources in Bergametti et al. (1992) was about 0.2 kg ha^{-1} yr^{-1}. The average atmospheric concentration of P aerosols was about 20 ng m^{-3}; Graham and Duce (1979) report an average value of 90 ng m^{-3} for continental regions between 60°N and 60°S.
- Assume that all tropical forest areas are receiving an average of 0.2 kg ha^{-1} yr^{-1} more P than in pre-industrial times, and that these forests are limited by P. Then:

  Area of forests = 2100 Mha

  Average C:P ratio = 1320 (Vitousek et al. 1988)

  Potential Gross C uptake = $(21 \times 10^8 \text{ ha})\times(1320)\times(0.2 \text{ kg ha}^{-1}) = 0.6$ Pg C yr^{-1}

Obviously, this calculation relies on a suite of assumptions that may be wrong; it is intended for illustrative purposes only. Most of the factors that will influence the response of tropical regions to P inputs are poorly understood. It is certainly an overestimate to assume that all tropical forests will convert all P inputs into vegetation C, even in the areas that are strongly P limited. There is little information on rates of P deposition, and furthermore, unlike N, much of the P that is deposited may be unavailable to the vegetation because it is tightly adsorbed to particle surfaces. Bergametti et al. (1992) showed that only about 10% of P inputs from crustal (soil, sand, etc.) sources were soluble in seawater, whereas industrial and/or fertiliser inputs were highly soluble. Thus, to estimate the response of the C cycle, one must know not only total P inputs, but the relative contribution of different sources of P aerosols that vary in P availability to vegetation. As well, many of the factors listed above for N (e.g., allocation, spatial gradients in deposition) will be important here, and again, it is the net flux of C that must be calculated. Finally, although some sources of atmospheric aerosols have increased in strength in recent decades, others—at least in many temperate areas—have weakened, so that deposition rates of other crustal elements such as Ca and Mg have also declined (Hedin et al. 1994).

Nevertheless, the potential effects of changes in P on tropical C dynamics should be investigated more thoroughly. Several studies have shown significant inputs of Saharan dust in the Amazon Basin, and some of those studies reported the P content of the dust aerosols (Artaxo et al. 1990; Andreae et al. 1990). Rates of Saharan particle flux to the atmosphere have increased in recent years (Bergametti et al. 1989); whether or not the P they contain is available for uptake by vegetation is not known. In contrast, fertiliser aerosols are a highly available form of P, and rates of fertiliser use in tropical regions are increasing dramatically. If P in dust inputs is relatively unavailable, then in regions where fertiliser-derived inputs are increasing, the effects on forest C balance are likely to be far greater than implied by the change in total deposition alone.

Table 3.1. Hypothesised changes in forest carbon (C) balance in response to increases in four factors: CO_2, temperature, nitrogen (N) deposition and frequency of disturbance. The responses listed are those that appear most likely based on recent studies, but none are certain.

PERTURBATION	INITIAL NET EFFECTS ON GLOBAL FOREST C STOCKS	POSSIBLE LONG-TERM RESPONSE
Increasing Atmospheric CO_2	*increase:* -increased carboxylation capacity -reduced water stress	decreasing C uptake due to acclimation to higher CO_2 levels
Increasing Temperature	*increase:* -transfer of nutrients from soil to plants increases system C:N so that ↑NPP > ↑Respiration	increased N mineralization will also increase N losses, which may decrease C uptake
Increasing N deposition	*increase:* -stimulation of growth in N limited ecosystems	decreasing C uptake due to: -N saturation, less N limitation -strong tropical uptake unlikely
Increase in Disturbance Frequency	*decrease:* -younger stands = less wood -increased loss of nutrients	continued losses of C likely due to losses of nutrients and soil organic matter

Interactions among global changes

Table 3.1 summarises the effects on the global forest C pool of each of the individual perturbations discussed above. In each case, the response listed is the overall change that appears most likely from recent modelled and experimental analyses, but substantial variation has been shown for each factor. In other words, we are not certain of how change in a single component will affect the C cycle, let alone how it will respond to the entire suite of current and projected perturbations. In many cases, the interactive effects of these changes are not likely to be a simple additive combination of the individual responses. For example, nutrient deposition and/or elevated temperatures may enhance the effects of CO_2 fertilisation, as nutrient limitation appears to constrain plant response to elevated CO_2 in many cases (Comins and McMurtrie 1993). Increased temperatures may increase nutrient availability (Rastetter et al. 1992), but if the warming is accompanied by increasing moisture stress, plant response to the extra nutrients may be constrained. On the other hand, the elevated CO_2 levels may compensate for the decrease in moisture.

Table 3.2 outlines some possible effects of change in two factors at once. Here, the responses are even less certain than those in Table 3.1, and do not include the additional complication of changes in moisture. These interactions, however, are extremely important to understand. Although analyses of single factor sensitivities are useful, more attempts to integrate multiple perturbations into analyses of ecosystem dynamics are needed to truly predict the response of ecosystems to the environmental changes of the coming centuries.

Table 3.2. Possible effects of a second environmental change on each of the four changes summarised in Table 3.1. The table is read as the effects of the factors in column 1 on those in columns 2–5.

	CO_2	TEMPERATURE	N-DEPOSITION	DISTURBANCE
CO_2	–	increase by easing water stress and increasing C:N ratios in vegetation	increase by increasing C:N ratios in vegetation and overall C uptake capacity	may decrease C losses by accentuating regrowth
Temperature	increase by increasing nutrient availability	–	both increase N; = larger initial response but more rapid saturation	may increase C losses by increasing N losses, but will also increase soil N mineralization
N deposition	increase by increasing nutrient availability	both increase N; = larger initial response but more rapid saturation	–	may decrease C losses by accentuating regrowth and by re-supplying lost N
Disturbance	decrease by removing vegetation and depleting nutrients	may shift T effects from net C gain to net loss	decrease by removing vegetation and increasing N losses	–

ACKNOWLEDGEMENTS

We thank Rob Braswell, Beth Holland, Robert McKane, Joyce Penner, Gus Shaver and Dave Schimel, and two anonymous reviewers for assistance with this manuscript.

REFERENCES

Aber JD, Melillo JM (1991) Terrestrial ecosystems. Saunders Pub., Philadelphia, PA, 430 pp.

Aber JD, Nadelhoffer KJ, Steudler P, Melillo JM (1989) Nitrogen saturation in northern forest ecosystems. *BioScience* 39(6): 378–386.

Andreae MO, Talbot RW, Berresheim H, Beecher KM (1990) Precipitation chemistry in central Amazonia. *J. Geophys. Res.* 95(D10): 16986–16999.

Artaxo P, Maenhaut W, Storms H., Van Grieken R (1990) Aerosol characteristics and sources for the Amazon Basin during the wet season. *J. Geophys. Res.* 95(D10): 16971–16985.

Bazzaz FA, (1990) The response of natural ecosystems to the rising global CO_2 levels. *Ann. Revs. Ecol. and Systematics* 21: 167–196.

Bergametti GL, Gomes E, Remoudaki, Desbois M, Martin D, Buat-Menard P (1989) Present transport and deposition patterns of African dusts to the northwestern Mediterranean. In: Leinen M, Sarnthein M (eds.) Paleoclimatology and paleometeorology: modern and past patterns of global atmospheric transport. Kluwer, Dordrecht, The Netherlands, pp 227–252.

Bergametti G, Remoudaki E, Losno R, Steiner E, Chatenet B (1992) Source, transport and deposition of atmospheric phosphorus over the northwestern Mediterranean. *J. Atm. Chem.* 14: 510–513.

Bolin B, Cook RB (eds) (1983) The major biogeochemical cycles and their interactions. Wiley, Chichester, UK, 532 pp.

Boring LR, Swank WT, Waide JB, Henderson GS (1988) Sources, fates and impacts of nitrogen inputs to terrestrial ecosystems: review and synthesis. *Biogeochem.* 6: 119–159.

Bormann FH, Likens GE (1967) N cycling. *Science* 155: 424–429.

Cole DW, Rapp M (1981) Element cycling in forest ecosystems. In: Reiche DE (ed) Dynamic Properties of Forest Ecosystems. Cambridge University Press, London, pp 341–409.

Comins HN, McMurtrie RE (1993) Long–term response of nutrient limited forests to CO_2 enrichment: equilibrium behavior of plant-soil models. *Ecol. Appl.* 3: 666–681.

Edwards NT, Johnson DW, McLaughlin SB, Harris WF (1989) C dynamics and productivity. In: Johnson DW, Van Hook RI (eds) Analysis of biogeochemical cycling processes in Walker Branch Watershed, Springer-Verlag, New York, pp 197–232.

Enting IG, Mansbridge JV (1991) Latitudinal distribution of sources and sinks of CO_2: results of an inversion study. *Tellus* 43: 156–170.

Galloway JN, Levy H II, Kasibhatla PS (1994) Year 2020: consequences of population growth and development on deposition of oxidized nitrogen. *Ambio* 23(2): 120–123.

Graham WF, Duce RA (1979) Atmospheric pathways of the phosphorus cycle. *Geochimica et Geophysica Acta* 43: 1195–1208.
Hedin LO, Granat L, Likens GE, Buishand TA, Galloway JM, Butler TJ, Rodhe H (1994) Steep declines in atmospheric base cations in regions of Europe and North America. *Nature* 367: 351–354.
Hobbie JE, Cole J, Dungan J, Houghton RA, Peterson B (1984) Role of biota in global CO_2 balance: the controversy. *BioScience* 34: 492–498.
Johnson DW (1992) Nitrogen retention in forest soils. *J. Environ. Qual.* 21: 1–12.
Johnson DW, Henderson GS (1989) Terrestrial N cycling In: Johnson DW and Van Hook RI (eds) Analysis of biogeochemical cycling processes in Walker Branch Watershed. Springer-Verlag, New York, pp 233–300.
Johnson DW, Van Hook RI (eds) (1989) Analysis of biogeochemical cycling processes in Walker Branch Watershed. Springer-Verlag, New York, 401 pp.
Jordan CF, (1985) N cycling in tropical forest ecosystems. Wiley, Chichester, UK, 190 pp.
Kurz, WA, Apps MJ (1993) Contribution of northern forests to the global C cycle: Canada as a case study. *Water, Air, Soil Pollut.* 70: 163–176.
Logan JA, Prather MJ, Wofsy SC, McElroy MB (1981) Tropospheric chemistry: a global perspective. *J. Geophys. Res.* 86: 7210–7254.
Melillo JM, Gosz JR (1983) Interactions of the biogeochemical cycles in forest ecosystems. In: Bolin B and Cook RB (eds) The major biogeochemical cycles and their interactions. Wiley, New York, pp 177–222.
Penner JE, Atherton CS, Dignon J, Ghan SJ, Walton JJ, Hameed S (1991) Tropospheric nitrogen: a three-dimensional study of sources, distributions, and deposition. *J. Geophys. Res.* 96: 959–990.
Peterson BJ, Melillo JM (1985) The potential storage of carbon caused by eutrophication of the biosphere. *Tellus* 37B: 117–127.
Rastetter EB, McKane RB, Shaver GR, and Melillo JM (1992) Changes in C storage by terrestrial ecosystems: How C–N interactions restrict responses to CO_2 and temperature. *Water, Air, Soil Pollut.* 64: 327–344.
Rastetter EB, Ryan MG, Shaver GR, Melillo JM, Nadelhoffer KJ, Hobbie JE, Aber JD (1991) A general model describing the responses of the C and N cycles in terrestrial ecosystems to changes in CO_2, climate, and N deposition. *Tree Physiology* 9: 101–126.
Rastetter, EB, Shaver GR (1995) A model of multiple element limitation for acclimating vegetation. *Ecology* 73 (In press).
Schimel DS, Braswell BH Jr, Holland EA, McKeown R, Ojima DS, Painter TH, Parton WJ, Townsend AR (1994) Climatic, edaphic and biotic controls over storage and turnover of carbon in soils. *Global Biogeochem. Cycles* 8(3): 279–293.
Schindler DW, Bayley SE (1993) The biosphere as an increasing sink for atmospheric carbon: estimates from increased nitrogen deposition. *Global Biogeochem. Cycles* 7(4): 717–733.
Schulze ED (1989) Air pollution and forest decline in a spruce (*Picea abies*) forest. *Science* 244: 776–783.
Townsend AR, Vitousek PM, Holland EA (1992) Tropical soils could dominate the short-term carbon cycle feedbacks to increased global temperatures. *Climatic Change* 22: 293–303.
Townsend AR, Braswell BH, Holland EA, Penner JE (1995a) Spatial and temporal patterns in terrestrial carbon storage due to deposition of fossil fuel derived nitrogen. *Ecological Applications* (In press).
Townsend AR, Vitousek PM, Trumbore SE (1995b) Soil carbon dynamics along gradients in temperature and land-use on the island of Hawai'i. *Ecology* 76: 721–733.
Trumbore SE (1993) Comparison of carbon dynamics in tropical and temperate soils using radio carbon measurements. *Global Biogeochem. Cycles* 7: 275–290.
Ungemach H (1972) Regenwasseranalysen aus Zentral Amazonien ausgefuhrt, Manaus, Amazonas. *Amazoniana* 3: 186–198.
Vitousek PM (1984) Litterfall, nutrient cycling, and nutrient limitation in tropical forests. *Ecology* 65(1): 285–298.
Vitousek P, Fahey T, Johnson D, Swift M (1988) Element interactions in forest ecosystems: succession, allometry and input–output budgets. *Biogeochem.* 5: 7–34.
Vitousek PM, Gosz JR, Grier GC, Melillo JM, Reiners WA, Todd RL (1979) Nitrate losses from disturbed ecosystems. *Science* 204: 469–474.
Vitousek PM, Howarth RW (1991) Nitrogen limitation on land and sea: how can it occur? *Biogeochem.* 13: 87–115.

4 Estimated extent of forested peatlands and their role in the global carbon cycle

Stephen C Zoltai and Pertti J Martikainen[1]

Natural Resources Canada, Canadian Forest Service,
Northwest Region, Northern Forestry Centre,
5320 - 122 Street, Edmonton AB T6H 3S5, Canada

INTRODUCTION

Wetlands are areas that are transitional between terrestrial and aquatic systems, where the water table is usually at or near the surface or the land is covered by shallow water (Cowardin et al. 1979). The definition of wetlands in Canada expresses a similar concept: wetlands are lands that have the water table at, near or above the land surface or which are saturated for a long enough period to promote wetland or aquatic processes as indicated by hydric soils, hydrophytic vegetation, and various kinds of biological activity that are adapted to the wet environment (Tarnocai 1980).

Although the general characteristic of perennial or seasonal water-logging is a common feature of wetlands, many fundamentally different kinds of wetlands have developed in response to climatic, hydrologic (quantity, quality) and biotic conditions. An important distinction can be made between those wetlands where the deposition rate of organic materials exceeds the rate of decomposition, resulting in an accumulation of waterlogged organic matter in various states of preservation: peat. Non peat-forming wetlands occur mainly where seasonal lowering of the water table allows aerobic decomposition of the organic debris to take place. In peat-forming wetlands (peatlands, mires) only a portion of the ground biomass is decomposed under aerobic conditions, the rest becomes waterlogged. Such organic matter under continuously water saturated conditions (peat) is subject to anaerobic decomposition which proceeds at a much slower rate than the aerobic process. The result is a continuous accumulation of peat, where carbon (C) is being stored, and has been at least since the beginning of the Holocene Period.

The same processes that drive the development of different kinds of wetlands also influence the formation of different kinds of peatlands. The quality of the water is important: peatlands that are fed by rainwater only (bogs) are poor in nutrients and cations, and are acidic in reaction. Peatlands that are influenced by groundwater that has been in contact with mineral soil (fens) are richer in cations, and are usually circumneutral in reaction. In some fens where the water table is consistently 20–30 cm below the surface (swamps), the ground layer is well decomposed and can support dense tree growth.

---

1 National Public Health Institute, Department of Environmental Microbiology, Kuopio, Finland

NATO ASI Series, Vol. I 40
Forest Ecosystems, Forest Management and the Global Carbon Cycle
Edited by Michael J. Apps and David T. Price

The subject of this paper is the C cycle of forested peatlands. These peatlands have relatively large standing biomass, sometimes approaching that of upland forests. On the other hand, the waterlogged substrate results in the accumulation of peat due to incomplete decomposition of the forest debris and associated ground vegetation. The forested peatlands are therefore different from both the upland forests and the non-forested fens and bogs, in that they support relatively large forest biomass and sequester a part of this biomass as peat.

EXTENT OF FORESTED PEATLANDS

Forested peatlands are defined for the purpose of this paper as poorly drained lands with a peat accumulation of >30 cm, where closed canopy tree cover (>5 m) form the dominant vegetation, or, if immature stands, have the potential to become closed canopy forests. This definition distinguishes the forested peatlands from treed, or wooded peatlands with sparse, often stunted trees occurring on many bogs and fens.

The determination of the global extent of peatlands and forested peatlands is fraught with difficulties. In some areas, such as in Europe, accurate data exist on the areas covered by peatlands. However, the breakdown into forested or treed peatlands can only be estimated from peatland ecosystem descriptions.

In other areas, such as Canada and the former USSR, only estimates of the peatland extent are available, but even these estimates are highly variable. Botch and Masing (1983) estimate that the total peatland area of the former USSR is 83 Mha, but Vompersky (1994) indicates that there are 139 Mha of mires in Russia where the peat thickness exceeds 30 cm. Such large differences in the estimates of peatland extent are undoubtedly due to inconsistencies in the definition of terms. In Canada, initial estimates have shown 10 Mha peatlands (Tibbetts 1969), but now the estimates are up to 111.3 Mha (Glooschenko 1990), as the estimate by Tibbetts (1969) referred only to peatland exploitable for horticulture.

The estimates of peatlands for other countries is even less reliable. This uncertainty is reflected in the estimates of the global extent of peatlands in the literature (Table 4.1). These estimates range from 150 to 422 Mha, but some (e.g., Post et al. 1982; Adams et al. 1990)

Table 4.1. Estimates of the global extent of peatlands/wetlands.

TYPE OF WETLAND[a]	AREA (Mha)	SOURCE
Bogs	314.9	Rodin et al. 1975
Bogs	297.4	Matthews and Fung 1987
Bogs	120.8	Esser 1984
Swamps	359.4	Rodin et al. 1975
Swamps	209.5	Matthews and Fung 1987
Swamps	214.4	Esser 1984
Wetlands	280	Post et al. 1982
Peatlands	500	Franzén 1994
Peatlands	422	Kivinen and Pakarinen 1981
Peatlands	388–408	Rieley et al. 1992
Peatlands	238	Taylor 1983
Bog, peatland	150	Ajtay in Armentano 1980
Swamp and marsh	200	Schlesinger 1984
Wetlands and peatlands	280	Adams et al. 1990

a Wetland/peatland type as identified by the original authors.

include all wetlands into their estimate. The current estimate of global peatland areas is about 352 Mha (Table 4.2).

Considering the lack of precise data of the global extent of peatlands, estimating the proportion of forested peatlands is even more difficult. Figures of forested wetlands quoted in the literature (Matthews and Fung 1987; Rodin et al. 1975; Esser 1984) vary from 316.4 to 124.9 Mha, possibly due to a more inclusive definition of "forests" and the inclusion of non-peaty wetlands. Without accessing the detailed forest inventories of dozens of jurisdictions, one can make only a rough estimate that will show the order of magnitude, but not the actual extent of forested peatlands.

The estimates are presented in Table 4.2. Reliable information is available only from Finland, all other figures were estimated from the description of peatland ecosystems of various authors. The main source for the estimate of the proportion of forested peatlands was the information given by the authors. Where such information was missing, experience from climatically similar areas in Canada and Finland were used to reach reasonable estimates.

For Canada, the procedure of Ketcheson and Jeglum (1972) was used, who examined the forest inventory data for the Province of Ontario and identified the areas of peatlands classed as productive forest land. They found that 18% of the total peatlands are forested. Similar estimates for the provinces of Alberta, Saskatchewan and Manitoba gave the proportions of forested peatlands as 15.5%, 15% and 17%, respectively. For Quebec, it was assumed that 18% of the peatlands are forested, giving a figure that is compatible with the estimated extent of potentially productive forested peatlands (Haavisto and Jeglum 1991). The subarctic and oceanic regions of Canada have negligible amounts of forested peatlands.

The resulting global estimate of 96.9 Mha is an approximation of the extent of forested peatlands (Figure 4.1). It does show that about 20% to 25% of the global peatlands support forest growth that is potentially exploitable for wood fibre production.

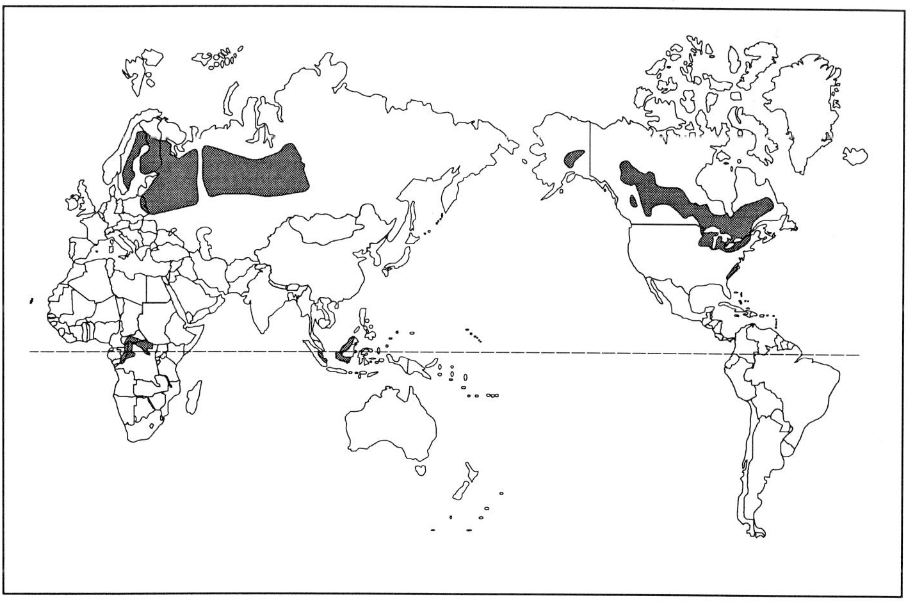

Figure 4.1. Main concentration of forested peatlands.

Table 4.2. Total area of peatlands and estimated extent of forested peatlands in different countries (Mha).

Region/country	Total peatlands	Forested peatlands	Source
Russia	139		Vompersky 1994
		29	Based on Botch and Masing 1983; Glebov 1979; Vompersky 1989
Canada	111.3		Glooschenko et al. 1993
		19.3	This paper
Malesia	29.3		Anderson 1983
		26.3	Rieley et al. 1992; Driessen et al. 1979
Africa	12.64		Denny 1993
Congo—Zaire		5.7	Denny 1993; Thompson and Hamilton 1983
USA (Alaska)	10.64	0.2	Rieger et al. 1979
(Contiguous)	6.8		Gorham 1990
		5.6	Cubbage and Flather 1993
Finland	9.7	2.7	Kivinen 1972
Sweden	6.3	1.8	Hånell 1991
Norway	3.0		Moen (Unpubl. ms)
Belarus	2.54	0.8	Ivkovich (Unpubl. ms)
United Kingdom	1.77		Lindsay (Unpubl. ms)
Poland	1.5		Szczepanski 1990
		0.2	Gore 1983
Ukraine	1.2	0.2	Based on Yurkovskaya 1980
Germany	1.26		Roderfeld and Kuntze (Unpubl. ms)
Ireland	1.18		Foss (Unpubl. ms)
Estonia	1.0	0.13	Ilomets and Kallas (Unpubl. ms).
Other Europe[a]	1.35		Lofroth (Unpubl. ms)
China	3.48		Kivinen and Pakarinen 1982
Japan	0.25		Robertson 1994
Central and South America	7.0	4.0	Robertson 1994
Australasia	0.16		Kivinen and Pakarinen 1982
Other tropical [b]	1.0	1.0	
Total	352.37	96.93	

a Czechoslovakia, Denmark, France, Holland, Iceland, Latvia, Switzerland.
b Caribbean, sub-equatorial Africa, etc.

The estimates of the global extent of peatlands and forested peatlands permit an approximation of the size of C pool in the peat and forest biomass of peatlands (Table 4.3). The extensive northern peatlands have relatively thin peat (average 2.3 m (Gorham 1991)), but the peat thickness in the tropics is greater (average 5 m (Anderson 1983; Bruenig 1990)). Similarly, the biomass of northern forested peatlands at 120 Mg C ha^{-1} is well below that of the tropical peatland forests, estimated at 500 Mg C ha^{-1} (Bruenig 1990). These estimates show that in addition to the C sequestered in peat, the forested peatlands store substantial amounts of C in their biomass.

The uncertainties associated with such global estimates have been emphasised throughout this discussion. The estimates of global extent of peatlands varies from 120.8 to 500 Mha, but the most reasonable estimates are in the range of 388–408 Mha (Immirzi et al. 1992, in Rieley et al. 1992), a range of ±2.5%. Our estimate of 352 Mha falls below the Immirzi et al. (1992)

Table 4.3. Estimate of global carbon (C) storage in peatlands.

	C STORAGE IN PEAT		C STORAGE IN FOREST BIOMASS	
	AREA (Mha)	C POOL (Pg)	AREA (Mha)	C POOL (Pg)
Northern peatlands	302	397	59.9	7.2
Tropical peatlands	50	144	37.0	18.5
Total peatlands	352	541	96.9	25.7

estimate, but we feel that the estimate range should be greater due to uncertainties of the survey; a range of ±5% may be warranted. This would place estimate of global peatlands in the range of 335–369 Mha.

Our estimate of the forested peatlands shows that 20–25% of the global peatlands are forested, ranging between a low estimate of 70 Mha and a high estimate of 88 Mha. The combined range of peatland and forested peatland extent uncertainty is about ±16%.

The estimation of the depth of peat accumulation is equally difficult. In Canada the depth of peat was estimated to range regionally mainly between 2 and 4 m (Tarnocai 1984), with regional depths up to 10 m. However, because of the dominance of shallower peat, the average is about 2.5 m. In Russia, 30% of the peatlands have a shallow (30–50 cm thick) peat layer (Vompersky 1994). In areas where few data exist, the determination of peat depth becomes largely speculative. In general, peat accumulation rate is between 0.5 and 1.0 mm per year in northern peatlands, which bounds the estimates of peat depths within limits of ±0.25 mm per year, with greater accumulation rates in climatically more favourable, but non-forested locations. For these reasons, peat depths can be estimated regionally within ±15%. A further uncertainty is introduced by estimating the bulk density of the peat. However, the bulk density of well decomposed woody peat, found in forested peatlands, is fairly uniform at about 0.14 g cm^{-3}, representing only a small source of error in the estimate.

The combined effect of uncertainties totals a maximum of ±33% of the estimated organic matter store in global forested peatlands. The actual error in the estimate may be considerably lower, as many estimates have a solid basis and the extension of such data is reasonable. Furthermore, not all the errors are additive.

LONG-TERM DYNAMICS OF FORESTED PEATLAND DEVELOPMENT

The development of peatland systems generally proceed through two basic processes: paludification and terrestrialization. Paludification refers to peatland formation over non-wetland areas (Gore 1983) due to autogenic processes (processes internal to the peatlands). Terrestrialization is a process of filling a body of water with organic remains (Gore 1983) until terrestrial or semi-terrestrial conditions are obtained (Tallis 1983). These concepts are useful in understanding the development of forested peatlands, but the classical processes may be interrupted or altered by forces external to the ecosystems, such as climate change.

Paludification generally occurs by the expansion of the wetlands into the marginal non-wetlands (uplands) as peat accumulates in the main basin of the wetland, raising the water table. Depending on the topography of the basin, such expansion may be restricted to a narrow band around the peatland or cover very large areas. The original vegetation is replaced by vegetation tolerant of high water table conditions. In many cases forests can adapt to such situations, as an aerated surface layer usually remains where much of the organic debris decomposes, supplying nutrients for tree growth. In the Canadian boreal forest the average depth to water table under dense conifer forest was 32 cm (Zoltai and Johnson 1987).

The peat of such forested peatlands consists of woody material that is more or less decomposed, generally within a humic matrix. In the boreal regions the peat is usually thin (< 1.5 m), thinner than in non-forested peatlands, but thicknesses over 10 m are common in the tropics (Rieley et al. 1992). It is not uncommon to encounter in the northern areas, layers of woody peat that are covered by fen peat, indicating that some forested peatlands have been overwhelmed by the fen at some point in the past.

Forested peatlands also occur on peatlands developed through the process of terrestrialization. They are found on peatlands where the surface drainage has improved through autogenic or allogenic (influences external to the peatland) factors. An autogenic process may be the development of a dome of peat which may provide sufficient surface drainage to allow tree growth, or the development of drainage ways in fens which can lower the water table in the surrounding peatland. Allogenic influences may be a change in climate or in the water supply of the wetland. The peat under the forest cover usually contains woody debris in a humic matrix. Under this forest peat one finds the peat deposited by the previous phases of peatland development: bog and fen phases, and even an initial lacustrine phase.

CARBON DYNAMICS

The present C store in northern peatlands is 455 Pg (Gorham 1991) and their average annual rate of C accumulation is 29 g m^{-2} yr^{-1} (0.1 Pg yr^{-1} for the whole northern peatlands) (Gorham 1991). The accumulation of C is thus only 7.5 % of the average annual net primary production of 307 g C m^{-2} yr^{-1} (Gorham and Janssens 1993). During summer, the total CO_2 produced in decomposition processes, root respiration and dark respiration of shrubs and ground vegetation on virgin (i.e., unaffected by human influences) northern peatlands has ranged from 200 to 500 mg CO_2 m^{-2} h^{-1} (Silvola and Heikkinen 1979; Moore 1989; Kim and Verma 1992; Silvola et al. 1992, 1994a; Martikainen et al., in press). The annual CO_2 production in these processes varies from 100 to 300 g C m^{-2} yr^{-1} (Silvola 1985; Moore 1986, 1989), depending on nutrient status and hydrological conditions (Silvola et al. 1994a). Nutrient poor ombrotrophic bogs have the lowest production rates (Silvola et al. 1994a). Temperature has a great influence on CO_2 production in peat (Svensson 1980); the log-transformed CO_2 production correlates well with temperature (Silvola et al. 1994a).

The C dynamics of a naturally forested peatland has been studied in the boreal region of Canada (Charman et al. 1994). The forest cover is formed by a 10–15 m tall closed canopy stand of black spruce (*Picea mariana*) (Rothwell 1991). Here the long-term rate of C accumulation in peat was 17 g C m^{-2} yr^{-1}. The approximate annual biomass increment in the stand can be calculated by using the biomass equations of Singh (1982), on the basis of mean diameter and height of young, mature and old black spruce trees growing on the site. The C accumulation in tree biomass was about 100 g C m^{-2} yr^{-1} exclusive of tree roots. The shrubs and ground vegetation could accumulate another 20 g C m^{-2} yr^{-1}.

A widespread allogenic influence on peatlands is the lowering of water table by ditching to enhance tree growth. In terms of peatland dynamics this practice mimics the effect of natural allogenic processes by promoting aerobic biological activities in peat. When the water table is lowered for forestry purposes by ditching, the C accumulation in peat can greatly decrease. Laine et al. (1995) showed that in a Finnish forest the peat can be a net source of CO_2 after draining. They estimated a loss of 14 g C m^{-2} yr^{-1} from peat whereas the developing tree stand accumulated 100 g C m^{-2} yr^{-1} above pre-drainage conditions. Similarly, Sakovets and

Germanova (1993) reported that in a forested peatland in southern Karelia (Russia) a net increase of 123 g C m^{-2} yr^{-1} occurred in the ecosystem, despite a loss of C from the peat (32 g C m^{-2} yr^{-1}). The measured loss of C from peat after draining supports the observations that lowering of the water table generally increases the CO_2 release from peat. Lowering the water table by 20–30 cm increases the CO_2 emission by 150 to 250 % (Silvola et al. 1985; Moore and Knowles 1989; Silvola et al. 1994a). This increase is caused by increased decomposition due to better oxygen availability. However, other studies indicate that the long-term changes in C storage in peat after drainage depend on peatland type and geographical location (Laine et al. 1994).

Peat soil is both a source (mineralization, plant respiration) and a sink (photosynthesis) for CO_2. If peat is accumulating, the average uptake of CO_2 has to be higher than its average release. Virgin peatlands with high water table generally release methane (CH_4). In addition, peat soil can be a potential source for nitrous oxide (N_2O). The latter trace gas has a far greater capacity to absorb infrared radiation than CO_2 or CH_4 and it also participates in reactions that destroy the ozone in the stratosphere. The following is a brief discussion of the fluxes of CH_4 and N_2O in virgin and drained northern peatlands.

METHANE

Measurements have shown large variations in methane release from virgin northern peatlands, the range being < 1 to 2000 mg CH_4 m^{-2} d^{-1} (Crill et al. 1993; Bubier et al. 1993; Roulet et al. 1994). There are wide seasonal and spatial variations, but the annual emissions for specific sites vary far less (Crill et al. 1993) The annual emissions of CH_4 have varied from 0 to 70 g CH_4 m^{-2} yr^{-1} (Crill et al. 1993; Bubier et al. 1993; Roulet et al. 1994) showing not only site-specific, but also regional variation. The total emission of CH_4 from the northern peatlands is not well evaluated; it may range from 0.024 to 0.038 Pg yr^{-1} (Crill et al. 1993). The present rather limited data from tropical wetlands suggest that their total annual CH_4 emission could be twice (0.066 Pg yr^{-1}) the emission from northern wetlands (Bartlett and Harriss 1993).

The nutrient status of peatlands together with water table levels are the two main factors affecting the CH_4 fluxes. In general, nutrient rich fens with high water table have higher CH_4 emissions than ombrotrophic bogs (Martikainen et al. 1994). However, minerotrophic mires with low water table also have a low CH_4 emission rate (Martikainen et al. 1994; Roulet et al. 1994).

Lowering the level of the water table has a very different effect on CH_4 release than on CO_2 emission. The lowering of water table generally greatly reduces the CH_4 emission (Moore and Knowles 1989, 1990; Moore et al. 1990; Pulliam and Meyer 1992; Roulet et al. 1992) and a drained fen with good tree growth can even be a sink for CH_4 (Martikainen et al. 1993a, 1994, in press; Crill et al. 1993). However, high levels of CH_4 release from ditches may increase the emissions after draining fens with low initial CH_4 release rates (Roulet 1994). Drained peat consumes CH_4 at rates up to 3 mg CH_4 m^{-2} d^{-1} (Crill et al. 1994) because of methane oxidation in the uppermost aerated peat profile (Crill et al. 1994; Lien et al. 1993; Svensson and Sundh 1993).

In general, the CH_4 release from northern peatlands increases with increasing temperature (Bartlett and Harriss 1993). However, the correlation between temperature and CH_4 release is weak in peatlands with low or highly fluctuating water tables. In such peatlands the total CH_4 emission to the atmosphere may be reduced, when some of the methane produced in the anaerobic peat is oxidised by a temperature dependent process in the aerobic peat (Roulet et al. 1992a; Martikainen et al. 1994).

NITROUS OXIDE

Nitrous oxide emissions from virgin northern and temperate region peatlands are low, below 0.03 g N_2O-N m^{-2} yr^{-1} (Urban et al. 1988; Freeman et al. 1993; Martikainen et al. 1993b). The effect of draining on N_2O emissions is dependent on the nutrient status of the peatlands. There is no increase in the N_2O emission when a nutrient poor ombrotrophic bog is drained, but the draining of fens increases their N_2O fluxes (Martikainen et al. 1993b). The N_2O-N release from drained boreal forested fens ranged from 0.05 to 0.14 g N_2O-N m^{-2} yr^{-1} (Martikainen et al. 1993b). One reason for low N_2O emission from virgin mires with a high water table is a lack of oxygen, but nitrification can take place in the peat profile after the water table has been lowered by drainage (Lång et al. 1994).

FORESTED PEATLANDS AND GLOBAL CHANGE

The conversion of forested peatlands to agriculture is a threat, especially in densely populated areas. The west Malesian peatland forests constitute a valuable resource that are being logged at the present (Anderson 1983). Agricultural development often follows, generally with little success, due to problems associated with the rapid oxidation/decomposition of the peat, the irreversible shrinkage of peat, poor hydraulic conductivity and high rate of subsidence (Anderson 1983; Armentano and Menges 1986).

The clearing of forested peatlands for other land uses, such as agriculture, horticulture, or industrial uses, with the attendant drainage, results in the destruction of the original wetland ecosystems. Depletion of the C resource both in the living and stored biomass follows. A very high proportion of the peatlands of all kinds have been exploited in western Europe (for example, over 80% in Germany, 90% in Poland, and virtually all in the Netherlands and Belgium (Gore 1983)). It is expected that such trends will continue in other lands as well, unless conscious efforts are made to preserve all peat-forming environments, as a measure for extracting C from the atmosphere and storing it for long periods. At the present and in the foreseeable future there is not a great need to convert northern peatlands to agricultural use. However, persistent droughts in the present important agricultural areas could create a pressure to increase cultivation activity in the north for the growing population. The use of northern peatlands for agriculture is estimated to cause a C loss of 1500 g m^{-2} yr^{-1} because of increased peat decomposition (Nykänen et al. 1995).

Ditching of forested peatlands to increase forest productivity is widespread in Fenno-scandia and Russia. As discussed above, the ditching generally increases C accumulation in the biomass, but decreases the peat formation, and even stimulates a C loss from peat. However, because of the increased tree growth, the net effect for the whole ecosystem can be positive, i.e., more C is stored, for hundreds of years (Laine et al. 1994). However, biomass removal can result in a large loss of C from the site (Trettin et al. 1992).

Soil moisture conditions in a changing climate are more important for the biogeochemical cycles in the northern peatlands than possible increases in temperature. Increasing temperature with no change in water table would enhance the CH_4 release from peatlands, but changes in C accumulation (net primary production (NPP) – decomposition) are difficult to predict. Peat accumulation may decrease because of an increase in decomposition in the uppermost peat profile (acrotelm), but this may be offset by greater productivity in a longer growing season. If the water table will be lower, as in peatlands drained for forestry, the increase in peat decomposition can greatly decrease the peat accumulation and can even lead to C loss (see above).

According to climatic models, the soils are expected to be drier during the summer. Such dry conditions would tend to lower the water table, counteracting the increase in CH_4 release induced by higher temperatures. On the other hand, warming can increase the unfrozen period and allow CH_4 emissions during the wet autumn and winter. The draining experiments suggest that drying would enhance the N_2O release from northern peatlands (Freeman et al. 1993; Martikainen et al. 1993b). The potential release of N_2O from forested northern peatlands in a changing climate would be 0.03–0.1 Tg yr^{-1} (Martikainen et al. 1993b). If the water table of peatlands will be lower, similar to drained peatlands, the increase in tree growth would increase the C accumulation in the biomass. This would compensate for the possible decrease in C accumulation as peat. The short-term (e.g., 100 years) net change would be even positive, as more C is accumulated in the ecosystem. Based on observations from drained forested peatlands (decrease in peat accumulation, loss of C, and increase in tree growth, see above), the maximum net increase in C store in northern peatlands that can support tree growth (192 Mha) would be 0.1 Pg yr^{-1}. However, the fluctuating water table will very likely limit the increase in tree growth to be lower than in the drained peatlands.

Melting of permafrost would increase the CH_4 emissions from the north, as the thawed peatlands subside below the water table. Presently there are about 140 Mha of peatlands affected by permafrost (in Russia 95 Mha (Vomperski 1994), in Canada 40 Mha). The annual increase in CH_4 emission would be in the order of 0.01 Pg (7 g CH_4 m^{-2} yr^{-1}). On the other hand, C accumulation would increase if the peatlands remain waterlogged after thawing, providing new areas for peat formation. The increase would be about 0.03 Pg C yr^{-1} (14×10^{11} $m^2 \times 23$ g C m^{-2} yr^{-1}).

Carbon and nitrogen fluxes in northern peatlands can react in many different ways to the combined effects of human impact and climate change. There are interactions among temperature, precipitation, evapotranspiration, soil moisture, vegetation development and gas exchange in the natural ecosystems. These interactions, especially when combined with human interference in a changing climate are not known, causing great uncertainties when predicting the changes in C and nitrogen dynamics and organic matter storage in northern peatlands.

The most effective C management under a changing climate is the preservation of the forested peatland ecosystems as functioning units. Conversion of forested peatland ecosystems to any other land use would not only result in the loss of the C stored in the peat, but the potential for peat formation would be destroyed. The forested peatlands would become attractive targets for agricultural or other development under a warming climate scenario. The conservation of these ecosystems would preserve a valuable C sequestering resource.

REFERENCES

Adams JM, Faure H, Faure-Denard L, McGlade JM, Woodward FI (1990) Increases in terrestrial carbon storage from the Last Glacial Maximum to the present. *Nature* 348: 711–714.

Anderson JAR (1983) The tropical peat swamps of western Malesia. In: Gore AJP (ed) Ecosystems of the World. 4B, Mires: Swamp, Bog, Fen and Moor, Elsevier, Amsterdam, pp 181–199.

Armentano TV (1980) Drainage of organic soils as a factor in the World carbon cycle. *BioScience* 30: 825–830.

Armentano TV, Menges ES (1986) Patterns of change in the carbon balance of organic soil—wetlands of the temperate zone. *J. Ecol.* 74: 755–774.

Bartlett K, Harriss RC (1993) Review and assessment of methane emissions from wetlands. *Chemosphere* 26: 261–320.

Botch MS, Masing VV (1983) Mire ecosystems of the USSR In: Gore AJP (ed) Ecosystems of the World. Vol. 4B. Mires: Swamp, Bog, Fen and Moor. Elsevier, Amsterdam, pp 95–152.

Bruenig EF (1990) Oligotrophic forested wetlands in Borneo. In: Lugo AE, Brinson M, Brown S (eds) Ecosystems of the World. Vol 15. Forested Wetlands. Elsevier, Amsterdam, pp 299–334.

Bubier JL, Moore TR, Roulet NT (1993) Methane emissions from wetlands in the midboreal region of northern Ontario, Canada. *Ecology* 74: 2240–2254.

Charman DJ, Aravena R, Warner BG (1994) Carbon dynamics in a forested peatland in north-eastern Ontario, Canada. *J. Ecol.* 82, 55–62.

Cowardin LM, Carter V, Golet FC, LaRoe ET (1979) Classification of wetlands and deepwater habitats of the United States. U.S. Dept. Interior, Fish and Wildlife Serv., FWS/OBS-79/31, Washington DC, 103 pp.

Crill P, Bartlett K, Roulet N (1993) Methane flux from boreal peatlands. *Suo* 43: 173–182.

Crill PM, Martikainen PJ, Nykänen H, Silvola J (1994) Temperature and N fertilization effects on methane oxidation in a drained peatland soil. *Soil Biol. and Biochem.* 26: 1331–1339.

Cubbage FW, Flather CH (1993) Forested wetland area and distribution. *J. Forestry* 91: 35–40.

Denny P (1993) Wetlands of Africa: Introduction. In: Whigham D, Dykyjova D, Hejny S (eds) Wetlands of the World I: Inventory, Ecology and Management. Kluwer, Dordrecht, The Netherlands, pp 1–31.

Driessen PM, Soepraptokardjo M, Pons L (1979) Formation, properties, reclamation and agricultural potential of Indonesian ombrogenous lowland peats. In: Proc. Intl. Symp. Peat in Agriculture and Horticulture. International Peat Society, pp 67–108.

Esser G (1984) Significance of biospheric carbon pools and fluxes for the atmospheric CO_2: A proposed model structure. In: Lieth H, Fantechi R, Schnitzler H (eds) Interactions Between Climate and Biosphere, Progress in Biometeorology, Swets and Zeitlinger BV, Lisse, Netherlands, pp 253–294.

Foss PJ (1994) Mires in Ireland. In: European mires; Distribution and conservation situation (ed. M Lofroth). International Mire Conservation Group, Trondheim, Norway, 1994, pp 87–93. (Unpublished manuscript).

Franzén LG (1994) Are wetlands the key to the ice-age cycle enigma? *Ambio* 23: 300–308.

Freeman C, Lock MA, Reynolds B (1993) Fluxes of CO_2, CH_4 and N_2O from a Welsh peatland following simulation of water table draw-down: Potential feedback to climate change. *Biogeochem.* 19: 51–60.

Glebov FZ (1979) Classification of space—time changes of forest and mire in the taiga zone of West Siberia. In: Kivinen E, Heikurainen L, Pakarinen P (eds) Proc. Internat. Symp. on Classification of Peat and Peatlands. Intl. Peat Society, Helsinki, pp 249–259.

Glooschenko WA, Tarnocai C, Zoltai S, Glooschenko, V (1993) Wetlands of Canada and Greenland. In: Whigham D, Dykyjova D, Hejny S (eds) Wetlands of the World I: Inventory, Ecology and Management. Kluwer, Dordrecht, The Netherlands, pp 415–514.

Gore AJP (1983) Introduction. In: Gore AJP (ed) Ecosystems of the World. Volume 4A. Mires: Swamp, Bog, Fen and Moor. Elsevier, Amsterdam, pp 1–34.

Gorham E (1990) Biotic impoverishment in northern peatlands. In: Woodwell GM (ed.) The Earth in Transition. Patterns and Processes of Biotic Impoverishment, Cambridge University Press, Cambridge, UK, pp 65–98.

Gorham E (1991) Northern peatlands: role in the carbon cycle and probable responses to climatic warming. *Ecol. Applic.* 1: 182–195.

Gorham E, Janssens JA (1993) The paleorecord of geochemistry and hydrology in northern peatlands and its relation to global change. *Suo* 43: 117–126.

Haavisto VF, Jeglum JK (1991) Peatland potentially available for forestry in Canada. In: Jeglum JK, Overend RP (eds) Peat and Peatlands: Diversification and Innovation. Quebec City, QC. Canadian Society of Peat and Peatlands, Echo Bay, ON, pp 30–37.

Hånell B (1991) Peatland forestry in Sweden. In: Jeglum JK, Overend RP (ed) Peat and Peatlands: Diversification and Innovation. Quebec City, P.Q. Canadian Society of Peat and Peatlands, Echo Bay, ON, pp 19–25.

Ilomets M, Kallas R (1994) Mires in Estonia. In: Lofroth M (ed) European mires; Distribution and Conservation Situation. International Mire Conservation Group, Trondheim, Norway, 1994, pp 39–52. (Unpublished manuscript).

Immirzi CP, Maltby E, Clymo RS (1992) The Global Status of Peatlands and their Role in the Carbon Cycle. Friends of the Earth, London, 145 pp.

Ivkovich V (1994) Mires in Belarus. In: Lofroth M (ed.) European mires; Distribution and conservation situation. International Mire Conservation Group, Trondheim, Norway, 1994, pp 20–23. (Unpublished manuscript).

Ketcheson DE, Jeglum JK (1972) Estimates of black spruce peatland areas in Ontario. Can. For. Serv. Ont. Inf. Rep. O-X-172, Canadian Forestry Service, Sault Ste. Marie, ON, 29 pp.

Kim J, Verma S (1992) Soil surface CO_2 flux in a Minnesota peatland. *Biogeochem.* 18: 37–51.

Kivinen E (1972) Area, distribution and ownership of peatlands. In: Päivänen J (ed) Finnish Peatlands and their Utilization. Finnish Peatland Society, Helsinki, pp. 7–9.

Kivinen E, Pakarinen P (1981) Geographical distribution of peat resources and major peatland complex types in the world. *Annales Academiae Scientiarum Fennicae*, Ser. A, III. *Geologica—Geographica*, No. 132, 28 pp.

Laine J, Minkkinen K, Puhalainen A, Jauhiainen S (1994) Effect of forest drainage on the carbon balance of peatland ecosystems. In: Kanninen M, Heikinheimo P (ed.) The Finnish Research Programme on Climate Change, Second Progress Report. Publications of the Academy of Finland 1/94. Painatuskeskus, Helsinki, pp 303–308.

Laine J, Minkkinen K, Sinisalo J, Savolainen I, Martikainen PJ (1995) Greenhouse impact of a mire after drainage for forestry. In: Ecology and Management of Northern Wetlands, August 1994, Traverse City, MI. (In press).

Lång K, Lehtonen M, Martikainen PJ (1994) Nitrification potentials at different pH values in peat samples from various layers of a drained mire. *Geomicrobiol.* 11: 141–147.

Lien T, Martikainen PJ, Nykänen H, Bakken L (1993) Methane oxidation and methane fluxes in two drained peat soils. *Suo* 43: 231–236.

Lindsay R (1994) Mires in the United Kingdom. In: Lofroth M (ed) European mires; Distribution and Conservation Situation. International Mire Conservation Group, Trondheim, Norway, 1994, pp 178–188. (Unpublished manuscript).

Lofroth M (ed) (1994) European mires: Distribution and Conservation Situation. International Mire Conservation Group, Trondheim, Norway, 1994, 188 pp. (Unpublished manuscript).

Martikainen PJ, Nykänen H, Crill P, Silvola J (1993a) The effect of changing water table on methane fluxes at two Finnish mire sites. *Suo* 43: 237–240.

Martikainen PJ, Nykänen H, Crill P, Silvola J (1993b) Effect of a lowered water table on nitrous oxide fluxes from northern peatlands. *Nature* 366: 51–53.

Martikainen PJ, Nykänen H, Lång K, Alm J, Silvola J (1994) Emissions of methane and nitrogen oxides from peatland ecosystems. In: Kanninen M, Heikinheimo P (eds) The Finnish Research Programme on Climate Change, Second Progress Report, Publ. Acad. Finland 1/94, Painatuskeskus, Helsinki, pp 279–284.

Martikainen PJ, Nykänen H, Alm J, Silvola, J (1995) Changes in fluxes of carbon dioxide, methane and nitrous oxide due to forest drainage of mire sites of different trophy. *Plant and Soil* (In press).

Matthews E, Fung I (1987) Methane emission from natural wetlands: global distribution area, and environmental characteristics of sources. *Global Biogeochem. Cycles* 1: 61–86.

Moen A (1995) Mires in Norway. In: Lofroth M (ed) European mires: Distribution and Conservation Situation. International Mire Conservation Group, Trondheim, Norway, 1994, pp 119–126. (Unpublished manuscript).

Moore TR (1986) Carbon dioxide evolution from subarctic peatlands in eastern Canada. *Arct. Alp. Res.* 18: 189–193.

Moore TR (1989) Plant production, decomposition, and carbon efflux in a subarctic patterned fen. *Arct. Alp. Res.* 21: 156–162.

Moore TR, Knowles R (1989) The influence of water table levels on methane and carbon dioxide emissions from peatland soils. *Can. J. Soil Sci.* 69: 33–38.

Moore TR, Knowles, R (1990) Methane emissions from fen, bog and swamp peatlands in Quebec. *Biogeochem.* 11: 45–61.

Moore T, Roulet N, Knowles R (1990) Spatial and temporal variations of methane flux from subarctic/northern boreal fens. *Global Biochem. Cycles* 4: 29–46.

Nykänen H, Alm J, Lång K, Silvola J, Martikainen PJ (1995) Emissions of CH_4, N_2O and CO_2 from a virgin fen and a fen drained for grassland in Finland. *Journal of Biogeography* 22: 1149–1155.

Post WM, Emanuel WR, Zinke PJ, Stangenberger AG (1982) Soil carbon pools and world life zones. *Nature* 298: 156–159.

Pulliam W, Meyer JL (1992) Methane emissions from floodplain swamps of the Ogeechee River: long-term patterns and effects of climate change. *Biogeochem.* 15: 151–174.

Rieger S, Schoephorster DB, Furbush CE (1979) Exploratory Soil Survey of Alaska. U.S. Dept. Agriculture, Soil Conservation Service. U.S. Gov. Printing Office, 1979-247-478/68. 213 pp.

Rieley JO, Siefferman RG, Page SE (1992) The origin, development present status and importance of the lowland peat swamp forests of Borneo. *Suo* 43: 241–244.

Robertson A (1994) Nature and distribution of peat and peatlands. *IPS Newsletter* No. 2 (April 1994) International Peat Society, Jyskä, Finland, pp 8–9.

Roderfeld H, Kuntze H (1995) Mires in Germany. In: Lofroth M (ed) European Mires; Distribution and Conservation Situation. International Mire Conservation Group, Trondheim, Norway, 1994, pp 66–76. (Unpublished manuscript).

Rodin LE, Bazilevich NI, Rozov NN (1975) Productivity of the world's main ecosystems. In: DE Franklin JF, Goodal DW (eds), Productivity of World Ecosystems, Reichle, National Academy of Science, Washington DC, pp 13–26.

Roulet NT (1995) The effect of forestry drainage practices on the emission of methane from northern peatlands. Abstract. In: Ecology and Management of Northern Wetlands. Technical Program, August 1994, Traverse City, MI.

Roulet NT, Ash R, Moore TR (1992a) Low boreal wetlands as a source of atmospheric methane. *J. Geophys. Res.* 97: 3739–3749.

Roulet NT, Jano A, Kelly CA, Klinger LF, Moore TR, Protz R, Ritter A, Rouse WR (1994) Role of Hudson Bay lowland as a source of atmospheric methane. *J. Geophys. Res.* 94: 1439–1454.

Roulet N, Moore T, Bubier J, Lafleur P (1992b) Northern fens: methane flux and climatic change. *Tellus* 448: 100–105 .

Rothwell RL (1991) Substrate environments on drained and undrained peatlands, Wally Creek experimental drainage area, Cochrane, Ontario. In: Jeglum JK, Overend RP (eds) Proceedings, Symposium '89, Peat and Peatlands, Volume I. Canadian Society for Peat and Peatlands, Echo Bay, ON, pp 103–108.

Sakovets VV, Germanova NI (1992) Changes in the carbon balance of forested mires in Karelia due to drainage. *Suo* 43: 249–252.

Schlesinger WH (1984) Soil organic matter: a source of atmospheric CO_2. In: Woodwell GM (ed) The Role of Terrestrial Vegetation in the Global Carbon Cycle: Measurement by Remote Sensing SCOPE 23, Wiley, pp 111–127.

Silvola J, Heikkinen S (1979) CO_2 exchange in the *Empetrum nigrum–Sphagnum fuscum* community. *Oecologia* 37: 273–283.

Silvola J, Valijoki J, Aaltonen H (1985) Effect of draining and fertilization on soil respiration at three ameliorated peatland sites. *Acta Forestalia Fennica* 191: 1–32.

Silvola J, Alm J, Ahlholm U (1992) The effect of plant roots on CO_2 release from peat soils. *Suo* 43: 259–262.

Silvola J, Alm J, Ahlholm U, Nykänen H, Martikainen PJ (1994a) CO_2 fluxes in peatlands under varying temperature and moisture conditions. In: Kanninen M, Heikinheimo P (eds) The Finnish Research Programme on Climate Change, Second Progress Report, Publ. Acad. Finland 1/94, Painastuskeskus, Helsinki, pp 272–276.

Silvola J, Alm J, Martikainen PJ, Nykänen H (1994b) Temporal and spatial variations of CO_2, CH_4 and N_2O fluxes in a boreal minerotrophic pine fen. In: Kanninen M, Heikinheimo P (eds) The Finnish Research Programme on Climate Change, Second Progress Report, Publ. Acad. Finland, Painastuskeskus, Helsinki, pp 277–278.

Singh T (1982) Biomass equations for ten major tree species of the prairie provinces. Env. Can. Inf. Rep. NOR-X-242, Can. For. Serv., Northern Forest Research Centre, Edmonton, AB, 35 pp.

Svensson B (1980) Carbon dioxide and methane fluxes from the ombrotrophic parts of a subarctic mire. In: Soneson M (ed) Ecology of a subarctic mire. Ecological Bulletin 30, Stockholm, pp 235–250.

Svensson BH, Sundh I (1993) Factors affecting methane production in peat soils. *Suo* 43: 183–190.

Szczepanski AJ (1990) Forested wetlands of Poland. In: Lugo AE, Brinson M, Brown S (eds) Ecosystems of the World. Vol. 15. Forested Wetlands. Elsevier, Amsterdam, pp 437–446.

Tallis JH (1983) Changes in wetland communities. In: Gore AJP (ed) Ecosystems of the World. Vol. 4B. Mires: Swamp, Bog, Fen and Moor, Elsevier, Amsterdam, pp 311–347.

Tarnocai C (1980) Canadian wetland registry. In: Pollett FC (ed) Workshop on Canadian Wetlands. CDA Rubec and Lands Directorate, Environment Canada. Ecological Land Classification Series No. 12, Ottawa, pp 453–470.

Tarnocai C (1984) Peat Resources of Canada. Natl. Res. Coun. Canada, Div. Energy, Peat Energy Progr. NRCC No. 24140, 17 pp.

Taylor JA (1983) The peatlands of Great Britain and Ireland. In: Gore AJP (ed) Ecosystems of the World. Vol. 4B. Mires: Swamp, Bog, Fen and Moor. Elsevier, Amsterdam, pp 1–46.

Thompson K, Hamilton AC (1983) Peatlands and swamps of the African continent. In: Gore AJP (ed) Ecosystems of the World. Vol. 4B. Mires: Swamp, Bog, Fen and Moor. Elsevier, Amsterdam, pp 331–373.

Tibbetts TE (1969) Canada: a brief review of peat in Canada, 1968. In: Lafleur C, Butler J (eds) Proc. Third Intl. Peat Congress. Quebec, pp 9–10.

Trettin CC, Gale MR, Jurgensen MF, McLaughlin JW (1992) Carbon storage response to harvesting and site preparation in a forested mire in northern Michigan, USA. *Suo* 43: 281–284.

Urban NR, Eisenreich SJ, Bayley SR (1988) The relative importance of denitrification and nitrate assimilation in mid-continental bogs. *Limnol. Oceanog.* 33: 1611–1617.

Vompersky SE (1989) Current status of forest drainage in the USSR and problems in research. In: Jeglum JK, Overend RP (eds) Proceedings '89, Peat and Peatlands, Volume 1, Canadian Society for Peat and Peatlands, Echo Bay, ON, pp 13–18.

Vompersky SE (1994) Biospheric role of mires in the carbon cycle. *Priroda* 7: 44–50.

Yurkovskaya T (1980) Mires. In: Gribovoy SA, Isachenko TI, Lavrenko EM (eds) The Vegetation of the European Portion of USSR. Nauka, Leningrad, pp 300–345.

Zoltai SC, Johnson JD (1987) Relationships between nutrients and vegetation in peatlands of the Prairie Provinces. In: Rubec CDA, Overend RP (eds) Proc. Symposium '87, Wetlands/Peatlands, Edmonton, AB, pp 535–542.

5 Sequestration of carbon in the Finnish boreal forest ecosystem managed for timber production

Seppo Kellomäki and Timo Karjalainen

University of Joensuu, Faculty of Forestry
PO Box 111, Fin-80101 Joensuu, Finland

INTRODUCTION

Carbon in forest ecosystems

The increasing concentration of carbon dioxide (CO_2) in the atmosphere alone contributes by over 50% to the enhancing greenhouse effect (Houghton et al. 1990, 1992). The major anthropogenic sources of carbon (C) entering the atmosphere are combustion of fossil fuels and land-use changes. Enhancement of the greenhouse effect can be mitigated by decreasing the emission of C or by increasing the capacity of terrestrial C sinks. Terrestrial C sinks can, in turn, be strengthened by increasing the C in vegetation or soil on the current land area or by increasing the area covered by vegetation. In this context, the role of forest ecosystems is crucial, since they occupy one third of the Earth's land area and contain approximately 45% of the total terrestrial C (915 Pg C of 2050 Pg C) (Bolin 1986; Houghton et al. 1990).

The amount of C storage in boreal forests, 300 Pg (110 Pg in vegetation and 190 Pg in soil), is larger than that of temperate forests (190 Pg total, 75 Pg in vegetation and 115 Pg in soil), but smaller than that of tropical forests (425 Pg total, 220 Pg in vegetation and 205 Pg in soil) (Bolin 1986). It has been estimated that in the 1980s, boreal and temperate forests in the Northern Hemisphere acted as a C sink of the order of 1.6–2.4 Pg C yr^{-1} (Tans et al. 1990), which is fairly large compared to the net average emission from changes in land use, 1.6 ±1.0 Pg C, in 1980–89 (Houghton et al. 1990) or to emission from fossil fuels, 6.0 ±0.5 Pg C yr^{-1}, in 1989 and 1990; but there is still a 1.2 ±0.5 Pg annual imbalance between the known sinks and sources (Houghton et al. 1992). Other chapters in this book (e.g., Chapters 2 by Fung, and 10 by Houghton) also discuss this issue from a global perspective.

Sequestration of carbon in forests and the forest sector

Preservation of the current forests and increasing the size of forested areas seem to be the most effective way to mitigate the increase in the build-up of atmospheric C since, despite international agreements, the use of fossil fuels, and thus C emissions, continues to increase (Houghton et al. 1992; UNCED 1992). However, only a part of the existing forests can be set aside, since forests are widely used for timber production or other benefits obtainable from the forest ecosystem, as is the case for Finland. On the other hand, the use of stemwood in the forest industry makes it possible to move C into storage outside the forest ecosystem in the form of wood-based products—in some cases this can be rather long-term storage (see other chapters in this book, including Matthews et al., Chapter 24).

NATO ASI Series, Vol. I 40
Forest Ecosystems, Forest Management and the Global Carbon Cycle
Edited by Michael J. Apps and David T. Price

Regular management maintains the net growth of forests, and thus preserves the sequestration capacity of the forest ecosystem, which would otherwise be lost through maturation of trees. Nevertheless, forests can provide only temporary C storage to compensate for the human-induced release of C. This is due to the dynamics of C cycling through the forest ecosystem and forest sector; C sequestered by forest vegetation is released—sooner or later—back to the atmosphere through respiration, decay of litter and soil organic matter, burning and decay of products. Usually forests are able to store C for a longer time than products can. However, prolongation of product lifespan and increased recycling could, to some extent, delay the unavoidable return of C back to the atmosphere.

This paper reviews some studies on the capacity of forests in Finland to sequester C and to supply timber under the influence of regular management. This is an attempt to demonstrate the potential for combining C sequestration and timber production in boreal conditions. In this context, attention is also paid to the sequestration potential of the forest sector (forestry and forest industry together) to set aside C from its natural cycle under current and changing climatic conditions.

CARBON IN FINNISH FORESTS

Finland is situated between latitudes 60°N and 70°N, where the forests form part of the boreal zone. Since the first national forest inventory in 1921–24, the proportion of Scots pine (*Pinus sylvestris*) has decreased from 49% to 45% of the growing stock, while the proportion of Norway spruce (*Picea abies*) has increased from 30% to 38%, and that of deciduous species (mainly birches, *Betula pendula* and *B. pubescens*) decreased from 21% to 15% (Ilvessalo 1956; Yearbook of Forest Statistics 1993–94). The age-class distribution has also become even-aged, implying that the management of forests for timber production has modified the forest structure, even though forests were previously under the influence of shifting cultivation (Kuusela 1990). Currently the total area of forest lands in Finland is approximately 23.2 Mha, of which one third are peatlands.

In 1990, the total amount of C in Finnish forests was about 2800 Tg including soil organic matter to 30 cm depth (Karjalainen and Kellomäki 1993). The aboveground components of trees and other vegetation contained about 21% of the total C in forests, the rest (79%) being in soils including the roots of trees. In the vegetation, the stemwood represents the largest C pool with 403 Tg, which is about 10% of the C stored in the growing stock in Europe. Since 1953, C storage of the growing stem-wood stock has increased by 83 Tg, due mainly to intensive management and the consequent growth, which since the early 1970s has continuously been greater than removal (Karjalainen and Kellomäki 1993; see also Kauppi, Chapter 16). The fertilisation effect of airborne nutrients and CO_2 may also have enhanced forest growth (Acidification Research in Finland 1991). Although a substantial proportion of the enhanced C sequestration is the result of forest drainage, drainage of half of the peatlands in the country may have enhanced C emissions more than sequestration. Detailed and accurate assessments are lacking for the whole country.

CAPACITY OF FINNISH FOREST SECTOR TO SEQUESTER CARBON

Forests under current climatic conditions

Calculations applying the IPCC (Intergovernmental Panel on Climate Change) guidelines for inventory of greenhouse gases (IPCC 1993) indicate that in Finland the sequestration of C by growing stock (whole tree biomass) increased substantially between 1970 and 1990, from

0.5 Tg C yr^{-1} to 8.3 Tg C yr^{-1} (Karjalainen and Kellomäki 1994a). This took place under management with cuttings of about 50×10^6 m^3 yr^{-1}, which was about 30×10^6 m^3 yr^{-1} less than the annual growth in 1990. Under this management regime, the amount of C could increase substantially in the future with no major reduction in the sequestration capacity of the forest ecosystem (Figure 5.1), but in the long run the sequestration capacity could decrease due to an over-mature forest structure. Enhanced use of forest resources could allow increased use of wood (9.3–13.9 $\times 10^6$ m^3 yr^{-1}) and in bioenergy production, in addition to the conventional use of by-products. (See also Section 3 of this book for several chapters dealing with bioenergy and wood products.) The combined effect could decrease emissions of fossil fuels by 0.14–1.56 Tg C yr^{-1} (Karjalainen and Kellomäki 1994a).

Interaction between tree species composition, management and climatic conditions

The sequestration capacity of the forest ecosystem is highly dependent on the tree species composition, site fertility and climatic conditions. In Finnish conditions the competitive capacity of Scots pine and pendula birch could increase under changing climatic conditions, with a subsequent increase in the productivity of the forest ecosystem, especially in northern Finland (north of 64°N) (Kellomäki and Kolström 1992, 1994). Also the growth of Norway spruce will increase in northern Finland, but may decrease in southern Finland, if the temperature increase is close to the predicted maximum of 0.6 °C per decade. However, the effects of changing climatic conditions on C sequestration are variable with regard to tree species and time horizon (Table 5.1). As the temperature increases, the C sequestration in unmanaged pendula birch stands could be enhanced also in the long run, while in stands composed of other tree species sequestration could decrease.

Better growth under conditions of increasing temperature implies that rotations could be shortened and thinnings performed earlier or more intensively than currently practised, particularly in northern Finland (Kellomäki and Kolström 1993; Karjalainen and Kellomäki 1994b). According to simulation studies, tree species, management and climatic conditions interact in the C sequestration, as demonstrated in Table 5.2. In all cases, C sequestration tended to be greater under changing climatic conditions than under current conditions. The forest management options were ranked in a different order depending on the measure used.

Karjalainen and Kellomäki (1994a) estimated that, in Finland as a whole, under the predicted changing climate, C uptake by trees could increase by about 10% (40-year average); with the current use of forest resources this would result in a 21.5 Tg C yr^{-1} positive balance (40-year average) under changing conditions compared to 17.6 Tg C yr^{-1} (40-year average) under the current climatic conditions. These estimates are based on growth enhancements similar to that found in stand-level simulations.

Effects of increasing CO_2 on the C balance and storage in forests were not considered in the above calculations, but one can expect increasing CO_2 to further enhance forest growth. Rastetter et al. (1991) simulated the effects of changes in CO_2 concentration, temperature and nitrogen deposition on C storage in temperate hardwood forests. A 50-year exposure to doubling of the CO_2 concentration increased the C storage in vegetation by only 4%, but if a 5 °C temperature increase was assumed, the increase was 22%. A combined CO_2 and temperature treatment caused an increase of 29% in C storage, and an increase in nitrogen deposition of 10 kg m^{-2} yr^{-1} over the growing season caused an increase of 1%. With an increased rate of mineralization, however, the C storage of the soil in the above-mentioned treatments changed by -1%, -6%, ±0%, and +4%, respectively.

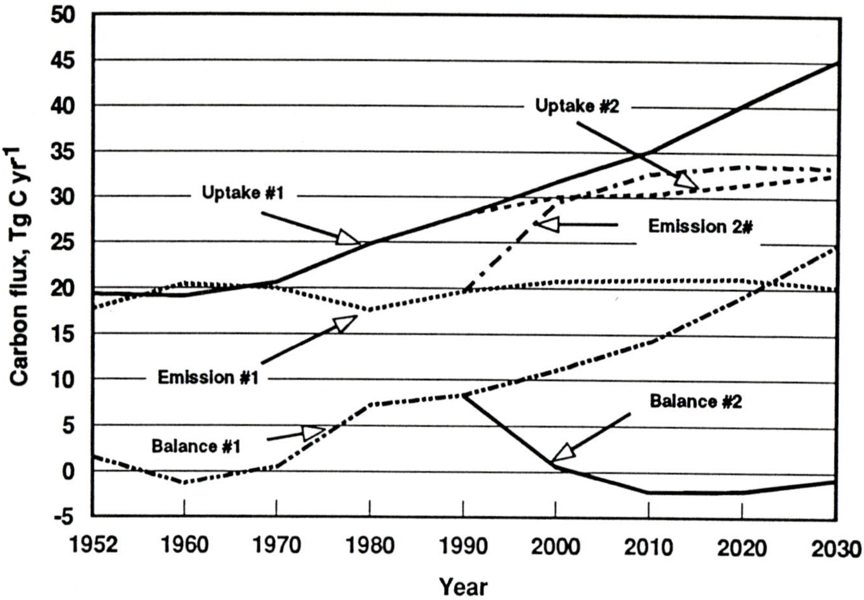

Figure 5.1. Development of the C balance of growing stock (whole tree biomass) in terms of uptake (growth) and emission (removal, logging and silvicultural waste, natural mortality) in the period 1952–2030 (Karjalainen and Kellomäki 1994a). #1 indicates current use of forest resources, and #2 shows the largest sustainable cuttings (growth and removal in balance in the long run).

The effect of CO_2 enrichment on forest growth could, however, be smaller than the effect of increasing temperature (Rastetter et al. 1991; Väisänen et al. 1994). Furthermore, the response to rising levels of atmospheric CO_2 could be short-lived, with the consequence that individual trees may not sustain higher growth rates for longer periods (Rastetter et al. 1991; Pitelka 1992; Wang et al. 1995) due to acclimation to the elevated level of CO_2 (Wang et al. 1995). New insects and fungal diseases which are not currently present can also cause unexpected attacks along with an increased risk of wind- and fire-induced damage assumed to be associated with the changing climate in boreal conditions (see Chapters 7, Volney, and 8, Stocks et al.).

CAPACITY OF MANAGED FORESTS TO SEQUESTER CARBON

Regardless of the management regime, the simulated net production of 3–4 Mg C ha^{-1} yr^{-1} of forest ecosystems in the conditions in southern Finland under current climatic conditions (Karjalainen 1994) is close to that (3.6–5.8 Mg C ha^{-1} yr^{-1}) reported by Woodwell et al. (1978) for boreal and temperate forests. However, C storage is lower in managed than in unmanaged forests. Over the rotation, the average amount of C in managed forests is only about one third of that in unmanaged forests, even though the aim in multi-purpose management is to reach maximum sustainable yield (Cooper 1983). Though differences in C sequestration between unmanaged stands and managed stands plus wood products seem to become even in the long run (Karjalainen and Kellomäki 1994b).

Table 5.1. Effect of increasing temperature on C sequestration in unmanaged stands of different tree species over 100- and 200-year perspectives, based on results with a gap-type simulation model (Karjalainen 1994). The calculations are for unmanaged stands in southern Finland (62°N). In each column, figures to the left are for current climatic conditions and figures to the right are for changing climatic conditions. All values include vegetation, litter and soil organic matter.

Tree Species	<C> Balance over:				C Storage after:				<C> Storage over:			
	100 Years ($Mg\ C\ ha^{-1}\ yr^{-1}$)		200 Years ($Mg\ C\ ha^{-1}\ yr^{-1}$)		100 Years ($Mg\ C\ ha^{-1}$)		200 Years ($Mg\ C\ ha^{-1}$)		100 Years ($Mg\ C\ ha^{-1}$)		200 Years ($Mg\ C\ ha^{-1}$)	
Scots pine	2.12	2.16	0.91	0.87	212	216	182	173	163	169	180	181
Norway spruce	2.39	2.34	0.90	0.27	240	234	180	53	183	192	196	166
Pendula birch	2.01	2.24	0.58	0.81	201	224	117	163	154	171	156	179
Pubescent birch	1.34	1.30	0.43	0.36	134	130	86	73	115	118	108	105
Mixed species	2.45	2.57	1.07	0.90	245	257	213	180	182	197	209	205

Table 5.2. Effect of changing climate and stand management on three key C sequestration measures of managed stands and wood products, and on the ranking order (Karjalainen 1994). For each group, the first column indicates increase (+) or decrease (-) in the quantity; the second column gives the ranking order under the current climate, and the third column gives the ranking order under changing climatic conditions.

Tree species and management[a]	<C> Balance over rotation			C Storage after clearcut			<C> Storage over rotation		
	1	2	3	1	2	3	1	2	3
Based on Scots pine									
Basic thinning[a], short rotation[b]	+	4	2	+	6	1	+	5	4
Intensive thinning, short rotation	+	2	1	+	4	2	+	6	6
Light thinning, short rotation	+	1	3	+	1	4	+/-	3	3
Basic thinning, long rotation	+	3	4	+	2	3	+	2	2
Intensive thinning, long rotation	+	5	5	+	5	5	+	4	5
Light thinning, long rotation	+	5	6	-	3	6	-	1	1
Based on Norway spruce									
Basic thinning, short rotation	+	4	2	+	4	4	+	4	4
Intensive thinning, short rotation	+	1	1	+	3	2	+	6	6
Light thinning, short rotation	+	5	5	-	5	6	-	3	3
Basic thinning, long rotation	-	2	4	-	2	3	+	2	2
Intensive thinning, long rotation	+	6	6	+	6	5	+	5	5
Light thinning, long rotation	+	3	3	+	1	1	-	1	1
Based on pendula birch									
Basic thinning, short rotation	+	3	3	+	3	4	+	4	4
Intensive thinning, short rotation	+	1	2	+	2	5	+	5	6
Light thinning, short rotation	+	2	1	+	1	1	+	2	2
Basic thinning, long rotation	+	5	4	+	5	2	+	3	3
Intensive thinning, long rotation	+	6	6	+	6	6	+	6	5
Light thinning, long rotation	+	4	5	+	4	3	+	1	1

a Basic Thinning: the current thinning regime applied in Finland.
Intensive Thinning: the current thinning regime with 10% increase in intensity.
Light Thinning: the current thinning regime with 10% reduction in intensity.

b Rotation length is defined in terms of the maximum mean overbark diameter limit for clearcutting:
Short rotation: Scots pine: 29 cm; Norway spruce and pendula birch: 26 cm.
Long rotation: Scots pine: 31 cm; Norway spruce and pendula birch: 28 cm.

Nabuurs and Mohren (1993) compared the expected C sequestration capacity and the probability of stand establishment at several sites world-wide. Based on the criterion *long-term average stock of C in biomass and products*, selectively-logged lowland evergreen tropical rainforest and Douglas-fir forest in the United States sequestered the largest amount of C. In both cases, C stocks of about 200 Mg C ha^{-1} were possible. If the organic matter in the soil was included, the figures rose as high as 309 and 339 Mg C ha^{-1}, respectively. Short-rotation forestry gave relatively small C storages at the time of harvest and high net fluxes during the rotation. The highest net annual fixation of C in the first rotation was achieved in Caribbean pine plantations in Brazil, where the storage rate was 5.1 Mg C ha^{-1} yr^{-1}. By comparison, under the current climate in the boreal conditions in southern Finland, the highest average net annual C balance over the rotation was 1.8 Mg C ha^{-1} yr^{-1} (Karjalainen 1994).

Young forests accumulate C more intensively than do mature forests, even though mature forests are major sites of C storage. However, conversion of old-growth forests to young fast-growing forests will not decrease the amount of atmospheric C in general; i.e., C accumulation will increase but the existing storage will be reduced. The major part of C sequestered in the forests and wood products is set aside in the forest ecosystem rather than in wood-based products (Table 5.3). Therefore sequestration of C in forests and wood products is a matter of optimising the size of different C pools keeping in mind limits set by the selected time horizon. For example, when Dewar and Cannell (1992) estimated the C storage in plantation forests in the United Kingdom applying varying growth rates, thinning regimes and species characteristics, they found that about 10–20% of the total C was in wood-based products. Similarly, Nabuurs and Mohren (1993) found that the boreal Norway spruce forests in the former Soviet Union contained more than 90% of the total C, which is similar to the findings of Karjalainen (1994). On the other hand, the C storage of wood-based products, 40–45%, found by Burschel et al.(1993) was clearly higher than that in the above-mentioned studies. The differences within and between those studies are mainly due to differences in stand productivity (e.g., tree species, site fertility, climatic conditions).

MITIGATION OF THE GREENHOUSE EFFECT

Role of forestry

Burschel et al. (1993) concluded that silvicultural management does not, to any great extent, improve the CO_2-mitigating characteristics of existing forests. Measures such as longer rotation, underplanting and change of species may in the short term lead to additional C mitigation, but in the long term the increase in the C storage will be less than that achieved through afforestation. Schroeder and Ladd (1991) concluded that if the objective is to remove

Table 5.3. Average amount of C storage (Mg C ha^{-1}) in the forest ecosystem and wood-based products as estimated in some recent studies in temperate and boreal zones.

C storage (Mg C ha^{-1})			
Forest	Products	Total	Source
125–195	20–40	145–235	Dewar and Cannell 1992
176–186	9–17	185–203	Nabuurs and Mohren 1993
82–103	60–91	142–194	Burschel et al. 1993
85–119	5–16	90–132	Karjalainen 1994

CO_2 from the atmosphere, then the emphasis should be on long-term C storage. Consideration of the short-term average annual rates of C fixation, without consideration of the growth dynamics of forest stands over time, can be misleading. Shroeder and Ladd's analysis showed that it is feasible to plant trees for storing C, but the large areas of land needed to do this could limit the potential of afforestation to sequester atmospheric C. Thus, stopping deforestation would be an immediate and direct contribution toward solving the CO_2 problem.

Role of the forest sector

In 1990, the total Finnish emissions of greenhouse gases caused by human activity amounted to 80.5 Tg CO_2-equivalents, of which CO_2 contributed 58.6 Tg (Carbon Dioxide Committee II 1994). Emissions of CO_2 from energy production and consumption amounted to 14.9 Tg C (54.5 Tg CO_2), of which traffic accounted for 3.3 Tg. If there are no changes in energy policy aimed at restricting emissions, CO_2 emissions from the production and consumption of energy would be about 21.3 Tg C yr^{-1} in 2000 and about 24.8 Tg C yr^{-1} in 2010 (Carbon Dioxide Committee II 1994). Since Finland has a low population density and substantial forest resources, the forest sector is able to sequester substantial proportions of the energy-related C emissions (see Figure 5.1).

Carbon emissions from the forest sector can only be delayed slightly because all C once bound by trees will later return to the atmosphere. By keeping forest structure optimal from the standpoint of C sequestration, emissions from forests can be delayed. In wood products, emissions are dependent on the production processes (conversion efficiency: proportion of sawn goods, chemical and mechanical pulp, use of by-products), use of products (lifespan, terminal use: proportion of energy conversion, recycling and landfill dumping), and structure of the forest industry (proportion of mechanical and chemical forest industry).

Release of C from wood products can be delayed by prolonging the lifespans of products, by increasing recycling, and by dumping abandoned products in conditions where they will decay slowly (Karjalainen et al. 1994, 1995; See also previously referenced chapters in this book). To have an effect on C balance, changes in product lifespans or recycling should be substantial (Figure 5.2). Landfills seem to store abandoned products effectively (see e.g., Chapter 23, Price et al.), but are sensitive to changes in decay rate. In fact, landfills may store C better than forests, but shortage of land and aesthetic reasons may restrict the use of landfills as C storage. Furthermore, replacing goods and energy based on fossil fuel by sustainably produced wood products would give additional benefits for C sequestration.

Stand-level assessments indicate that wood products should be included in assessments of C balance (Figure 5.3), although there are several uncertainties related to wood product lifespans and decomposition rate in landfills (Karjalainen 1994; Karjalainen et al. 1994). Of the long-term average C storage more than half could be in wood products. Nevertheless, C sequestration potentials in the long-term are rather low if comparing gross production and net C storage (e.g., 5096 Mg compared to 304 Mg in Figure 5.3).

In Finnish conditions, forest management seems to play a more significant role in determining forest C balance than, for example, forest fires or insect and wind damage, since the annual cutting areas are larger than the areas affected by fires. The average annual cutting area in 1970–1993 was 422 000 ha yr^{-1}, which is less than 2% of the total area of forest and scrub land (Yearbook of Forest Statistics 1993–94). The proportion of thinnings on the cutting area was 37% and clear cutting 26%, while about 30% of the harvest area was related to cuttings serving natural regeneration, 7% was other cuttings and cuttings on scrub land. The average burnt area in 1952–1992 was 1880 ha yr^{-1}, while during the last 20 years only about 500 ha have burnt annually (Yearbook of Forest Statistics 1993–94). In some other parts of the boreal

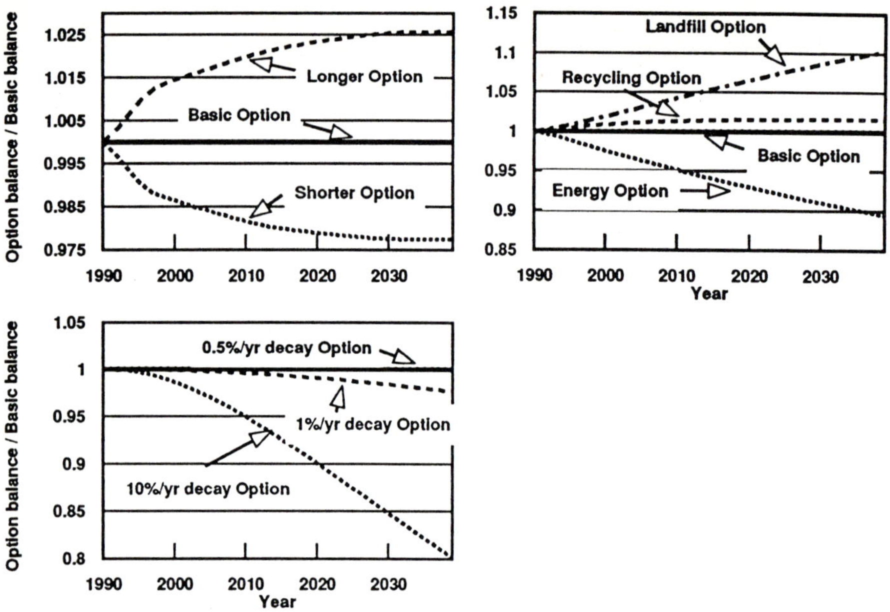

Figure 5.2. C balance of the wood product reservoir affected by changes in product lifespan (±10% change in product lifespan), by changes in terminal use (Basic Option: one third recycled, one third burned to generate energy, and one third disposed in landfills; in other options proportions changed to one half and one fourth), and by changes in landfill decay rate (Karjalainen et al. 1993).

region, however, forest fires could have a significant effect on the C balance (Auclair and Carter 1993; see also Chapters 8, Stocks et al., and 14, Kurz and Apps). Regular forest management and the dense network of forest roads have led to a decrease in forest fires in Finland.

CONCLUSIONS

The potential of managed forest to sequester C was discussed in this paper. However, the potential of the forest ecosystem cannot be treated without reference to the C in wood-based products, since according to stand-level assessments the amount of C in wood-based products may make up more than half of the total C sequestered in the long run. Problems related to estimates of product life spans and decomposition rates in landfills should be taken into account (Kurz et al. 1992; Karjalainen et al. 1994, 1995; Chapters 18–24 of this book). On a global scale, an increasing amount of wood-based products has been produced during this century. Part of this product pool remains in long-term storage, either in use or transferred to permanent disposal sites, thus affecting the global C balance. So far, the C in wood-based products is usually included in C budget assessments by merely assuming instant decay for the products. This is valid only when the production of products is stable and changes in product lifespan or in terminal use are excluded. Furthermore, the effects of disturbances such as forest fire, insect attacks and wind damage on the C cycle, the C emissions from harvesting and management operations and manufacturing of wood-based products are poorly documented, which leads to uncertainty as to how well the current calculations account for

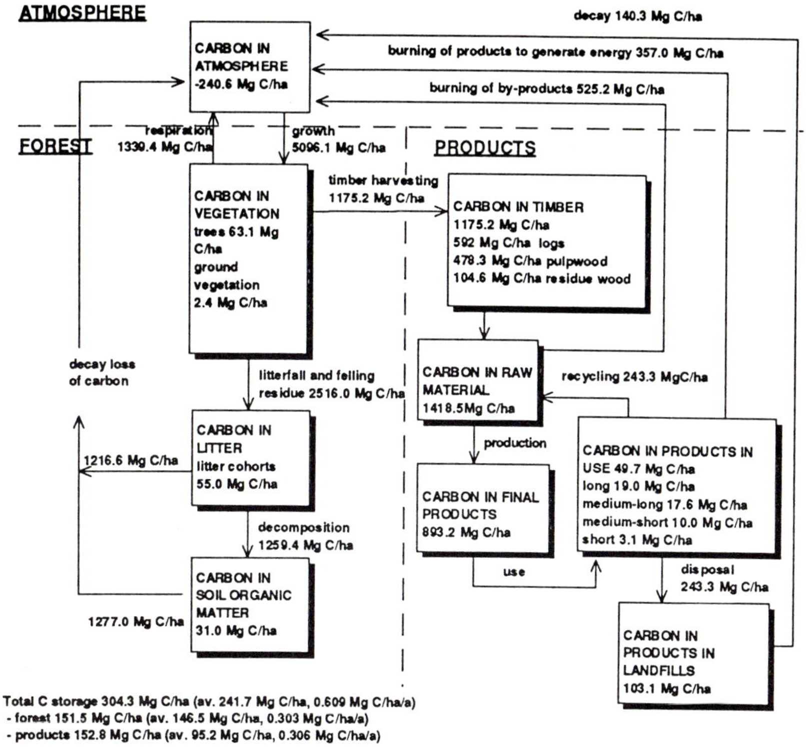

Figure 5.3. Example of total C flows and storages after 500 years simulation under changing climatic conditions in southern Finland (Karjalainen and Kellomäki 1994b). Storages in boxes represent the situation after 500 years simulation; flows are cumulative values.

fluxes of C that affect the global C balance. These problems are among the major obstacles when the C balances for different countries are compared in terms of their contribution to the global C cycle. Guidelines developed by the IPCC (IPCC 1993) for conducting and reporting national greenhouse gas inventories are most welcome, but need to be further developed (Karjalainen and Kellomäki 1994a).

REFERENCES

Aarne M (ed) (1994) Yearbook of Forest Statistics 1993-94. The Finnish Forest Research Institute. Official Statistics of Finland, Agriculture and forestry 1994:7, Helsinki, 348 pp.

Auclair AND, Carter TB (1993) Forest wildfires as a recent source of CO_2 at northern latitudes. *Can. J. For. Res.* 23: 1528–1536.

Bolin B (1986) How much CO_2 will remain in the atmosphere? The carbon cycle and projections for the future. In: Bolin B, Döös BR, Jäger J, Warrick RA (eds) The Greenhouse Effect, Climatic Change, and Ecosystems. SCOPE 29, Chichester, UK, pp 93–155.

Burschel P, Kürsten E, Larson BC (1993) Die Rolle von Wald und Forstwirtschaft im Kohlenstoffhaushalt-Eine Betrachtung für die Bundesrepublik Deutschland. *Forstliche Forschungsberichte* 126/1993. München, Deutschland, 135 pp.

Carbon Dioxide Committee II (1994) Committee report 1994:2. Ministry of Environment, Finnish Publishing Centre Ltd., Helsinki, 145 pp.
Cooper CF (1983) Carbon storage in managed forests. *Can. J. For. Res.* 13: 155–166.
Dewar RC, Cannell MGR (1992) Carbon sequestration in the trees, products and soils of forest plantations: an analysis using UK examples. *Tree Physiol.* 11: 49–71.
Houghton JT, Jenkins GJ, Ephraums JJ (eds) (1990) Climate Change. The IPCC Scientific Assessment. Cambridge University Press, Cambridge, UK, 364 pp.
Houghton JT, Callander BA, Varney SK (eds) (1992) Climate Change 1992. The Supplementary Report to the IPCC Scientific Assessment. Cambridge University Press, Cambridge, UK, 200 pp.
Ilvessalo Y (1956) The forests of Finland from 1921–24 to 1951–53. A survey based on three national forest inventories. *Commun. Instit. For. Fenn.* 47(1): 227 pp.
IPCC (1993) Intergovernmental Panel on Climate Change, Draft Guidelines for National Greenhouse Gas Inventories. Vol 1–3.
Karjalainen T (1994) Simulation of carbon budget for forests and wood-based products; application into boreal conditions in Finland. Thesis for the Licentiate degree in Silviculture. University of Joensuu, Faculty of Forestry, Joensuu, Finland, 123 pp.
Karjalainen T, Kellomäki S (1993) Carbon storage in forest ecosystems in Finland. Proceedings of the IPCC AFOS workshop Carbon Balance of World's Forested Ecosystems: Towards a Global Assessment. *Publ. Acad. Fin.* 3/93, Joensuu, Finland, 11–15 May 1992. pp 40–51.
Karjalainen T, Kellomäki S, Pussinen A (1994) Role of wood-based products in absorbing atmospheric carbon. *Silva Fenn.* 28(2): 67–80.
Karjalainen T, Kellomäki S (1994a) Greenhouse gas inventory for land use change and forestry in Finland based on international guidelines. *Mitig. and Adapt. Strat. for Glob. Chan.* (Unpublished manuscript).
Karjalainen T, Kellomäki S (1994b) Simulation of long-term forest sector carbon budget; stand level assessment under a changing climate in Finland. *Water, Air, Soil Pollut.* 82: 309–320.
Karjalainen T, Kellomäki S, Pussinen A (1995) Carbon balance in the forest sector in Finland during 1990–2039. *Climatic Change*, 28 pp. (In press).
Kellomäki S, Kolström M (1992) Simulation of tree species composition and organic matter accumulation in Finnish boreal forests under changing climatic conditions. *Veget.* 102: 47–68.
Kellomäki S, Kolström M (1993) Computations on the yield of timber by Scots pine when subjected to varying levels of thinning under a changing climate in southern Finland. *For. Ecol. Manage.* 59: 237–255.
Kellomäki S, Kolström M (1994) The influence of climate change on the productivity of Scots pine, Norway spruce, pendula birch and pubescent birch in southern and northern Finland. *For. Ecol. Manage.* 65: 201–217.
Kentlamies K (ed) (1991) Acidification Research in Finland. Review of the Results of the Finnish Acidification Research Programme (HAPRO) 1985–1990. Brochure 39/1991. Ministry of the Environment, Helsinki, Finland, 48 pp.
Kurz, WA, Apps MJ, Webb TM, McNamee PJ (1992) The Carbon Budget of the Canadian Forest Sector: Phase 1. Forestry Canada Information Report NOR-X-326. Forestry Canada, Northwest Region, Edmonton, Alberta, 93 pp.
Kuusela K (1990) The dynamics of boreal coniferous forests. Finnish National Fund for Research and Development, 112. Gummerus, Jyväskylä, 172 pp.
Nabuurs GJ, Mohren GMJ (1993) Carbon fixation through forestation activities. A study of the carbon sequestration potential of selected forest types, commissioned by the FACE Foundation. *IBN Res. Rep.* 93/4. FACE/Institute for Forestry and Nature Research (IBN–DLO), Arnhem, Wageningen, The Netherlands, 205 pp.
Pitelka L (1992) Forest response to carbon dioxide. *EPRI J.* January/February 1992: 38–41.
Rastetter EB, Ryan MG, Shaver GR, Melillo JM, Nadelhoffer KJ, Hobbie JE, Aber JD (1991) A general biogeochemical model describing the responses of the C and N cycles in terrestrial ecosystems to changes in CO_2, climate, and N deposition. *Tree Physiol.* 9: 101–126.
Schroeder P, Ladd L (1991) Slowing the increase of atmospheric carbon dioxide: a biological approach. *Climatic Change* 19: 283–290.
Tans P, Fung IY, Takahashi T (1990) Observational constraints on the global atmospheric CO_2 budget. *Science* 247: 1431–1438.
UNCED (1992) United Nations Conference on Environment and Development. Adoption of agreements on Environment and Development. 97 pp.
Väisänen H, Strandman H, Kellomäki S (1993) A model for simulating the effects of changing climate on the functioning and structure of the boreal forest ecosystem: an approach based on object-oriented design. *Tree Physiol.* 14: 1081–1095.
Wang K, Kellomäki S, Laitinen K (1995) Effect of needle age and long-term temperature and CO_2 treatment on the photosynthesis of Scots pine. *Tree Physiol.* 15: 211–218.
Woodwell GM, Whittaker RH, Reiners WA, Likens GE, Delwiche CC, Botkin DB (1978) The biota and the world carbon budget. *Science* 199: 141–146.

6 Carbon storage and climate change in Swedish forests: a comparison of static and dynamic modelling approaches

Martin T Sykes and I Colin Prentice

Global Systems Group, Ecology
Lund University, 223 62 Lund, Sweden

INTRODUCTION

Changes in the global carbon (C) cycle caused by human activities have focused the attention of environmental scientists on where and how C is distributed through the terrestrial biosphere. Forests are the largest land reservoir for C (e.g., see Kellomäki and Karjalainen, Chapter 5). They also have the potential to be a C sink in the future. However, their future role in this respect depends not only on present and future management practices, but also on how the vegetation responds to climate changes that may already be underway.

We compare two approaches to the estimation of forest responses to climate change, and we consider their implications for C sequestration or loss. One approach uses a static model, in which the state of the vegetation is uniquely determined by the climate and a single value for C storage (representing a mature condition) is attributed to each vegetation type (biome). The other approach uses a dynamic model, in which the processes of regeneration, growth, competition, mortality and disturbance are modelled explicitly and the C balance emerges through the opposing effects of different processes. We use regional forest inventory data as a standard against which to compare the models' estimates of present C storage, and we examine simulated forest composition and C storage responses to an illustrative climate change scenario derived from four general circulation model experiments. We focus on boreal and mixed forests along the north–south climatic gradient in Sweden. Our approach parallels that of Price et al. (1993) for Canadian forests.

The primary purpose of the Chapter is to compare approaches from a methodological standpoint, and not to present a complete picture of the effects of climate change on C storage in Swedish forests. Thus we do not consider the below-ground component of C storage, which includes soil C fractions with slower dynamics that need to modelled separately. Nor do we consider the direct effects of increasing CO_2, although the FORSKA2 modelling framework would allow such effects to be incorporated.

METHODS

Biome simulations

The static model BIOME1.1 (Prentice et al. 1993b) is a minor variant of the biome model of Prentice et al. (1992). It predicts potential natural vegetation, defined as the vegetation that would be expected (given the prevailing climate and soil conditions) in the absence of land use. The model uses a small set of plant functional types to predict the distribution of biomes on a global scale. The model is driven by a small number of bioclimatic variables that determine the potential presence or absence of plant functional types. These types interact

NATO ASI Series, Vol. I 40
Forest Ecosystems, Forest Management and the Global Carbon Cycle
Edited by Michael J. Apps and David T. Price

through a simple dominance hierarchy. BIOME1.1 differs from the original biome model of Prentice et al. (1992) only in distinguishing three additional high-latitude biomes (wooded tundra, northern taiga, northern cold deciduous forest). These are distinguished because they have different C densities from their parent biomes. BIOME1.1 has been used to simulate changes in global C storage between the last glacial maximum and the present (Prentice et al. 1993b) as well as possible global changes in C storage capabilities in the future (Prentice and Sykes 1995).

The environmental constraints on plant functional type distribution in BIOME1.1 are implemented as limiting values of bioclimatic variables: mean temperature of the coldest month, GDD (growing degree days) on a 5 °C base, and the Priestley–Taylor coefficient (annual actual evapotranspiration/annual equilibrium evapotranspiration) to represent annual plant-available moisture. All of these variables are calculated from monthly mean temperature, precipitation and sunshine data. Prentice et al. (1992) used the Leemans and Cramer (1991) 0.5° gridded monthly mean climate data for global simulations. As we planned a more regional study and also to compare our results with FORSKA2 at specific locations, we required higher resolution and accuracy. Results presented here are based on a 10' topographic grid (Sykes et al. 1995), with values computed using three-dimensional spline interpolation over the range 12.0–20.5°E and 55.0–69.0°N (W Cramer, Potsdam Institute of Climate Impact Research, 1994, pers. comm.).

An illustrative future climate scenario was calculated based on the temperature and precipitation anomalies (differences between simulated 2×CO_2 and control climates) from four general circulation models (GCMs): GFDL (Manabe and Wetherald 1987), OSU (Schlesinger and Zhao 1989), UKMO (Mitchell 1983) and GISS (Hansen et al. 1988). These anomalies were smoothly interpolated and added to the current climate data-set. This method was adopted in preference to using untransformed GCM output because it implicitly corrects for bias in the GCM simulation of present-day climate. The mean of the four modified climate data-sets was then calculated to give a new data-set of monthly mean temperature and precipitation for each 10' cell. We refer to this data-set as the *mean 2×CO_2 climate*. Sunshine data were kept as present because cloud cover anomalies were not available from the GCM experiments. Maximum soil water storage was set at 150 mm (Prentice et al. 1993a) because of the lack of a suitable soil texture data set that could be used to modify this parameter. Sensitivity experiments have indicated that moderate variations in cloudiness and soil texture do not have major effects in most of northern Europe, although soil texture becomes important in some low-rainfall areas.

Standard values for C in biomass were prescribed as in Prentice et al. (1993b) and Prentice and Sykes (1995), based on estimates by Olson (1983). In comparisons with FORSKA2 and with forest inventory data, these values were converted into estimates of aboveground stem C assuming a harvest index of 0.65 (Jarvis and Leverenz 1983).

FORSKA2 simulations

The dynamic model FORSKA2 (Prentice et al. 1993a) simulates changes in the species composition and biomass of forested landscapes through explicit growth responses and competitive interactions. Species-specific information is needed on dynamic characteristics such as shade tolerance as well as climatic responses. Output includes time-dependent estimates of species composition and biomass under specified scenarios, with or without climate change (Prentice et al. 1991). The model structure allows effects of limited migration rates or changes in disturbance regime to be explored by themselves and in interaction with climate change (Sykes and Prentice 1995b).

FORSKA2 is a gap model according to the definition of Shugart (1984), i.e., it follows the basic concept of the JABOWA model (Botkin et al. 1972), although most specific processes are formulated differently. At the core of FORSKA2 is the original FORSKA model of Prentice and Leemans (1990). FORSKA has successfully predicted forest composition (Price et al. 1993) and successional patterns (Prentice and Leemans 1990) in boreal forests, and in temperate and tropical deciduous forests (Leemans 1992; Desanker and Prentice 1995). In FORSKA2, establishment and growth rates in the FORSKA core model are modified according to species-specific environmental drivers. Landscape-scale forest composition and stem biomass are estimated by Monte Carlo simulation, assuming a stochastic disturbance regime in which the probability of aboveground biomass removal on any one patch increases with time. We routinely use a patch size of 1000 m^2 and a sample of 200 patches to represent the landscape. Note that these patches are not supposed to be spatially contiguous but rather are a statistical sample of a larger area. Applied in this "standard landscape mode", FORSKA2 has been shown to give a reasonable approximation to forest composition and biomass both regionally in different climatic zones (Prentice et al. 1991) and at particular locations (Sykes 1994; Sykes and Prentice 1995a) in Sweden. The version of FORSKA2 used here incorporates the revised set of species-specific climate response parameters estimated by Sykes et al. (1995).

We ran FORSKA2 in standard landscape mode at ten locations scattered through Sweden, covering the major biomes and climatic regions. A root mean square disturbance return time of 100 years was assumed. An initial 400-year period at today's climate allowed forest composition and biomass to equilibrate. The climate was then changed linearly over 100 years, each month's temperature and precipitation being progressively adjusted towards the mean $2\times CO_2$ climate. A final 500-year period in this warmer climate allowed a new equilibrium to develop. Biomass estimates (averages across patches) for each species were archived at 50-year intervals.

C storage estimates were obtained by applying species-specific conversion factors to the sum over all trees of an aboveground stem volume index (D^2H, where D is diameter at breast height and H is top height) to obtain total stem biomass. This in turn was corrected to stem C, for comparison with forest inventory data, by assuming a value of 50% for the C content of dry matter.

RESULTS

Current climate

Contemporary biome distributions simulated with BIOME1.1 were found to be in good agreement with potential vegetation zones as mapped e.g., by Sjörs (1967). Note that the classifications of potential natural vegetation as used by Sjörs and others are physiognomic and floristic, with no connotation of biomass; these physiognomic/floristic zones are simulated well by biome models (e.g., Prentice et al. 1992). Current species composition and dominance simulated with FORSKA2 also accurately reflected the vegetation of the test locations.

Stem C estimates based on BIOME1.1 (characteristic values for biomes) and FORSKA2 (values computed at the test locations) were compared with estimates based on forest inventory data for the surrounding regions (Skogsdata 1990; Table 6.1). BIOME1.1 estimates were *systematically too high*, by factors ranging from 1.5 in southern Sweden to 4.5 at some locations in northern Sweden. These biases probably reflects a tendency for the Olson et al.

(1983) C densities to reflect atypical (uniform, mature or old-growth) forest stands rather than normal (heterogeneous, managed) forest landscapes. Other studies have indicated that these standard C densities are also too high for unmanaged boreal forests (Botkin and Simpson 1990; Price et al. 1993), where fire and other natural factors restrict biomass. The standard C density value for cool conifer forest, in particular, is much too high because it mainly reflects measurements in the climatically distinct, massive conifer forests of lower latitudes in the Pacific Northwest (USA). We infer that the standard C densities for these biomes do not apply to most of the landscape and that the error is especially large in the cool-conifer and boreal forests.

FORSKA2 estimates are much closer to the forest inventory data (Figure 6.3). In contrast with the standard values, the FORSKA2 estimates tend to be somewhat *lower* than the forest inventory estimates. This downward bias may be due to imprecision in simulating growth or logging rates. The comparison is only rough because the FORSKA2 calculations were carried out at particular locations whereas the forest inventory data are aggregates for the surrounding regions. We infer that the C storage values simulated by FORSKA2 are approximately correct (a more exacting test would have to be defined in order to quantify the extent of any possible disagreement). They are clearly more reliable than those obtained by imputing the Olson et al. (1983) values to biomes.

Table 6.1. Comparison of stem carbon (C) estimates derived from FORSKA 2, BIOME1.1, and forest inventory data. The FORSKA2 estimates are for test locations within the specified forestry regions (lat./long. are given in Table 6.3). The BIOME1.1 estimates are "standard" values for the biome simulated at each location. The data-based estimates are average values for the regions.

REGION	SIMULATED BIOME	FORSKA2 kg C m^{-2}	BIOME1.1 kg C m^{-2}	DATA kg C m^{-2}
Malmöhus	Temperate Deciduous Forest	2.9	6.5	4.4
Kristianstad	Cool Mixed Forest	3.7	6.5	4.0
Halland	Cool Mixed Forest	3.3	6.5	3.7
Kalmar	Cool Mixed Forest	2.9	6.5	3.2
Örebro	Cool Mixed Forest	2.8	6.5	3.0
Stockholm	Cool Mixed Forest	1.7	6.5	3.2
Västmanland	Cool Conifer Forest	2.6	10.9	3.2
Kopparberg	Cool Conifer Forest	2.3	10.9	2.4
Västernorrland	Taiga	2.1	5.7	2.9
Norrbotten	Taiga	1.7	5.7	1.3

Future climate

Simulated biome distributions from BIOME1.1 (Figure 6.2, Table 6.2) under the 2×CO_2 climate show remarkable changes over the current distribution (Figure 6.1). Montane tundra is largely replaced by cold deciduous (*Betula* spp.) and cold mixed (*Pinus sylvestris–Betula* spp.) forests. Temperate deciduous forest is predicted to cover most of Sweden, replacing cool mixed and cool conifer forests and most of the taiga. Warm-temperate evergreen/mixed forest could grow in southern and coastal areas. The main driving force behind these changes is the extreme winter warming that the GCM simulations show for high latitudes in general, and Scandinavia in particular. In these simulations, boreal evergreen conifers (e.g., *Picea abies*) are

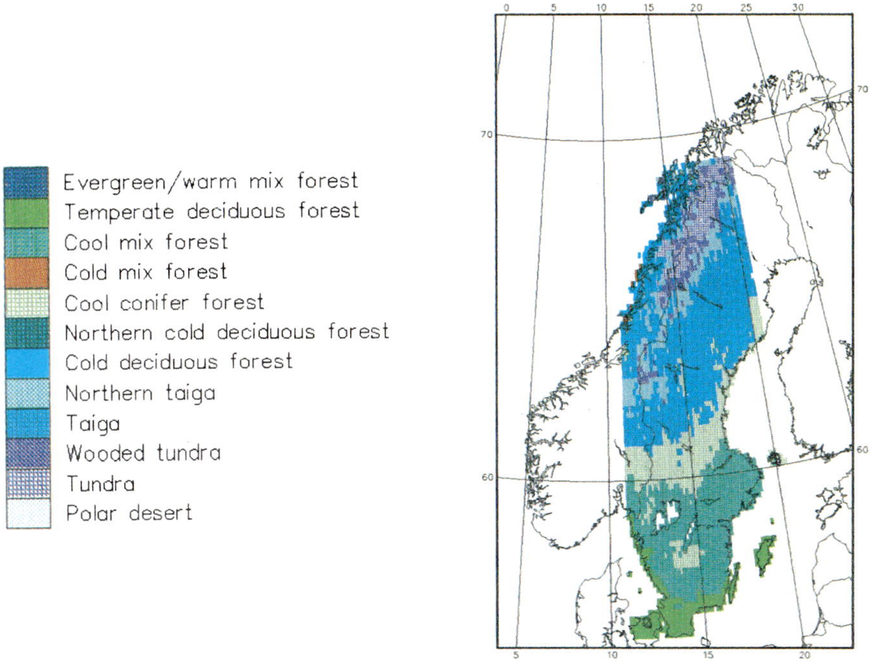

Figure 6.1. Present day biomes, simulated with BIOME1.1 and based on present climate data.

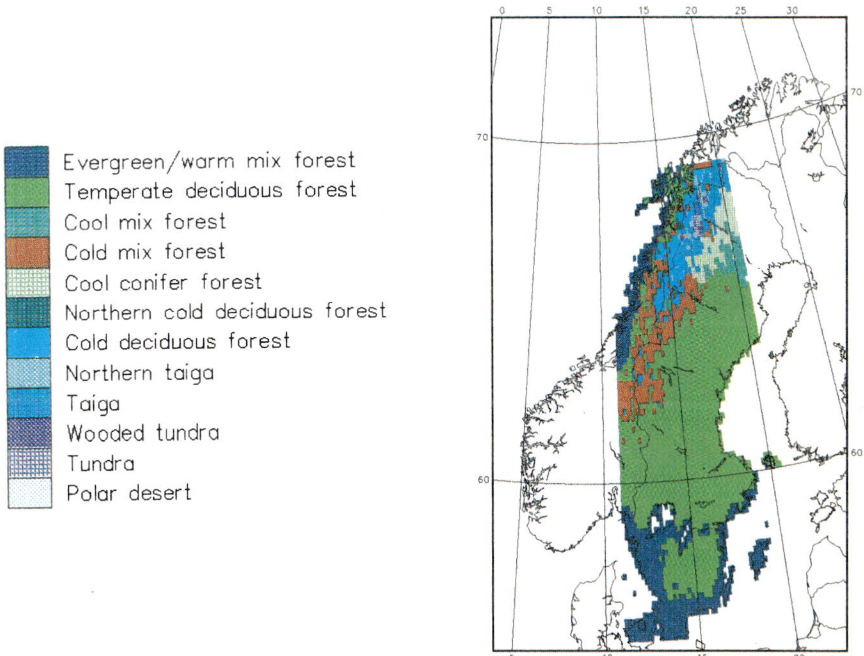

Figure 6.2. Future (2×CO_2) biomes, simulated with BIOME1.1 and based on the average of four scenarios derived from GCM experiments.

driven to more continental regions, while temperate deciduous trees (e.g., *Quercus robur* and *Fagus sylvatica*) take over in central Sweden. Warm-temperate evergreen trees (broad-leaved trees such as *Quercus ilex*, and conifers such as *Pinus pinaster*) potentially extend into southern Sweden, far to the north of their present distribution areas in southern and south-western Europe.

Table 6.2. Comparison between present and 2×CO_2 climate: changes in biome area and total carbon (C) in biomass simulated with BIOME1.1. Values for both area and biomass are totals for the window for which results are given in Figure 6.1.

	PRESENT		2×CO_2	
BIOME	AREA (10^6 km^2)	C (Pg)	AREA (10^6 km^2)	C (Pg)
Polar Desert	0.000	0.00	0	0
Tundra	0.028	0.02	0.001	0.00
Wooded Tundra	0.020	0.04	0.002	0.00
Taiga	0.186	1.62	0.018	0.16
Northern Taiga	0.053	0.27	0.006	0.03
Cold Deciduous Forest	0.006	0.05	0.025	0.22
N. Cold Deciduous Forest	0.000	0.00	0.001	0.01
Cool Conifer Forest	0.081	1.36	0.020	0.34
Cold Mixed Forest	0.001	0.01	0.050	0.50
Cool Mixed Forest	0.120	1.20	0.010	0.10
Temperate Deciduous Forest	0.042	0.42	0.294	2.94
Evergreen/Warm mixed Forest	0	0	0.110	1.10
Totals	0.537	4.99	0.537	5.40

These spatial changes imply changes in biome area and C storage (Table 6.2). The total area of taiga in the region shown in Figure 6.1 is reduced by a factor of ten, while the area of temperate deciduous forest is increased by a factor of seven. Potential stem C storage in warm-temperate evergreen/mixed forest could be greater than 1 Pg C. Total C storage in the study area increases by about 8%.

Simulations using FORSKA2 confirm the general directions of change in vegetation composition indicated by BIOME1.1 but show a far more complex picture. As an example, we show the case of cool mixed forest at a location in Halland, SW Sweden (Figure 6.4). Here the winter warming drives out *Picea abies* and the other major conifer, *Pinus sylvestris*, is largely outcompeted by deciduous species. *Fagus sylvatica*, which is present in the area today, increases substantially in abundance during the transient phase as thermal conditions for *Fagus* growth improve. But later this species declines rapidly again, as the climatic becomes extremely oceanic. The simulated decline is due to failure of the long winter chilling period that *Fagus sylvatica* requires (Sykes et al. 1995). Overall, biomass declines from 2.9 Pg C to 1.9 Pg C (Table 6.3). This decline is not due to drought. Our results agree with earlier findings (Prentice et al. 1991) that the GCM-simulated increase in precipitation is sufficient to cancel out any increased drought effect due to the warming; measured by the Priestley–Taylor coefficient, there is almost precise compensation. Rather, the biomass decline appears to be due to lack of suitable species with a high growth potential. We speculate that it could be avoided by the introduction of climatically better adapted, warm-temperate species that have not been included in these simulations.

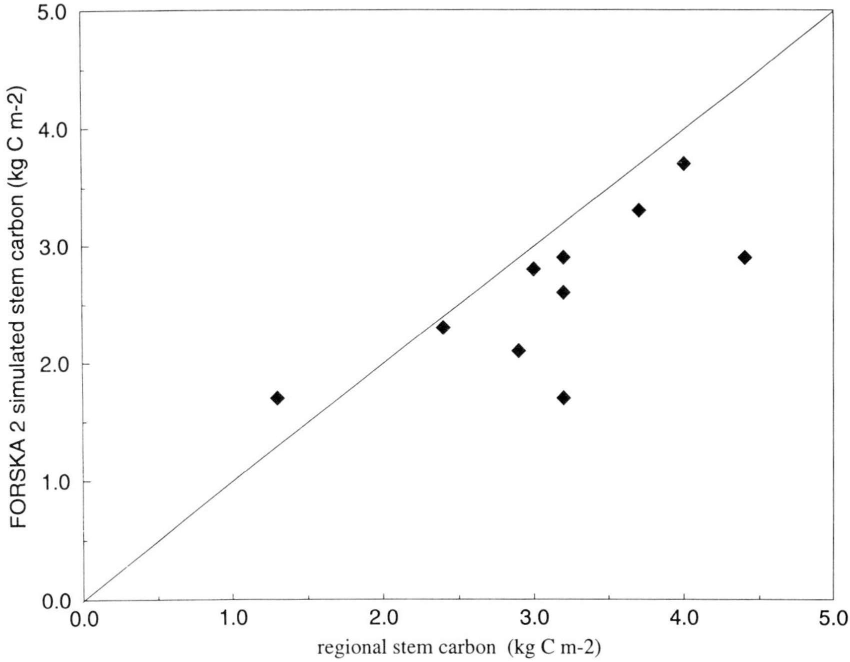

Figure 6.3. Comparison of FORSKA2 estimates of stem carbon (C) at the test locations with figures for the surrounding forestry region, based on forest inventory data for 1989/90.

Table 6.3. Comparison between present and 2×CO_2 climate: changes in steady-state stem carbon at selected locations, simulated with FORSKA 2.

	PRESENT-DAY		2×CO_2	
LAT./LONG.	BIOME DESCRIPTION	C (kg m^{-2})	BIOME DESCRIPTION	C (kg m^{-2})
55.67/13.33	Temperate deciduous forest	2.9	Evergreen/warm forest	1.8
56.02/13.23	Cool mixed forest	3.7	Evergreen/warm forest	2.4
57.08/12.77	Cool mixed forest	3.3	Evergreen/warm forest	1.9
57.77/15.58	Cool mixed forest	2.9	Temperate deciduous. forest	1.9
58.72/14.63	Cool mixed forest	2.8	Evergreen/warm forest	2.6
59.17/18.25	Cool mixed forest	1.7	Evergreen/warm forest	1.9
59.63/15.45	Cool conifer forest	2.6	Temperate deciduous forest	1.6
60.95/13.13	Cool conifer forest	2.3	Temperate deciduous forest	2.6
62.25/16.23	Taiga	2.1	Temperate deciduous forest	2.2
65.78/19.13	Taiga	1.7	Temperate deciduous forest	2.8
65.97/15.97	Tundra	0.1	Cold deciduous forest	2.7

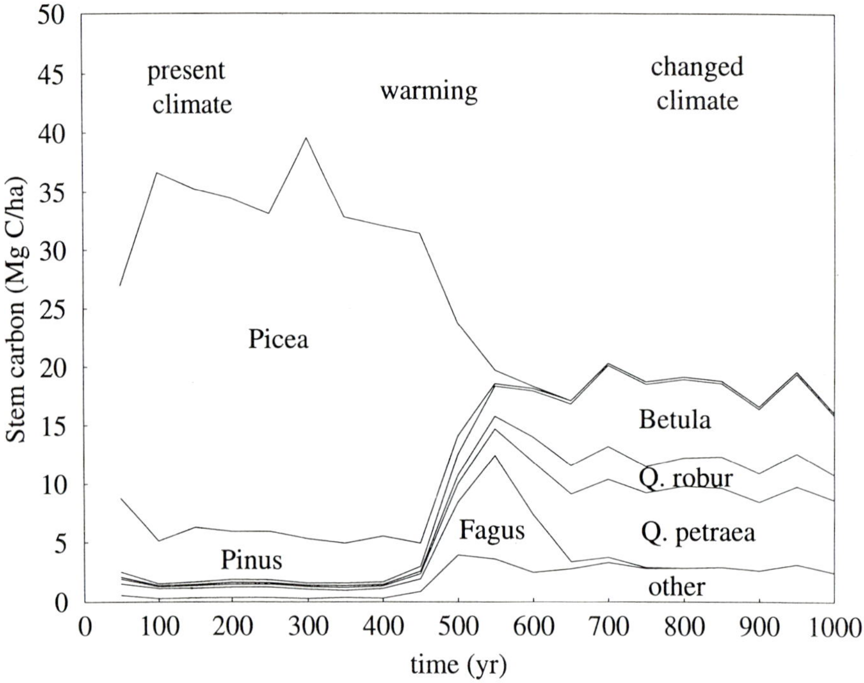

Figure 6.4. Simulation of a forest landscape at Boa Berg in Halland region, SW Sweden 12.77°E, 57.08°N. A linear change in climate was imposed during the 100-year period marked "warming". After that, the climate remained constant at the mean 2×CO_2 climate for this location.

Biomass increases are shown only in northern Sweden, where the lengthening growing season promotes forest growth (including *Picea abies*) and potential compositional changes are less drastic. The greatest increases in biomass are near the northern and montane tree limit. Our results are consistent in this respect with those of Kellomäki and Kolström (1992) for neighbouring Finland. Using another gap model, they found that CO_2-induced warming would favour *Picea abies* invasion in northern Finland, whereas in southern Finland (on the latitude of central Sweden) *Picea abies* would decline in competition with deciduous trees. They did not include more warmth-demanding species, such as *Quercus* species or *Fagus sylvatica*, in their simulations.

DISCUSSION

Both models provide insight into how rapid climate could affect C storage in Swedish forests. BIOME1.1 simulations indicate major changes in the selection of tree species climatically suited to different regions. Simple book-keeping calculations imply that C storage would in the long run increase due to these biome shifts; however, this conclusion is rendered suspect by the fact that standard values of biomass C in each biome are apparently far too high (here as elsewhere), and by the results of transient simulations with FORSKA2. FORSKA2, which does a much better job of simulating present (landscape-average) biomass, shows that many regions in the south of Sweden could actually decline in biomass. We speculate that the (natural or intentional) introduction of warm-temperate species that have not been considered as part of the natural regeneration in our simulations, e.g., evergreen species from southern Europe, might avert this decline.

We have not considered the possible direct effects of increased CO_2 on forest growth. If CO_2 fertilisation occurs then we would expect a general increase in biomass, or smaller declines. CO_2 increase may also modify species composition, although little information is available yet about species-specific effects. CO_2 effects could thus modify the patterns simulated here, but the major simulated changes in forest composition (including complex transient effects) would be expected to occur anyway in response to the change in climate.

ACKNOWLEDGEMENTS

We thank Wolfgang Cramer for providing the baseline climate data set, and climate data interpolated to the test locations. This study was funded by the Swedish Environmental Protection Board (SNV).

REFERENCES

Botkin DB, Simpson LG (1990) Biomass of the North American boreal forest. A step towards accurate global measures. *Biogeochem.* 9: 161–174.

Botkin DB, Janak, JF, Wallis JR (1972) Some consequences of a computer model of forest growth. *J. Ecol.* 60: 849–873.

Desanker and Prentice (1995) MIOMBO—a vegetation dynamics model for the Miombo woodlands of Zambesian Africa. *For. Ecol. Manage.* 69: 87–95.

Hansen J, Fung I, Lacis A, Rind D, Russell G, Lebedeff S, Ruedy R (1988) Global climate changes as forecast by the GISS-3-D model. *J. Geophys. Res.* 93: 9341–9364.

Jarvis PG, Leverenz JW (1983) Productivity of temperate, deciduous and evergreen forests. In: Lange OL, Nobel PS, Osmond CB, Ziegler H (eds) Physiological Plant Ecology IV, Ecosystem Processes: Mineral Cycling, Productivity and Man's Influence, Springer-Verlag, Berlin, pp 233–261.

Leemans R, Cramer W (1991) The IIASA Climate Database for Monthly Mean Values of Temperature, Precipitation and Cloudiness on a Terrestrial Grid. RR-91-18, IIASA, Laxenburg, Austria, 62 pp.

Kellomäki S, Karjalainen T (1995) Sequestration of carbon in the Finnish boreal forest ecosystem managed for timber production. (Chapter 5, this volume).

Kellomäki S, Kolström M (1992) Simulation of tree species composition and organic matter accumulation in Finnish boreal forests under changing climatic conditions. *Vegetatio* 102: 47–68.

Manabe S, Wetherald RT (1987) Large scale changes in soil wetness induced by an increase in carbon dioxide. *J. Atmos. Sci.* 44: 1211–1235.

Mitchell JFB (1983) The seasonal response of a general circulation model to changes in CO_2 and sea temperature. *Quart. J. Roy. Met. Soc.* B109: 113–152.

Olson JS, Watts JA, Allison LJ (1983) Carbon in live vegetation of major world ecosystems. ORNL-5682, Oak Ridge National Laboratory, Oak Ridge, TN, 152 pp.

Prentice IC, Leemans R (1990) Pattern and process and the dynamics of forest structure: a simulation approach. *J. Ecol.* 78: 340–355.

Prentice IC, Sykes MT, Cramer W (1991) The possible dynamic response of northern forests to global warming. *Global Ecol. Biogeogr. Lett.* 1: 129–135.

Prentice IC, Cramer W, Harrison SP, Leemans R, Monserud R, Solomon AM (1992) A global biome model based on plant physiology and dominance, soil properties and climate. *J. Biogeog.* 19: 117–134.

Prentice IC, Sykes MT, Cramer W (1993a) A simulation model for the transient effects of climatic change on forest landscapes. *Ecol. Modelling* 65: 51–70.

Prentice IC, Sykes MT, Lautenschlager M, Harrison SP, Denissenko O, Bartlein PJ (1993b) Modelling global vegetation patterns and terrestrial carbon storage at the last glacial maximum. *Global Ecol. Biogeogr. Lett.* 3: 67–76.

Prentice IC, Sykes MT (1995a) *Climatic Change* (In press).

Prentice IC, Sykes MT (1995b) Vegetation geography and global carbon storage changes. In: Woodwell GM, Mackenzie FT (eds) Biotic Feedbacks in the Global Climate System. Will the warming speed the warming? Oxford University Press, New York, pp 304–312.

Price DT, Apps MJ, Kurz WA, Prentice IC, Sykes MT (1993) Simulating the carbon budget of the Canadian Boreal forest using an integrated suite of process-based models. In: Huor-Ung C (ed) Forest Growth Models and their Uses. Canadian Forest Service, Québec, pp 251–264.

Schlesinger ME, Zhao ZC (1989) Seasonal climatic changes induced by doubled CO_2 as simulated by the OSU atmospheric GCM/mixed-layer ocean model. *J. Clim.* 2: 459–495.

Sjörs H (1967) *Nordisk Växtgeografi.* Svenska Bokförlaget, Bonniers, Stockholm, 239 pp.

Shugart HH (1984) A Theory of Forest Dynamics. Springer, New York, 278 pp.

Skogsdata (1990) *Aktuella uppgifter om de svenska skogarna från riksskogstaxeringen, taxering 89/90.* Institutionen för skogstaxering, Sveriges lantbruksuniversitet, Umeå, Sweden, 70 pp.

Sykes MT (1994) Modelling the effects of climate change on forest diversity. *Biolog.* 2–3: 17–24. Oslo Norway.

Sykes MT, Prentice IC (1995a) Boreal forest futures: Modelling the controls on tree species range limits and transient responses to climate change. *Water, Air, Soil Pollut.* 82: 415–428.

Sykes MT, Prentice IC (1995b) Climate change, tree species distributions and forest dynamics: a case study in the mixed conifers/northern hardwoods zone of northern Europe. (Unpublished manuscript).

Sykes MT, Prentice IC, Cramer W (1995) A bioclimatic model for the potential distribution of northern European tree species under present and future climates. *J Biogeogr.* (In press).

7 Climate change and management of insect defoliators in boreal forest ecosystems

W Jan A Volney

Natural Resources Canada, Canadian Forest Service,
Northwest Region, Northern Forestry Centre,
5320 - 122 Street, Edmonton, AB T6H 3S5, Canada

INTRODUCTION

Insects are the most diverse group of terrestrial animals. Not surprisingly, they influence several ecosystem processes and thus have a profound influence in most terrestrial ecosystems by occupying a myriad of niches. The reciprocal interaction of insects with forest plants includes some of the most finely tuned examples of mutualism known as well as some of the most destructive interactions between insects and plant populations. These interactions have important consequences to the development of forest ecosystems and insects may regulate forest productivity (Mattson and Addy 1975). Insects also have a more indirect role in forested ecosystems in that they are often important regulators of herbivore populations, may be important food items for vertebrate populations, and perform a significant function in nutrient cycling in concert with micro-organisms. (See also Chapter 15, Galinski and Witowski). However, both insects and plants are poikilotherms and thus their activities in forest ecosystems are regulated to a large extent by climate.

Insects are a major disturbance factor in many northern and temperate forests. Insect outbreaks in northern latitudes often occur over vast areas and over long periods. There is thus a direct effect on the productivity of forest stands through growth loss, tree deformity and ultimately tree mortality. Outbreaks may reduce the accumulation of biomass in forest stands for up to a decade.

Boreal forests are dominated by disturbances. The most notable disturbance is wildfire (see Chapter 8, Stocks et al.). Fires periodically burn large areas but in the process regenerate many forest stands and re-initiate secondary successional processes. The nature of the stand and the history of stand development immediately prior to fire are important determinants of the extent and intensity with which these forests burn. An important factor that determines the flammability of forests is the amount of pest-caused damage to trees in forest stands.

Both the direct effects of insect outbreaks, and their indirect effects on fire extent and intensity, liberate vast amounts of CO_2 to the atmosphere. This biogeochemical cycling of C can be accelerated if the frequency and intensity of disturbances increases. On a global scale, the net result is an elevated level of CO_2 which can adversely affect climate and result in more disturbance prone forests: a positive feedback loop. Thus it is possible that the global effect of these disturbances is to increase the frequency of disturbances as a consequence of the effects of elevated CO_2 levels on climate (Kurz et al. 1995; Chapter 14, Kurz and Apps). This can only be offset by the accelerated removal of CO_2 from the atmosphere. Forest tree growth is one such process.

NATO ASI Series, Vol. I 40
Forest Ecosystems, Forest Management and the Global Carbon Cycle
Edited by Michael J. Apps and David T. Price

Forest management often targets insect outbreaks. To the extent that management actions are successful in mitigating the effects of outbreaks, some retardation of the accumulation of CO_2 in the atmosphere will result. Pest management thus offers one opportunity for managing C flux on a global scale.

Although insects affect most terrestrial ecosystem processes, their most visible impact in boreal forests is through tree growth reduction and tree mortality associated with outbreaks. My objective in this paper is to describe how weather may influence outbreaks of select insect species and thus affect boreal forest productivity. A second objective is to discuss how outbreak behaviour may be altered by changes at the landscape and global levels. Included in these changes are both inadvertent and deliberate anthropogenic-induced modifications to landscapes. Pest management activities and the opportunities they present to mitigate reduced forest productivity are discussed.

The spruce budworm *Choristoneura fumiferana* (Clem.) is a pest of boreal spruce–fir forests found on mesic sites. The closely related jack pine budworm (*C. pinus* Free.) feeds on jack pine and is most damaging in stands growing on well drained sites. Together these species have a significant impact on the productivity of forests in Canada. Their activities thus provide a sampling of outbreak effects that may be evaluated in relation to global climate change.

DEFOLIATOR BIOLOGY

The spruce budworm is native to North America and has been found wherever pheromone traps have been deployed in the boreal forest. This species has a transcontinental distribution but its range in north-western Canada and adjacent Alaska is uncertain (Harvey 1985). The range of the principal host tree, white spruce (*Picea glauca* (Moench) Voss.), is transcontinental, extending from the Atlantic coast, west to the Chukchi sea and north to the Arctic Ocean along the Mackenzie River (Little 1971). This is probably also the range of the spruce budworm. If this is the case,. then the insect can be found in the entire range of climates that its host occupies. Furthermore, there are no known insurmountable reproductive barriers among the spruce budworm and the closely related western budworm species (*C. occidentalis* Free., *C. orae* Free., *C. retiniana* (Wlshm.) and *C. biennis* Free.) (Silk and Kuenen 1988). The western budworm species also feed on members of the Abietoidea—the subfamily that includes the spruces (Volney 1989). The conditions under which this pool of genotypes can be found and the climatic range over which a budworm type is adapted thus include sites ranging from latitudes in the tropics to the Arctic.

With the exception of *C. biennis*, the life cycle of the budworms are similar; varying only in detail. The life cycle of the spruce budworm as described by McGugan (1954) begins with the hatching of eggs in late July, the first instars finding hibernation sites, spinning hibernacula and moulting to second instars before over-wintering. In spring, second instars emerge from hibernacula, locate suitably developed buds to mine and initiate feeding or mine needles if buds are not available. There are six instars each apparently adapted to a particular stage of host shoot development. Upon the completion of feeding, the larvae pupate in late June. Moths eclose in early July, mate and lay eggs. The pattern is slightly different for the jack pine budworm, where the insects emerge later in the season and have as many as seven instars (Nealis 1987). *Choristoneura biennis*, which feeds in forests at high elevation, enters diapause in the fourth instar and re-emerges the following spring, taking 2 years to complete its life cycle (Harvey 1985). This 2-year life cycle is found to a varying degree in most budworm species.

The seasonal variation in the development of these spruce budworms is remarkable. Within the range of the spruce budworm, western populations require fewer heat units (degree days above some threshold) to complete development (Volney and Cerezke 1992) than do populations in eastern Canada (Lysyk 1989). This east–west difference is associated with changes in forest types as well as differences in latitudes. Yet when populations from a variety of locations along a latitudinal gradient were examined, no difference in their basic development rates could be detected (Weber 1994). Clearly the variation in local populations is high. There is evidence that the genetic systems of these species conserve this variation and that this is immune to directional selection over the short-term (Volney et al. 1983). Synchrony between emergence of larvae and host tree bud development to a stage suitable for larvae to establish feeding sites is critical to budworm survival in spring. Diapause in over-wintering spruce budworm larvae is effectively over by March; there is little geographical or genetic variation in this trait and photoperiod has no influence on development at this stage (Régnière 1990). Régnière (1990) also found that following diapause termination, budworms' sensitivity to warm temperatures increases gradually, effectively synchronising insects' development with host-tree bud development. The pattern of variation observed in these populations suggests that directional changes in climate experienced to date are well within the ability of the insect to cope. Furthermore, the genetic and environmental mechanisms which control development in these populations are such that they are unlikely to be barriers to these insects exploiting forests growing in altered climates.

There is evidence that warming may be associated with elevated mean budworm densities in boreal forest systems. Most spruce budworm outbreaks last for several years and the interval between outbreaks varies from 25 to 45 years. Superimposed on this long-term trend are higher frequency fluctuations probably driven by weather (Royama 1992). At the peak of one of these episodes several Mha of eastern North American boreal forest may be severely defoliated (Hardy et al. 1986). The dynamics of the spruce budworm in the western boreal forest of Canada differ in that host stands are limited in extent and thus outbreak areas are smaller. Nevertheless, outbreaks in the west persist for decades only to be terminated by severe late spring frosts (Cerezke and Volney 1995). Jack pine budworm outbreaks, by contrast, are periodic in west–central Canada and appear to be associated with drought prior to the beginning of outbreaks (Volney 1988). To the south, outbreaks are more frequent and this frequency is associated with edaphic conditions (Volney and McCullough 1994). Jack pine budworm outbreaks are of much shorter duration than those of the spruce budworm. Outbreak initiation requires mature, flowering jack pine trees (Nealis and Lomic 1994). Flowering in jack pine is associated with warm periods and drought followed by the subsequent recovery of trees. The intensity, extent and frequency of western spruce budworm outbreaks in British Columbia are often associated with the synchrony of larval emergence with bud development in the spring (Thomson et al. 1984). Thus weather modulates many aspects of population dynamics and is suspected to be involved in the initiation, maintenance and termination of outbreaks in different parts of the budworms' range.

The damage caused by budworm feeding depends on the characteristic age structure of foliage retained by host trees, their ability to produce epicormic shoots when defoliated, and the duration of outbreaks. Jack pine (*Pinus banksiana* Lamb.) retains needles for three years and produces few, if any, epicormic shoots when defoliated. Thus local jack pine budworm outbreaks that persist for more than two years can cause widespread tree mortality. However, even when outbreaks are short, there is a reduction in volume increments and some stem deformity is induced in the host tree by jack pine budworm feeding (Kulman et al. 1963). Balsam fir (*Abies balsamea* (L.) Mill.) is the most susceptible to damage by feeding of the spruce budworm. Although several age classes of foliage are retained on fir trees, later in life they lose their ability to recover from defoliation by initiating the growth of latent buds. As a consequence, outbreaks of five or more years can kill entire stands of this species (MacLean 1985). White spruce, like balsam fir, retains several age classes of foliage but continues to produce epicormic shoots after several consecutive years of defoliation even on old trees. Consequently, white spruce stands are capable of sustaining outbreaks of long duration without widespread tree mortality. Tree mortality does occur, however, and up to 40% mortality has been reported for white spruce stands (MacLean 1985). Finally, the damage sustained in stands depends on the general health of the stand: it would appear that the degree to which individual trees are damaged is influenced by the incidence of infections by pathogens prior to, during, or immediately following, the outbreak (Mallett and Volney 1990). Forest type, stand history and forest health thus combine to determine the extent to which stands can be damaged by budworm outbreaks.

The relative magnitude of the C flux effected by insects and diseases can be assessed from the forest depletions compiled by the Canadian Forest Service (1994). Between 1982 and 1987, the annual wood volume loss attributed to insects and diseases in Canada amounted to 102.8 million m^3 (half of which were caused by insects). The corresponding losses for fire were 36 million $m^3$and harvesting accounted for 160 million m^3. Of the total insect caused depletions, the spruce budworm caused 27.3 million m^3 and jack pine budworm caused an average of 5.2 million m^3. These losses are not constant but vary depending on whether outbreaks occurred in the reporting period or not (see also Chapter 14, and Figure 14.2). During this (1982–1987) period, spruce budworm outbreaks declined but the mortality from an earlier outbreak persisted into the reporting period. Nevertheless, these two species damaged 10% of the annual net forest productivity of Canada as measured by the annual allowable cut (Canadian Forest Service 1994).

BUDWORM/FOREST DYNAMICS AND GLOBAL CHANGE

The spruce budworms

Tree-ring studies provide the best insight into long-term changes in budworm dynamics. The oldest records are for the western spruce budworm feeding on Douglas-fir (*Pseudotsuga menziesii* var. *glauca* (Beissn.) Franco) and white fir (*Abies concolor* (Gord. and Glend.) Lindl.) in the American south-west (Swetnam and Lynch 1994). Since 1600, outbreaks have lasted from 5 to 25 years. In the longest record of 750 years, there were 20 outbreaks. Swetnam and Lynch (1994) suggest that spring precipitation may drive population change. Wet periods are associated with release and outbreak years. In addition, there is evidence that there are endogenous factors regulating outbreaks that are related to the response and recovery of the forest canopy from defoliation.

The tree-ring record for the spruce budworm is not as lengthy as that for the western spruce budworm. An analysis of tree-ring data suggest that outbreaks in eastern North America have become more frequent in the 20th century in contrast to those in the two previous centuries (Blais 1983). Whereas the period between outbreaks was commonly 60 to 100 years, intervals are now 19 to 37 years. The limited observational records of outbreaks in western Canada suggest even shorter intervals in the 20th century (Cerezke and Volney 1995). In western Canada where the forest merges with the grassland biome over a fairly steep gradient, several disjunct white spruce forests on the southern margin of the boreal forest support chronic spruce budworm infestations. Tree ring observations suggest that the duration and frequency of outbreaks have increased in western Canada in the last century (Volney, unpublished data). The evidence from the geographical and temporal outbreak patterns suggest a major shift in the dynamics of this insect throughout North America in the 20th century.

A detailed understanding of the processes which underlie the cause of outbreaks must be presented if this change in outbreak pattern is to be attributed to any change in global climate. There are confounding effects such as the introduction of chemical insecticides and fire protection in the mid-20th century, and the clearing of land for settlement and forest harvesting following colonisation by Europeans. It is reasonable to expect that changes in the pattern of outbreaks of the spruce budworm would be associated with changes in climate because so many of the important activities related to budworm survival are related to weather parameters. This is particularly true for the association between the insect and its host in synchronising larval development with shoot development and the recurrent association between late spring frosts and the collapse of outbreaks. Nevertheless, this attractive explanation remains only as a tantalising hypothesis.

Theories of climatic release proposed by Wellington et al. (1950) suggested that warm dry summers favoured outbreaks whereas cool wet periods favoured their collapse. Greenbank (1963) provided a correlation between 5-year moving averages of temperature and precipitation in June and July with budworm densities in support of this climatic release theory. Based on the relatively short observational data base for the spruce budworm in eastern Canada, Royama (1992) provides an alternative interpretation for the causes of outbreaks. He concluded that the 5-year moving average correlation obtained by Greenbank between July weather and budworm densities was spurious. Rather, the nature, frequency and amplitude of the oscillations observed in spruce budworm densities are the result of the interaction of the complex food web to which the pest belongs. At best, weather functions to influence recruitment and results in a vertical perturbation of the population trajectory. However, the observations on which this synthesis was developed were made far from the region in which climatically induced stress on host trees is a chronic problem.

The jack pine budworm

Observations on jack pine budworm outbreaks support the hypothesis that eruptions are associated with climatic release. Although the insect is found throughout the range of the host, outbreaks have only been documented at the southern extremity of this distribution. Predictably, this is also the driest part of the host's range. Observations of stands damaged by the jack pine budworm suggest that drought stressed trees, whether stressed physiologically by root disease, root deformities, or by low precipitation, were prone to develop damaging budworm populations. The severity of infection by *Armillaria* root rotting fungi is strongly correlated with the severity of damage sustained by individual trees following an outbreak (Mallett and Volney 1990). Considerably less is known about the dynamics of the jack pine budworm populations than those of the spruce budworm, yet analysis of the areas defoliated in the prairie provinces of Canada (Volney 1988) and population data from Wisconsin strongly implicate an association with climate (Clancy et al. 1980). The Wisconsin data further suggest that outbreaks are more frequent on drier sites (Volney and McCullough 1994). Furthermore, the elevated survival of early instars in pollen bearing conelets is a necessary condition for outbreak initiation (Nealis and Lomic 1994) but the control of cone production in conifers is known to be associated with weather (Puritch 1972).

The three budworm species, for which we have fairly detailed information, all seem to react similarly to drought. This is evident in the geographical regions in which chronic outbreaks occur, the association of xeric site conditions with elevated budworm densities, and the condition of individual trees damaged by these insects. The food web in which these insects exist may in fact determine the nature of population change. However, a major element of these webs is the host tree and it is through the effect of weather on the host that the effects of climate change will be expressed in altered population behaviour. The types and numbers of buds produced by trees, the timing of bud-break, and the phenological development of shoots, are processes that are affected by weather and known to influence budworm populations. The intensity of defoliation and the risk of tree mortality is also affected by weather. Whereas the increase in populations may not be affected proximally by weather, the overwhelming body of evidence suggests that the local climate has a profound effect on the dynamics of the insects and the consequent damage to forest stands.

MANAGEMENT AND ECOLOGICAL CONSTRAINTS

One might wonder if the intensity, extent and severity of outbreaks can be changed. This change may be effected either through direct intervention or inadvertently through management intervention intended to achieve some other objective. Efforts at direct control of the spruce budworm have not been implemented on a sufficiently wide scale to suppress populations with much success except at a very local level. Spraying in eastern North America was directed at conifer-dominated stands after they had become defoliated. There has been no analysis to suggest that the trajectory of the outbreak was altered. In any case, the area sprayed during the outbreak was a small proportion of the total area involved. One could envision a policy in which population suppression over the insect's range was the objective. Although this strategy may not be feasible for the spruce budworm now, the European spruce sawfly (*Gilpinia hercyniae* (Hartig)) is an example where direct control agents were successfully introduced to maintain a forest insect at endemic levels in eastern North America (Magasi and Syme 1984).

Inadvertent change in the dynamics of the western spruce budworm on a regional scale has been implicated in the American south-west (Swetnam and Lynch 1993). Swetnam and Lynch (1993) found that outbreaks were more synchronous over the whole study area in the 20th century than they had been previously. This change came about as a response of the forest to the impacts of Anglo–American settlement in the area. Logging, grazing and fire reduced the density and age-class diversity of the forest cover in the region as settlement progressed. With the relaxation of these pressures in the 20th century and the regrowth of the forest, a fairly uniform, dense coniferous forest reoccupied much of the land. These conditions have resulted in synchronous region-wide outbreaks. Land-use patterns thus have an impact on the C flux activated by uncontrolled insect populations. Forest management policies could therefore be evaluated for their effects on the pattern and frequency of outbreaks.

A further concern is that climate change on a global scale may influence the pattern of outbreaks (Kurz et al. 1995). Temperatures in the western boreal forest extending from the western shore of Lake Manitoba to the Mackenzie Valley appear to have risen in excess of 0.75 °C in the last 70 years (Oechel and Vourlitis 1994). The recent chronic spruce budworm outbreaks in this region may be associated with this warming. Certainly the warmest sites in the south, particularly the isolated forest disjuncts, have experienced more frequent damage.

So far, the western boreal forests have not been subjected to extensive forest harvesting or anthropogenically altered fire regimes. They thus offer the opportunity to investigate the impacts of warming on the frequency of outbreaks. The global warming/increased outbreak frequency conjecture is consistent with similar regional comparisons in the jack pine budworm system. At present, the frequency of outbreaks, as evidenced by defoliation records, decreases as one goes north-west in the western boreal forest of west–central Canada. The earliest records of infestations in stands north of the line of discontinuous permafrost are for the 1960s. The first outbreaks expanded into Alberta in the 1980s. The net result is that the area in which defoliation has occurred has increased exponentially in the last 5 decades (Volney 1988). Population surveys in north-western Wisconsin suggest that outbreak frequency is higher in the more xeric sites in a region where climatic differences are relatively minor compared to edaphic differences among sites (Volney and McCullough 1994). One could thus hypothesise that if global change results in moisture deficit, the sites in Wisconsin that currently experience outbreaks every 10 years will switch to an outbreak frequency of 5 to 6 years currently experienced at the more xeric sites. Similarly, the sites in western Canada may experience elevated jack pine budworm outbreak frequencies under these new climates. As elevated insect densities accelerate tree mortality there would be a further increase in the release of C to the atmosphere which would either result in a greater area involved in these frequent outbreaks or a reduction in the area supporting jack pine stands. The synergistic effects of fires would accelerate this process further. In such a feedback loop, management of anthropogenically produced greenhouse gases on a global scale should have a dramatic effect on insect outbreaks and *vice versa*.

CONCLUSIONS

Many of the process known to affect the survival of the budworms are subject to control or modulation by weather parameters. Among the most critical of these processes are the seasonality of insect emergence, the initiation of feeding and the relationship between insect development and that of the host tree shoot systems. From what we know, there is no known genetic or environmental process that would limit the success of budworms in boreal forest systems subjected to a warming climate. The same applies to ecological processes. In the case of the jack pine budworm, the production of pollen conelets is a necessary (but not sufficient) condition for elevated survival of larvae. There is evidence that drought is involved in triggering the production of flower crops in most conifers, thus the survival of jack pine budworm larvae is subject to the vagaries of weather several seasons in advance of the outbreak. The termination of spruce budworm outbreaks in the western boreal forest by late spring frosts also argues strongly for the association of weather parameters in affecting outbreak duration and frequency. Therefore climate, to the extent that it represents the mean occurrence of these somewhat extreme weather events, can have an influence on the frequency of outbreaks and ultimately on the C flux between forests and the atmosphere.

Analyses of biogeoclimatic gradients within the range of the insects further support the hypothesis that there is a connection between climate and insect outbreaks. The temporal changes in the frequency of outbreaks witnessed in the 20th century need explanation. One hypothesis is that elevated temperatures, and presumable elevated moisture deficits, have caused this change. Confounded with this change, however, are the activities of humans in altering the temporal and spatial patterns of forest occupancy in the landscape. The greatest impact of resource management on population behaviour has been inadvertent. The tree-ring evidence from the south-west implicates these anthropogenic changes with the 20th century outbreak pattern. Attempts at direct control of spruce budworms in eastern Canada may not have altered the outbreak dynamic because they did not treat a large proportion of the forest prone to the spruce budworm. In the western boreal forest, where white spruce stands are restricted to the more mesic sites along water courses, the opportunity exists to alter the course of outbreaks either by direct intervention or inadvertently through the positive feedback loop involving climate change induced by greenhouse gases, insects and/or fire, and tree mortality. Unlike other animal species in this region, such as whooping crane and bison populations which were driven to virtual extinction through hunting, and the rocky mountain grasshopper which became extinct as a result of habitat modification, however, it is probable that budworm populations will become more problematical for resource managers. These insects may, in fact, be among the agents which inadvertently but ultimately extirpate their host trees from landscapes affected by an altered climate.

REFERENCES

Blais JR (1983) Trends in the frequency, extent, and severity of spruce budworm outbreaks in eastern Canada. *Can. J. For. Res.* 13: 539–547.

Canadian Forest Service (1994) Forest depletions caused by insects and diseases in Canada 1982–1987. Inf. Rep. ST-X-8. Nat. Resour. Can., Ottawa, 14 pp.

Clancy KM, Giese RL, Benjamin DM (1980) Predicting jack–pine budworm infestations in northwestern Wisconsin. *Environ. Entomol.* 9: 743–751.

Cerezke HF, Volney WJA (1995) Forest insect pests in the Northwest Region. In: Armstrong JA, Ives WGH (eds) Forest Insect Pests in Canada. Nat. Resour. Can., Can. For. Serv., Ottawa, pp 59-72.

Greenbank DO (1963) The development of the outbreak. In: Morris RF (ed) The dynamics of Epidemic Spruce Budworm Populations. *Mem. Entomol. Soc. Can.* 31: 19–23.

Hardy Y, Mainville M, Schmitt DM (1986) An atlas of spruce budworm defoliation in eastern North America, 1938–1980. US Dept. Agric., For. Serv., Coop. State Res. Serv., Misc. Publ. No. 1449. USDA Washington, DC. 52 pp.

Harvey GT (1985) The taxonomy of coniferophagous *Choristoneura* (Lepidoptera: Tortricidae): a review. In: Sanders CJ, Stark RW, Mullins EJ, Murphy J (eds) Recent advances in Spruce budworms research. Proceedings of the CANUSA Research Symposium, Bangor ME, 16–20 Sept. 1984. Canadian Forestry Service, Ottawa, pp 16–48.

Kulman HM, Hodson AC, Duncan DP (1963) Distribution and effects of jack pine budworm defoliation. *For. Sci.* 9: 146–157.

Kurz WA, Apps MJ, Stocks BJ, Volney WJA (1995) Global climate change: disturbance regimes and biospheric feedbacks of temperate and boreal forests. In: Woodwell GM, Mackenzie FT (eds) Biotic feedbacks in the global climate change system will the warming speed the warming? Oxford University Press, New York, pp 119–133.

Little EL Jr. (1971) Atlas of United States Trees. Vol. 1. Conifers and important hardwoods. USDA For. Serv., Misc. Publ. No. 1146. USDA, Washington, DC.

Lysyk TJ (1989) Stochastic model of eastern spruce budworm (Lepidoptera: Tortricidae) phenology on white spruce and balsam fir. *J. Econ. Entomol.* 82: 1161–1168.

McGugan BM (1954) Needle-mining habits and larval instars of the spruce budworm. *Can. Entomol.* 86: 439–454.

MacLean DA (1985) Effects of spruce budworm outbreaks on forest growth and yield. In: Sanders CJ, Stark RW, Mullins EJ, Murphy J (eds) Recent advances in spruce budworms research. Proceedings of the CANUSA Research Symposium, Bangor, ME, 16–20 Sept. 1984. Can. For. Serv., Ottawa, pp 148–175.

Magasi LP, Syme PD (1984) In: Kelleher JS, Hulme MA (eds) *Gilpinia hercyniae* (Hartig), European spruce sawfly (Hymenoptera: Diprionidae). Biological control programmes against insects and weeds in Canada 1969–1980. Commonwealth Agric. Bureaux, Farnham Royal, UK, pp 295–297.

Mallett KI, Volney WJA (1990) Relationship between jack pine budworm damage, selected tree characteristics, and *Armillaria* root rot in jack pine. *Can. J. For. Res.* 20: 1791–1795.

Mattson WJ, Addy ND (1975) Phytophagous insects as regulators of forest primary production. *Science* 190: 515–522.

Nealis VG (1987) The number of instars in jack pine budworm, *Choristoneura pinus pinus* Free. (Lepidoptera: Tortricidae), and the effect of parasitism on head capsule width and development time. *Can. Entomol.* 119: 773–777.

Nealis VG, Lomic PV (1994) Host-plant influences on the population ecology of the jack pine budworm, *Choristoneura pinus* (Lepidoptera: Tortricidae). *Ecol. Entomol.* 19: 367–373.
Oechel WC, Vourlitis GL (1994) The effect of climate change on land-atmosphere feedbacks in arctic tundra regions *Trends in Ecol. and Evol.* 9: 324–329.
Puritch GS (1972) Cone production in conifers. Can. For. Serv. Inf. Rep. BC-X–65, Pacific For. Res. Centre, Victoria BC, 94 pp.
Régnière J (1990) Diapause termination and changes in thermal responses during post-diapause development of the spruce budworm, *Choristoneura fumiferana. J. Insect. Physiol* 36: 727–735.
Royama T (1992) Analytical Population Dynamics. Chapman and Hall, London, 371 pp.
Silk PJ, Kuenen LPS (1988) Sex pheromones and behavioral biology of the coniferophagous *Choristoneura. Ann. Rev. Entomol.* 33: 83–101.
Swetnam TW, Lynch AM (1993) Multicentury, regional-scale patterns of western spruce budworm outbreaks. *Ecol. Monogr.* 63: 399–424.
Thomson AJ, Shepherd RF, Harris JWE, Silversides RH (1984) Relating weather to outbreaks of western spruce budworm, *Choristoneura occidentalis* (Lepidoptera: Tortricidae), in British Columbia. *Can. Entomol.* 116: 375–381.
Volney WJA (1988) Analysis of historic jack pine budworm outbreaks in the prairie provinces of Canada. *Can. J. For. Res.* 18: 1152–1158.
Volney WJA (1989) Biology and dynamics of North American coniferophagous *Choristoneura* populations. *Agric. Zool. Rev.* 3: 133–156.
Volney WJA, Waters WE, Akers RP, Liebhold AM (1983) Variation in spring emergence patterns among western *Choristoneura* spp. (Lepidoptera: Tortricidae) populations in southern Oregon *Can. Entomol.* 115: 199–209.
Volney WJA, Cerezke HF (1992) The phenology of white spruce and the spruce budworm in northern Alberta. *Can. J. For. Res.* 22: 198–205.
Volney WJA, McCullough DG (1994) Jack pine budworm population behaviour in northwestern Wisconsin. *Can. J. For. Res.* 24: 502–510.
Weber JD (1994) Latitude, physiological time and the spruce budworm. Dept. Entomology, University of Alberta, MSc Thesis, Edmonton. p. 84 (Unpublished manuscript).
Wellington WG, Fettes JJ, Turner KB, Belyea, RM (1950) Physical and biological indicators of the development of outbreaks of the spruce budworm. *Can. J. Res.* 28: 308–331.

8 Some potential carbon budget implications of fire management in the boreal forest

Brian J Stocks, Bryan S Lee[1] and David L Martell[2]

Canadian Forest Service, Ontario Region, 1219 Queen Street East, Sault Ste. Marie, Ontario, Canada P6A 5M7

INTRODUCTION

Forest fire is a natural force that has shaped boreal and temperate forest ecosystems in the northern hemisphere for millennia, to the point that these forests have not only adapted to, but are dependent upon, periodic stand-replacement wildfires to maintain their existence. Over the past century people throughout northern forest ecosystems have coexisted, at times somewhat uneasily, with this important natural force, as fire management agencies attempted to balance public safety concerns and the industrial and recreational use of these forests, with costs, and the need for natural forest cycling through forest fires. In recent years fire management has been complicated by a growing recognition of the need to reconcile attempts to minimise fire losses with the growing cost of fire management and the beneficial impacts of fire. Canadian, Russian, and American fire managers have always designated parts of the boreal zone, usually in northern regions, as "lower priority" zones that receive little or no fire protection, since fires that occurred there were thought to have little or no significant detrimental impact on public safety and forest values. The realisation that total fire exclusion is neither possible nor ecologically desirable, initiated a gradual move toward the widespread adoption of fire management strategies that prioritise protection of high-value resources while permitting burning in more remote areas. This is particularly true in the boreal forest regions of Canada, Russia, and Alaska where lower population densities and forest use allow more flexible fire management strategies.

The area burned in the world's boreal zone increased dramatically during the 1980s, the warmest decade on recent record, primarily due to short-term extreme fire weather situations, but also due in part to long-term drought in some regions and changes in fire suppression strategies (Stocks 1991). The looming possibility of global warming, which is expected to have the most significant impact at northern latitudes, raises legitimate concerns that boreal zone fire activity will continue to increase in future decades, with potentially serious implications for carbon (C) fluxes and stores in northern forests.

Two-thirds of the world's boreal forests are located in Russia, and both historical (Shostakovitch 1925) and recent (Cahoon et al. 1994) investigations indicate that large areas burn annually in this country, particularly in previously-unmonitored remote regions of Siberia. Although a major proportion of the Earth's large boreal forest fires occur in Siberia, past Russian forest fire statistics are incomplete and unreliable and, while this situation is

---

1 Canadian Forest Service, Northwest Region, 5320-122 Street, Edmonton, Alberta T6H 3S5, Canada
2 Faculty of Forestry, University of Toronto, 33 Willcocks Street, Toronto, Ontario M5S 3B3, Canada

NATO ASI Series, Vol. I 40
Forest Ecosystems, Forest Management and the Global Carbon Cycle
Edited by Michael J. Apps and David T. Price

expected to improve in the near future, no comprehensive statistics from this country are currently available for analysis.

Conversely, detailed Canadian forest fire statistics have been archived since 1920, permitting a thorough analysis of trends in this country. Annual fire occurrence in Canada has steadily increased, without much fluctuation, over the past seven decades, to approximately 10 000 fires yearly during the 1980s. Over the same interval annual area burned—which fluctuates dramatically from year-to-year—actually decreased over the first four decades of record (although poorer detection capabilities during this period undoubtedly resulted in many fires not being recorded), only to increase over the next three decades, culminating during the 1980s when initial estimates revealed an average of more than 2.5 Mha were burned annually (Stocks 1991).

The objectives of this paper are:

1. to demonstrate, by means of highly aggregated burned area data, that fire plays a significant role in boreal forest ecology, and that the impact varies by ecoclimatic parameters (defined below) and the level of fire protection,
2. to stress the importance of integrated approaches that will encourage ecologists to consider the ecological implications of fire management policies, and fire managers to consider the potential fire management implications of global warming.

METHODOLOGY

The database used for this analysis is a portion of a larger, GIS-based data set (which is updated annually) of all Canadian fires larger than 200 ha in area that have occurred since 1980. Individual wildfire reports have been collected from all Canadian fire management agencies (ten provincial, two territorial, and Parks Canada), and analysed in detail to construct this database. Statistical records (Ramsey and Higgins 1982, 1986, 1991; Higgins and Ramsey 1992) show that during the 1980–89 period a total of 96 870 fires burned 25 311 485 ha in Canada. However, those statistics did not include many large fires that burned in northern Quebec during the early 1980s, and a more realistic estimate of the total Canadian area burned in this decade, based on this study, would be 27–28 Mha. Only 3.5% of these wildfires grew larger than 200 ha in size, undoubtedly influenced by aggressive initial attack strategies in high-value areas, but these fires accounted for approximately 97% of the annual area burned in the country, resulting in a small database (3362 fires) which characterises the most significant impacts of fire on the Canadian landscape.

This database is not without some limitations. These include: some fire location errors, some large fires in remote regions (particularly the Arctic) perhaps being overlooked, and the fact that burned area in the Canadian prairies is underestimated because many fires occur in areas that are not under the jurisdiction of provincial forest fire management agencies and therefore are not recorded.

In this paper large 1980–89 fires are analysed by 10 Canadian Ecoclimatic Provinces (Ecoregions Working Group 1989). Ecoclimatic Provinces (EPs) are groupings of Ecoclimatic Regions (ERs) and are representative of broad vegetation physiognomy at a global scale (see Table 8.1). For the purposes of this analysis, three EPs were subdivided into aggregations of somewhat similar ERs: the Subarctic was partitioned into High and Low, the Boreal into High, Mid and Low, and the Cordilleran into High and Low. The Subarctic, Boreal and Cordilleran (Subarctic and High) EPs used in this study correspond to the Boreal Region described by Rowe (1972). As a result 14 spatial analysis zones are used in this study.

Table 8.1. Vegetation physiognomy by ecoclimatic province (Ecoregions Working Group 1989).

ECOCLIMATIC PROVINCE	VEGETATION DEVELOPMENT
Arctic	Treeless, with tundra, polar semi-desert, or polar desert.
Subarctic	Open-canopied conifer woodlands, with tundra patches.
Boreal	Closed-canopied forests of conifer or mixed conifer-hardwood.
Cool Temperate	Mixed forests of shade-tolerant hardwood-conifer.
Moderate Temperate	Deciduous forests.
Grassland	Grassland with or without small groves of hardwood trees.
Subarctic Cordilleran	Open-canopied conifer woodland and alpine tundra in elevation zones.
Cordilleran	Closed-canopied conifer or mixedwood forest, open-canopied conifer Woodland, and alpine tundra in elevational zones.
Interior Cordilleran	Grassland (with or without scattered trees), closed-canopied conifer or mixedwood forest, open-canopied conifer woodland, and alpine tundra in elevational and rain shadow zones.
Pacific Cordilleran	Closed-canopied conifer forest, open-canopied conifer woodland, and alpine tundra in elevational zones.

RESULTS AND DISCUSSION

Figure 8.1 illustrates the preponderance of large fires across the Canadian boreal zone, with the greatest concentration of these fires in the boreal region of west-central Canada, particularly in north-western Ontario, Northern Manitoba and Saskatchewan. Table 8.2 summarises large fires by size class within each EP, while Table 8.3 presents data on the total area burned in each EP during the 1980s and the percentage of each EP burned annually. Both tables accentuate the fact that more and larger fires occur within the boreal zone. The highest percentage of burned area is observed in the boreal zone, with percent annual area burned averaging 0.90%, 0.58%, and 0.24% in the High, Mid, and Low Boreal EPs respectively. A more detailed picture emerges from Figure 8.2, which illustrates the distribution of large fires within the High and Low Subarctic EPs; the High, Mid and Low Boreal: and the Subarctic EPs and the Low Cordilleran EP. Specific ER figures show that percent annual area burned generally decreases from north to south and west to east across the boreal zone, dropping, within the High Boreal EP, from a high of 1.53% in northern Manitoba and Saskatchewan to 0.34% in northern Ontario and Quebec. The same pattern occurs for the Mid Boreal EP, with a high of 1.00% in central Saskatchewan and Manitoba and north-western Ontario dropping to less than 0.1% in central Ontario and Quebec. The Low Boreal EPs pattern is similar, with the exception of the southern portion of north-western Ontario where approximately 1.0% burned annually. Within the Low Subarctic EP, large fire occurrence was generally consistent from the Northwest Territories through Labrador, with a percent annual area burned of 0.53%. In the mountainous regions of western Canada percent annual area burned is highest within the High Cordilleran EP (0.36%), declining quickly with movement north into the Subarctic Cordilleran EP and south into the Low Cordilleran EP.

There are perhaps four major factors which are collectively responsible for the prominent impact of fire in the Canadian boreal zone: fuels, weather, fire occurrence potential, and level of protection. A brief review of each of these factors explains in large part their collective influence on the manner in which large forest fires are distributed across the Canadian landscape.

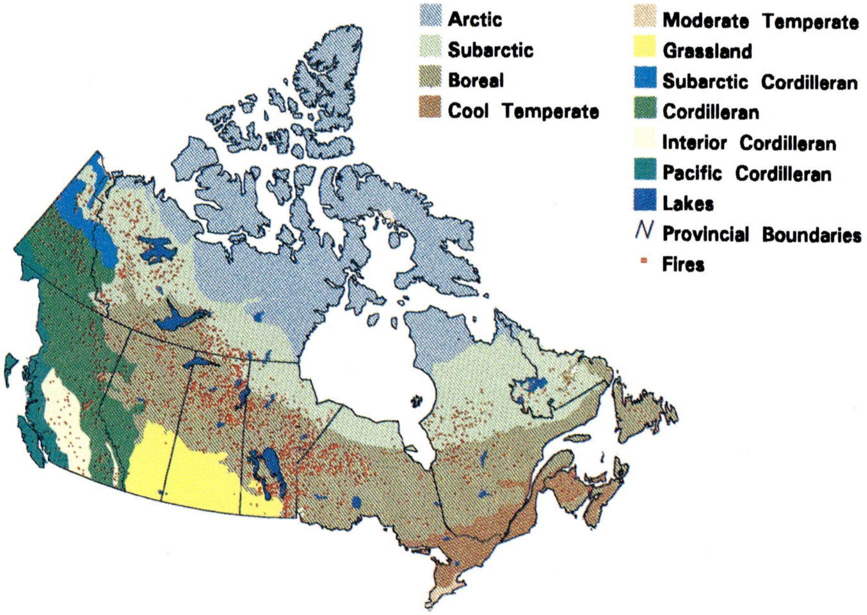

Figure 8.1. Map illustrating the distribution of 1980–89 fires larger than 200 ha across Canadian EPs.

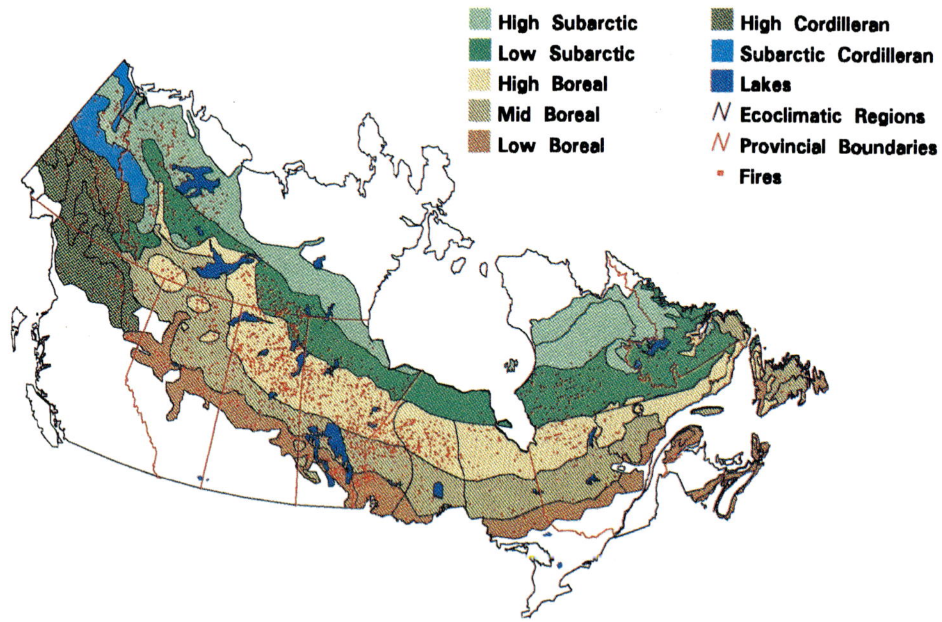

Figure 8.2. Map of 1980–89 Canadian fires >200 ha within the Subarctic, Boreal, Subarctic Cordilleran, and High Cordilleran EPs. Boundaries of ERs are also shown.

Table 8.2. Number of fires by ecoclimatic province by fire size class for Canada for the period 1980 to 1989.

		NUMBER OF FIRES IN SIZE RANGE (× 1000 ha)				
ECOCLIMATIC PROVINCE		0.2–1.0	1.0–10	10–100	100+	TOTAL
Arctic		0	0	0	0	0
Subarctic	High	122	108	24	0	254
	Low	195	207	99	10	511
Boreal	High	507	482	179	19	1187
	Mid	248	206	88	13	555
	Low	223	85	17	2	327
Cordilleran	High	113	106	23	2	244
	Low	95	35	6	0	136
Subarctic Cordilleran		15	7	0	0	22
Pacific Cordilleran		13	4	0	0	17
Interior Cordilleran		35	15	2	0	52
Grasslands		15	12	1	1	29
Cool Temperate		19	8	1	0	28
Moderate Temperate		0	0	0	0	0
Totals		1600	1275	440	47	3362

Table 8.3. EP by land area, area burned, number of fires greater than 200 ha (N > 200 ha), number of fires greater than 200 ha per Mha. yr^{-1}, and average annual area burned as a percentage for Canada for the period 1980 to 1989.

ECOCLIMATIC PROVINCE		LAND AREA (ha)	BURNED AREA (ha)	NUMBER OF FIRES (N) > 200 ha	N > 200 ha PER Mha yr^{-1}	MEAN ANNUAL AREA BURNED (%)
Arctic		244 413 318	0	0	0.00	0.00
Subarctic	High	91 437 379	1 184 677	254	0.28	0.13
	Low	97 336 888	5 295 112	511	0.53	0.54
Boreal	High	114 547 536	10 327 064	1187	1.04	0.90
	Mid	103 154 363	5 956 341	555	0.54	0.58
	Low	56 083 368	1 363 378	327	0.58	0.24
Cordilleran	High	37 327 905	1 353 738	244	0.65	0.36
	Low	40 374 722	354 285	136	0.34	0.09
Subarctic Cordilleran		13 997 724	27 103	22	0.16	0.02
Pacific Cordilleran		22 198 683	13 593	17	0.08	0.01
Interior Cordilleran		19 361 579	79 733	52	0.27	0.04
Grasslands		46 395 330	512 784	29	0.06	0.11
Cool Temperate		37 622 286	52 789	28	0.07	0.01
Moderate Temperate		2 208 438	0	0	0.00	0.00
Totals		926 459 519	26 520 597	3362		

The boreal forest, consisting primarily of broad expanses of contiguous even-aged conifer stands with large quantities of fire-prone fuels, is ideally suited to large, high-intensity, stand-replacing wildfires. Conversely, Cordilleran and temperate forests often contain more deciduous, less-flammable tree species, and more discontinuous fuel complexes due to human impacts. Although people-caused fires are more common in temperate, more heavily populated areas, lightning fire occurrence potential is generally higher in the boreal zone than in many other regions of Canada (e.g., Stocks and Hartley 1979). People-caused fires also occur near more heavily populated areas, increasing the likelihood of early detection and successful suppression action. Lightning fires often occur in more remote areas, slowing detection and response times. Lightning also commonly causes multiple fire starts in a short period, overwhelming fire management resources and resulting in larger fires. In addition, fire weather patterns, and resulting fire danger conditions, are not similar across Canada. Generally, fire danger conditions are much more severe west of the Great Lakes than in eastern Canada (Simard 1973; Harrington et al. 1983; Stocks and Lynham 1995), as the climate in this region is much more continental. Finally, the level of protection applied across Canada is not uniform, with high levels of protection being afforded the more heavily utilised and populated regions of the country. These regions (e.g., Interior Cordilleran and Cool Temperate EPs) have many more fires, particularly people-caused ignitions, but aggressive protection contributes to relatively small areas being burned. Generally, the level of protection decreases with movement northward into the boreal and subarctic regions, although intensive protection is still the policy where values-at-risk warrant the expense. Although ER boundaries do not coincide with the jurisdictional boundaries upon which levels of protection are implemented, this data set suggests level of protection affects area burned, particularly in the Boreal and Subarctic ERs of west-central Canada where more severe fire weather and modified suppression policies combine to create large fire impacts and the shortest mean fire-return intervals.

The less intensive protection of selected regions within the Canadian boreal zone, which creates large areas where a form of modified suppression is practised and fires generally are allowed to assume their natural role in ecosystem dynamics, has existed since the advent of organised fire management in Canada. Recent studies comparing fire sizes relative to levels of protection indicate that, on average, fires in the unprotected regions of the boreal zone are much larger than fires in intensively protected regions (Stocks 1991; Ward and Tithecotte 1993). Martell (1994), in an extensive study of the impact of fire on timber supply in Ontario, conducted an analysis of the effect of level of protection on area burned. Data for the 1976–88 period were used to investigate fire occurrence and area burned in the intensive (fires attacked until suppressed), measured (fires attacked but re-evaluated if not contained by the initial attack force), and extensive (fires only attacked if threatening communities or property) zones of Ontario. Results for north-western Ontario show that, although extensive zone fire occurrence (number of fires $km^{-2} yr^{-1}$) was only 12% and 28% as frequent as fire occurrence in the intensive and measured zones respectively, they result in an annual percent area burned approximately twice that of these much more heavily protected zones. Detailed level of protection measures have not been compiled for other provinces and territories where modified protection is practised, but the effect is almost certainly very similar.

The 1980s Canadian fire data reflect the dominant role of fire in the boreal zone. Although fire strongly influences the cycling and storage of C in boreal forests, it has probably not, to date, resulted in a net loss of C from the boreal biosphere (see however, Chapter 14). While boreal fires release large quantities of carbonaceous aerosols and trace gases directly into the atmosphere, with resultant impacts on atmospheric chemistry, C sequestration begins almost immediately due to prompt regeneration. Therefore boreal fires will only become important to the C budget if fires occur more frequently than at present, shortening fire return intervals to the point where the forest age-class distribution and net C sequestration are affected (see Kurz and Apps 1995a, 1995b). There are, however, two potentially strong forces that could bring about a significant increase in the impact of boreal fires on the C budget in the near future: economic constraints and global warming.

Russia and Canada share the vast majority of the boreal zone and have had somewhat similar fire management strategies for much of this century. Extreme economic difficulties in Russia make more effective fire protection in that country unlikely in the foreseeable future, making increased boreal fire impacts in that country a virtual certainty. Economic realities also dictate that there will be less, rather than more, financial resources for Canadian fire management agencies in the future. Increased budget scrutiny, in response to growing public demands and shrinking resources, will likely result in a gradual erosion of real funds and much more selective fire suppression. Severe pressure to reduce expenditures, along with the growing recognition that fire is a natural component of boreal ecosystems, will probably encourage Canadian agencies to gradually modify their fire management policies. This will result in a careful reconsideration of the social, economic, and ecological implications of fire management, and a focusing of efforts aimed at enhancing the value of forest resources in a broader sense. Current intensive and modified suppression strategies will be refined as forest fire management agencies shift from traditional widespread fire exclusion policies and adopt fire impact management strategies characterised by the selective suppression of fires based on social, economic, and ecological considerations. Considerable thought must be devoted to determining how best to move from a state of fire exclusion to one of fire impact management. Martell (1983) discusses some of the many complex practical issues to be dealt with during such a transition.

Projected climate change scenarios all predict warmer and somewhat drier conditions for the boreal zone, particularly in Siberia and north-central Canada. This would quickly result in increased fire danger (Flannigan and Van Wagner 1991) and a significant increase in fire activity in these areas and, as fire management agencies currently operate with constrained budgets and a small margin for error, a large increase in area burned and shorter fire-return intervals (Stocks 1993). This raises the possibility of positive biospheric feedbacks to global climate change resulting from increased boreal fire activity (Kurz et al. 1994; Kurz and Apps 1995a), as C would be released from forest ecosystems much faster than it could accumulate (Kasischke et al. 1995). As discussed earlier, economic constraints will likely ensure fire managers do not receive the large budget increases that would be necessary to maintain burned area at current levels under a warmer climate.

A realistic conclusion from the preceding analysis is to expect generally more and larger boreal forest fires in future. The result will be a re-evaluation of protection priorities, increased protection of smaller, high-value areas, and a concurrent return to more natural fire regimes over large portions of the world's boreal zone. Such prognoses should be considered when estimating future C storage and fluxes.

REFERENCES

Cahoon DR, Stocks BJ, Levine JS, Cofer WR, Pierson, JM (1994) Satellite analysis of the severe 1987 forest fires in northern China and southeastern Siberia. *J. Geophys. Res.* 99 (D9): 18627–18638.

Ecoregions Working Group (1989) Ecoclimatic regions of Canada, first Approximation Ecoregions Working Group, Canadian Committee on Ecological Land Classification. Ecological Land Classification Series No 23, Sustain. Dev. Br., Can. Wild. Serv., Env. Can., Ottawa, 119 pp. and map.

Flannigan MD, Van Wagner CE (1991) Climate change and wildfire in Canada. *Can. J. For. Res.* 21: 66–72.

Harrington JB, Flannigan MD, Van Wagner CE (1983) A study of the relation of components of the Fire Weather Index to monthly provincial area burned by wildfire in Canada 1953–80. Env. Can., Can. For. Serv. Inf. Rep. PI-X-25, Petawawa Natl. For. Inst., Chalk River, ON, 65 pp.

Higgins DG, Ramsey GS, (1992) Canadian forest fire statistics: 1988–1990 For. Can. Inf. Rep. PI-X-107, Petawawa Natl. For. Inst., Chalk River, ON, 76 pp.

Kasischke ES, Christensen NL, Stocks BJ (1995) Fire, global warming, and the carbon balance of boreal forests. *Ecol. Appl.* 5(2): 437–451.

Kurz WA, Apps MJ, Stocks BJ, Volney WJA (1994) Global climate change: disturbance regimes and biospheric feedbacks of temperate and boreal forests. In: Woodwell GM, Mackenzie FT (eds), Biotic Feedbacks in the Global Climate System: Will the Warming Speed the Warming? Oxford Univ. Press, New York, pp 119–133.

Kurz WA, Apps MJ (1995a) An analysis of future carbon budgets of Canadian boreal forests. *Water Air and Soil Pollution:* 82: 321-331.

Kurz WA, Apps MJ (1995b) Retrospective assessment of carbon flows in Canadian boreal forests. (Chapter 14, this volume).

Martell DL (1983) Fire impact management in the boreal forest region of Canada. In: Wein RW, Riewe RR, Methven IR (eds) Resources and Dynamics of the Boreal Zone. Association of Canadian Universities for Northern Studies, Ottawa, pp 526–533.

Martell, DL (1994) The impact of fire on timber supply. *For. Chron.* 70: 164–173.

Ramsey GS, Higgins DG (1982) Canadian forest fire statistics: 1980 Dep. Env., Can. For. Serv. Inf. Rep. PI-X-17, Petawawa Natl. For. Inst., Chalk River, ON, 38 pp.

Ramsey GS, Higgins, DG (1986) Canadian forest fire statistics: 1981, 1982, 1983 Dep. Env., Can. For. Serv. Inf. Rep. PI-X-49, Petawawa Natl. For. Inst., Chalk River, ON, 148 pp.

Ramsey GS, Higgins DG (1991) Canadian forest fire statistics: 1984–87 For. Can. Inf. Rep. PI-X-74, Petawawa Natl. For. Inst., Chalk River, ON, 163 pp.

Rowe JS (1972) Forest regions of Canada. Can. For. Serv. Publ. No. 1300, 173 pp.

Shostakovitch VB (1925) Forest conflagrations in Siberia. *J. For.* 23: 365–371.

Simard AJ (1973) Forest fire weather zones of Canada. Env. Can., Can. For. Serv., For. Fire. Res. Inst., Ottawa. (Unpublished manuscript).

Stocks BJ (1991) The extent and impact of forest fires in northern circumpolar countries In: JS Levine (ed) Chapman Conf. on Global Biomass Burning: Atmospheric, Climatic, and Biospheric Implications, MIT Press, Cambridge, MA, pp 197–202.

Stocks BJ (1993) Global warming and forest fires in Canada. *For. Chron.* 69: 290–293.

Stocks BJ, Hartley GR (1979) Forest fire occurrence in Ontario. Env. Can., Can. For. Serv., Sault Ste Marie, ON. (Unpublished manuscript).

Stocks BJ, Lynham TJ (1995) Fire weather climatology in Canada and Russia. In: Goldammer JG, Furyaev VV (eds) Fire in Ecosystems of Boreal Eurasia. Springer-Verlag, Heidelberg. (In press).

Ward PC, Tithecott AG (1993) The impact of fire management on the boreal landscape of Ontario. Ont. Min. Nat. Res. AFFMB Pub. No. 305, 10 pp.

9 WG1 Summary: Natural and anthropogenically-induced variations in terrestrial carbon balance

Alan R Townsend[1,2], Martin T Sykes[1,2], Michael J Apps[2], Inez Fung[2], Seppo Kellomäki[2], Pertti J Martikainen[2], Edward B Rastetter[2], Brian J Stocks[2], W Jan A Volney[2] and Stephen C Zoltai[2]

[1] Co-Chairs of Working Group 1.
[2] Addresses are provided in the participant list.

INTRODUCTION

Combustion of fossil fuels and changes in land use have caused a substantial increase in atmospheric CO_2 over the past century, yet the actual atmospheric change is only about half of total fossil fuel emissions alone. This difference is accounted for by carbon (C) uptake into terrestrial and oceanic systems (Table 9.1). CO_2 dissolves into surface waters of the ocean, and can then be stored for centuries via transport into deep waters. Terrestrial ecosystems act as a sink for CO_2 when their net ecosystem productivity (NEP) is positive, in other words when net primary productivity (NPP) exceeds heterotrophic respiration. Recent estimates suggest that significant C sinks are operating in both land and ocean systems (Tans et al. 1990; Keeling et al. 1993; Siegenthaler and Sarmiento 1993), but considerable uncertainty exists about the relative sizes of these sinks, about the mechanisms behind terrestrial storage, and about the potential for the terrestrial biosphere to continue moderating the accumulation of CO_2 in the atmosphere.

Resolution of these uncertainties is critical to projections of future CO_2 concentrations and climate. Storage of C in the oceans is likely to be more stable and persistent than storage on land (IPCC 1990). In turn, alternative mechanisms for terrestrial sinks have very different implications for the biosphere's long-term response to change. For example, C fixed into leaves and easily decomposable litter will return to the atmosphere in a matter of months, whereas C in wood and recalcitrant soil organic matter (SOM) can remain on land for centuries, therefore equal increases in gross C uptake into wood versus leaves will result in very different changes in net C uptake.

This chapter draws upon the individual papers of Working Group 1 (WG1), and summarises some of the group discussions. We discuss ways in which forested biomes may respond to and influence anthropogenic perturbations to the global C cycle, with a focus on "natural" processes in the C cycle and on the response of those processes to "indirect" anthropogenic perturbations. By indirect, we mean human-induced changes such as elevated atmospheric CO_2, climate change, and high rates of N deposition. The "direct" effects of land use changes are discussed in Part II of this volume.

There is roughly 1.5 times more C in the vegetation and soils of the world's forests than is in the atmosphere—*ca.* 1000 Pg C *versus ca.* 700 Pg C (Schlesinger 1991; Dixon et al. 1994)—and annual exchanges of CO_2 between forest ecosystems and the atmosphere are several times larger than yearly fossil fuel emissions. Thus, a relatively small imbalance in the global fluxes of C to and from forests could have significant effects on atmospheric CO_2 levels, and where that imbalance is due to changes in wood or SOM, the effects may last for decades to centuries. We discuss several factors which could cause variability in global forest NEP, including:

NATO ASI Series, Vol. I 40
Forest Ecosystems, Forest Management and the Global Carbon Cycle
Edited by Michael J. Apps and David T. Price

Table 9.1. Average annual budget of CO_2 perturbations for 1980–1989 (Pg C yr^{-1}). Reprinted from Schimel et al. (IPCC 1995) with author's permission.

Reservoir	Mean annual reservoir change (Pg C yr^{-1})
Emissions from fossil fuel combustion and cement production	-5.5 ±0.5
Storage in atmosphere	3.2 ±0.2
Oceanic uptake	2.0 ±0.8
Emissions from changes in tropical land use	-1.6 ±1.0
Uptake by Northern Hemisphere forest regrowth	0.5 ±0.5
Additional net terrestrial sinks	1.4 ±1.5

fire, insect and other pest outbreaks, climate, atmospheric CO_2 levels, and the deposition of N from the atmosphere. We also devote separate sections to the role of carbon-rich peatlands, and to the potential changes in C storage caused by the movement of forest biomes in response to a changing environment.

Natural *versus* anthropogenic effects on global C balance

A host of modelling studies exist in the literature in which quantitative ranges are given for terrestrial C gain or loss in response to a given anthropogenic perturbation (Jenkinson et al. 1991; Townsend et al. 1992; Melillo et al. 1993; Schimel et al. 1994). One uncertainty in all of these estimates, though, is that they assume the C cycle is in steady state prior to the perturbation being investigated. In contrast to the modelled estimates, recent field data on changes in C stocks (Kauppi et al. 1992; Sedjo 1992; Dixon et al. 1994) represent the combined effects of human perturbation *and* natural "background" variability. If background fluctuations in regional to global exchanges of C are non-trivial, then they must be quantified to allow comparisons of modelled and data-based estimates of human perturbations to the C cycle (Fung, Chapter 2).

The key question is: how large and frequent are these background departures from steady state? Fluctuations in the global environment and in terrestrial C pools have occurred throughout the history of the planet (Webb and Bartlein 1992); northern peatlands, for example, have been accumulating C since the end of the last ice age (Schlesinger 1991). Recent attention to global change phenomena is due to concern that human activities may alter the earth at a rate and/or magnitude that is unmatched over, at the very least, human history. For atmospheric CO_2, that appears to be the case (Figure 9.1). Ice core records show that CO_2 has varied by more than 80 ppm over the last 160 000 years (Barnola et al. 1987), rising and falling several times, but the recent rate of increase and current concentration appear greater than at any other point in the record. This change in CO_2 is predicted to change climate, but the evidence for a human-induced shift in climate is much less clear than that for CO_2. A long-term reconstruction of climate for the same period as the ice core records of CO_2 does not show as clear a departure in recent decades from background variability (Figure 9.1).

The atmospheric CO_2 record also shows that land and/or ocean systems have taken up and stored roughly half of anthropogenic emissions (Fung 1993). However, there is considerable uncertainty over how the total sink is partitioned among land and ocean, in part because of potential variations in natural sources and sinks (Fung 1995). A variety of inherently variable environmental factors can alter the balance between NPP and respiration, including fire, pest outbreaks, and climate (both short- and long-term). Soils, for example, have been accumulating C since the last deglaciation at a rate of about 0.04 Pg yr^{-1} (Schlesinger 1991).

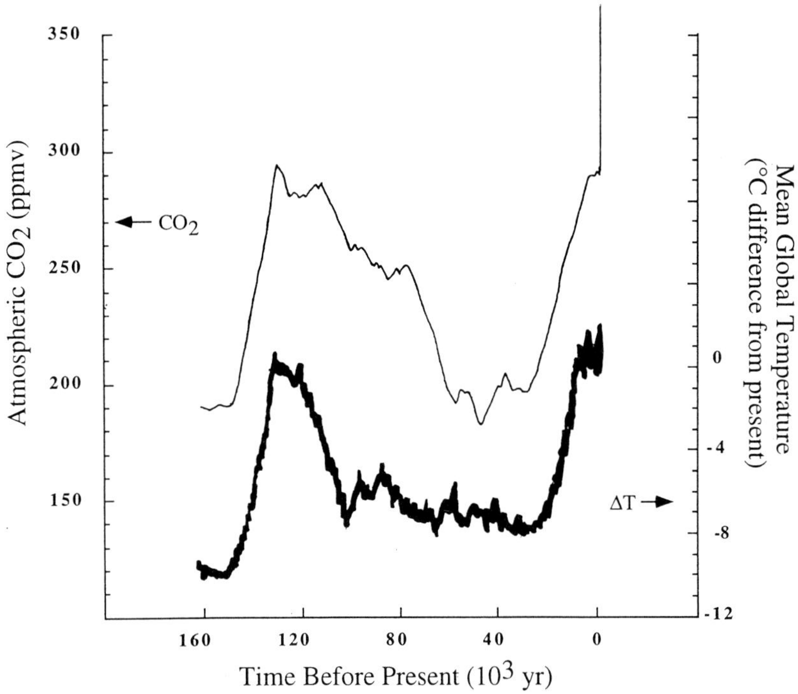

Figure 9.1. Atmospheric CO_2 concentrations and mean global temperatures over the last 160 000 years. CO_2 concentrations are determined by analyses of air bubbles trapped in the Vostok ice core (Antarctica) and from direct measurements at the Mauna Loa Observatory, and temperatures are deduced from ^{18}O values in the core. Redrawn from Barnola et al. (1987).

Fung (Chapter 2) gives a quantitative summary of the current global C budget from the perspective of the atmosphere. She clearly illustrates how background variability in atmosphere–biosphere C exchange could confuse our understanding of how much anthropogenic CO_2 may be stored on land. The "missing" terrestrial sink is commonly thought of as simply an increased flux of C into land ecosystems "induced" by human activity, but Fung points out that the complete equation for the missing sink must include not only this induced flux, but also any imbalance in the background state (see Equation 11, Chapter 2).

Fire

The net effect of fire on the C balance of a forested region is the balance between short term losses of C during and just after burning and gains of C during forest regrowth. In any one year, this is a difference in the dynamics of two very different areas: relatively small C gain for much of the region minus large losses from a small fraction of the area. If integrated over hundreds of years, the net exchange may be a slight sink due to the formation of highly recalcitrant charcoal, but at shorter time scales, a region must vary between net uptake and loss of C. Variability in total C uptake should be small, since it is the product of small gains over a large area, therefore short-term departures from a regional NEP equal to zero will be over a large area, therefore short-term departures from a regional NEP equal to zero will be driven by variations in fire frequency. In particular, a series of intense fire years could cause the region to be a relatively large short-term source of C to the atmosphere, followed by a longer period in which it is a C sink. See Chapter 14 (Kurz and Apps) for a Canadian example.

The area burned in boreal and temperate regions can fluctuate dramatically on an annual basis. In some regions fire suppression policies are likely to have changed the frequency and variability of fire, but year-to-year variability for the entire region is primarily due to short term weather (Stocks et al., Chapter 8). Based on 1980s data, the area burned in northern temperate and boreal latitudes varied between about 3–12 Mha (Stocks and Simard 1993). Assuming a combustion loss of 1.5 kg C m^{-2} (which is somewhat conservative), this translates into a range in C efflux of 0.14 Pg. This is a rough estimate at best of the potential variability in NEP that is due to fire, and the added contribution of tropical regions is unknown, but it suggests that "natural" variability in global fire frequency could at most cause a short term (several years) source of no more than a few tenths of a Pg. Periods of net uptake should last longer than those of net release, but account for a smaller per year departure from steady state.

More detail on forest fires and the global C cycle is provided by Stocks et al. in Chapter 8. They point out that any attempt to estimate the effects of a naturally varying fire regime are confounded recently by fire management. This effect is much larger in southern boreal and temperate zones, where industrial forestry dominates. The longer past is also confounded by land use change, especially in the U.S. and southern Canada. In general, they state that we can assume much more burned in the past in the temperate zone, and that for the boreal region, the area burned seems to be primarily driven by shifts in climate (see also Working Group 2 summary).

Stocks et al. also suggest that in the future, economic constraints will likely translate into larger burned areas in the boreal, and that a climate change toward warmer and drier conditions would mean increases in boreal fires. Fire management agencies with decreasing budgets would protect only high-value regions, allowing more fire over a larger area. Climate change may also shorten fire return intervals, altering the forest age-class structure and therefore C storage. Finally, they point out that fire may accelerate biome migration.

Pest outbreaks

Periodic outbreaks of forest pests such as jack pine budworms (*Choristoneura pinus*) or spruce budworms (*C. fumiferana*) also cause the release of forest C to the atmosphere. In the circumpolar boreal zone, the average area per year affected by pest outbreaks is roughly equivalent to the area burned (Kurz et al. 1995). Like fire, the effects on NEP come from balancing C lost during infestation against C gained during regrowth. Although a severe outbreak may cause the release of as much C as a fire, the average loss per area is much lower, and any release of C will occur much more slowly than during combustion. As a result, short-term variability in C balance due to insect outbreaks should be less than that for fire. Based on Canadian data for C losses (Volney 1988), we derived a rough estimate of variability in NEP of the circumpolar boreal zone of 0.01–0.03 Pg yr^{-1}.

A couple of major unknowns, however, limit our ability to estimate the full effects of pest outbreaks. One, the quantitative estimates of area affected and C lost for boreal and temperate zones are for insects only; the regional effects of chronic diseases such as root rot are poorly understood. Two, very little is known about pest outbreaks in tropical forests. Therefore, values for C fluctuations based on the northern insect data alone may underestimate the global variability induced by all forest pests. Finally, although natural variability in pest outbreaks may not appear significant to large scale C balance, human activities may cause pronounced increases in pest outbreaks, and as Kurz and Apps (Chapter 14) point out, it is "easier to kill than grow". In other words, an increase in the frequency of disturbance is likely to cause a decrease in the capacity for C storage.

Volney (Chapter 7) outlines the interactions between forest pests and C balance in Chapter 7. He proposes that the impacts of pests on forests will increase dramatically if changes in climate result in lower soil moisture regimes. This speculation is based on observations of conditions in areas increasingly affected by pests. Historically, the most intense pest outbreaks were restricted to the southern portions of the boreal forest, but Volney points out that a changing climate could allow intense outbreaks to spread to all productive forests in the biome. In addition, he suggests that the introduction of exotic pest species into forested areas will certainly increase the frequency and intensity of outbreaks, and that C losses from these introductions could be large unless managed by effective biological control programs.

Climate

It is quoted widely that the 1980s were the warmest years on record, while the 1950s, 1960s and 1970s were relatively cool; the range in mean global temperature across this period is more than 0.5 °C (Folland et al. 1990). At the same time, global precipitation was somewhat above average in the 1950s and the 1970s, while the 1980s were relatively dry (Eischeid et al. 1991). How do these relatively short-term variations in climate affect terrestrial NEP? Using simple relationships between climate, soil respiration and NPP, Dai and Fung (1993) estimate that climate variability over the last 60 years may have caused the biosphere to alternate between net uptake and release of C, with a total range of more than 1 Pg and an annual variability of 0.3 Pg yr^{-1}. They suggest that the effects of climate alone may account for about half of the unexplained residual in the C budget between 1950 and 1984.

Simple relationships between climate, NPP and decomposition tend to show that warming will cause a net loss of C, while cooling will cause net C uptake (Woodwell 1990; Townsend et al. 1992; Dai and Fung 1993). The effects of climate on NEP, however, may be more complex than these relationships suggest. Increasing decomposition rates not only cause greater respiratory C losses, they also cause an increase in N mineralization, and N limitation of plant growth is widespread in forest ecosystems (Vitousek and Howarth 1991). Because of a dynamic equilibrium between vegetation and soil processes, about 80–90% of forest N is in SOM and litter, and only 10–20% is in vegetation. Rastetter et al. (1992) suggest that this equilibrium distribution will shift as the rate of N cycling changes. If the rate of turnover is increased, the dynamic equilibrium will shift so that a higher percentage of the N will be in vegetation. Thus, a shift toward more optimal temperature and moisture for soil processes will result in a redistribution of N from soils to vegetation. Because the C:N ratio of forest vegetation is about an order of magnitude higher than that of soils (100–200 *vs.* 10–20), any redistribution of N from soils to vegetation will result in a net increase in total ecosystem C. That is, any loss of C resulting from SOM decomposition should be more than made up by the stimulation of vegetation growth at a much higher C:N ratio.

This mechanism for increasing total ecosystem C will work as long as NPP is limited by N or some other nutrient that is supplied largely through the decomposition of SOM (e.g., P). In forests where NPP is limited by light or water, an increase in nutrient availability resulting from an increase in SOM decomposition will not stimulate production and therefore will not stimulate C storage. Under such conditions, nutrients released from the SOM will likely be lost from the ecosystem altogether. Thus, C released by the increased decomposition will not be made up by production in the vegetation and there should be a net loss of C from the ecosystem as a whole.

The Dai and Fung analysis did not account for these potential nutrient effects on NEP. In fact, if most terrestrial systems are primarily nutrient limited (rather than by moisture or light), then the warming period during the eighties might be expected to increase terrestrial net C uptake, rather than shift the biosphere towards the net source suggested by Dai and Fung. If, however, this nutrient effect dominates the terrestrial response, then climate variability should have caused the net C flux in recent decades to oppose, rather than track, temporal estimates of the missing sink.

The effects of a warmer climate on both C and nutrient cycles, and the overall effect on forest C balance, are discussed by Townsend and Rastetter (Chapter 3). For a given biome, whether warming leads to C gain or loss should depend on the response of vegetation growth rates to increasing rates of nutrient mineralization balanced against the increases in respiration. Townsend and Rastetter suggest that net uptake of atmospheric C should prevail, at least in the short term, although Fung (1995) points out that a positive feedback, in which warming leads to C loss from the biosphere, is more consistent with the recent C budget. The interactive effects of changes in moisture add to the uncertainty, as does the likelihood that a warmer climate will initiate biome redistribution.

Summary of background variability in carbon balance

There is still much uncertainty over how the net C flux of an undisturbed biosphere might vary both in magnitude and in time. Of the factors discussed above, variations in climate appear to produce the greatest potential range in NEP, but the net effects of climate are still unclear, as are the total potential influences of fire, pest outbreaks and other disturbances in forested regions.

Estimates of global variability derived from scaling up regional data are subject to large errors; it may be that the range decreases significantly with greater area considered. None of the numbers given in the preceding sections are robust; instead they are meant to show the potential importance of natural variability in terrestrial ecosystems—especially in forests—to estimates of recent anthropogenic influences on C balance. It does appear that total background fluctuations in the biosphere may be of the same order of magnitude as the missing C sink, therefore the typical baseline assumption of zero net C exchange prior to human disturbance may introduce significant errors into the current C budget (see also WG2 summary, Chapter 17). On the other hand, it seems unlikely that natural variability in the global C cycle could have been predominantly responsible for the apparent steady increase in terrestrial C storage over the last two centuries.

Atmospheric CO_2

In contrast to the climate record, it is clear that atmospheric CO_2 concentrations are much higher today than they were in the centuries preceding the industrial revolution (Figure 9.1). In addition, decadal variability in CO_2 was a tiny fraction of the change brought on by human activities (Keeling et al. 1989). Therefore, any recent changes in terrestrial C storage due to changes in CO_2 may be safely attributed to human influence. These changes, however, are extremely difficult to estimate, as the effects of CO_2 have been shown to vary substantially with plant species, resource availability, and time (Mooney et al. 1991; Bazzaz and Fajer 1992; Comins and McMurtrie 1993). Based on the immediate response and several model computations, Kellomaki and Karjalainen (Chapter 5) suggest that boreal forest growth should be enhanced under the combination of elevated temperature and CO_2. The duration of this response, however, is not certain. They describe field experiments which indicate that the photosynthetic response of coniferous species subjected to long-term doubled CO_2 is much smaller than the initial change. Similarly, they suggest that long term temperature elevation will reduce C uptake. Finally, any enhancement of growth will speed successional processes, thereby causing an earlier reduction in a forest's capacity to sequester C than under pristine conditions. In the long run, Kellomaki and Karjalainen propose that the C uptake by boreal forest ecosystems subjected to higher temperatures and CO_2 concentrations will probably not be higher than current levels.

Nitrogen deposition

Like atmospheric CO_2, recent effects of anthropogenic activity on the N cycle are much larger than background variability; human activity has caused the global rate of N_2 fixation to more than double in just three decades (Vitousek 1994; Figure 9.2). Such disturbance of the N cycle has led to much higher rates of N deposition on to land and ocean. Galloway et al. (1994) estimate that deposition has increased by more than an order of magnitude in parts of the temperate zone, and since most temperate ecosystems appear to be N limited, it has been suggested that this increase may stimulate C uptake and storage (Peterson and Melillo 1985; Schindler and Bayley 1993; Hudson et al. 1994; Townsend et al. 1985). However, the magnitude and duration of such a sink are not well known.

In Chapter 3, Townsend and Rastetter suggest that forests are likely to dominate the global response to N deposition due to their capacity to produce woody tissue with both high C:N ratios and long turnover times. They also point out that a global estimate of the effects of N deposition must take into account a number of factors which vary strongly in time and space.

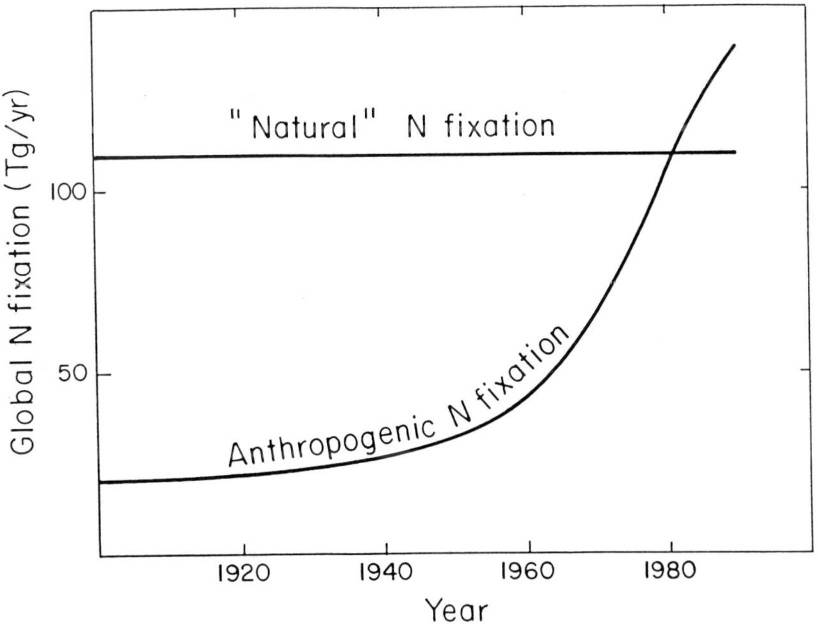

Figure 9.2. Natural and anthropogenic N fixation over the twentieth century. The natural line is biological fixation in natural terrestrial systems plus fixation by lightning; the anthropogenic line is the sum of fixation for fertilisers, fossil fuel combustion and leguminous crops. Reprinted from Vitousek (1994) with permission of the author.

Uptake in any given year will be equal to the net primary production stimulated by the current year's deposition minus the respiration from excess C fixed in previous years. In other words, C storage in the present depends upon N inputs in the past. In addition, deposition is not evenly distributed across the terrestrial landscape; it varies by more than an order of magnitude among world biomes (Penner et al. 1991), and different ecosystem types will vary greatly in their response to that N. For example, N which falls on cultivated areas will not contribute significantly to any terrestrial sink.

Townsend and Rastetter identify three major variables as critical in determining the global effects of N deposition on C storage. The first is the distribution of N-limited forest ecosystems relative to the distribution of N deposition. The second is the fraction of N inputs which are allocated to the production of woody tissue. Finally, both current and future C storage will depend on the fraction of N inputs which are retained in the ecosystem. Aber et al. (1989) suggest that as elevated N inputs continue, the extent of N limitation will decline and therefore that N retention will decline. In the extreme case, N inputs may become so high that they are damaging, causing ecosystem decline and C loss. Forest dieback in some areas of Europe and the north-eastern United States has been attributed to long-term exposure to high N and sulphur inputs (Schulze 1989). Townsend and Rastetter propose that even if these inputs do not prove harmful, the global extent of N limitation must be declining, and therefore the long-term persistence of any N-derived sink is inherently limited.

Peatlands

Peatlands are a large reservoir of C, and we address them here for several reasons. Firstly, many of the world's peatlands also have a significant amount of tree biomass. Secondly, the rate of C accumulation in peatlands may be important in establishing the natural variability of terrestrial NEP. Finally, changes in climate may cause large scale conversion of open and sparsely wooded peatlands to more dense forest.

The present average accumulation of C in northern peatlands is estimated at 29 g C m^{-2} yr^{-1} (Zoltai and Martikainen, Chapter 4). This value leads to an annual C sink of about 0.1 Pg yr^{-1} in all northern peatlands (330 Mha). Total C in this region has been estimated to be between 192–455 Pg, and NPP averages about 300 g m^{-2} yr^{-1}, thus roughly 8% of annual NPP is accumulated as peat. The ~0.1 Pg yr^{-1} rate of accumulation appears to be relatively stable, at least in recent centuries. Peatlands are therefore unlikely to have caused any significant inter-decadal variability in the terrestrial C budget, but their steady C uptake should be accounted for when establishing a background against which recent perturbations are to be measured. The additional influence of the large area of tropical peatlands is not well known.

Zoltai and Martikainen (Chapter 4) outline how the effects of peatlands on the C cycle and climate depend not only on CO_2 exchange, but also on the production and emission of methane. For example, a drying of peatlands may increase CO_2 efflux, but it may also decrease CH_4 production, and since on a per molecule basis CH_4 has a greater impact on radiative balance than does CO_2, the net effect could be a negative feedback to warming (Billings et al. 1984; Oechel et al. 1993). On the other hand, they point out that a warmer climate will both increase microbial activity and allow greater emissions in fall and spring by shortening the number of months in which surface water is frozen. Finally, they propose that the practice of converting peatlands to forested areas via draining will result in a net loss of C, due to an increase in decomposition rates as conditions switch from anaerobic to aerobic that exceeds any C gain from the accretion of woody biomass.

Biome migration

Present day ecosystem types, or biomes, are not fixed in space, nor have they been at any time in the past (Prentice and Fung 1990; Friedlingstein et al. 1995). Changes in both the composition of the biomes and in their distribution over the globe have been driven by changes in climate. In the past, changes in climate have been relatively slow, so that the movement of biomes has usually occurred on the time scale of centuries. Future changes, however, are predicted to occur at an unprecedented rate, which may well lead to rapid changes in the present boundaries of biomes, and even to complete breakdown of some biomes and a consequential reduction in their ability to store C.

Possible equilibrium changes in the distribution of biomes and C storage under an altered climate have been examined both on a global scale (Prentice and Sykes 1994) and on a regional scale (Sykes and Prentice, Chapter 6) using the equilibrium model of potential vegetation, BIOME 1 (Prentice et al. 1992). The model uses a small number of plant functional types, a dominance hierarchy and five environmental parameters to project the distribution of biomes. Prentice and Sykes (1994) used the model to project changes in biome distribution and C storage under doubled CO_2 scenarios from two general circulation models GFDL (Manabe and Wetherald 1987) and OSU (Schlesinger and Zhao 1989). Generally, their projections show a poleward shift of the taiga, cool conifer, cool mixed, temperate deciduous and warm temperate evergreen/warm mixed forest belts in northern high- and mid-latitudes, an eastward shift of forests in Western Eurasia, and a poleward shift of warm (C_4 dominated) grass and shrublands. They also a project a slight increase in tropical rain and seasonal forests. However, there are regional variations to this general pattern.

On a global scale, the net change in C is not clear as there are both regional gains and losses, and large differences depending on the climate scenario used (Prentice and Sykes 1994; Friedlingstein et al. 1995). In the Prentice and Sykes study, the OSU scenario estimated that terrestrial C storage would increase by 90 to 100 Pg C, while the GFDL scenario produced an increase of only 50 Pg C down to no change at all. Prentice and Sykes (1994) also suggested that the terrestrial biosphere may respond in a non-linear fashion to rapid changes in climate, and that beyond a certain point the shifts in biomes may turn the biosphere from a sink into a source of C if taiga and tundra regions are reduced sufficiently. All current estimates of C storage under a new biome distribution are equilibrium simulations, and the transient responses may be much different, but a dynamic vegetation model is yet to be developed which could answer these questions on a global scale. Sykes and Prentice discuss the effects of biome shifting in Chapter 6, and describe some regional attempts at estimating transient changes in C storage through the use of a dynamic simulation model.

SUMMARY

In summary, the members of WG1 discussed several factors which influence the C balance of the terrestrial biosphere. These discussions focused on the role of forest ecosystems, as we felt that forest dynamics may dominate the terrestrial response to many current and future changes in the environment. The potential effects of fire, forest pests, climate, atmospheric CO_2, N deposition, and biome redistribution were all addressed, and each is presented in some detail by individual members of WG1 in the following chapters. Finally, we identified five major unknowns that arose during the group discussions:

- How much of recent net C uptake into terrestrial systems might be due to natural variability?
- How important are tropical regions in all of the factors discussed?

- What is the long-term persistence of any current C sinks stimulated by elevated N deposition and atmospheric CO_2?
- What is the most likely net effect of a warmer climate on terrestrial C pools?
- What are the transient responses of biome areas and C pools to climate and atmospheric change?

ACKNOWLEDGEMENTS

We thank P Vitousek, E Holland, R Braswell and D Schimel for contributing figures, tables and/or ideas, and C Prentice and R Sedjo for critical reviews of an earlier version.

REFERENCES

Aber JD, Nadelhoffer KJ, Steudler P, Melillo JM (1989) Nitrogen saturation in northern forest ecosystems. *BioScience* 39(6): 378–386. .

Barnola J-M, Raynaud D, Korotkevitch YS, Lorius C, (1987) Vostok ice core provides 160,000 year record of atmospheric CO_2. *Nature* 329: 408–414.

Barnola J-M., Pimenta P, Raynaud D, Korotkevich TS (1991) CO_2-climate relationship as deduced from the Vostok ice core: a re-examination based on new measurement and on a re-evaluation of the air dating. *Tellus* 43B: 83–90.

Bazzaz FA, Fajer ED (1992) Plant life in a CO_2-rich world. *Scientific American* 266: 68–74.

Billings WD, Lukeen JO, Mortensen DA, Peterson KM, 1984. Arctic tundra: a source or sink for atmospheric carbon dioxide in a changing environment? *Oecologia* 53: 7–11.

Comins IIN, McMurtric RE (1993) Long-term response of nutrient limited forests to CO_2 enrichment: equilibrium behavior of plant–soil models. *Ecological Applications* 3: 666–681.

Dai A, Fung IY (1993) Can climate variability contribute to the "missing" CO_2 sink? *Global Biogeochemical Cycles* 7: 599–609.

Dixon RK, Brown SA, Houghton RA, Solomon AM, Trexler MC, Wisniewski J (1994) Carbon pools and flux of global forest ecosystems. *Science* 263: 185–190.

Eischeid JK, Bradley RS, Jones PD (1991) A comprehensive precipitation data set for global land areas. Tech. Report TR051, DOE ER-69017T-H1, US Dept. of Energy CO_2 Research Division, Washington, DC, 81 pp.

Folland CK, Karl TR, Vinnikov KY (1990) Observed climate variations and change. In: Houghton JT, Jenkins GJ, Ephraums JJ (eds) Climate Change, the IPCC Scientific Assessment WMO/UNEP, Cambridge University. Press, New York, pp 201–238.

Friedlingstein P, Prentice KC, Fung IY, John JG, Brasseur GP (1995) Carbon-Biosphere-Climate interactions in the LGM climate. *J. Geophys Res.* (In press).

Fung IY (1993) Models of oceanic and terrestrial sinks of anthropogenic CO_2: A review of the contemporary carbon cycle. In: Oremland R (ed) Biogeochemistry of global change: radiatively active trace gases: selected papers from the Tenth International Symposium on Environmental Biogeochemistry, San Francisco, 19–24 August, 1991. Chapman and Hall, New York.

Fung IY (1995) Perturbations to the biospheric carbon cycle: uncertainties in the estimates. In: Woodwell GM, MacKenzie FT (eds) Biotic Feedbacks in the Global Climate System: Will Warming Speed the Warming? Oxford University Press, New York, 416 pp.

Fung IY (1995) The global carbon cycle and the atmospheric record: "the problem definition". (Chapter 2, this volume).

Galloway JN, Levy II H, Kasibhatla PS (1994) Year 2020: consequences of population growth and development on deposition of oxidized nitrogen. *Ambio* 23(2): 120–123.

Hudson RJM, Gherini SA, Goldstein RA (1994) Modeling the global carbon cycle: nitrogen fertilization of the terrestrial biosphere and the missing CO_2 sink. *Global Biogeochemical Cycles* 8(3): 307–333.

IPCC (1990) Climate change: the IPCC scientific assessment. Cambridge University Press, Cambridge, UK, 365 pp. .

Jenkinson DS, Adams DE, Wild A (1991) Model estimates of CO_2 emissions from soil in response to global warming. *Nature* 351: 304–306.

Kauppi PE, Mielikäinen K, Kuusela K (1992) Biomass and carbon budget of European forests, 1971–1990. *Science* 256: 70–74.

Keeling CD, Bacastow RB, Carter AF, Piper SC, Whorf TP, Heimann M, Mook WG, Roeloffzen H (1989a) A three-dimensional model of atmospheric CO_2 transport based on observed winds: 1. Analysis and observational data. In: Peterson DH (ed), Aspects of Climate Variability in the Pacific and Western Americas. *Geophysical Monograph* 55, American Geophysical Union, Washington, DC, pp 305–363.

Kellomaki S, Karjalainen T (1995) Sequestration of carbon in the Finnish boreal forest ecosystem managed for timber production. (Chapter 5, this volume).

Kurz WA, Apps MJ (1995) Retrospective assessment of C flows in Canadian boreal forests. (Chapter 14, this volume).

Kurz WA, Apps MJ, Beukema SJ, Lekstrum T (1995) Twentieth century carbon budget of Canadian forests. *Tellus* 47B: 170–177.

Manabe S, Wetherald RT (1987) Large scale changes in soil wetness induced by and increase in carbon dioxide. *J. Atmos. Sci.* 44: 1211–1235.

Melillo JM, McGuire AD, Kicklighter DW, Moore III B, Vorosmarty CJ, Schloss AL (1993) Global climate change and terrestrial net primary production. *Nature* 363: 234–240.

Mooney HA, Drake BG, Luxmoore RJ (1991) Predicting ecosystem responses to elevated CO_2 concentrations. *BioScience* 41: 96–104.

Oechel WC, Hastings SJ, Vourlitis G, Jenkins M, Riechers G, Grulke N (1993) Recent change of Arctic tundra ecosystems from a net carbon dioxide sink to a source. *Nature* 361: 520–523.

Penner JE, Atherton CS, Dignon J, Ghan SJ, Walton JJ, Hameed S (1991) Tropospheric nitrogen: a three-dimensional study of sources, distributions, and deposition. *J. Geophys. Res.* 96: 959–990.

Peterson BJ, Melillo JM (1985) The potential storage of carbon caused by eutrophication of the biosphere. *Tellus* 37B: 117–127.

Prentice KC, Fung IY (1990) The sensitivity of terrestrial C storage to climate change. *Nature* 346: 48–51.

Rastetter EB, McKane RB, Shaver GR, Melillo JM (1992) Changes in C storage by terrestrial ecosystems: how C–N interactions restrict responses to CO_2 and temperature. *Water, Air, and Soil Pollution* 64: 327–344.

Schimel DS, Braswell Jr BH, Holland EA, McKeown R, Ojima DS, Painter TH, Parton WJ, Townsend AR (1994) Climatic, edaphic and biotic controls over storage and turnover of carbon in soils. *Global Biogeochemical Cycles* 8(3): 279–293.

Schindler DW, Bayley SE (1993) The biosphere as an increasing sink for atmospheric carbon: estimates from increased nitrogen deposition. *Global Biogeochemical Cycles* 7: 717–734.

Schlesinger ME, Zhao ZC (1989) Seasonal climate change introduced by doubled CO_2 as simulated by the OSU atmospheric Gcm/mixed-layer ocean model. *J. Climate* 2: 429–495.

Schlesinger WH (1991) Biogeochemistry: An Analysis of Global Change. Academic Press. San Diego, CA, 443 pp.

Schulze ED (1989) Air pollution and forest decline in a spruce (*Picea abies*) forest. *Science* 244: 776–783.

Sedjo RA, (1992) Temperate forest ecosystems in the global carbon cycle. *Ambio* 21: 274–277.

Sellers P, Schimel D (1993) Remote sensing of the land biosphere and biogeochemistry in the EOS era: science priorities, methods and implementation—EOS land biosphere and biogeochemical cycles panels. *Global and Planetary Change* 7: 279–297.

Siegenthaler U, Sarmiento JL (1993) Atmospheric carbon dioxide and the ocean. *Nature* 365: 119–125.

Stocks BJ, Simard AJ (1993) Forest fire management in Canada. *Disaster Management* 5: 21–27.

Stocks BJ, Lee BS, Martell DL (1995) Some potential carbon budget implications of fire management in the boreal forest. (Chapter 8, this volume).

Sykes MT, Prentice IC (1995) Carbon storage and climate change in Swedish forests: a comparison of static and dynamic modelling approaches. (Chapter 6, this volume).

Tans PP, Fung IY, Takahashi T (1990) Observational constraints on the global atmospheric CO_2 budget. *Science* 247: 1431–1438.

Townsend AR, Vitousek PM, Holland EA (1992) Tropical soils could dominate the short-term carbon cycle feedbacks to increased global temperatures. *Climatic Change* 22: 293–303.

Townsend AR, Braswell BH, Holland EA, Penner JE (1995) Spatial and temporal patterns in potential terrestrial carbon storage due to deposition of fossil fuel derived nitrogen. *Ecological Applications* (In press).

Townsend AR, Rastetter EB (1995) Nutrient constraints on carbon storage in forested ecosystems. (Chapter 3, this volume).

Vitousek PM, Howarth RW (1991) Nitrogen limitation on land and in the sea: how can it occur? *Biogeochemistry* 13: 87–115.

Vitousek PM (1994) Beyond global warming: ecology and global change. *Ecology* 75(7): 1861–1876.

Volney WJA (1988) Analysis of historic jack pine budworm outbreaks in the Prairie provinces of Canada. *Can. J. For. Res.* 18: 1152–1158.

Volney WJA (1995) Climate change and management of insect defoliators in boreal forest ecosystems. (Chapter 7, this volume).

Webb TL, Bartlein PJ (1992) Global changes during the last three million years: climatic controls and biotic responses. *Ann. Rev. Ecology and Systematics* 23: 141–174.

Woodwell GM (1990) The effects of global warming. In: Leggett J (ed) Global Warming, The Greenpeace Report. Oxford University Press, New York.

Zoltai SC, Martikainen PJ (1995) Estimated extent of forested peatlands and their role in the global carbon cycle. (Chapter 4, this volume).

Part II

The global carbon cycle and forest land use: past, present and future

II The global carbon cycle and forest land use: past, present and future

Introduction

David T Price and Michael J Apps

Part II reviews the available information and contemporary understanding of past and present changes in global and regional terrestrial ecosystem carbon (C) pools, and of the C fluxes associated with forest management and land-use activities. Tropical, temperate and boreal forests are assessed separately, in recognition of the profound differences in their ecological dynamics and socio-economic pressures. The possibilties for managing existing forest ecosystems sustainably, and for reclamation of degraded forest lands, as C stores are examined. Some projections of future changes (under future climate and socio-economic scenarios) are also reviewed.

The Working Group discussions tended to overlap with those of Working Group 3 (Part III), particularly in recognising the need for standardised methods for determining regional C budgets. In contrast however, Working Group 2 took a "top-down" view of the role of forests in the global C cycle, concentrating on the interaction of forest dynamics with disturbances and human perturbations. This was manifested in a focus towards regional comparisons of past and future trends in land-use change, (whereas the focus of Working Group 3 places more emphasis on the opportunities for, and practicalities of, using forests and forest products, including energy substitution, as tools for increasing C sequestration). Areas of uncertainty requiring further research or better data were identified, with a concluding discussion on the need for a global monitoring system to track C fluxes resulting from changes in vegetation cover, in land use and forest land management.

Review of past and current rates of carbon loss

At the global scale, Houghton (Chapter 10), sets the scene by reviewing the impacts of human land-use and land-use change on global C emissions from forests during the period 1850–1980. The rate of land-use change has quadrupled since 1925, with approximately 5 million km^2 of tropical forest removed during 1850–1980. In the same period, commercial logging operations have been practiced on more than twice this area in the boreal and temperate regions, although only a relatively small portion of this land-use practice has resulted in land-use change and net release of terrestrial carbon. Houghton estimates the net loss of C from terrestrial ecosystems during 1850–1980 to be about 97 Pg C, with net additions to storage (by 1980) of approximately 15 Pg C in wood products and 10 Pg C in forest floor slash.

NATO ASI Series, Vol. I 40
Forest Ecosystems, Forest Management and the Global Carbon Cycle
Edited by Michael J. Apps and David T. Price

Conversions of forest to agriculture (croplands and pastures) released about 73 Pg C, while logging (after accounting for regeneration) released a further 23 Pg C.

Brown (Chapter 11) summarises progress in estimation of C losses from tropical land-use and land-use changes. Owing to poor inventory data in many tropical regions, precise estimates of soil and biomass C densities and net fluxes are difficult to obtain. Recent developments in remote sensing and mapping, however, now allow improved monitoring of changes in forest area. Significant deforestation occurs in tropical America and this area has been widely thought to be the largest contributor to the tropical C flux. Forest management and regeneration may be compensating for deforestation losses in some areas, notably Brazil and India. Brown et al. (Chapter 17) also report that forest degradation, caused by various human land-use practices and other pressures, is widespread and contributes considerable C releases in addition to those resulting from deforestation. The area of degraded forests is much greater than the deforested area, but since about 1950, it appears that the C released due to deforestation has significantly exceeded that from degradation, at least in tropical Asia.

Regional C budgets and the search for the missing sink

Houghton emphasises that the term *missing sink* refers only to the global C imbalance between anthropogenic emissions and the observed net increases in atmospheric CO_2 after accounting for *known* oceanic and terrestrial sinks. Within the period of the global land-use changes reported by Houghton, several authors report regional studies which permit the comparison and contrasting of large-scale C dynamics of forests in different geographic regions. As noted by Brown et al. (Chapter 17), estimates of C balances tend to be more precise in the temperate and boreal regions because inventory data are more complete. Other Chapters addressing factors influencing regional forest C budgets are provided by Kellomäki and Karjalainen (Chapter 5), Kurz and Apps (Chapter 14) and Maclaren (Chapter 21).

For the boreal regions, Shvidenko et al. (Chapter 12) report that recent analyses of age-class dynamics in the 7.35 million km^2 of forests in the Former Soviet Union indicate that they have collectively been a strong C sink of the order of 1 Pg C yr^{-1}, although they recognise that the data upon which this estimate is based require improvement. A striking conclusion however was the possible disappearance of this sink in the last 10 years: both the 'official' statistics and their own reconstruction of recent biomass dynamics, indicate that in the period 1983—1993, the Russian forests did not sequester any carbon—primarily due to losses of growing stock in the Asian parts. Kurz and Apps (Chapter 14) reached similar conclusions for the Canadian boreal forests. Using forest inventory data and records of disturbances, they show that the average age of boreal forest in Canada has increased from about 61 yr in 1920 to 82.5 yr in 1969. From 1970, a doubling in the average rate of disturbances has caused the mean age to decrease to about 76.4 yr by 1989. Using a dynamic C budget model, they find that Canadian forests therefore accumulated 8.3 Pg C in the period 1920–1970, but lost about 0.6 Pg C during 1970–1989.

Kauppi (Chapter 16) describes causal mechanisms affecting C reservoirs in European temperate zone forests for 1851–2050. For much of the temperate zone, he concludes, logging has been sustainable. In fact, he argues, annual volume removals have been significantly lower than mean annual increment. This is due largely to low harvesting rates but other reasons include fire protection, forest preservation initiatives and the virtual disappearance of firewood collection. Productivity may also have increased though the underlying mechanisms are still uncertain. It seems that CO_2 fertilisation can currently be rejected as a significant factor. Many European forest sites are nutrient-limited however, so it is likely that nitrogen pollution has increased productivity on some sites. This pollution-fertilisation effect, when combined with silvicultural enhancements in many European countries, appears to have resulted in increased average rates of forest growth, since the recorded volume of wood harvests has remained more-or-less constant since 1965 while the forest inventories have shown a general increase in wood volume.

While at large scales, air pollution impacts to date may have enhanced rather than caused a general decline in productivity, Kauppi anticipates that deleterious effects may occur in future. Indeed Galinski and Witowski (Chapter 15) report a case study for the heavily industrialised border regions of Poland and Czechoslovakia, where pollution reduced the natural resistance of local forests to insect attack and evidently resulted in significant losses of C from the region (to some extent offset by transfers to products through salvage logging). Galinski and Witowski's findings act as a warning about the possible C implications of such forest die-back if pollution loadings were to increase in Europe as a whole.

Role of natural disturbances

In his assessment of the biospheric C flux, Houghton (Chapter 10) considers only land-use practices and changes in these, ignoring ecosystems not cleared, cultivated or otherwise modified. He considers it unlikely that management will eliminate natural disturbances, but assumes that natural disturbances maintain a steady-state C balance over long periods of time and large regions and do not contribute significantly to the long-term global C imbalance. Kurz and Apps (Chapter 14) argue that the assumption of a steady-state C balance in natural forests does not hold when the frequency of natural disturbances changes, e.g., in response to changes in climate occurring over periods of several decades. Analyses that do not make the assumption of steady-state are required, they argue, in order to assess correctly the possible historical and future contributions to the long-term C budget, as affected by both natural and anthropogenic disturbances.

Brown et al. (Chapter 17) also conclude that it is inappropriate to assume that all "natural" forest ecosystems are in a state of equilibrium with respect to C exchange with the atmosphere. They conclude that shifts in age-class structure resulting from changes in disturbance regimes, particularly in the boreal regions, could substantially alter regional C budgets. As a general observation applicable to all regions, contributions to the missing C sink may or may not be attributed to the presence of relatively large areas of younger actively regrowing forests. Where such forests exist as the result of logging, or deforestation and subsequent afforestation, then this has been included in estimates of land-use change (e.g., Houghton and Brown), and cannot therefore contribute to the observed global imbalance. If the result of natural disturbances, however, changes in average C uptake have not been accounted for in the land-use flux, and may very well help to explain the observed imbalance (particularly for the boreal regions addressed by Kurz and Apps and Shvidenko et al.).

Bringing a halt to deforestation, especially in the tropics

In general, Houghton observes, it seems that abandoned agricultural land in the tropics does not return naturally to forest, or else shifting cultivation methods are becoming less sustainable due to increased population pressures. He argues therefore that tropical deforestation can be reduced only by making agriculture more sustainable, while unproductive lands must be managed to sequester C.

Brown (Chapter 11) on the other hand points out that biomass estimates for tropical forests are often based on samples that misrepresent the contributions of mature forest—due to past human activities in degrading these forests. She therefore believes that the C sequestration potential of many tropical forests may be underestimated. Some tropical forests may take centuries to reach maximum size and C storage (a feature shared with the massive Pacific coastal forests of North America).

In conclusion, Brown et al. suggest that because tropical forests have been subjected to human disturbance and changes in land use over many centuries, significant C storage gains are likely to occur in areas previously disturbed but now abandoned. For example, a recent C budget analysis for Brazil has shown that C gains in vast areas of secondary forest almost offset C losses from current deforestation.

Managing forests to sequester more carbon

Houghton (Chapter 10) proposes three main approaches to C sequestration in forests: (1) increase area; (2) increase utilisation efficiency and longevity of products; (3) manage the climate (i.e., prevent climate warming) to avoid net losses of ecosystem C.

Fischlin (Chapter 13) addresses the question: is it possible to maximise C sequestration in forests while still harvesting wood? He uses a model to focus on the effects of harvesting on forest-level C storage and shows that intensive management, with low harvesting rates, or selection cutting (which maintains the forest at high mean age with short periods between harvests), is the best possible. Fischlin concludes that maximum C storage is obtainable by avoiding clear-cutting. This contrasts with the general view expressed by Working Group 3, although much of the difference can be explained by considering the potential value of direct and indirect energy substitution.

Brown et al. (Chapter 17) assess the future potential for using forests to sequester C:

> *"...forests will continue to be critical to the global C balance, and will be used to sequester C emissions."*

They recognise however, that it is unrealistic to expect forests to be managed solely for this purpose; other forest management objectives must be considered, such as sustainability, timber and fuel production, traditional uses, protection, recreation, etc. Three main management strategies will assist with C sequestration: (1) *conservation management*, i.e., conserving C pools by controlling deforestation and applying protection; (2) *storage management*, e.g., forestation to increase C storage densities in vegetation and soil (limited largely by economic constraints); and (3) *substitution management*, i.e., increasing the transfer of C to wood products and fuel, to reduce fossil fuel use. It is estimated that such options have the potential to conserve and sequester ~90 Pg C over the next 50 years—offsetting up to 15% of total fossil fuel emissions. Approximately 80% of this increased storage could potentially occur in the tropics, in addition to any net gains or losses resulting from projected changes in existing cover and land use.

Dealing with the possible effects of global change

Houghton observes that climate change is likely to occur, and that there appears to be an inverse relationship between the residual C flux (i.e., the "inadvertent" natural net C flux for terrestrial ecosystems inferred from modelling) and temperature. The implication is that with increasing temperature, ecosystem uptakes of CO_2 will be reduced, resulting in a gradual shift from net sinks to net sources. The results of Shvidenko et al., and Kurz and Apps, indicate this may already be occurring in parts of the boreal zone. The conclusion is that the rate of warming must be reduced if we are to reduce C emissions from terrestrial ecosystems.

Brown et al. (Chapter 17) draw attention to the fact that variations in human pressures cause changes in forest biomass C densities. The factors include the intensity of local human pressure, the accessibility of the land, and the relative roles of natural and anthropogenic disturbances. Current projections of the effects of changes in land use on C emissions in tropical regions assume no change in climate or demography. If climate warming occurs, the tropical zone is expected to be substantially reduced due to increased pressure for agriculture as well as a harsher climate. In temperate and boreal zones, the area of forest land area may be constant or even increase, but the outcome under a warmer climate is still unclear.

The need for standardised procedures for C budget accounting

In Chapter 10, Houghton draws attention to the importance of the C accounting methodology used to assess regional contributions to the global estimate. For example, wood products may be regarded as a C sink or as a source, depending on definition. The presence of the global C imbalance (missing sink) implies the net effect that modified lands are generally sources, while undisturbed ecosystems are generally sinks. He suggests that it is important to distinguish forests regrowing following harvest from those regrowing following natural disturbances. Only the former are included in his analysis of land-use changes affecting global C emissions.

Kurz and Apps (Chapter 14), on the other hand, consider the influences of changes in natural disturbance regimes which affect forest age-class structure and hence C dynamics—particularly in the boreal regions in addition to the effects of land-use change. Shvidenko et al. (Chapter 12) recognise many problems and contradictions in the data for boreal Asia. Because of the large extent of these forests, these data problems limit the accuracy with which the contributions of the boreal zone can be established.

In their summary, Brown et al. (Chapter 17) state that the best source of objective data for estimation of biomass C densities is forest inventories, as compared to data obtained from ecological studies, because the former are generally more comprehensive with respect to variations in stand age and site quality while the latter tend to be biased toward "mature" stands on productive sites. Improvements in estimates of regional C fluxes are possible with better spatial data, particularly when biomass-C density data can be coupled with geographic information systems (GIS) and remote sensing information. In general, the contributions of belowground biomass and other forest components are poorly known. More effort is needed in compiling comprehensive data bases containing both forest inventory data and supporting ecological information.

The methodology adopted for C accounting may be selected on the basis of available data. In the tropics, fires, insect, disease and logging are all generally followed by regrowth; hence net changes in C storage are small (when spatially averaged) relative to those resulting from permanent conversion. Outside the tropics however, net changes in C are determined mainly by the balance of disturbances (both natural and anthropogenic) and forest regrowth. Most countries in the boreal and temperate zones can provide inventory data and at least some historical disturbance records needed for reconstructing past exchanges. Galinski and Witowski (Chapter 15) in particular, document the steps needed to record forest inventory data, and the changes therein due to harvesting, to obtain estimates of C losses due to harvesting.

Brown et al. recognise three terms with different implications for monitoring changes in area and the related impacts on C storage and flux. These are Land Cover (LC), Land Use (LU) and Land Management (LM). They then outline the need for a system to track net national C gains and losses due to past, present and near future projected changes in LC/LU/LM, as well as natural disturbances, climate and atmospheric composition. This monitoring system must account for the effects of growth, respiration and decomposition over large areas as well as episodic disturbances.

10 Land-use change and terrestrial carbon: the temporal record

Richard A Houghton

The Woods Hole Research Center
P.O. Box 296, Woods Hole, MA 02543 USA

INTRODUCTION

Understanding changes in the terrestrial pools of carbon (C) over the last two centuries helps identify management strategies that might be used to enhance the storage of C on land in the future. Three general types of management can be recognised. The first two: policies affecting the area of forests; and policies affecting the amount of C per unit area of forest, including the amount stored in forest products; are directly related to the management of land, particularly forests. The third type of management pertains to maintaining the climate of the Earth so as, again, to conserve and enhance both the area of forests and the amount of C stored per unit area of forest. Although the latter type of management is not related to land use or land-use change, *per se*, knowing the contribution of land-use change to the global C cycle over the last two centuries helps identify the critical role that forests may play in the future.

More specifically, the current imbalance in the global C budget, defined in part by the net emissions of C from changes in land use, is thought to be explained by an accumulation of C in terrestrial ecosystems, primarily forests. The mechanisms responsible are not known (indeed, the increased storage has not been measured), and those mechanisms with the potential to have been important in the past, yield wildly different futures under similar projections of global change. If increasing concentrations of CO_2 have been responsible for increasing the storage of C on land, for example, the storage might be expected to continue and even increase in the future. If small changes in temperature and precipitation have been important, future changes in climate could either continue or reverse the accumulation of C. However, whatever mechanisms may have stored C in the past, their future capacity is being eroded by changes in land use, that is, the global loss of forests. Management strategies that enhance the storage of C on land should involve not only policies of land use but policies affecting the rate and extent of climatic change.

APPROACH

The approach used here to estimate long-term changes in terrestrial C as a result of changes in land use has been described previously (Houghton et al. 1983, 1987, 1991a, b; Houghton and Hackler 1994) and is only briefly summarised here. Annual rates of land-use change (expansion or abandonment of croplands, pastures, and shifting cultivation; harvest of wood; establishment of new forests) are used in a model to start the progression of a specified area of land along a temporal gradient of C. The gradient depends on the type of ecosystem and

NATO ASI Series, Vol. I 40
Forest Ecosystems, Forest Management and the Global Carbon Cycle
Edited by Michael J. Apps and David T. Price

the type of management or land use. Cleared forests lose C as dead plant material and soil organic matter decay; growing forests accumulate C in live vegetation and soil. With the initiation of a change in land use, portions of the live vegetation are assigned in the model to slash (roots, stumps, and other material left on site at the time of clearing), that decays exponentially, and to products that last an average of one year (fuelwood), 10 years (pulp and paper products), 100 years (timber products), or 1000 years (elemental C or charcoal formed with burning). Rates of decay and regrowth vary with the type of land use, the geographical region (9 regions globally), and the type of ecosystem (3 to 8 types per region).

Data on land-use change since 1850 have been obtained from agricultural and forestry statistics, and a broad range of studies that include anthropology, remote sensing, and historical geography. Rates of deforestation and biomass of tropical forests for the last two decades were obtained from the FAO Tropical Forest Assessments (FAO/UNEP 1981; FAO 1993). Data on C stocks and rates of decay and growth were generally obtained from the ecological and forestry literature. Specific references for different regions are given in the analyses cited above. Detailed data on rates of land-use change, average C stocks, and annual fluxes of C are available elsewhere (Houghton and Hackler, in press).

The mathematical model used to calculate annual changes in C is conceptually simple but provides the potential for an enormous volume of output. Besides keeping track of the areas and C contents of specific types of ecosystems under different types of land use within each region, the model also tracks individual age classes and different compartments of an ecosystem, such as live vegetation, slash, products, and soils. The C content of these compartments, again, is determined for each combination of land use, ecosystem, and region. Gross and net fluxes of C from changes in land use can be calculated for individual regions or types of ecosystems. The calculated fluxes are only those resulting from use of land. Ecosystems that are not cleared, cultivated, logged, planted, or otherwise deliberately modified, were ignored in this analysis.

THE PAST

Humans have modified the landscape and changed the amount of C stored in forests since they were capable of managing fire (Perlin 1989). Collectively the modifications have reduced the total amount of C held in terrestrial ecosystems globally, but before the last few centuries they were probably not important in affecting atmospheric concentrations of CO_2. Indeed, net annual emissions as large as 1.7 Pg C yr^{-1} (see below) might be insignificant today if there were not simultaneously an additional source of 6 Pg C yr^{-1} from combustion of fossil fuels. Even a long-term release of 1 to 2 Pg C yr^{-1} might not raise atmospheric concentrations appreciably because the rate is slow enough for the oceans to absorb it. The build-up of CO_2 over the last century has resulted from rates of emissions that are high relative to the rate of oceanic mixing and, hence, uptake. Thus, changes in terrestrial C stocks probably did not become important for atmospheric CO_2 until the industrial revolution.

1850 to 1980

Rates of land-use change

Changes in land use responsible for the greatest per ha change in terrestrial C are those changes that involve the transformation of forests to cleared land, such as croplands or pastures, and, conversely, those that involve the establishment of forests on cleared land.

Figure 10.1 shows the net annual rates of transformation for the world's boreal, temperate, and tropical forests, and all other ecosystems combined.

Over the 140-year period between 1850 and 1990, most of the lands converted to agriculture were grasslands of the temperate zone. Clearing of forests accounted for only about 27% of the net increase in agricultural area. The proportion taken from forests increased over the period, however. Between 1850 and 1925 the net rate at which non-forests were converted to agriculture was approximately equal to the rate at which forests of all types were lost. From 1925 to 1965, rates of non-forest conversion were about equal to rates of tropical deforestation. After 1965, clearing of non-forest ecosystems decreased, while clearing of tropical forests increased (Figure 10.1).

Throughout the period, net deforestation in the tropics (508 Mha) exceeded net deforestation in the temperate zone (91 Mha). Clearing of boreal forests for agriculture was always minor (4 Mha total). In contrast, about 1633 Mha of grasslands and desert scrub were converted to croplands or pastures.

The total increase in agricultural lands (2228 Mha) includes the net increase in cultivated land and the net conversion of forest to pasture. The increase is difficult to compare directly with agricultural statistics. Adding it to the 289 Mha already in farmlands in 1850 (Table 10.1) yields a total of 2517 Mha, which is greater than the FAO (1991) estimate of cropland area (1444 Mha) and lower than their estimate of grazing areas (3402 Mha). The discrepancy results from the fact that conversion of natural grasslands to pastures was ignored in this analysis, and a large fraction of the lands grazed today were derived from natural grasslands. The conversion of natural grasslands to pastures was ignored here because introducing cattle to grasslands or replacing wild grazers with domesticated ones is thought to have little effect on C stocks.

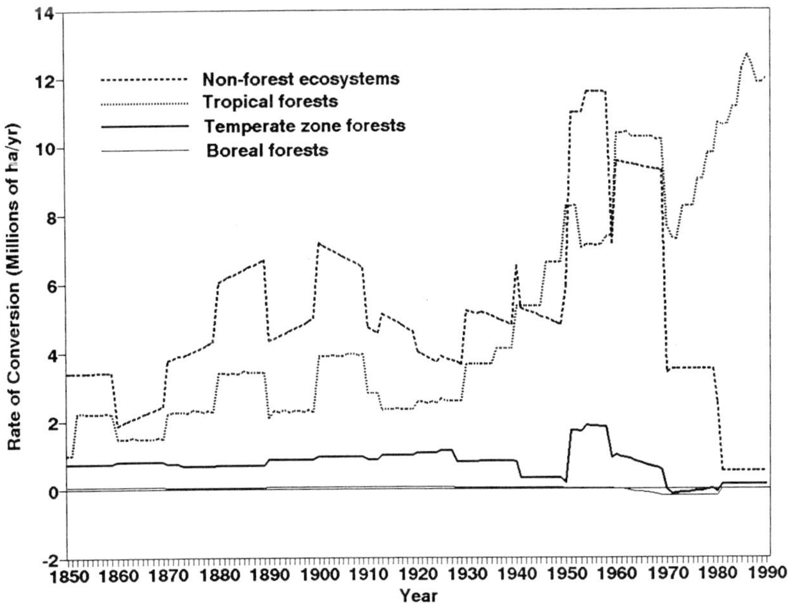

Figure 10.1. Annual rates of conversion of boreal, temperate, and tropical forests and other natural ecosystems to agricultural lands (conversion of grasslands to pastures not included).

Between 1850 and 1980 about 1069 Mha of forests were logged, an area about 77% greater than the area of forest converted to agriculture. In contrast to the geographical distribution of deforestation over this period, most logging occurred in forests of the temperate and boreal zones (Figure 10.2).

Total stocks of carbon (and changes) in live vegetation and soils

Forests in 1980 are estimated to have held about 87% of the 583 Pg C in living terrestrial vegetation and about 40% of the 1440 Pg C in the organic C in soils (Table 10.1). The greatest changes in total C storage between 1850 and 1980 occurred in forests, especially tropical forests. About 150 Pg C (including vegetation and soils) were lost from forests, 66% from tropical forests. Another 150 Pg C were lost from temperate grasslands, 88% from soils. These estimates of loss are somewhat misleading because they include both the C released to the atmosphere and the apparent losses of C associated with the transfer of land from natural ecosystems to agricultural ecosystems. Agricultural ecosystems *gained* C as a result of their increase in area. Taking into account this transfer of C from one ecosystem to another, net losses of C from live vegetation (90 Pg C) and soils (30 Pg C) totalled about 120 Pg C (Table 10.1).

Stocks of carbon in slash (dead vegetation) and products

Some of the C lost from live vegetation was not released to the atmosphere but accumulated in slash and wood products. The total amount of slash and wood products increased by about 25 Pg C between 1850 and 1980 (Figure 10.3). The increase in slash was about 10 Pg C (Figure 10.3a), and the increase in products was about 15 Pg C (Figure 10.3b). In contrast to the net increase, total additions to slash and wood products were 180 and 68 Pg C, respectively, over this 140-year period.

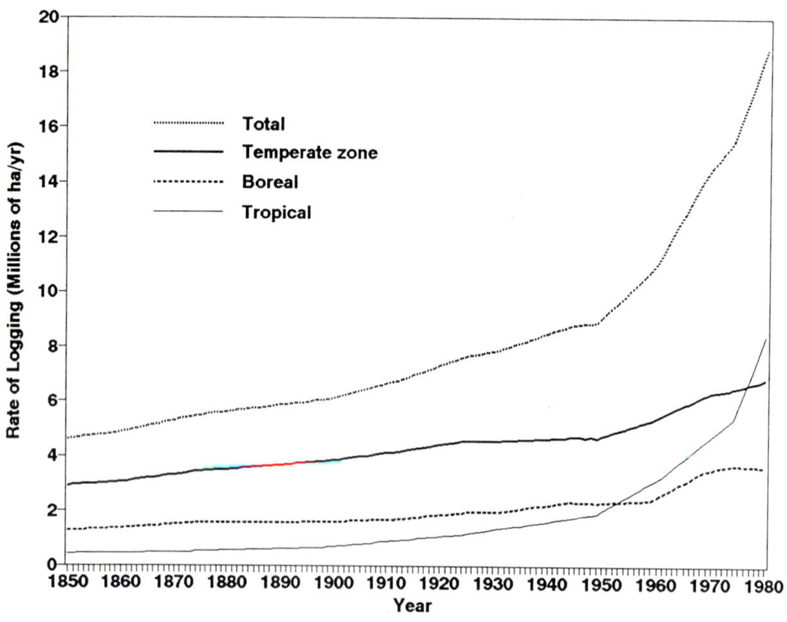

Figure 10.2. Annual rates of logging in boreal and temperate zone forests.

Table 10.1. Areas and carbon (C) content of major terrestrial ecosystems in 1850 and 1980, and changes over the 130-year period.

	AREA (Mha)			CARBON CONTENT (Pg C): VEGETATION			CARBON CONTENT (Pg C): SOIL		
	1850	1980	Change	1850	1980	Change	1850	1980	Change
Tropical evergreen forest	608	540	-68	116	100	-15	67	59	-8
Tropical seasonal forest	1647	1345	-302	208	173	-36	149	124	-25
Tropical open forest/woodland	420	282	-138	23	5	-8	28	19	-9
Subtotal tropical forests	2675	2167	-508	347	288	-59	245	203	-42
Temperate evergreen forest	558	564	6	83	71	-12	73	67	-6
Temperate deciduous forest	461	398	-63	55	41	-14	61	51	-9
Temperate woodland	564	530	-34	15	14	-1	39	37	-2
Subtotal temperate forests	1583	1492	-91	153	127	-26	173	155	-17
Boreal forest	1172	1167	-4	102	96	-6	240	237	-3
Subtotal all forests and woodlands	5430	4827	-603	601	510	-91	657	595	-62
Desert scrub	1120	849	-271	3	2	-1	65	49	-15
Tropical fallows (shifting cultivation)	199	229	30	8	8	1	17	19	2
Tropical grassland and pasture	459	545	86	8	9	1	20	28	8
Temperate grassland and pasture	2150	2010	-140	34	26	-8	310	270	-41
Cultivated, temperate zone	261	762	501	2	5	2	49	105	56
Cultivated, tropical zone	28	417	389	0	4	4	2	24	22
Other[a]	4080	4088	8	18	18	0	351	351	0
TOTAL	13 727	13 727	0	673	583	-90	1471	1440	-31

a Includes tundra, alpine meadow, marsh, rock, ice, and sand; from Whittaker and Likens (1973) and Schlesinger (1984).

Net flux of carbon

The sum of changes in live and dead biomass, soil organic C, and wood products represents the net exchange of C with the atmosphere (Figure 10.4). According to these analyses the clearing of lands for agriculture and the harvest and regrowth of forests caused a net release to the atmosphere of about 97 Pg C over the 130-year period. The annual net release increased from about 0.3 Pg C yr^{-1} in 1850 to about 1.5 Pg C yr^{-1} in 1980.

Deforestation in the tropics released the largest amount of non-fossil C. Between 1850 and 1980 the net release from tropical forests was 52 Pg C. The annual net flux from the tropics began to exceed all other non-fossil sources combined after 1950. Temperate zone forests, including their clearing, logging, and regrowth, released a net of 27 Pg C. This net loss of C from temperate zone forests includes the uptake of C resulting from the abandonment of agricultural land. Boreal forests have had the smallest net loss (4 Pg C), and human activities in ecosystems other than forests have caused a net release of about 14 Pg C.

The expansion of croplands has been responsible for the largest release of C (63 Pg), followed by logging and regrowth of forests (23 Pg), and the conversion of forests to pastures (10 Pg) (Figure 10.5). Again, these estimates of flux are net fluxes. They include the uptake of C in forests regrowing after logging and after abandonment of agriculture.

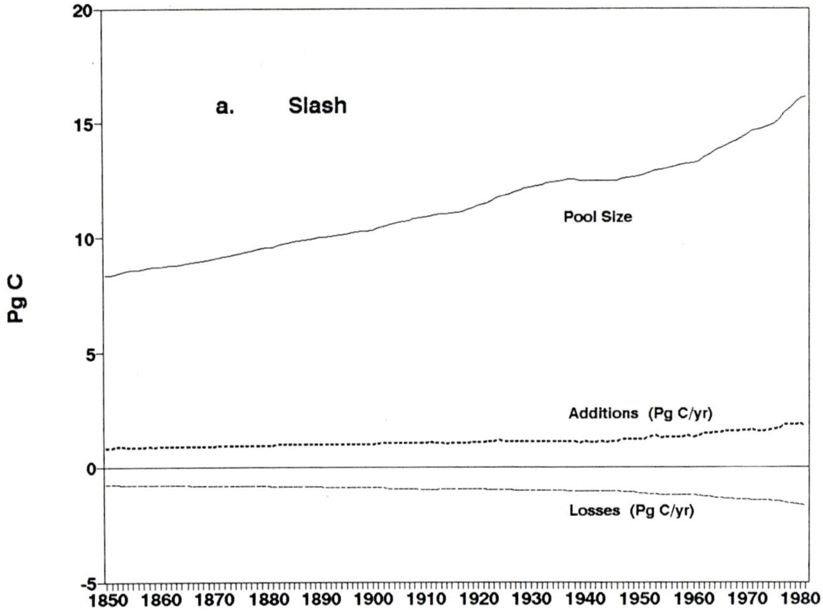

Figure 10.3a. Carbon in slash (stumps, roots, and logging debris), and annual additions to, and losses from, this pool.

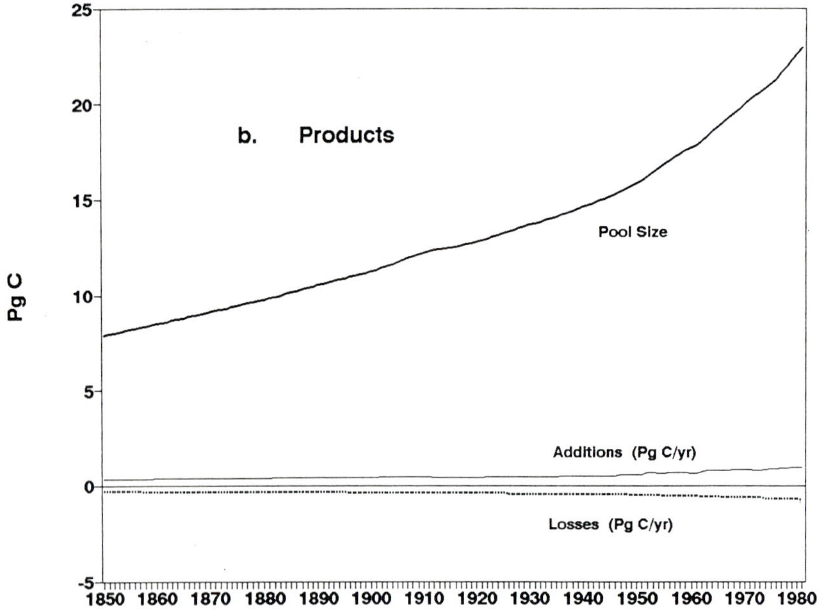

Figure 10.3b. Amount of carbon in wood products, and annual additions to, and losses from, this pool.

1980 to 1990

The net flux of C from the tropics between 1980 and 1990 resulting from changes in land use, is estimated here to have averaged 1.6 ±0.5 Pg C yr^{-1}. Estimates for the individual regions are 0.6 ±0.3, 0.7 ±0.3, and 0.3 ±0.2 Pg C yr^{-1} for Latin America, tropical Asia, and tropical Africa, respectively. Errors are estimated to be within approximately ±50%. Outside the tropics, the net flux is estimated to have ranged between a net source of 0.1 Pg C yr^{-1} (Houghton 1993) and a net sink of 0.5 to 0.9 Pg C yr^{-1} (Dixon et al. 1994, based on recent estimates for individual regions by Kauppi et al. 1992; Birdsey et al. 1993; Apps and Kurz 1994; and others), for an overall sink of 0.5 ±0.5 Pg C yr^{-1}. The error associated with the flux from temperate and boreal zones appears as large as it is for the tropics. On the other hand, the high estimates of accumulation outside the tropics are based on changes not necessarily related to land-use change (see section on Uncertainties, below). The flux associated with land-use change, alone, may be close to zero (Houghton et al. 1987; Melillo et al. 1988; Houghton 1993).

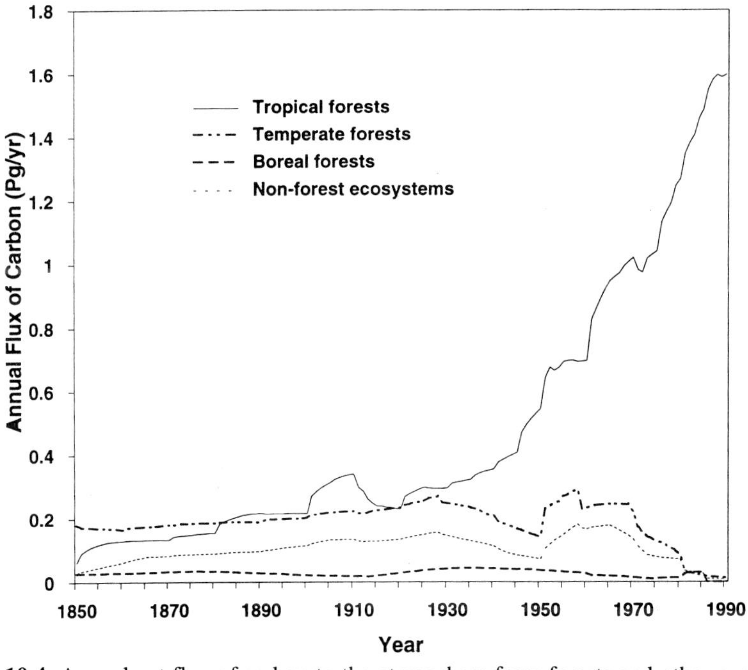

Figure 10.4. Annual net flux of carbon to the atmosphere from forests and other ecosystems as a result of changes in land use (logging, clearing, and abandonment) and decay of slash and wood products.

The estimates of net flux reported here are comprised of gross rates of uptake and release that are more than twice the net flux (Tables 10.2 and 10.3). Gross rates in the tropics are underestimated here because logging and shifting cultivation were not included in the analysis of tropical Africa (Table 10.2). Nevertheless, regrowing vegetation accounted for about 80% of the gross uptake, globally, and soils, for about 20%. The processes contributing to the gross

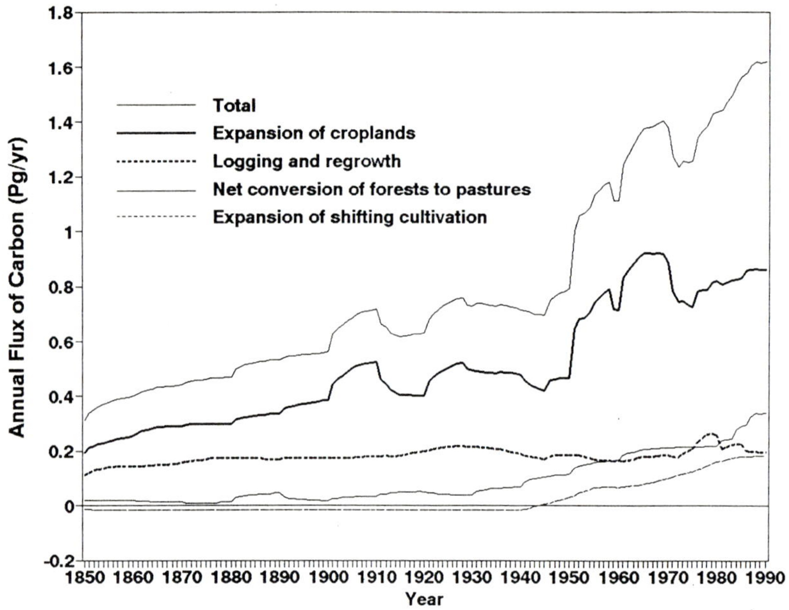

Figure 10.5. Annual flux of carbon to the atmosphere from different types of land-use change.

release were decay of slash (42%), burning associated with clearing for agriculture, including some use of fuelwood (22%), oxidation of wood products (17%), and oxidation of organic C in soils as a result of cultivation (19%). Emissions of C from fuelwood not obtained in the process of clearing land were assumed to be balanced by uptake of C in recovering forests. These emissions from burning include only those resulting from land-use change; they are less than gross emissions associated with total biomass burning (Crutzen and Andreae 1990).

THE FUTURE

The results of these analyses suggest at least three broad approaches to sequestering C in forests: first, by increasing the area of forests; second, by reducing the production of logging residues and increasing the fraction of products that are placed in long-term storage; and, third, by managing the Earth's climate such that rates of warming do not result in a net loss of the C stored in forest ecosystems.

Area of forests

Between 1850 and 1980, the area of forests and woodlands decreased by 11% (603 Mha), world-wide (Table 10.1). About 84% of this loss was in the tropics and about 15% in the temperate zone. The changes in forests and other ecosystems caused a net release of about 97 Pg C into the atmosphere and increased the amount of C in dead organic matter (slash) and wood products by about 25 Pg C. Clearly, the most effective way to sequester C on land is to

Table 10.2. Average annual rates of accumulation (–) and release (+) of carbon (C) from changes in the major forested regions of the tropics and the global total for the years 1975–1980 (Pg C yr^{-1}).

	TROPICAL ASIA	LATIN AMERICA	TROPICAL AFRICA[a]	TOTAL[a] TROPICS	GLOBAL TOTAL
A. Gross accumulations of C in regrowing forests					
Vegetation	-0.65	-0.54	0	-1.19	-2.10
Soil	-0.15	-0.09	0	-0.24	-0.42
Total accumulation	-0.80	-0.64	0	-1.44	-2.52
B. Gross releases of C					
Decay of slash	0.57	0.41	0.05	1.04	1.62
Burning[b]	0.38	0.26	0.07	0.70	0.86
Oxidation of wood products	0.12	0.29	0.04	0.44	0.65
Oxidation of soils	0.22	0.16	0.06	0.44	0.75
Total of gross releases	1.28	1.13	0.21	2.62	3.89
C. Net flux of C					
(A+B)	0.48	0.49	0.21	1.18	1.36

a Because neither logging nor shifting cultivation were included in Africa, there is no accumulation of C in growing forests, and estimates are not of gross flux.

b Burning associated with logging, clearing for agriculture, shifting cultivation, and fuelwood. Emissions do not include those resulting from wild fires.

reverse this trend in forest loss. If the world's forest area could be expanded, an amount of C similar to that released over the last centuries might be withdrawn again and accumulated on land.

Such an expansion may not be as unrealistic as it seems. The surplus of agricultural production in Europe, for example, is leading to a reduction in farmlands and an increase in forests. An estimated 50 to 100 Mha may be added to forest areas there. Similarly, in the U.S., as many as 100 Mha of marginal agricultural lands might be converted to forests. Large areas are being reforested in New Zealand, as well (Maclaren, this volume).

In the tropics, an expansion of forest area seems more problematic. Growing populations require an increased supply of food, and some of the increase in food production has always come from increasing the area under production. Between 1980 and 1985 the areas of croplands and pasture increased at average annual rates of 2.7 and 1.8 Mha, respectively (FAO 1991). The area of forests cleared during this period, however, was more than twice as large, implying that the addition of new land to the agricultural base was, in part, offset by the loss of old agricultural land (Houghton 1994). The abandoned lands are probably lands that no longer support production because of nutrient loss, erosion, salinization, or other form of impoverishment. These abandoned lands seem not to be returning to forest (FAO 1991).

An alternative interpretation is that half the tropical deforestation was for shifting cultivation, which is not included by the FAO (1991) as either cropland, permanent pasture, or forest. But shifting cultivation, itself, is becoming less sustainable. Periods of fallow have been shortening, with the result that the cropping cycles are less productive (Myers 1980). The end result may be the same as impoverishment, with large areas of fallow unable or only slowly able to recover to forest. Whatever the actual pathway, the increasing areas of unproductive lands without forest cover seem likely to be a consequence of unsustainable agriculture. Deforestation can only be reduced in the tropics by making agriculture sustainable.

The positive aspect of these unproductive lands is that they might be managed to accumulate C. Estimates of the areas available and suitable for reforestation vary, as do estimates of the costs involved to establish forests. Several hundred Mha may be available in

Table 10.3. Average annual rates of accumulation (–) and release (+) of carbon (C) from changes in the major forested regions of the northern temperate and boreal zones for the years 1975–1980 (Pg C yr^{-1}).

	North America	Europe	Former Soviet Union	Subtotal	Other[a]	Total temperate and boreal
A. Gross accumulations of C in regrowing forests						
Vegetation	-0.28	-0.25	-0.14	-0.67	-0.24	-0.91
Soil	-0.05	-0.05	-0.03	-0.13	-0.04	-0.18
Total accumulation	-0.33	-0.30	-0.17	-0.80	-0.28	-1.08
B. Gross releases of C						
Decay of slash	0.15	0.17	0.07	0.39	0.20	0.59
Burning[b]	0.02	0.02	0.05	0.08	0.08	0.16
Oxidation of wood products	0.09	0.05	0.03	0.17	0.04	0.21
Oxidation of soils	0.06	0.05	0.08	0.19	0.12	0.32
Total of gross releases	0.32	0.29	0.23	0.84	0.43	1.26
C. Net flux of C (A+B)	-0.02	-0.01	0.06	0.04	0.14	0.18

a Includes China, Japan, Korea, and countries in the middle east and northern Africa (north of the Sahara).
b Burning associated with logging, clearing for agriculture, and consumption of fuelwood. Emissions do not include those resulting from wildfires.

the tropics and might be used to accumulate 100–200 Pg C over the course of decades (Houghton et al. 1993). But, while the storage of C in forests is permanent, the annual accumulation of C diminishes as forests mature. The only long-term solution to reducing global emissions of C with forest management is replacement of fossil fuels with sustainably produced wood fuels. Forests should not be thought of as major solution for stabilising the concentration of CO_2 in the atmosphere, however, even if the land area available for wood fuel production were adequate to supply the world's current demand for energy. The predominant solution must lie with reducing the dominant source of CO_2, that is, combustion of fossil fuels. The reasons are both scientific and political. Scientific reasons include the numerous services provided by forests that might not be provided by lands managed solely for wood production or C storage. Political reasons are related to the north–south differences in energy consumption and development. Developing countries, largely in the tropics, are unlikely to want to manage their forests so as to accommodate the C emissions of the north. Lands suitable for C storage will also be suitable for production of food, and the competition is likely to be intense.

Management of forests and wood products

A second strategy for accumulating C on land is to store more C in long-lived products (buildings and furniture, for example). Increased storage might be achieved in at least two ways. First, wastage can be reduced. Wastage occurs at all steps, from the mortality of trees damaged during selective logging, to wastage of those parts of the tree not used in roundwood, to wastage at sites of construction. According to the analyses reported here, the amount of C released annually from decay of slash is about 1.6 Pg C, world-wide (Figure 10.3a, Table 10.2). It is the largest component of the current net flux and might be reduced significantly through management. The result would be a net accumulation of C on land as long as the average lifetime of C in products was longer than that of the slash.

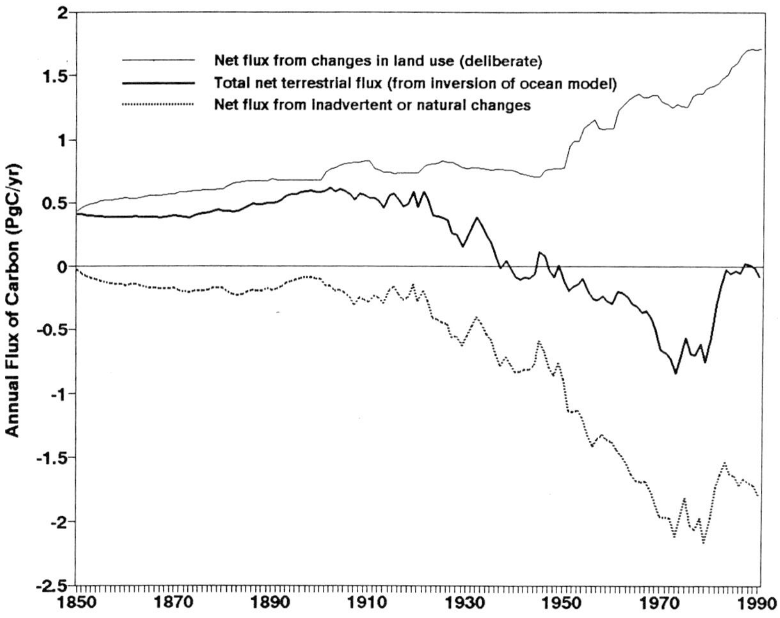

Figure 10.6. The total net terrestrial flux of carbon (determined from inverse calculations; Sarmiento et al. 1992), the net flux from changes in land use, and the difference between the two, assumed here to result from inadvertent or natural changes in terrestrial ecosystems (from Houghton 1995).

A second approach for storing more C on land is to use more wood in the construction of buildings and other long-lasting structures. Wood was the dominant building material until the middle of this century. It is still used extensively in the U.S., but its expense has reduced its use elsewhere. Financial mechanisms might be used to restore its use. However, the storage of wood in long-lasting structures is not permanent. As the total amount of C in products increases, annual losses from fire and decay also increase (Figure 10.3b). With time, a new balance is reached, where additions approximate losses. Long-lasting products are similar to mature forests in this respect.

A third approach to accumulating C on land is to enhance the storage of C in secondary forests. Large areas of secondary forests might be protected from fire or logging and thereby accumulate significant amounts of C. But, while protection may enhance the storage of C temporarily, the natural fire cycle or disturbance cycle limits the storage of C in vegetation and soil, and it seems unlikely that management practices will eliminate disturbances. On the contrary, projected changes in climate will probably decrease the storage of C rather than enhance it.

Inadvertent changes to carbon held in forests

In addition to the changes in terrestrial C that result from deliberate changes in land use, the amount of C stored in vegetation and soils is affected also by natural disturbances and changes in climate and, perhaps, atmospheric concentrations of CO_2. In contrast to changes in land use, these other changes are inadvertent and difficult to measure globally. Evidence for

Table 10.4. Comparison of the 1980 forest areas and carbon (C) stocks used in this study (upper values) with those reported by Dixon et al. (1994) (lower values).

		TOTAL C (Pg C)		AVERAGE C STOCKS (Mg C ha^{-1})	
	AREA (Mha)	VEGETATION	SOIL	VEGETATION	SOIL
Tropical forests	2167	288	203	133	94
	1755	212	216	121	123
Temperate forests	1492	127	155	85	104
	1038	59	100	57	96
Boreal forests	1167	96	237	82	203
	1372	88	471	64	343
Total forests	4827	510	595	106	123
	4165	359	787	86	189

such changes in terrestrial ecosystems comes from geochemical analyses of the global C cycle. According to inverse methods using historic CO_2 concentrations and models of oceanic uptake, the net terrestrial flux of C has decreased since about 1920 (Sarmiento et al. 1992) (Figure 10.6). The magnitude and temporal pattern of this net flux are clearly different from those of the flux attributable to changes in land use. If both the land-use analyses and the inverse modelling analyses are correct, the difference between the total net terrestrial flux (from inversion) and the flux from changes in land use, alone, defines an inadvertent or natural flux (Houghton 1995) (Figure 10.6).

Over the long term, this inadvertent flux seems to have accumulated C, presumably on land, in proportion to industrial activity. Rising CO_2 concentrations, rising temperatures, or increasing nitrogen mobilisation, are suggested as possible causes for the accumulation. Over the decades since 1940, more of the variation in the residual flux is explained by variations in temperature than by CO_2 concentrations, suggesting an inverse relationship between temperature and C storage (Houghton 1995). The results are equivocal. They demonstrate the strong possibility, however, that a global warming will increase the emissions of C from terrestrial ecosystems (Woodwell et al. 1995), and that management strategies aimed at sequestering C will not succeed unless the strategies include limiting the rate and extent of the warming.

DISCUSSION

Uncertainties

The total amount of C estimated to be in the world's terrestrial vegetation (583 Pg) (Table 10.1) is somewhat higher than estimates of Ajtay et al. (1979) and Olson et al. (1983), 560 Pg and 562 Pg, respectively. The estimate for soils (1440 Pg C) is consistent with recent estimates of global soil C stocks (Eswaran et al. 1993; Post et al. 1982; Schlesinger 1984, 1986; Zinke et al. 1986).

The estimates for forest areas and C stocks are different, however, from estimates summarised recently by Dixon et al. (1994) (Table 10.4). The total area of forests reported here is about 15% higher. The largest difference is for tropical and temperate zone forests, which occupy areas 23% and 44% greater, respectively, according to the estimates given here. Estimates of average per ha biomass are 10% higher for the tropics; estimates for tropical soils average 24% less here. The other major discrepancies involve the C in temperate forest biomass (more than 50% greater per ha here) and boreal forest soils (about 40% lower here).

Only some of the differences in estimates would modify the estimates of change or the estimates of C flux reported here. Data can be divided into three types: areas or rates of land-use change, initial stocks of C, and changes in C stocks associated with land-use change.

Areas or rates of land-use change

Changes in the area of forests are known better than total areas. Total areas are dependent on definition, and areas of forest are generally not well documented historically. *Changes* in agricultural area, on the other hand, are reasonably well defined and recorded. Furthermore, if the areas of forest shown in Table 10.1 are overestimates, correcting them will not affect the calculated flux. Adjustments will modify the areas estimated to have existed in 1850 and 1980 and the total stocks of C at both dates, but not the change or the net flux of C over the interval. The emphasis of this analysis has been on change.

Initial stocks of carbon and changes in stocks without deforestation

If the estimates of average per ha biomass and soil C used here are higher than the actual stocks of C, one might expect the calculated flux to be overestimated. Forests with low stocks release less C when cleared than forests with high stocks of C. But an overestimate is obtained only if the stocks have always been low (or, for these calculations, low since 1850). If not, if the stocks were high in 1850 and have been declining over the period since 1850, then the decline itself represents a loss of C in addition to that lost from deforestation. Evidence for such a decline comes from throughout the tropics, especially where selective logging is important, as in Asia (Brown et al. 1994; Flint and Richards 1994; Houghton and Hackler 1994). Degradation of C stocks results in a lower release of C from subsequent deforestation (calculated and actual), but degradation itself releases C as a result of the burning and decay associated with use of the wood removed. As much as 40–50% of the total net flux calculated for tropical Asia is calculated to have resulted from a loss of biomass not associated with deforestation (Flint and Richards 1994; Houghton and Hackler 1994). Knowledge of C stocks in the past is important, yet difficult, to obtain. The distribution of biomass today is probably determined more from human activities than by climatic and edaphic conditions. Furthermore, stocks of C keep changing as a result of repeated disturbance and recovery. Much of the discussion concerning estimates of forest biomass may reflect the uncertainties introduced by human activities (Brown et al. 1989; Brown and Lugo 1992; Fearnside 1992; Houghton and Hackler 1994).

Changes in carbon stocks associated with land-use change

Changes in C per ha are well defined in some cases and poorly known in others. Those changes associated with clearing forests or with reforestation are generally well known. They involve large changes per ha and are thought to account for most of the C lost to the atmosphere. On the other hand, changes in C stocks within forests, as a result of degradation or enhanced growth, are poorly documented. In the absence of information about land-use change, ecosystems were assumed here to be in balance with respect to C. But some types of land-use change are under-reported or not reported at all. These types include some domestic and illegal logging, shifting cultivation, fuelwood extraction, grazing, and deliberate burning. These chronic uses of forests may reduce C stocks, but the rates are difficult to quantify. Forests believed to be degraded today may have been degraded slowly over the last century, rapidly over the last decade, or may already have been degraded (and the C lost) prior to 1850. In contrast to changes in forest area which are known better than the areas themselves, the changes in C stocks are confounded with initial conditions, and neither is known well.

Current estimates of flux

For the tropics, current estimates of the flux of C from changes in land use range between 0.6 and 1.6 Pg C yr^{-1} (Hall and Uhlig 1991) and between 1.1 and 2.1 Pg C yr^{-1} (this analysis). The difference between the high and low estimates of Hall and Uhlig (1992) and between their low estimates and the estimates from this study result largely from different estimates of biomass used. The different estimates of biomass, in turn, result from different assumptions about the extent of historic degradation. As discussed above, although the lower estimates of biomass may be good estimates of current biomass, they are inadequate for calculation of C emissions unless the calculations explicitly include degradation as well as deforestation (Houghton and Hackler 1994).

Outside the tropics, current estimates of flux range between a net release close to zero (Melillo et al. 1988; Houghton 1993) and a net accumulation of 0.74 ±0.19 Pg C yr^{-1} (Dixon et al. 1994). The difference seems to be due to a difference in methods. Those studies calculating a small net flux considered only changes in land use. Those analyses showing larger accumulations of C (for example, Kauppi et al. 1992; Birdsey et al. 1993; Kolchugina and Vinson 1993) considered not only changes in land use but changes in growing stocks reported by forest inventories. The latter include processes such as natural disturbance, CO_2 fertilisation, or climatically-driven change that may be independent of changes in land use. Thus, different estimates of flux, depending on their methods, may include different processes.

Conceptual issues

Deliberate vs. inadvertent changes in terrestrial ecosystems

It is useful to distinguish between changes in terrestrial C that result from changes in land use and those that result from other factors, such as environmental change. The former are more easily quantified because the areas affected are generally documented and the per ha changes in C are generally large. The latter may be spread over wide, undefined areas, and the changes are generally small relative to standing stocks. Distinguishing between the two allows them to be compared, or allows the inadvertent changes to be calculated by difference, as shown in Figure 10.6. Distinguishing between the two may help identify specific mechanisms and seems essential for a resolution of the global C imbalance and for determining how terrestrial ecosystems will respond to future changes in climate and atmospheric CO_2.

Terrestrial carbon sinks versus missing sinks

The difference between sinks and *missing* sinks of C has led to some confusion in the literature. Sinks can be defined as the net uptake or net accumulation of C. The world's oceans and growing forests are sinks of C. The missing sink refers to the imbalance in the terms of the global C equation: emissions of C from combustion of fossil fuels (5.4 ±0.5 Pg C yr^{-1} during the 1980s) and from changes in land use (1.6 ±0.5 Pg C yr^{-1}) are not balanced by the accumulations of C in the atmosphere and oceans (3.4 ±0.2 and 2.0 ±0.8 Pg C yr^{-1}, respectively). The imbalance (or missing C) is approximately of the same magnitude as, but of opposite sign to, the net release of C from changes in land use. The imbalance is thought to be explained by the accumulation of C on land in undisturbed or incorrectly understood disturbed ecosystems. Indirect evidence for a terrestrial sink comes from geochemical data and models (Siegenthaler and Oeschger 1987; Keeling et al. 1989; Tans et al. 1990; Sarmiento et al. 1992). The findings are not inconsistent with analyses of land-use change because the latter assume that ecosystems not deliberately modified by human activities (about 80% of the forest lands in this analysis) are in balance with respect to C. Thus, the net effect of modified lands may be to release C, while undisturbed ecosystems may be accumulating it.

Calculation of the net flux of C from changes in land use includes the sinks of C in forests recovering from logging and agricultural abandonment. This sink is more than 2.6 Pg C yr^{-1} in the tropics (Table 10.2) and about 1.3 Pg C yr^{-1} in forests of the temperate and boreal zones (Table 10.3). Thus, forest inventories showing increased growing stocks do not necessarily help explain the missing sink unless: (1) the sink calculated from inventories is greater than the gross sink calculated from land use; or (2) the sink calculated from inventories refers to forests not included in the analyses of land use. It should be possible, for example, to distinguish forests regrowing after harvest from those regrowing after natural disturbance. Both will be included in forest inventories; only the former will be included in analyses of land use. This distinction seems not to have been made with respect to fluxes of C, although it should be possible to do so in some regions (Kurz et al. 1995).

To date, sinks of C calculated from forest inventories seem not to be higher than gross sinks calculated from changes in land use (Houghton 1993). The average annual rate at which C must be accumulating on land to account for all of the missing sink (about 1.5 Pg C yr^{-1}), is 1.4 Mg C ha^{-1} yr^{-1} if all of it were accumulating in forests of the temperate zone, 0.65 Mg C ha^{-1} yr^{-1} if all of it were accumulating in forests of both temperate and boreal zones, and 0.34 Mg C ha^{-1} yr^{-1} if it were accumulating in all forests (boreal, temperate, and tropical) (areas from Table 10.4). By comparison, the average rates of C accumulation determined from forest inventories were 0.4–0.6 Mg C ha^{-1} yr^{-1} in European forests (Kauppi et al. 1992) and less in the forests of the former Soviet Union, Canada, and the U.S. (Table 10.5).

The estimates in Table 10.5 were obtained from ECE/FAO (1992). They are based on exploitable forests only, which account for only about half of the total forest area in Canada and the former USSR. The average net change per ha shown in the last column of Table 10.5 was calculated by subtracting fellings from the net increment in growth. These estimates are crude, but the difference between observed rates of accumulation (Kauppi et al. 1992; and Table 10.5) and those required to balance the global C cycle suggests that little of the missing sink is to be found in the trees of northern forests. There are few forest inventories in the tropics, and thus no way to ascertain whether rates of accumulation there are as great as in northern regions. In general, if the missing sink is to be found on land, it must be considerably larger than the sink calculated from forest inventories. The difference between the estimates of net uptake in the former Soviet Union (Kolchugina and Vinson (1993) compared to ECE/FAO (1992) in Table 10.5) require further analysis.

Table 10.5. Areas of exploitable forest in northern temperate and boreal zones and volumes of wood accumulated through growth and removed through logging (from ECE/FAO 1992).

	Exploitable forest area (Mha)	Fraction of total forest area (%)	Net[a] annual increment (Mg C ha^{-1} yr^{-1})	Fellings (Mg C ha^{-1} yr^{-1})	Net accumulation[b] (Mg C ha^{-1} yr^{-1})
Europe	133	89	1.3	0.9	0.4
Former USSR	414	55	0.5	0.4	0.1
Canada	112	45	0.6	0.4	0.2
United States	196	93	1.2	1.0	0.2

a *Net* refers here to growth and natural mortality. Wood volumes were assumed to have an average density of 0.6 Mg m^{-3} and an average C content of 50%.

b Net accumulation is net increment *minus* fellings.

Should products be considered sources or sinks of carbon?

There are at least two different ways to describe the role of wood products in the global C balance. Many studies refer to products as C sinks, while, for the approach used here, products can only be sources. Part of the difference is definition. The approach used in this paper defines sources and sinks with respect to the atmosphere. Wood products can only release C to the atmosphere. Live vegetation can only remove C from the atmosphere. If the total C in products (or slash) has increased over some interval, that increase represents an accumulation of C on land but not, by this definition, a C sink.

In contrast, those who refer to wood products as sinks use a different definition. However, one must be careful to distinguish between production of wood products (or additions to the pool) and net increments (or changes) in the total pool. An *increasing pool* of wood products can be logically defined as a sink for C. The *production* of wood products cannot be, because the production of wood products does not address the decay of wood products. A consistent method must consider either (1) net changes in the pool size of products or (2) both additions to *and* losses from that pool. Either approach (definition) is appropriate, but additions (or losses) alone should not be referred to as a sink (or a source).

ACKNOWLEDGEMENTS

The author acknowledges with pleasure the able assistance of Mr. Joseph Hackler with the analyses and graphics and the helpful suggestions of two anonymous reviewers. Research was supported by the U.S. Department of Energy, Office of Health and Environmental Research (Grant No. DE-FG02-90ER61079).

REFERENCES

Ajtay GL, Ketner P, Duvigneaud P (1979) Terrestrial primary production and phytomass. In: Bolin B, Degens ET, Kempe S, Ketner P (eds) The Global Carbon Cycle. Wiley, New York, pp 129–182.

Apps MJ, Kurz WA (1994) The role of Canadian forests in the global carbon budget. In: Kanninen M (ed) Carbon Balance of World's Forested Ecosystems: Towards a Global Assessment. Publ. Acad. Finland 3/93, Helsinki, pp 14–39.

Birdsey RA, Plantinga AJ, Heath LS (1993) Past and prospective carbon storage in United States forests. *Forest Ecol. Manage.* 58: 33–40.

Brown S, Lugo AE (1992) Aboveground biomass estimates for tropical moist forests of the Brazilian Amazon. *Interciencia* 17: 8–18.

Brown S, Gillespie AJR, Lugo AE (1989) Biomass estimation methods for tropical forests with applications to forest inventory data. *Forest Sci.* 35: 881–902.

Brown S, Iverson LR, Lugo AE (1994) Land use and biomass changes of forests in Peninsular Malaysia from 1972 to 1982: A GIS Approach. In: Dale VH (ed) Effects of Land Use Change on Atmospheric CO_2 Concentrations: South and Southeast Asia as a Case Study. Springer-Verlag, New York, pp 117–143.

Crutzen PJ, Andreae MO (1990) Biomass burning in the tropics: Impact on atmospheric chemistry and biogeochemical cycles. *Science* 250: 1669–1678.

Dixon RK, Brown S, Houghton RA, Solomon AM, Trexler MC, Wisniewski J (1994) Carbon pools and flux of global forest ecosystems. *Science* 263: 185–190.

ECE/FAO (1992) The Forest Resources of the Temperate Zones. Main Findings of the UN-ECE/FAO 1990 Forest Resource Assessment. ECE/TIM/60, United Nations, New York, 32 pp.

Eswaran H, Van Den Berg E, Reich PF (1993) Organic carbon in soils of the world. *Soil Sci. Soc. Amer. J.* 57: 192–195.

FAO (1990) (1991) Production Yearbook. FAO, Rome, 306 pp.

FAO (1993) Forest Resources Assessment 1990 Tropical Countries. FAO Forestry Paper No. 112 FAO, Rome, 61 pp.

FAO/UNEP (1981) Tropical Forest Resources Assessment Project. FAO, Rome, 1598 pp.

Fearnside PM (1992) Forest biomass in Brazilian Amazonia: comments on the estimate by Brown and Lugo. *Interciencia* 17: 19–27.

Flint EP, Richards JF (1994) Trends in carbon content of vegetation in south and southeast Asia associated with changes in land use. In: Dale VH (ed) Effects of Land Use Change on Atmospheric CO_2 Concentrations: South and Southeast Asia as a Case Study. Springer-Verlag, New York, pp 201–299.

Hall CAS, Uhlig J (1991) Refining estimates of carbon released from tropical land-use change. *Can. J. For. Res.* 21: 118–131.

Houghton RA (1993) Is carbon accumulating in the northern temperate zone? *Global Biogeochemical Cycles* 7: 611–617.

Houghton RA (1994) The worldwide extent of land-use change. *BioScience* 44: 305–313.

Houghton RA (1995) Effects of land-use change, surface temperature, and CO_2 concentration on terrestrial stores of carbon. In: Woodwell GM, Mackenzie FT (eds) Biotic Feedbacks in the Global Climatic System: Will the Warming Feed the Warming? Oxford University Press, New York, pp 333–350.

Houghton RA, Hackler JL (1994) The net flux of carbon from deforestation and degradation in south and southeast Asia. In: Dale VH (ed) Effects of Land Use Change on Atmospheric CO_2 Concentrations: South and Southeast Asia as a Case Study. Springer-Verlag, New York, pp 301–327.

Houghton RA, Hackler JL (1995) Continental Scale Estimates of the Biotic Carbon Flux from Land Cover Change: 1850–1980. ORNL/CDIAC-79, Oak Ridge National Laboratory, Oak Ridge, TN. (In press).

Houghton RA, Boone RD, Fruci JR, Hobbie JE, Melillo JM, Palm CA, Peterson BJ, Shaver GR, Woodwell GM, Moore B, Skole DL, Myers N (1987) The flux of carbon from terrestrial ecosystems to the atmosphere in 1980 due to changes in land use: geographic distribution of the global flux. *Tellus* B 39: 122–139.

Houghton RA, Hobbie JE, Melillo JM, Moore B, Peterson BJ, Shaver GR, Woodwell GM (1983) Changes in the carbon content of terrestrial biota and soils between 1860 and 1980: A net release of CO_2 to the atmosphere. *Ecol. Monogr.* 53: 235–262.

Houghton RA, Lefkowitz DS, Skole DL (1991) Changes in the landscape of Latin America between 1850 and 1980. I. A progressive loss of forests. *Forest Ecol. Manage.* 38: 143–172.

Houghton RA, Skole DL, Lefkowitz DS (1991) Changes in the landscape of Latin America between 1850 and 1980. II. A net release of CO_2 to the atmosphere. *Forest Ecol. Manage.* 38: 173–199.

Houghton RA, Unruh JD, Lefebvre PA (1993) Current land use in the tropics and its potential for sequestering carbon. *Global Biogeochemical Cycles* 7: 305–320.

Kauppi PE, Mielikainen K, Kuusela K (1992) Biomass and carbon budget of European forests, 1971–1990. *Science* 256: 70–74.

Keeling CD, Bacastow RB, Carter AF, Piper SC, Whorf TP, Heimann M, Mook WG, Roeloffzen H (1989) A three-dimensional model of atmospheric CO_2 transport based on observed winds: 1. Analysis of observational data. In: Peterson DH (ed) Aspects of Climate Variability in the Pacific and the Western Americas. *Geophysical Monograph* 55, American Geophysical Union, Washington DC, pp 165–236.

Kolchugina TP, Vinson TS (1993) Carbon sources and sinks in forest biomes of the former Soviet Union. *Global Biogeochemical Cycles* 7: 291–304.

Kurz WA, Apps MJ, Beukema SJ, Lekstrum T (1995) Twentieth century carbon budget of Canadian forests. *Tellus* 47B: 170–177.

Maclaren, JP (1995) Plantation forestry: its role as a carbon sink. (Chapter 21, this volume).

Myers N (1980) Conversion of Tropical Moist Forests. National Academy of Sciences Press, Washington DC, 205 pp.

Olson JS, Watts JA, Allison LJ (1983) Carbon in live vegetation of major world ecosystems. TR004, U.S. Department of Energy, Washington DC, 152 pp.

Perlin J (1989) A Forest Journey: The Role of Wood in the Development of Civilization. Harvard University Press, Cambridge, MA, 445 pp.

Post WM, Emanuel WR, Zinke PJ, Stangenberger AG (1982) Soil carbon pools and world life zones. *Nature* 298: 156–159.

Sarmiento JL, Orr JC, Siegenthaler U (1992) A perturbation simulation of CO_2 uptake in an ocean general circulation model. *J. Geophys. Res.* 97: 3621–3645.

Schlesinger WH (1984) The world carbon pool in soil organic matter: A source of atmospheric CO_2. In: Woodwell GM (ed) The Role of Terrestrial Vegetation in the Global Carbon Cycle: Measurement by Remote Sensing, SCOPE 23, Wiley, New York, pp 111–124.

Schlesinger WH (1986) Changes in soil carbon storage and associated properties with disturbance and recovery. In: Trabalka JR and Reichle DE (eds) The Changing Carbon Cycle: A Global Analysis. Springer-Verlag, New York, pp 194–220.

Siegenthaler U, Oeschger H (1987) Biospheric CO_2 emissions during the past 200 years reconstructed by deconvolution of ice core data. *Tellus* B 39: 140–154.

Tans PP, Fung IY, Takahashi T (1990) Observational constraints on the global atmospheric CO_2 budget. *Science* 247: 1431–1438.

Whittaker RH, Likens GE (1973) Carbon in the biota. In: Woodwell GM and Pecan EV (eds) Carbon and the Biosphere. US Atomic Energy Commission, Symposium Series 30, National Technical Information Service, Springfield, Virginia, pp 281–302.

Woodwell GM, Mackenzie FT, Houghton RA, Apps MJ, Gorham E, Davidson EA (1995) Will the warming speed the warming? In: Woodwell GM and Mackenzie FT (eds) Biotic Feedbacks in the Global Climatic System: Will the Warming Feed the Warming? Oxford University Press, New York, pp 393–411.
Zinke PJ, Stangenberger AG, Post WM, Emanuel WR, Olson JS (1986) Worldwide Organic Soil Carbon and Nitrogen Data. ORNL/CDIC-18. Oak Ridge National Laboratory, Oak Ridge TN, 141 pp.

11 Tropical forests and the global carbon cycle: estimating state and change in biomass density

Sandra Brown[1]

Department of Forestry, University of Illinois
W-503 Turner Hall, 1102 S Goodwin
Urbana, IL 61801 USA

INTRODUCTION

Tropical forests have an important role in the global carbon (C) cycle because of their existing large areal extent, high rates of deforestation, large C pool in vegetation and soil, and high rates of C emissions resulting from conversion to other uses (equivalent to between 22 and 37% of current fossil fuel C emissions) (Table 11.1). Tropical forests currently account for about 43% of the global forest area (Dixon et al. 1994), most of which is in tropical America (52%), followed by tropical Africa (30%) and tropical Asia (18%). They occur mostly as lowland formations where 88% are at an elevation of 1000 m or less. Within the lowlands, 47% are in the rain forest ecological zone, 38% in the moist deciduous zone, and 15% in the dry to very dry zone (Food and Agriculture Organization (FAO) 1993).

During 1981–90, the average rate of deforestation was about 15.4 Mha yr^{-1} (Table 11.1). The area deforested in tropical America was almost twice as high as in the other two continental regions, but tropical Asia had the highest rate of deforestation relative to its remaining forest area. In addition to deforestation, logging affects another 5.6 Mha, with over 80% of this from mature forests, and the remainder from previously logged forests (Table 11.1). Therefore, deforestation and harvesting alone affect a total of 21 Mha yr^{-1}. In addition tropical forests are subject to various forms of degradation, including decrease in canopy density (e.g., transformed from closed to open forest formations), fragmentation (partial deforestation), and conversion to other wooded classes such as shrubs or short-fallow agriculture (Brown et al. 1993; FAO 1993). It is estimated that rates of conversion to degraded forests are several times higher than rates of deforestation (FAO 1993).

Some of the loss in natural forest cover is compensated for by plantation establishment, currently estimated as about 1 ha planted for every 6 ha deforested (Table 11.1). Tropical Asia, with the largest percentage loss in forest cover also has the highest rate of plantation establishment, about 1 ha planted for every 2 ha deforested. However, in the rest of the tropics, the rate of plantation establishment is slow.

The C pool in tropical forest vegetation is more or less equal to that in the soil (Table 11.1). Consequently, tropical forests contain over 60% of the world's forest C pool in vegetation, but only 33% of the soil C pool (Dixon et al. 1994). Large uncertainties probably exist in both of these C pool estimates caused by uncertainties in the extent of forests, the proportions that are

[1] Current address: US Environmental Protection Agency, 200 SW 35th Street, Corvallis, OR 97333, USA

NATO ASI Series, Vol. I 40
Forest Ecosystems, Forest Management and the Global Carbon Cycle
Edited by Michael J. Apps and David T. Price

mature, degraded, or disturbed, and, consequently, the C densities of the corresponding forest vegetation and soil. The current estimate of the C flux from conversion of tropical forests in 1990 is about 1.6 Pg C yr^{-1} (Houghton, this volume). The uncertainty around this C flux is also high, caused not only by uncertainty associated with biomass and soil C densities, but also by uncertainty in the rates of forest conversion and forest degradation. Except for very few countries, remote sensing imagery is not used regularly to estimate change in forest cover or use.

Table 11.1. Characteristics of forests, by tropical region, with respect to state and change in area and C pools (data from Dixon et al. 1994; FAO 1993).

CHARACTERISTIC	AFRICA	AMERICA[a]	ASIA	TOTAL
Forest cover: 1990 (Mha)	527	918	310	1756
% of land area	24	56	35	37
Deforestation: 1981–90 (Mha yr^{-1})	4.1	7.4	3.9	15.4
% of forest cover	0.7	0.8	1.2	0.8
Logging: 1981–90 (Mha yr^{-1})	0.92	2.6	2.1	5.6
% newly logged lands	73.6	88.9	82.2	83.9
Plantation forests (Mha)	2.1	6.0	22.6	30.7
Plantations: 1981–90 (Mha yr^{-1})	0.13	0.37	2.10	2.61
Ratio planted:deforested	1:32	1:20	1:2	1:6
Carbon pools (Pg)				
Vegetation	52	119	48	219
Soil	63	110	43	216

a Includes the tropical parts of South and Central America and the Caribbean.

The extent to which the total tropical C flux includes C fluxes from forest degradation, regrowth and management of secondary forests, plantation establishment, or logging practices is unclear. The impact of these activities could increase or decrease the C flux depending upon their relative importance in a given country or region. For example, recent analyses indicate that the C flux from the atmosphere from management and regrowth of secondary forests in India and Brazil almost compensates for the C flux to the atmosphere from deforestation (Ravindranath et al. 1992, Schroeder and Winjum 1995).

The magnitude of the C pool and flux of tropical forests depends on two data bases: (1) state and change in cover, use, and management of forests and (2) the corresponding biomass densities (Mg C ha^{-1}). Estimates of the state and change in area of tropical forests are improving as a result of new efforts by FAO, the remote sensing community, and ongoing national forest inventories by some countries (e.g., Brown et al. 1994; FAO 1993; Skole and Tucker 1993). However, parallel efforts to improve, update, and map biomass densities of tropical forests have not proceeded at the same pace. Consequently, the uncertainty in biomass density estimates remains relatively large. The main goal of this paper is to summarise the progress and increased understanding of tropical forest biomass density estimation, including: (1) the main methods available and their application at scales suited for modelling the tropical terrestrial C cycle, (2) the role of large diameter trees in biomass density estimation, and (3) human activities that cause biomass densities of forests to change over time.

ESTIMATING BIOMASS DENSITIES

For calculating forest C budgets, the quantity of concern is the total above- and belowground live and dead mass of organic matter per unit area (i.e., biomass density). This includes the aboveground parts of trees (leaves, twigs, branches, bole, and bark), shrubs, herbs, vines, and palms; belowground mass of fine and large roots; fine litter; and lying and standing dead wood (woody debris). In practice, particularly in the tropics, aboveground biomass density of trees is often the only quantity estimated with any degree of reliability because either the mass of the other components is relatively insignificant, or it is of no commercial value, or reliable data are lacking. Estimating the biomass density of trees will be the main focus in this paper.

Some trends in the biomass of the other forest components are given here from the few data available. Data on belowground biomass density suggest that it can vary from about 10% to 50% of the aboveground biomass (Brown and Lugo 1982), probably as a function of species differences, soil type, soil fertility, and moisture regime. With such a limited data base, however, determining algorithms for estimating root biomass is difficult. From the few studies available, the biomass of the non-tree components in mature tropical forests is generally 3% or less of the tree biomass (Brown and Lugo 1992). However, in disturbed forests the biomass of the non-tree component is likely to be significantly higher, but by how much is largely unknown (Brown and Lugo 1990). Data on the mass of fine litter are relatively abundant and range from about 2 to 10 Mg ha^{-1} (Brown and Lugo 1982), but the mass and dynamics of coarse litter or woody debris are poorly known. It may account for 10% to 40% of the aboveground biomass and serve as a long-term C pool (Brown and Lugo 1992). It is clear that these other components could contain up to as much mass as the aboveground trees. An obvious area of future research is to develop methods to estimate the mass of these other components at the scale and accuracy necessary for landscape level analysis, particularly in disturbed or secondary forests which are becoming more important features of the tropics (Brown and Lugo 1990).

To estimate the biomass density of large expanses of forests poses a serious sampling problem. Many factors influence biomass density, including climatic, edaphic, and topographic conditions; age; land-use history; and human and natural disturbances. Rarely can ecological studies, which generally encompass small areas and are often designed to address topics other than biomass density, capture this large spatial heterogeneity because the data are not from the population of interest and thus do not represent the tropical landscape. For reliably estimating the biomass density of tropical forests, the sampling design must be random or stratified random from the population of interest. Data that generally meet this criterion are those obtained from national to sub-national forest inventories. Not all inventory data are suitable for this purpose, however, because they often include selected commercial species only or measure trees above some relatively large minimum diameter (>30 cm).

A *toolbox* of approaches has been developed to convert tropical forest inventory data into estimates of aboveground biomass density (Brown and Lugo 1992; Brown et al. 1989, 1991; Gillespie et al. 1992) starting with either stand tables (number of trees per hectare by diameter class) or commercial volumes. The preferred approach, which incorporates the fewest assumptions, is the use of stand tables in combination with an appropriate biomass regression equation selected on the basis of climate regime (dry, moist, or wet; Brown et al. 1989). Basically, the method involves estimating the biomass of trees whose diameters are the mid-point of each diameter class, and multiplying the biomass per tree by the number of trees in each class. This is then summed for all classes to obtain an estimate of the total aboveground biomass density.

This approach has been used with inventories from nine tropical Asian countries. The resulting estimates of aboveground biomass density range from less than 50 Mg ha^{-1} to about 500 Mg ha^{-1}, with an arithmetic mean of 225 Mg ha^{-1} (Brown et al. 1991; Figure 11.1). In Figure 11.1 biomass density is plotted against basal area as one means of showing the variation in the estimates, and the high correlation is expected because the biomass regression equation includes a diameter-squared term.

I applied the same approach to inventories for regions of ten countries in tropical America: Bolivia, Brazil, Ecuador, Guatemala, Guyana, Nicaragua, Panama, Peru, Surinam, and Venezuela (the data base and sources are available from author). These inventories encompassed a total forest area of 6.7 Mha. The resulting aboveground biomass density estimates range from less than 50 Mg ha^{-1} to about 850 Mg ha^{-1} (a mean of 235 Mg ha^{-1}) in the moist zone and to about 550 Mg ha^{-1} (a mean of 150 Mg ha^{-1}) in the wet zone (Figure 11.1). The national inventory for Panama encompassed the full range of biomass density estimates

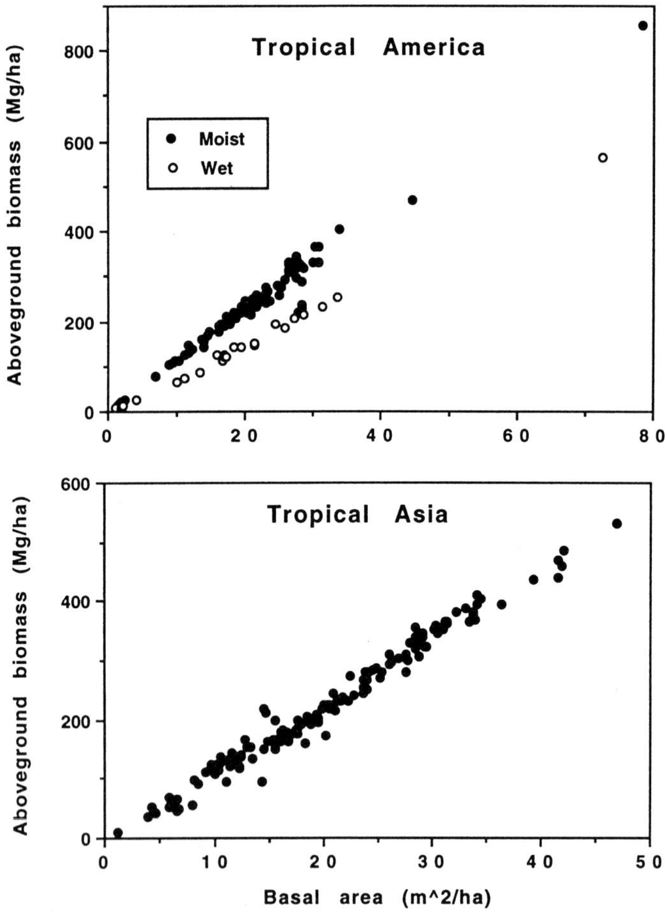

Figure 11.1. Estimates of aboveground biomass density of forests based on inventory data for tropical America (data and sources from author) and tropical Asia (based on data given in Brown et al. 1991). Biomass density estimates are plotted against basal area as one way of showing the range in values.

presented here. Forests in the wet zone of Panama tended to have lower biomass densities for a given basal area as has been shown before (Brown and Lugo 1982). Except for the one outlier (a forest dominated by *Campnosperma panamensis* in Panama), the range of biomass density estimates for moist tropical American forests is practically identical to that for moist tropical Asian forests (Figure 11.1). As was the case for the topical Asian forests (cf. Brown et al. 1991), many of the tropical American forests were identified as being disturbed (e.g., commercial harvesting, harvesting by indigenous communities, young to late secondary, shifting cultivation).

Aboveground biomass density estimates for mature moist forests in tropical America based on ecological studies typically range between 300 to 400 Mg ha^{-1} (Brown and Lugo 1982; Olson et al. 1983), values that are within the range of those based on inventories but are considerably higher on average than most of the values shown in Figure 11.1. Based on the analyses presented here and elsewhere (Brown and Lugo 1992), support for the continued use of aboveground biomass density estimates from ecological studies is weak and it is time to recognise that conversion of reliable forest inventory data (i.e., contain all tree species and measure trees down to a minimum diameter of at least 10 cm; cf. Brown and Iverson (1992)) to biomass density estimates is the preferred approach. Similar conclusions have been made for forests in the boreal and temperate zones of North America (e.g., Botkin and Simpson 1990; Simpson et al. 1993). What is needed to improve the robustness and accuracy of estimating tropical forest biomass densities from inventory data are additional studies to improve the tools, such as biomass measurements of more large diameter trees (> 100 cm) to add to the regression equations for converting stand tables to biomass and expansion factors for converting commercial volumes (m^3 ha^{-1}) to biomass density (Brown et al. 1989). Additional field studies are also needed for the other forest components as well as efforts to include appropriate new measurements in future forest inventories.

LARGE DIAMETER TREES AND BIOMASS DENSITY ESTIMATION

Part of the reason that many of the tropical forests in America and Asia have low biomass is the lack of large diameter trees, defined as those with DBH > 70 cm and a biomass of > 5 Mg per tree in the moist zone (Brown and Lugo 1992). In general, a decrease in the percent of the biomass in large diameter trees is associated with a decrease in the biomass density of a forest (Figure 11.2). Significant linear relations between these two variables, with $r^2 \geq 0.6$, were obtained for Myanmar (Figure 11.2a), Sri Lanka (moist) (Figure 11.2b), Peninsular Malaysia and the Philippines (Figure 11.2c), and Brazilian Amazon (Figure 11.2d). The reasons why these moist forests have low biomass densities and low proportions of biomass in large trees are either that the large diameter trees have been removed by sanctioned or unsanctioned selective harvesting (Brown et al. 1991; Callister 1992) or they are secondary forests which have not had time to grow large trees. Whatever the reason, this analysis shows that removal of a few large diameter trees can have a marked effect on the biomass density of moist forests. I expect similar, though probably less pronounced, trends in dry or wet forests, with the diameter limit of a large tree being smaller.

Forests with a biomass density of 300 Mg ha^{-1} or more generally contain 30% or more of the biomass in the large diameter trees. The exceptions to this are the forests in (1) the Philippines where aboveground biomass density of mature forests was 500 Mg ha^{-1} or more with 30% of this amount in large diameter trees and (2) Myanmar which has low biomass densities (about 200 Mg ha^{-1} or less) and yet up to 70% of the biomass is in large trees. The forests inventoried in Myanmar were located near townships along the Rangoon to Mandalay railway line, an area that has been under intense human pressure for years. Although the forests were disturbed, resulting in low biomass densities, large diameter trees appeared to have been

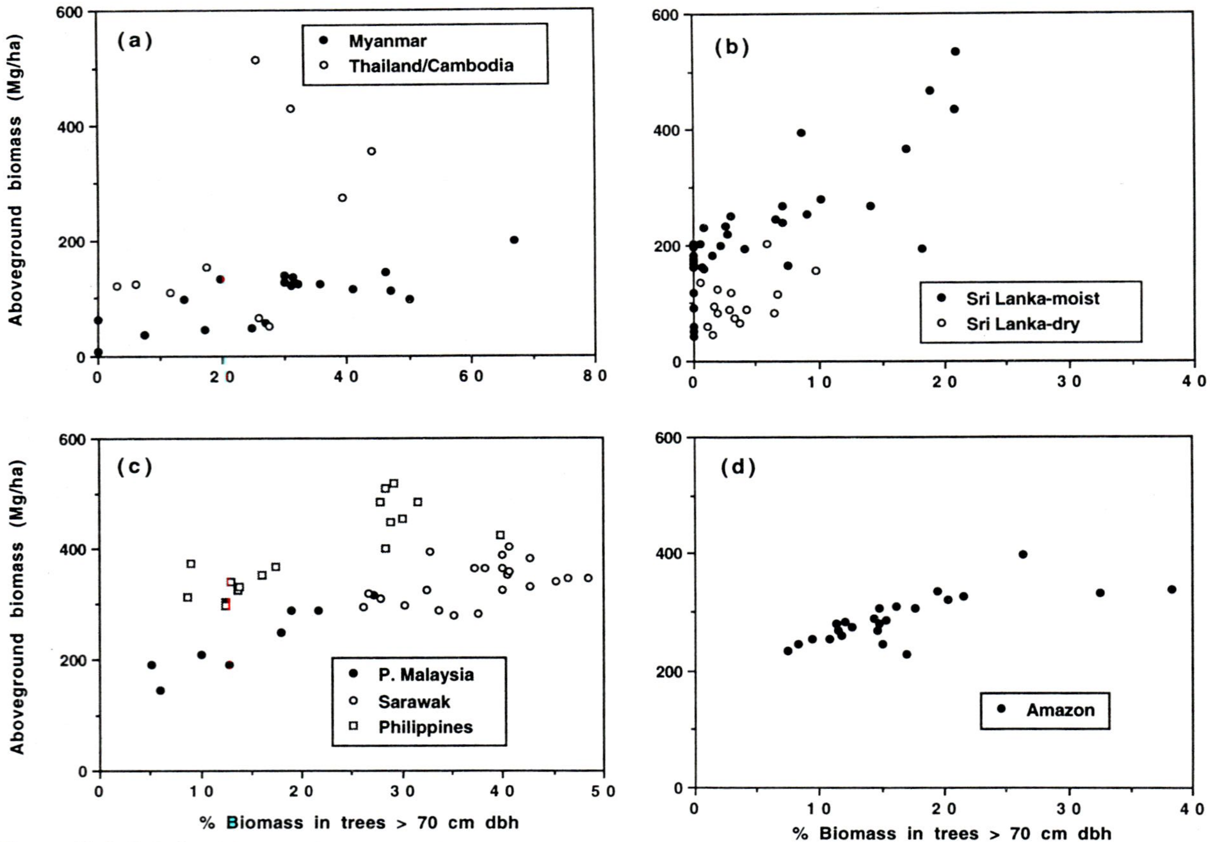

Figure 11.2. Relation between aboveground biomass density and percentage of this biomass in trees with diameters >70 cm. Data for (a–c) are from sources given in Brown et al. (1991) and for (d) from sources given in Brown and Lugo (1992).

selectively retained in the forest for unknown reasons. This example represents a case where the cutting regime obviously affects the proportion of the biomass found in large diameter trees.

The presence or absence of large-diameter trees may be a useful index for assessing degree of disturbance in tropical forests growing on zonal or typical soils. If the percentage of the aboveground biomass density in large diameter trees is greater than 30%–35%, then the forests could be considered relatively mature (but not necessarily undisturbed by humans; Brown et al. 1994; Iverson et al. 1994). This is supported by the inventory descriptions for most of the Sarawak forests, the high biomass Philippine forests (>400 Mg ha^{-1}), and a few Amazonian forests. However there may be exceptions to this trend, as is the case of Myanmar.

The influence of large diameter trees on biomass appears to largely explain why ecological studies may produce high biomass density estimates. Ecologists tend to select sample plots on the notion of what a mature forest should look like, i.e., one with numerous large diameter trees, thus creating a serious sampling bias in mature forests (McCune and Menges 1986). In inventories of moist forests, the number of trees with DBH > 70 cm is generally of the order of < 5% of the stem density (number per ha), whereas in the fewer and smaller plots of ecological studies, trees with DBH > 70 cm often comprise 10% or more of the total number of stems and up to 90% of the aboveground biomass density (Brown and Lugo 1992). Future ecological studies need to be cognisant of this problem and adjust their sampling scheme accordingly.

The reduced number of large diameter trees in many forests of the tropics, and the indication that this is caused by human activities, demonstrates that many of these forests are not at their maximum C storage. This has implications for mitigation of greenhouse gas emissions because if some of these forest areas could be adequately protected, they could sequester C. It is the large trees that can store significant quantities of C and it can take on the order of hundreds of years to reach their maximum size.

REDUCTION IN BIOMASS DENSITY BY HUMAN DISTURBANCE

That tropical forests are undergoing degradation in addition to deforestation is well documented, not only for the recent decades (Brown et al. 1993, 1994; FAO 1993) but also as far back as 1880, e.g., in tropical Asia (Flint and Richards 1994; Richards and Flint 1994). This degradation of forests is caused by several human activities including sanctioned and unsanctioned logging for fuel and timber, grazing by domestic animals, and slash-and-burn or shifting cultivation (Brown et al. 1993; Flint and Richards 1994). The net result is conversion of closed forest to open forests, increased forest fragmentation, a decrease in biomass densities, and a change in the temporal pattern of C flux to the atmosphere (Brown et al. 1994; FAO 1993; Houghton and Hackler 1994). However, there is evidence that some forest areas become less fragmented, e.g., Peninsular Malaysia (Brown et al. 1994), or are better managed and protected, e.g., parts of India (Ravindranath et al. 1992) and in these situations biomass density recovers.

Attempts have been made to quantify the effects of human activities on forest degradation in tropical Asia (Brown et al. 1993; Flint and Richards 1994; Iverson et al. 1994). These approaches relied heavily on population density as an index that encapsulates all the activities by humans in degrading the biomass density of forests. The approach by Iverson et al. (1994) and extended by Brown et al. (1993) assumed that the impact of humans on tropical forests is a function of population density, and that the intensity of the impact varies by climate (dry, seasonal, and moist). That is, the same population density causes relatively less forest degradation in moist than in dry forests because of the ability of the former to produce biomass more rapidly (Brown and Lugo 1982). We first defined a degradation ratio as the ratio of actual biomass density based on inventory data (e.g., results shown in Figure 11.1) to potential biomass density obtained from a rule-based model (Iverson et al. 1994). Then three significant regression equations were developed of degradation ratio versus the natural logarithm of population density at sub-national scales (Brown et al. 1993). Because of the semi-logarithmic nature of the function, degradation ratios decreased relatively steeply at low population densities and less steeply at higher population densities.

Although the relationship between population density and forest degradation that I present here seems to apply well to tropical forests, I do not mean to imply that it is universal. First, there are examples even in the tropics where through forest protection, biomass density increases (e.g., Ravindranath et al. 1992). Furthermore, biomass density of forests in many developed countries such as the U.S. and in Europe has increased for decades while population has also increased. I believe that when a country becomes more developed and industrialisation increases, population density continues to increase but it is distributed more in the urban rather than rural areas. As a consequence, forest degradation tends to slow down and in fact forests may recover as in many of the developed countries.

The approach of Flint and Richards (1994) was similar to that used by Brown et al. except their model related degradation to the inverse of population density and they did not stratify by climate. Instead, they added other modifiers to account for the relative importance of shifting cultivation in a sub-national region and the relative importance of subsistence use, export, or input of timber and agriculture products. Like the model of Brown et al. (1993), their model also exhibited steep degradation of biomass at low population densities with a levelling at high population densities.

Using either model with sub-national scale data on (1) rates of forest land-cover and land-use change in the tropical Asian region from 1880 to 1990, (2) population density for the same time period, and (3) estimates of potential biomass density (Brown et al. 1993), the change in forest C pools due to degradation was about equal to that of the change in forest C pools due to deforestation from 1880 to 1950 (Table 11.2). From 1950 to 1990, the change in C pool due to degradation became less and that due to deforestation became more important. The relatively large change in C from degradation at low population density suggests that it is the large diameter trees that are removed first (cf. Figure 11.2). Removal of a few of these trees significantly reduces the biomass density as discussed above. What then remains are the lower biomass forests that are typically found today (cf. Figure 11.1).

Table 11.2. Comparison of the change in vegetation C pools (Pg C per time period) from clearing forests to non-forests with the change in vegetation C pools from degradation of forests (i.e., caused by sanctioned and unsanctioned logging, fuelwood gathering, shifting cultivation, etc.) in tropical Asia (from ongoing research of S Brown and colleagues).

	CHANGE IN VEGETATION C–POOLS (Pg C time period^{-1})		
TIME PERIOD	CLEARING (CLEAR)	DEGRADATION (DEGRADE)	RATIO (CLEAR/DEGRADE)
1880–1920	5.9	5.2	1.1
1920–1950	6.2	4.9	1.3
1950–1970	6.5	4.8	1.4
1970–1980	4.4	1.9	2.3
1980–1990	5.3	1.4	3.8
1880–1990	28.3	18.2	1.6

These approaches appear to offer ways by which the effect of humans on the tropical forest of Asia can be quantified. Historical accounts of the forests of the region also support the analysis shown here. For example, many of the forests of the region were subjected to the selective removal of the largest trees by the middle of the 1800s (a historical summary of these activities is given in Flint and Richards 1994). How typical the relationships presented here are of other tropical regions is unknown, although preliminary analysis of tropical Africa suggests that the degradation of forests is similarly related to population density (S. Brown and G. Gaston, ongoing research). More research on the human dimension of forest area and biomass density change is clearly needed.

ACKNOWLEDGEMENTS

I thank Aji Perera for help with the analysis of the inventory data for tropical American forests and the librarians at FAO for help in locating the inventory reports. This research was supported by a grant from the US Department of Energy, DOE DEFGO2-90ER61081, to the University of Illinois, S Brown, principal investigator.

REFERENCES

Botkin DB, Simpson LG (1990) Biomass of the North American boreal forest: a step toward accurate global measurements. *Biogeochemistry* 9: 161–170.

Brown S, Iverson LR (1992) Biomass estimates for tropical forests. *World Resource Review* 4: 366–384.

Brown S, Lugo AE (1982) The storage and production of organic matter in tropical forests and their role in the global carbon cycle. *Biotropica* 14: 161–187.

Brown S, Lugo AE (1990) Tropical secondary forests. *Journal of Tropical Ecology* 6: 1–32.

Brown S, Lugo AE (1992) Aboveground biomass estimates for tropical forests of the Brazilian Amazon. *Interciencia* 17: 8–18.

Brown S, Gillespie AJR, Lugo AE (1989) Biomass estimation methods for tropical forests with application to forest inventory data. *Forest Science* 35: 881–902.

Brown S, Gillespie AJR, Lugo AE (1991) Biomass of tropical forests of south and southeast Asia. *Canadian Journal of Forest Research* 21: 111–117.

Brown S, Iverson LR, Prasad A, Liu D (1993) Geographical distributions of carbon in biomass and soils of tropical Asian forests. *Geocarto International* 4: 45–59.

Brown S, Iverson LR, Lugo AE (1994) Land-use and biomass changes of forests in Peninsular Malaysia from 1972 to 1982: a GIS approach. In: Dale VH (ed) Effects of land-use change on atmospheric CO_2 concentrations: south and southeast Asia as a case study. Springer-Verlag, New York, pp 117–144.

Callister DJ (1992) Illegal tropical timber trade: Asia-Pacific. TRAFFIC International, Cambridge, UK, 83 pp.

Dixon RK, Brown S, Houghton RA, Solomon AM, Trexler MC, Wisniewski J (1994) Carbon pools and flux of global forest ecosystems. *Science* 263: 185–190.

Flint ER, Richards JF (1994) Trends in carbon content of vegetation in south and southeast Asia associated with changes in land-use. In: Dale VH (ed) Effects of land-use change on atmospheric CO_2 concentrations: south and southeast Asia as a case study. Springer-Verlag, New York, pp 201–300.

Food and Agriculture Organization of the United Nations (1993) Forest resource assessment 1990: tropical countries. FAO Forestry Paper 112, Rome, 102 pp.

Gillespie AJR, Brown S, Lugo AE (1992) Tropical forest biomass estimation from truncated stand tables. *Forest Ecology and Management* 48: 69–87.

Houghton RA, Hackler JL (1994) The net flux of carbon from deforestation and degradation in south and southeast Asia. In: Dale VH (ed) Effects of land-use change on atmospheric CO_2 concentrations: south and southeast Asia as a case study. Springer-Verlag, New York, pp 301–328.

Iverson LR, Brown S, Prasad A, Mitasova H, Gillespie AJR, Lugo AE (1994) Use of GIS for estimating potential and actual forest biomass for continental south and southeast Asia. In: Dale VH (ed) Effects of land-use change on atmospheric CO_2 concentrations: south and southeast Asia as a case study. Springer-Verlag, New York, pp 67–116.

McCune B, Menges ES (1986) Quality of historical data on midwestern old-growth forests. *The American Midland Naturalist* 116: 163–172.

Olson JS, Watts JA, Allison LJ (1983) Carbon in live vegetation of major world ecosystems. DOE/NBB-0037 National Technical Information Service, Springfield, VA, 152 pp.

Ravindranath NH, Somashekhar BS, Gadgil M (1992) Carbon flows in Indian forests. In: Makundi W, Sathaye J (eds) Carbon emissions and sequestration in forests: case studies from seven developing countries Vol. 3: India and China, LBL-32759 UC-402, Energy and Environment Division, Lawrence Berkeley Laboratory, Berkeley, CA, pp 1–30.

Richards JF, Flint ER (1994) A century of land-use change in south and southeast Asia. In VH Dale (ed) Effects of land-use change on atmospheric CO_2 concentrations: south and southeast Asia as a case study. Springer-Verlag, New York, pp 15–66.

Schroeder P, Winjum JK (1995) Balancing Brazil's carbon budget: the potential importance of secondary forests. *Interciencia* (In press).

Simpson LG, Botkin DB, Nisbet RA (1993) The potential aboveground carbon storage of North American forests. *Water, Air and Soil Pollution* 70: 197–208.

Skole D, Tucker C (1993) Tropical deforestation and habitat fragmentation in the Amazon: satellite data from 1978 to 1988. *Science* 260: 1095–1910.

12 Carbon budget of the Russian boreal forests: a systems analysis approach to uncertainty

Anatoly Z Shvidenko, Sten Nilsson, Vjacheslav A Rojkov[1] and Valentin V Strakhov[2]

International Institute for Applied Systems Analysis,
A-2361 Laxenburg, Austria

INTRODUCTION

The total land area of the Russian boreal zone is 1527.6 Mha, including 1143.0 Mha of Forest Fund areas and 735.8 Mha of forested areas. These estimates are based on Forest State Account data (Goscomles SSSR 1990, 1991). Forest Fund areas include forest land and non-forest land. Forest land is in turn divided into forested areas, covered by closed forests, and unforested areas, designated for forests but temporarily without a forest (sparse forests, burnt areas and dead stands, grassy glades). Non-forest land is represented by unproductive land, such as bogs, rocks, sand, and glaciers, and by land with special uses (forest roads, water reservoirs, and relatively small areas of arable lands, pastures farms, etc., situated on Forest Fund areas). For a more detailed description see, e.g., Nilsson et al. (1992). Nearly 95% of all Russian closed forests are considered boreal. Thus, the Russian boreal forests play an important role in the global carbon (C) cycle.

Estimates of the C budget for the boreal zone, as well as for the Russian forests and landscapes, vary greatly (some estimates are listed in Table 12.1). Estimates of the total C flux for the Russian forests and the forests of the former Soviet Union during the last 20 years have varied, from showing them to be a slight net source to being a significant sink of about 1 Pg C yr^{-1}.

The disagreement in estimates concerning the role of the boreal forests as a whole (and the Russian forests as a part of the whole) in the C budget has not yet been resolved (Kauppi et al. 1992; Houghton 1993; Shvidenko et al. 1994a). Basic reasons for the disagreement in the estimates include the following:

- Differences in the approaches and models used, differences in the structuring of the C cycles, and differences in the measurement of the basic indicators of the C cycle.
- The lack of knowledge of some basic processes.
- The complicated structure of the problem.
- The scale of Russian territory, which encompasses extremely diverse climatic, orographic, and anthropogenic conditions.
- The incompleteness, uncertainties, and shortcomings of data needed for robust analyses.
- The disagreements between the classification schemes used.

---

1 VV Dokuchaev Soil Institute, Russian Academy of Agricultural Sciences, Pyzhevski Lyne 7, 109015 Moscow, Russia

2 Federal Forestry Service of Russia, All-Russian Informational and Research Center for Forest Resources, VNIITSlesresource, Novocheriomushkinskaya Str. 69A, 117877 Moscow, Russia

NATO ASI Series, Vol. I 40
Forest Ecosystems, Forest Management and the Global Carbon Cycle
Edited by Michael J. Apps and David T. Price

Table 12.1. Estimates of the role of boreal and temperate biota in the carbon (C) budget.

Zone, Region	Period	Pg C yr^{-1}	Source
TBZ[a], forests	1953–1975	+1.0 to +1.2	Armentana and Ralston (1980)
FSU[a], forests	1953–1975	+0.5	Armentana and Ralston (1980)
FSU, forests	1982	-0.199	Shaver et al. (1982)
TBZ, forests	–	+1.6 to +1.9	Johnson and Sharpe (1983)
TBZ, forests	–	-0.5	Houghton et al. (1983)
TBZ, forests	1980	+0.973	Melillo et al. (1988)[c]
TBZ, land	1980	-0.025	Melillo et al. (1988)
FSU, land	1980	-0.037	Melillo et al. (1988)[d]
FSU, forests	1980	+0.048	Melillo et al. (1988)
Russia, forests	1990	+1.02	Makarov et al. (1990)
Russia, forests	1990	+0.4	Zavarzin (1992)
Russia, forests	1988	+0.184 to +0.212	Isaev et al. (1993)[e]
BZ[a], forests	1966–1988	+0.76	Sedjo (1992)
Canada forests	1986	+0.0768	Kurz et al. (1992)
Europe, forests	1970–1980	+0.08 to 0.12	Kauppi et al. (1992)
FSU, forests	1978–1988	+0.416	Sedjo (1992)
FSU, forest biomes	mid-1980s	+0.475, +0.517	Kolchugina and Vinson (1993a, 1993b)
Russia, forests	1990s	+0.1 to +0.4	Kokorin et al. (1993)
FSU, forests	1966–1988	+0.46	Shvidenko et al. (1994a)
Russia, forests	1987–1990	+ 0.3 to +0.5	Dixon et al. (1994)

a TBZ = temperate and boreal zones; FSU = former Soviet Union; BZ = boreal zone.
b + (-) indicates a sink (source) of atmospheric C in the terrestrial biota;
c Regrowth and reforestation.
d Includes boreal forests, deciduous forests, and temperate grasslands.
e Sequestration in wood on forested areas (771 Mha).

The estimation of the C budget is complex, involving different types of variables, such as deterministic, stochastic, quantitative, and qualitative variables and expert estimates. Currently some important internal system interactions (e.g., many aspects of the post-disturbance processes in soil) can only be represented in the form of "black-box" models. There are rather restricted possibilities for a consecutive formalisation and structuring of the C cycle due to a lack of required information from previous decades or centuries. The problem as a whole and its components have a typical fuzzy character. Consequently, any investigations of the total forest C budget or of the Russian forest C budget must be greatly simplified.

In our analyses, we used simple, pragmatic principles of system (holistic) analysis (goals, objects, structure, integrity, partition, connection, etc.) to estimate the sources of basic uncertainties in the C budget of the Russian boreal forests and landscapes. The following features of the problem were given special attention:

- Stability and improvement of the C budget are essential parts of the sustainable development paradigm and forests are the frame of the boreal landscape's ecological sustainability.
- The forest C problem is dynamic, and forest succession is one of the most important processes of the system; also, the dynamics of the forests are, in turn, part of temporal and spatial changes in land use, land cover change, and land management change.
- Methods and models used should yield quantitative estimates of basic indicators and related uncertainties using available information.

There are different basic approaches and criteria available for investigations of the C budget for large scale regions (see Nabuurs 1994), but most of the current analyses fall somewhere between the following two approaches: making an inventory of the C pools both at the beginning and at the end of the period investigated, e.g., estimating the sequestered C in the biomass during some specified period; and estimating the fluxes of the C exchange between pools. From the system point of view the first approach is preferable, but for the study of Russian forests neither the first nor the second approach is independently sufficient: a combination of the above-mentioned approaches is required.

SOME SYSTEM FEATURES OF THE PROBLEM

Goals

> *"Far better an approximate answer to the right question, which is often vague, than the exact answer to the wrong question, which can always be made precise"* (Tukey 1962).

Generally, a system does not exist until its goals exist. The decision tree of the C problem has a multilevel hierarchical structure with three main goals:

1. evaluation of the current state of the interaction between forests, landscapes and the greenhouse gas composition of the atmosphere.
2. evaluation of the impacts of global change on the forest C cycle and identification of the dynamics of the basic C pools and fluxes.
3. identification of a system of forest management options for C conservation and for increased C sequestration.

A simplified version of the goal tree for the forest C budget is presented in Figure 12.1. The lower levels and any details of goals (2) and (3) mentioned above are not considered; this paper addresses only the most important areas closely connected with the problem, namely, bioproductivity, landscape structures, forest management regimes, and natural and anthropogenic disturbances of the forests, which are interlinked in a complicated manner.

Spatial and temporal scale

It is expedient to identify the appropriate basic land measurement units and scales for C analysis in order to meet the study goals (see Figure 12.1). There are three possible units or scales for C budget evaluation: forest ecosystems, boreal landscapes, and ecological regions (ecoregions).

The forest ecosystem (we use this term as a synonym for *forest biogeocoenosis,* as defined by Suchachev (1964); our use is also similar to Tansley's (1935) definition) is a natural (primary, elementary) unit for the evaluation of the forest C budget at a local level. Forest ecosystems are the primary units of the Russian forest inventory system and they coincide with the primary management units in the managed forests. This means that any parametric aggregation (such as species, site indices, and forest management regimes) or account of results of management activities can be used at the regional level without additional collection of information.

The boreal forests constitute an ecological framework for the boreal landscapes; for a thorough analysis, the impact of forests and adjunct territories on the C budget must be considered. Anthropogenic impact, changes in land use and natural successions lead to

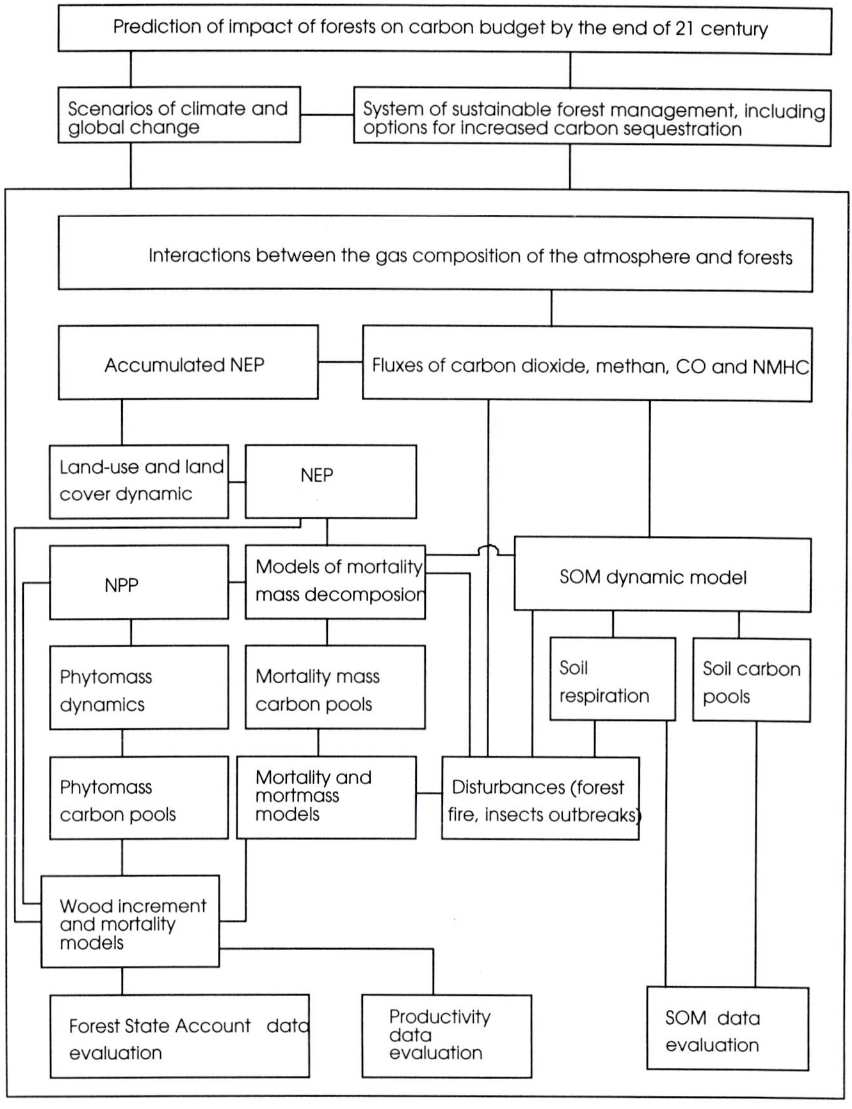

Figure 12.1. Conceptual scheme of the analytical system. NEP = Net Ecosystem Production; NPP = Net Primary Production; SOM = Soil Organic Matter.

transformations of land cover; therefore, estimations of C budget dynamics cannot be restricted to forested areas. Sustainable management of natural resources also requires consideration of the landscape as a whole. Consequently, the boreal landscape is also a logical unit for forest C evaluation.

One of the Russian landscape classification systems uses *facies* as a primary unit for the landscape division. Facies can be aggregated into *stows* ("urochische"); stows can be aggregated into *terrain* ("mestnost"); and terrains can be aggregated into *landscapes*. The average transect of landscape is around 10 to 20 km for Russia. This is probably the best unit for C evaluation, at least theoretically, because it makes it possible to consider, to some extent, a closed turnover of matter and energy. Nevertheless, in principal, such an approach needs a new database, e.g., multilevel overlay of large-scale digitised maps. Development of such maps takes many years and has not yet begun.

There are two limits to the use of landscape as a basic unit for relevant large-scale C evaluations. First, the great number of landscapes in Russia (several thousand) precludes the gathering of the detailed information on each landscape necessary for C analysis. Second, important information is normally only available by administrative units which are aggregates of several landscapes. Thus, in our analyses we have replaced the more detailed landscape level with aggregations of landscapes, called *ecoregions*. We divided all of the Russian territory into ecological regions using the following criteria:

- Each ecoregion's impact on the global C cycle should have a similar magnitude. The level of productivity [annual net primary production (NPP)] is an example of a possible indicator for the division.
- The climate within an ecoregion should be homogenous enough to permit a description of the ecoregion's basic features and its impacts on the C budget and, consequently, on the forest associations/types, structure, state and production of the terrestrial biota. We used the existing climatic and vegetation subzones (Kurnaev 1973).
- The macro-relief must be considered, and the distinction between mountainous and flat areas has to be taken into account.
- Frozen areas must be taken into account. Therefore, the areas must be divided into permafrost, soils frozen for extended periods and unfrozen soils.
- The levels and basic features of anthropogenic loads and disturbances of the terrestrial biota and the ecological status of forests and landscapes must be homogeneous.
- Administrative boundaries should be used as guides for ecoregion division so that official data can be used.

A division of this kind allows C budget results to be compared at different levels such as the regional, national, or international level. Unfortunately, the principles for the ecoregional division mentioned above have built-in contradictions and we could not avoid some of these in the final division. As a result of this division, the European-Ural part of Russia has been divided into 78 ecoregions and Siberia and the Far East have been divided into 63 ecoregions. Although the average areas of the ecoregions in European and Asian Russia differ by about fivefold, basic C pools and fluxes are of similar magnitude in the two parts of Russia. However, the variation in the basic indicators of the C cycle is still high between ecoregions. For example, the total NPP, based on Bazilevich's estimates (1993), varies between 20 and 330 Tg for dry organic matter for the ecoregions, with an average of 113.4 Tg.

The temporal scale of the C budget is extremely important. Current C flows depend mainly on the rate of forest succession, land-use changes, turnover time of C in different C pools, time of decomposition of organic material (such as natural mortality in forests and disturbances) and ecoregional characteristics. In Russia, the temporal scale varies from a few years (for green parts in the south) to several hundred years (for decomposition of wood in the extreme north). Carbon budget estimates without accurate identification of the temporal scale and periods of indicators' averaging, are of limited use; nevertheless, some publications (including several listed in Table 12.1) do not consider this problem specifically.

Models

There are three basic types of approaches or models currently available for assessment of boreal (terrestrial) biota in the C budget:

1. Dynamic models, mostly of a global scale (e.g., Goudrian and Ketner 1984; Esser 1987, 1990; Kokorin et al. 1993; Hudson et al. 1994; Alcamo 1994) that take into account some rather common prerequisites and considerations, mainly the CO_2-fertilisation effect and the geographical distribution of productivity.
2. Models that attempt to estimate all essential items of the C budget for the system under evaluation—so-called book-keeping models (e.g., Woodwell et al. 1978; Houghton et al. 1983).
3. Modelling systems based on a unified conceptual approach using components of the two approaches mentioned above (Kurz et al. 1992; Trexler 1991).

Each of the three approaches tries both to solve some specific problem of the C budget and to get a general understanding of the problem. The third approach is the most relevant for C balance estimates for individual regions and countries. Probably the most complete conceptual approach to date was developed for the Canadian forest sector by Kurz et al. 1992). This approach also seems to be sufficient for international investigations and for comparisons of the C problem of the forests in the Northern Hemisphere. Basic features of this approach were used for these analyses concerning Russia, although numerous adaptations, modifications, and changes had to be made in classifications and in modelling concepts.

UNCERTAINTIES

There are different types and sources of uncertainties in the C budget estimates for Russia.

Definitions and classification schemes

Often there is an absence of effective standards in the Russian data and the definitions used change over time. Many of the identified difficulties are connected with the essentially inconsistent nature of administrative classification, such as land-use categories, definitions of forests and different forest subdivisions, and forest indicators such as stocking and site indices.

Shortcomings of available data

Some important indicators are not and have never been measured. This can strongly impact the final results of the C calculations. Some typical examples of this are addressed below.

Russian forest inventory does not detect net and gross growth (current increment) and mortality directly in the field; existing national estimates of total growth differ 1.5 to 2-fold. Different regional estimates of growth are often incomplete or are based on different, sometimes incorrect, models and methods.

The same is true for the detection of the type of age structure of stands (TASS). For Siberia and the Far East, this factor may influence the net ecosystem production (NEP) of forests by 15% to 20% due to the fact that 40–80% of the area is covered by major forest-forming species of different types of uneven-aged stands (Table 12.2). Based on numerous investigations of the TASS in Russian forests (e.g., Semetchkin 1970; Gusev 1964; Shanin 1965) the following classification of the TASS has been employed in the analyses:

- EVA = even-aged stands, for which at least 70% of the growing stock is produced by trees belonging to one age class. The age-class interval for coniferous species in Russia, as well as for deciduous species of seed regeneration, is 20 years; for cedar it is 40 years; and for the remaining species (with some minor exceptions) it is 10 years. (Variation coefficients for age frequency distribution within stands, Va < 8%; and variation coefficient of the diameter at breast height, Vd < 28%.)
- RUE = relatively uneven-aged stands, for which at least 70% of the growing stock is produced by trees belonging to more than one, but no more than two, age classes (Va = 8% to 15%; Vd = 28% to 35%).
- UEV = uneven-aged stands. The growing stock is produced by trees of more than two, and usually of all, age groups (the age groups are young, middle-aged, premature, mature, and overmature and generally contain several age classes) (Va >15%; Vd > 35%).
- GUE = gradually uneven-aged stands. The growing stock is composed of more than one canopy layer in which dominant species belong to different age groups or different TASS (EVA, RUE, or UEV).

Current forest fire statistics in Russia present data for three kinds of forest fires: crown, ground-based and belowground fires. Crucial for estimation of the C emissions is knowledge of areas and intensity of litter, turf and peat fires. More detailed data on the fire characteristics may substantially alter the results.

Table 12.2. Types of age structure of stands for Siberia and the Far East. The data have been calculated using different Russian publications and the International Institute for Applied Systems Analysis (IIASA) database on ecoregions for Russia.

	Distribution by age structure type[a] (% of forested areas)							
Species	EVA	RUE	UEV	GUE	EVA	RUE	UEV	GUE
	Middle taiga				South taiga			
Pine	23	51	25	1	38	49	12	1
Spruce	1	41	40	18	26	27	25	22
Fir	16	28	14	42	27	24	10	39
Larch	10	69	21	0	10	74	16	0
Cedar	19	28	28	25	33	20	22	25
	Mixed forests				Forest steppe			
Pine	-	-	-	-	80	17	2	1
Spruce	0	70	25	5	15	30	20	35
Fir	-	-	-	-	16	30	10	44
Larch	10	74	16	0	18	71	11	0
Cedar	-	-	-	-	17	29	14	40

a EVA = even-aged,; RUE = relatively uneven-aged; UEV = uneven-aged; GUE = gradually uneven-aged. See text for full description of these acronyms:

Unknown or insufficient accuracy of initial data

Part of the problem with accuracy of Russian data stems from Russia's former centrally planned system. Statistics on forest fires before 1988 are not reliable because large uncontrolled forest fires were regarded by former central authorities to be a result of inadequate local management. This situation caused the official statistics of the actual burnt areas to be underestimated two to tenfold, and sometimes more. An attempt in the 1980s to determine the real burnt area in the extreme north-eastern area of the Far East, including some areas with no fire protection, found an approximately 50-fold difference between official data and remote sensing data. The second problem connected with forest fires is the absence of statistics on fires in large areas with no fire protection (around 35% to 40% of the total Forest Fund area has no fire protection). No measured data are available for these areas, and different expert estimates vary two to threefold (Stocks 1991; Krankina 1993; Dixon and Krankina 1993; Shvidenko et al. 1994b).

Significant uncontrolled bias by transformation of local (point) measurements to aggregated characteristics

Crucial parameters for estimation of the C budget are productivity and especially its components such as phytomass, necromass and NPP which are used to estimate the NEP. There are several estimates on such parameters for both terrestrial biota and forests in Russia. However, only one thorough, systematic study has so far been carried out for all of Russia (Bazilevich 1993). In this study, more than 2000 direct measurements of these parameters were made and the results have been overlaid and digitised for ecoregions showing aggregated data on phytomass, necromass (also known as mortmass) and NPP (Table 12.3). In order to transform this information into C data, we used the conversion factor 0.5 for woody phytomass and 0.45 for the rest of the vegetational components. The share of the woody phytomass has been estimated based on the percentage of forest cover and average growing stocks of the forests.

Validation of these data is only possible in an indirect way. The average global NPP (g C m^{-2} yr^{-1}), estimated by 13 different studies during the last 20 years and cited by Goldwijk et al. (1994), is estimated to be 403 g C m^{-2} yr^{-1}, ranging from 324 to 511g C m^{-2} yr^{-1}. The two most recent model results available, by IMAGE 2.0 (Goldwijk et al. 1994), and by TEM (Melillo et al. 1993), vary from 418 to 434 g C m^{-2} yr^{-1}. If the areas of the most productive evergreen tropical forests are excluded, the average global NPP is in the range of 310 to 320 g C m^{-2} yr^{-1} (Goldwijk et al. 1994), which is close to our result of 311 g C m^{-2} yr^{-1} for Russia.

Table 12.3. Carbon content in terrestrial vegetation in Russia. From the digitised maps of Bazilevich (1993).

		PHYTOMASS		NECROMASS		PRODUCTION	
ZONE[a]	AREA (Mha)	TOTAL (Tg)	DENSITY (kg m^{-2})	TOTAL (Tg)	DENSITY (kg m^{-2})	TOTAL (Tg)	DENSITY (g m^{-2})
AD	24.8	139	0.560	388	1.565	19.2	77
T	268.2	4383	1.634	6550	2.442	434.0	162
FT	291.2	9200	3.159	10 536	3.620	544.3	187
NT	125.6	6096	4.853	8792	7.000	243.6	194
MT	385.9	25 747	6.672	14 197	3.679	944.1	245
ST	232.1	19 460	8.384	11 495	4.953	801.7	345
MF	78.1	9680	12.410	3353	4.293	404.4	518
DF	61.0	6289	10.310	2043	3.349	370.3	607
FS	103.6	4460	4.305	1868	1.803	837.1	808
S	80.0	908	1.135	628	0.785	521.9	652
SDD	24.3	186	0.765	72	0.296	83.4	343
Total	1674.8	86 548	5.168	59 922	3.579	5204	311

a AD: Arctic desert; T: tundra; FT: forest tundra + sparse taiga + meadow forests; NT, MT, and ST: northern, middle, and southern taiga, respectively; MF: mixed forests; DF: deciduous forests; FS: forest steppe; S: steppe; SDD: semi-deserts and deserts.

The estimate using IMAGE 2.0 for the Commonwealth of Independent States (CIS) is 354 $g\,C\,m^{-2}\,yr^{-1}$. Productivity of the Russian terrestrial biota estimated by means of a specific chlorophyll index (Mokronosov 1994) is lower (258 $g\,C\,m^{-2}\,yr^{-1}$). The estimate by Mokronosov for taiga biomes (303 $g\,C\,m^{-2}\,yr^{-1}$) is higher than our estimate based on the Bazilevich (1993) data (275 $g\,C\,m^{-2}\,yr^{-1}$). Estimates for subarctic areas (85 $g\,C\,m^{-2}\,yr^{-1}$) and for forest tundra (175 $g\,C\,m^{-2}\,yr^{-1}$) have been presented by Zamolodchikov (1994). Zamolodchikov (1994) also suggests an estimate of 144 $g\,C\,m^{-2}\,yr^{-1}$ for flat tundra, which can be compared with our estimate (162 $g\,C\,m^{-2}\,yr^{-1}$) based on the Bazilevich (1993) data.

The average global estimate of living phytomass based on eight global estimates is 4.64 $kg\,C\,m^{-2}$, with a range of 3.7 to 5.6 $kg\,C\,m^{-2}$ (Schlesinger 1977; Atjay et al. 1979; Goudrian and Ketner 1984; Olson et al. 1985; Polgase and Wang 1992; Smith et al. 1992; Cramer and Solomon 1993; Goldwijk et al. 1994). This is less than our result for Russia of 5.17 $kg\,C\,m^{-2}$. The IMAGE 2.0 (Goldwijk et al. 1994) estimate for the CIS is 4.9 $kg\,C\,m^{-2}$. Calculations for Russia based on data from Goldwijk et al. result in about 5.9 $kg\,C\,m^{-2}$ or a total of 98.8 Pg C. The average estimate of the global C in phytomass is about 640 Pg, and based on our calculations it can be estimated that terrestrial biota of Russia is about 14% of the global estimate. Dixon et al. (1994) estimated C densities for the Russian forests to be 8.3 $kg\,C\,m^{-2}$ in the vegetation.

The data presented above, when combined with aggregated data on decomposition of organic matter (Pastor and Post 1986; Esser 1990; Grishina et al. 1990; Bonan and Van Cleve 1992) can be used for large-scale NEP estimates. Accuracy of such regional applications is unknown due to the variety and complexity of the natural and anthropogenic mortality of forests and uncertainties in numerical estimations of C exchange in the forest-soil system. It is necessary to stress that aggregation into larger scales like the Bazilevich maps, based on numerous research projects carried out over a long time, can only be used for overall and aggregated calculations.

Another approach for the above calculations would be to use forest inventory data. Isaev et al. (1993) used the Russian Forest State Account (FSA) and employed conversion factors in order to transfer C stored in stemwood into C stored in the overall forest ecosystems. This study reports a total C amount in forested areas of Russia to be 39.8 Pg C (or 5.16 $kg\,C\,m^{-2}$) and to be 1.4 Pg (0.53 $kg\,C\,m^{-2}$) in non-forest and unforested lands. The latter value does not coincide with other estimates. Zamolodchikov (1994) for example, reports 0.96 Pg C for tundra, which is the least productive of the non-forest lands in the northern territories. The difference can probably be explained by the different methods and data employed in the analyses.

This contradiction requires a discussion on the accuracy of the FSA, and in particular on the reliability of the FSA data, for estimates of changes of C sequestered in wood between inventory periods. Numerous investigations (Antanaitis and Repshis 1973; Fedosimov 1986) have shown that the growing stock has been underestimated by the Russian ground-based inventory by at least 5% to 15%. A comparison of traditional inventory data with a special control inventory in Ivanovo *oblast* showed a systematic underestimate of 15.7% (Anuchin et al. 1970) and of 7% for all of Lithuania (Antanaitis and Repshis 1973).

There is also another source of systematic error in the Russian inventory data and in the FSA. By 1956 all the forests of the former Soviet Union had been inventoried but the majority of the taiga forests were inventoried by so-called *visual air surveys* with very low accuracy. By 1966, only 41% of the Forest Fund areas in the European part of Russia and only 22% in the Asian part, had been inventoried by ground-based methods. Additional areas were subsequently inventoried by ground-based methods. The control of differences between visual air survey data and later inventory data, provided by more reliable methods and gathered in the 1980s on large territories, showed a consistent overestimation by the visual air survey method.

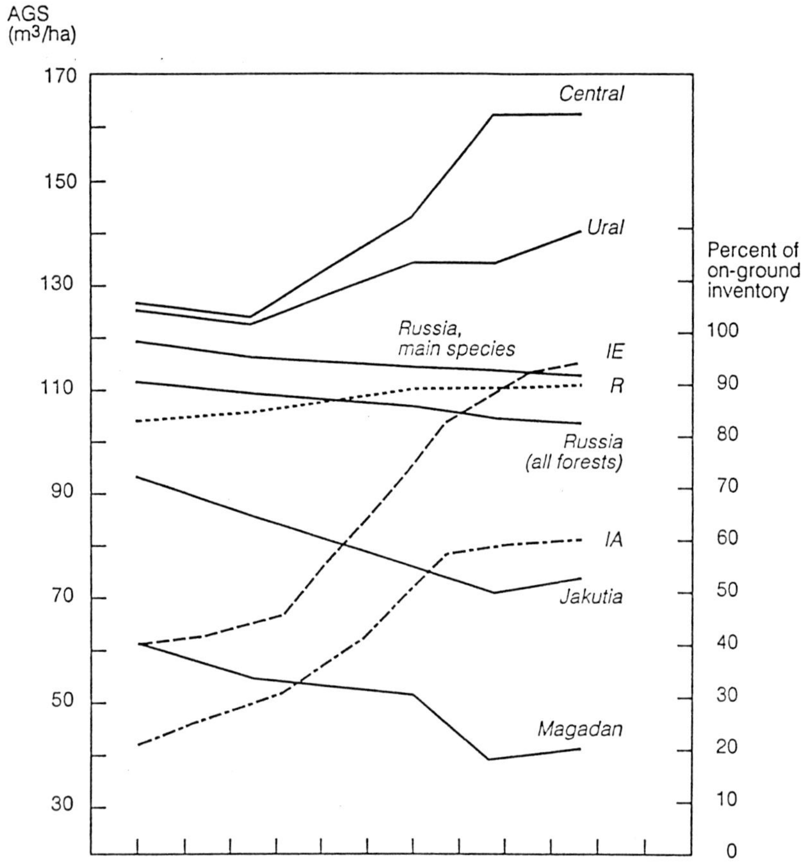

Figure 12.2. Average growing stock (m^3 ha^{-1}) for Russia and separate regions from 1966 to 1988. R = growing stock (m^3 ha^{-1}) by our reconstruction; IE and IA = accumulated percentage of Forest Fund areas inventoried by ground-based methods in European and Asian parts of Russia, respectively

In Yakutia, for four enterprises (about 20 Mha) the growing stock according to the up-to-date inventory was 23.6% lower than had been estimated using visual air surveys. For the Turukhansky *leskhoz* of Krasnojarsk *kraj* (14.2 Mha forested area) the difference was 37%. However, there is a rather long time span (some 30 years) between the visual air survey inventory and the more reliable inventory. For instance, in the enterprises in Yakutia, the forested areas increased by 17.9% during this time span and the average growing stock decreased by 34.7%. These inventory conditions also explain the reported development of the gross growth in some Russian regions. For example, in Yakutia and Magadan, an essential decline of the average growing stock occurred during the period from 1966 to 1988; this decline corresponds with an increase in the number of areas inventoried by reliable methods (Figure 12.2). This process cannot be explained by disturbances, e.g., by forest fire.

We tried to take these two sources of systematic errors into account when reconstructing the *real* dynamics of the growing stock in Russia during the period from 1966 to 1993. The correction factor for the ground-based inventory has been assumed to vary from +15% to +8% in Asian Russia and from +12% to +6% in European Russia. The calibration factors for the visual air survey data varied from -25% to -10% in Asian Russia and from -18% to -6% in European Russia. The official dynamics and the reconstructed dynamics of the Russian forests are presented in Table 12.4.

Table 12.4. Dynamics of Russian forests, 1966 to 1993. Growing stock (stemwood over bark) is presented for forested areas.

	YEAR					
INDICATOR[a]	1966	1973	1978	1983	1988	1993
Official statistics data						
FF (Mha)	1161.9	1161.4	1182.2	1187.2	1182.6	1180.9
FA (Mha)	705.6	729.7	749.5	766.6	771.1	763.5
GS (10^9 m^3)	77.0	78.7	80.7	81.9	81.6	80.7
ER (10^9 m^3)	17.0	17.4	18.7	19.3	20.3	21.1
AR (10^9 m^3)	60.0	61.3	62.0	62.6	61.4	59.6
OIAR (%)	22	30	38	52	59	60
OIER (%)	41	44	56	75	88	94
Reconstructed dynamics						
GS (10^9 m^3)	73.5	76.7	80.4	84.5	85.6	84.8
ER (10^9 m^3)	16.0	17.0	18.3	19.9	21.4	22.2
AR (10^9 m^3)	57.5	59.7	62.1	64.6	64.2	62.6

a FF = Forest Fund area; FA = forested area; GS = growing stock; ER = European Russia; AR = Asian Russia; OIAR and OIER = percentage of ground-based inventory in Asian and European Russian Forest Fund, respectively.

Reconstructed growing stock dynamics do not differ greatly from the official FSA data; there is a difference of -4.5% to +3.8%. However, the differences generate large impacts on the C budget. Without discussing the details, we present two conclusions.

1. Using an average ratio between the volume of whole tree wood and stemwood of 1.4, an average green moisture content of 50%, and C content of dry wood of 0.50, (territorial and seasonal variability of these ratios is an additional source of uncertainty) we can calculate the C fluxes. Based on the official statistics the C sequestered in wood during the period from 1966 to 1993 averaged 1.3 Pg or 50 Tg annually. If the reconstructed dynamics are used, the sequestered wood C is estimated to average 4 Pg or 152 Tg C yr^{-1} for the same period. The sequestration of C during the period from 1966 to 1983 was about 105 Tg C yr^{-1} based on the official inventory data and 236 Tg C yr^{-1} based on our reconstructed dynamics. In both cases, during the last 10 years the Russian forests did not sequester any C. The reason for this is a large decrease of the growing stock in Asian Russia. The forests of European Russia were still a C sink: 65 Tg annually (official data) and 84 Tg annually (reconstructed data) during the last 10 years.
2. We can only explain half of the disagreement between the C storage in Russian forest biomes reported by Isaev et al. (1993) and our results calculated from Bazilevich's data. The Isaev et al. (1993) estimate is about 15% lower than our estimate which is based on

the reconstruction of the growing stock. It should be noted that the differences considered in point 1 above deal only with wood. Often-used ratios to convert growing stock (or merchantable) wood to total forest phytomass or to C of entire forest ecosystems vary over a wide range. Sampson (1992) used 0.53 to convert timber volume (m^3) to total forest phytomass; the same ratio calculated on data by Isaev et al. (1993) for Russian forests is 0.49; Sedjo (1992) used the ratio 4.0 for boreal forests of the FSU to convert accumulated C of stemwood into C of forest ecosystems; the estimate given by Shvidenko et al. (1994a) for the FSU in 1966–1988 was 3.1, etc. The accuracy of these aggregated ratios is unknown.

Short time-series as a source of additional uncertainties

Estimation of post-disturbance biogenic C fluxes is an important part of C evaluation. As a rule, disturbances generate long-term C fluxes. For instance, post-fire biogenic flux (PFBF) can be expressed as follows (Shvidenko et al. 1994b):

$$PFBF{=}F\left\{t,\left[(DUW+DPFD)+SOC-(RF+BE)\right]-ChC\right\}_{i,j,k,q} \qquad [1]$$

where F is a function of time, t; DUW is decomposition of non-combusted wood; $DPFD$ is decomposition of post-fire dieback; SOC is soil organic changes; RF is regrowth of forests; BE is the boreal effect (increased productivity on post-fire areas in the boreal zone, especially on permafrost); ChC is charcoal stored in soil; and i, j, k, and q are indices of different classifications (i = climatic zones; j = C pools; k = types of fires; q = kinds of forest combustibles).

The *pure* fire impact is estimated without taking $\{(RF + BE) - ChC\}_{i,j,k,q}$ of the above function into account.

The function $F\{t, (DUW + PRFD)\}$ is defined in the following way: $O_{ij}(t)$ expresses the amount of organic matter originating from the decomposition at time t; the curve $G(t^*)$ (Figure 12.3) describes the decomposition of organic matter $DUW + DPFD$ which is in the decomposition pool in year t^* and will decompose during the period TD. Let the time for the decomposition of 95% of all the organic material be $T(0.95)$, then

$$\tau = t^* - t', \qquad \text{for} \quad 0 \le \tau \le \text{int}[T(0.95)+1] \qquad [2]$$

where τ is the period between t^* and t' during which the C emission is considered. Usually an exponential character of $G(t)$ is anticipated for homogenous pools:

$$G(t) = e^{-\alpha t} \qquad [3]$$

from which $T(0.95) = \ln[20/\alpha]$. It can be shown that the total amount of decomposed organic material in year t' from the annual inputs during the period $\{0, \text{int}[T(0.95)] + 1\}$, int($x$) is the integer part of x and can be calculated according to

$$G_{ij}(t) = (e^{\alpha_{ij}} - 1)\sum_{\tau=0}^{\Phi+1} O_{ij}(t-\tau)e^{-\alpha_{ij}\tau} \qquad [4]$$

where $\Phi = \text{int}[T(0.95) + 1]$.

Duration of the decomposition periods for different C pools in boreal forests varies from several months to about 200 years.

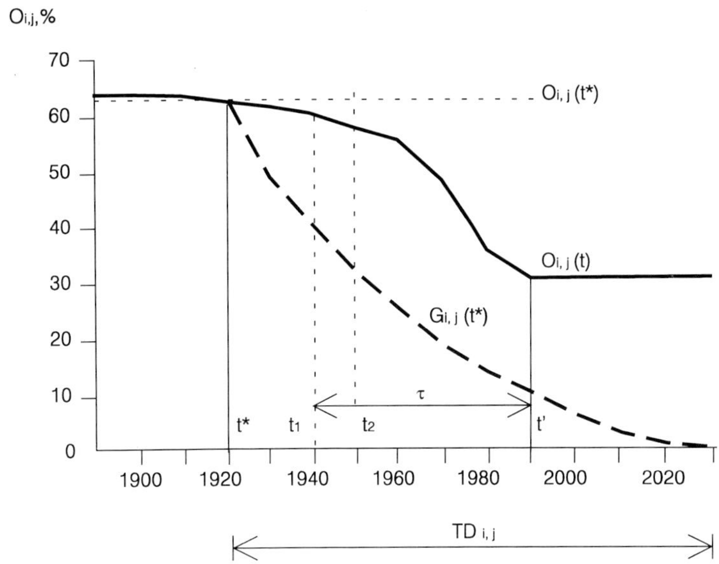

Figure 12.3. Decomposition of forest-fire dieback (explanation in the text).

The accuracy of the results depends on the curves $O_{ij}(t)$, which can be reliably reconstructed for the Russian conditions only for the last few years. We have assumed the following: the total input of organic matter for decomposition in year t is in proportion to the area burnt and the areas of dead stands at that point in time; the ratio of different kinds of forest fires and the geographical distribution of fires are the same for the whole period studied; and during the period from 1900 to 1960, the areas of burnt forests were of the same magnitude (Figure 12.3). Beginning from the 1960s, the areas decreased due to forest-fire suppression.

The reliable identification of the curve $G(t)$ is possibly based on experimental data on decomposition rates for litter. However, the data for the woody C component are very poor, and the best that can be done is to use expert estimates based on ecoregional climatic conditions and dominant tree species. The indirect post-fire fluxes for Russia during the 1990s are estimated to be 1.5 to 2 times higher than the direct ones. Nevertheless, the specific form of the curve $G(t)$ provided relatively stable estimates with changes of the values of $O_{ij}(t)$ up to 30%. For other types of disturbances (e.g., final harvesting) the errors may be much higher.

Simplification by the modelling approach and difficulties with verification

There are two types of uncertainties caused by the modelling approach that cannot be avoided. One of them is the impossibility of achieving an exact reflection of the complicated natural processes using models. The second is the problem of uncertainties of forecasts which in the frame of the C analyses can generate difficulties (e.g., quantitative impact of CO_2 fertilisation, modified fire frequency, etc.).

Another difficulty is dealing with long-term forecasts of forests' reactions to climate change. Attempts to formulate the most probable regional climate change scenario for the boreal zone in Russia have been made by Ganopolsky (1994) on the basis of analysis of the latest general circulation model scenarios produced for the Intergovernmental Panel on

Climate Change (IPCC) (Greco et al. 1994). These scenarios show common features of the climate change predicted by the different models on a global scale; however, on the regional level the differences between the examined projections are significant. The impact of extreme values of climatic characteristics (temperature, precipitation) and their specific distribution (very hot and dry periods) on forest fire frequency and intensity, and outbreaks of insect and disease may be more important for boreal forests than is the direct change of the climatic trend. Currently there are no models that can provide this information.

Evidently, only partial models of some processes of current interactions between forests and gas composition of the atmosphere can be verified by quantitative methods. No direct verification of forecasted futures by the models is possible.

FOREST MANAGEMENT REGIMES ORIENTED TOWARD INCREASED CARBON SEQUESTRATION

In contrast to the limited capability for predicting reliable impacts of climate change on the forest C budget, the impacts of forest management regimes are more evident. The most significant source of uncertainties in this area is the incomplete knowledge of how the proposed management regimes will be implemented in practice which is crucial for the results of the C budget. For instance, there are about 107 Mha of forests constituted by soft deciduous species in Russia (about 39 Mha in European Russia). The majority of the European soft deciduous forests are growing on sites of high productivity and have been replacing the original coniferous species following fires and clearcuts. The life span of the soft deciduous species is relatively short and the production of soft deciduous species, with a rotation period of 40 to 60 years, can provide higher C sequestration than can coniferous forests which have a longer rotation. However, the current level of harvest of soft deciduous stands in Russia is only 40% of what is regarded as a sustainable annual allowable cut. The soft deciduous stands are even-aged. This means that large areas of soft deciduous stands will degenerate and decompose during the next 40 to 50 years which will increase the CO_2 flux into the atmosphere.

Clearly, any predictions are the result of prerequisites used; changes in these assumptions can alter the conclusions in essential ways. This may be demonstrated with an example: the following assumptions about the implementation of forest operations were used to develop scenarios of possible C sequestration by Russian forests. First, the cost of the sequestration of 1 Mg C should not exceed US$3 (1991 value), in order to be competitive in the framework of a global program of large scale reforestation (Nilsson 1993). Second, the forest actions included in the program should be realised during the next 40 years. The time horizon for the analyses of the impact of the scenario is 100 years. One main issue of the scenario is the protection of existing forests. Analyses of the future C-sequestration potential of Russian forests (Shvidenko et al. 1994a, b) show that implementation of different forestry operations within the framework of the scenarios discussed above (such as large-scale reforestation, human replacement of low productivity stands, better fire protection, improved health status of forests, and more efficient wood utilisation) may result in an additional C sink of between 28.5 and 62.0 Pg C during a period of 100 years, depending on whether a "realistic" or "optimistic" scenario is achieved (Table 12.5). In the optimistic scenario, a set of upper estimates of indicators was used, no risk functions were considered, and all wood removed from forests was considered as a substitution of fossil fuel, i.e., was included in the *sink* estimate.

Table 12.5. Possible increase of carbon sequestration due to improved forest activities in Russia.

MEASURE TAKEN	MAX40[a] (Mha)	RATE (Mha yr^{-1})	CADD[b] (Pg)	TOTAL C SINK (Pg)		AVERAGE SINK (Mg ha^{-1} yr^{-1})	
				REAL[c]	OPTIMUM[d]	REAL	OPTIMUM
Large-scale reforestation on unforested areas	64	1.6	9.79	6.87	13.98	1.34	2.73
Reforestation of clear-cut and burnt areas	20	0.5	3.06	2.14	4.37	1.34	2.73
Replacement of low productive stands	60	1.5	6.72	3.98	12.07	0.83	2.52
Reconstruction of climax stands	20	0.5	2.16	1.25	3.85	0.78	2.41
Replacement of soft deciduous stands	25	0.6	2.70	2.20	5.14	1.09	2.57
Implementation of thinnings	75	4.0	-	-	2.00	-	5.00
Improvement of fire protection	1100	-	-	6.89	10.50	-	-
Mitigation of insect/disease impact	-	-	-	2.10	5.40	-	-
Efficient utilisation of wood	-	-	-	2.30	4.70	-	-
Total			24.4	27.7	62.0		

a MAX40 is the area that can be improved by the end of the 40 year period (see text).
b CADD is additional C fixation at the end of a 100 year period.
c Realistic scenario (see text).
d Optimistic scenario (see text).

This estimate is based on the assumptions that climate and land use will not change in the future. The accuracy of projections is dependent upon the following: the reliability of existing climate change scenarios; the uncertainties of up-to-date estimates of direct and indirect impact of climate change on forests; the reliability of economic and demographic forecasts for Russia; and the real rate of implementation of the paradigm of sustainable development in Russia. In the current situation there are limited possibilities for validating the accuracy of this estimate for C sequestration by Russian forests.

RESEARCH NEEDS

Based on the results of this study, we can identify a number of topics that are crucial to further research concerning the C sequestration potential of Russian boreal forests and landscapes.

- Improved knowledge of bioproductivity of forests and terrestrial biota based on improved data for phytomass, detritus, growth (increment) and natural mortality of forest stands is required. A specific modelling system for the estimation of increment and mortality in Russia is needed.
- A monitoring system for the estimation of the impacts and real extent of natural and anthropogenic disturbances must be developed.
- The changes in soil organics caused by natural and post-disturbance processes must be investigated.
- Regional models for the dynamics of soil organic matter and vegetation in permafrost areas must be investigated.

- A set of models for the estimation of changes in productivity, areas and species composition due to climate change must be developed.
- A set of models for the analysis of indirect climate change impact on forests, such as forest fires and outbreaks of insects and disease, must be developed.

REFERENCES

Alcamo J (ed) (1994) IMAGE 2.0: Integrated modeling of global climate change. Kluwer, Dordrecht, The Netherlands, 318 pp.

Antanaitis VV, Repshis IN (1973) Experience of inventory of the Lithuanian forests by mathematic-statistical methods. Forest Industry, Moscow, 102 pp. [In Russian].

Anuchin NP, Fedosimov AN, Chuenkov VS (1970) Results of inventory of forest fund by mathematic-statistical methods in Ivanovo oblast. *VNIILM Pushkino*, 42 pp. [In Russian].

Armentana TV, Ralston CV (1980) The role of temperate zone forests in the global carbon cycle. *Can. J. For. Res.* 10: 53–60.

Atjay GL, Ketner P, Duvigneaud P (1979) Terrestrial Primary Production and Phytomass. Wiley, New York, 187 pp.

Bazilevich NI (1993) Biological productivity of ecosystems of Northern Eurasia. Nauka Publ., Moscow, 294 pp [In Russian].

Bonan GB, Van Cleve K (1992) Soil temperature, nitrogen mineralization, and carbon source-sink relationships in boreal forests. *Can. J. For. Res.* 22: 629–639.

Cramer W, Solomon AM (1993) Climatic classification and future global redistribution of agricultural lands. *Clim. Res.* 3(1–2): 97–110.

Dixon RK, Brown S, Houghton RA, Solomon AM, Trexler MC, Wisniewski J (1994) Carbon pools and flux of global forest ecosystems. *Science* 263: 185–190 pp.

Dixon RK, Krankina ON (1993) Forest fire in Russia: Carbon dioxide emission to the atmosphere. *Can. J. For. Res.* 23(4): 700–705.

Esser G (1987) Sensitivity of global carbon pools and fluxes to human and potential climatic impacts. *Tellus* B 39: 245–260.

Esser G (1990) Modelling global terrestrial sources and sinks of CO_2 with special reference to soil organic matter. In: Bouwman AF (ed) Soils and the Greenhouse Effect. Proceedings of an International Conference, Wiley, New York, pp 247–262.

Fedosimov AN (1986) Forest inventory by sampling methods. Forest Industry, Moscow, 191 pp. [In Russian].

Ganopolsky A (1994) Climate change scenario for boreal forest zone. International Institute for Applied Systems Analysis, Laxenburg, Austria. (Unpublished manuscript).

Goldwijk KK, Minnen van JG, Kreileman GJJ, Vloedfeld M, Leemans R (1994) Simulating the carbon flux between the terrestrial environment and the atmosphere. Kluwer, Dordrecht, pp 199–230.

Goscomles SSSR (1989) Dynamics of the state forest managed forests during 1966–1988 (basic tree species). The USSR State Committee of Forest, Moscow, 156 pp [In Russian].

Goscomles SSSR (before 1988—Gosleshoz) Forest fund of the USSR 1968 (state of 1.01.1966 744 pp), 1976 (of 1.01 1973 V1 600 pp, V2 561 pp, V3 800 pp), 1982 (of 1.01.1978 V1 601 pp, V2 683 pp), 1986 (of 1.01.1983 V1 892 pp, V2 973 pp), 1990 (of 1.01.1988 V1 1005 pp), 1991 (V2 1021 pp). The USSR State Committee of Forest, Moscow [In Russian].

Goudrian J, Ketner D (1984) A simulation study for the global carbon cycle, including man's impact on the biosphere. *Climatic Change* 6: 167–192.

Greco S, Moss RH, Viner D, Jenne R (1994) Climate scenarios and socioeconomic projects for IPCC WG II. Intergovernmental Panel on Climate Change, Geneva.

Grishina LA Koptick GH, Makarov MT (1990) Transformation of soil organic matter. Moscow State University, Moscow, 88 pp [In Russian].

Gusev II (1964) Structure and peculiarities of taxation of spruce stands in the north. forest industry, Moscow 75 pp [In Russian].

Houghton RA, Hobbie JM, Melillo JM, Moore B, Peterson BJ, Shaver BJ, Woodwell GM (1983) Changes in the carbon content of terrestrial biota and soils between 1860 and 1980: A net release CO_2 to the atmosphere. *Ecol. Monogr.* 53: 235.

Houghton RA (1991) Releases of carbon to the atmosphere from degradation of forests in tropical Asia. *Can. J. For. Res.* 21: 132–142.

Houghton RA (1993) Is carbon accumulating in the Northern temperate zone? *Global Biogeochem. Cycles* 7(3): 611–617.

Hudson RJM, Cherini SA, Goldstein RA (1994) Modeling the global carbon cycle: Nitrogen fertilization of the terrestrial biosphere and the "missing" CO_2 sink. *Global Biogeochem. Cycles* 8(3): 307–333.

Isaev AS, Korovin GN, Utkin AI, Priashnikov AA, Zamolodchikov DG (1993) Estimation of carbon pool and its annual deposition in phytomass of forest ecosystems in Russia. *Forest Science* (Lesovedenie), Moscow 5: 3–10 [In Russian].

Johnson WC, Sharpe CM (1983) The ratio of total to merchantable forest biomass and its application to the global carbon budget *Can. J. For. Res.* 13: 372–383.

Kauppi PE, Mielikainen K, Kuusela K (1992) Biomass and carbon budget of European forests, 1971 to 1990. *Science* 256: 70–74.

Kokorin AO, Nazarov IM (1993) The role of Russian forests in global CO_2 uptake from the atmosphere. Presented at the Int. Symp. Soil Processes and Management Systems: Greenhouse gas emissions and carbon sequestration. April 1993, Columbus, OH, 11 pp.

Kolchugina TP, Vinson TS (1993a) Equilibrium analysis of carbon pools and fluxes of forest biomes in the former Soviet Union. *Can. J. For. Res.* 23: 81–88.

Kolchugina TP, Vinson TS (1993b) Comparison of two methods to assess the carbon budget of forest biomes in the former Soviet Union. *Water, Air, Soil Pollut.* 70: 207–221.

Krankina ON (1993) Forest fires in the former Soviet Union: Past, present and future greenhouse gas contributions to the atmosphere. In: Vinson TS, Kolchugina TP (eds) Proceedings of the International Workshop on Carbon Cycling in Boreal Forest and Subarctic Ecosystems: Biospheric Responses and Feedback to Global Climate Change, EPA/600/r–93/084. Corvallis, OR, pp 179–186.

Kurnaev SF (1973) Forest growth division of the USSR. Nauka Publ., Moscow, 202 pp [In Russian].

Kurz WA, Apps MJ, Webb TM, McNamee PJ (1992) The carbon budget of the Canadian forest sector: Phase I Forestry Canada Information Report NOR-X-326, Forestry Canada, Northwest Region, Edmonton, AB, 93 pp.

Makarov AA, Bashmakov I (1990) The Soviet Union. In: Chandler, WU (ed) Carbon Emission Control Strategies: Case studies in International Cooperation. 263 pp.

Melillo JM, Fruci JR, Houghton RA, Moore B III, Scole DL (1988) Land-use change in the Soviet Union between 1850 and 1980: Causes of a net release to the atmosphere. *Tellus* B 40: 116–128.

Mokronosov AT (1994) Photosynthesis and change of CO_2 content in the atmosphere. *Nature (Priroda), Moscow* 7: 25–27. [In Russian].

Nabuurs CJ (1994) State-of-the-art in the field of forest sector carbon budget studies—with reference to the European situation. European Forestry Institute, Working Paper 2, Joensuu, Finland. 87 pp.

Nilsson S (1993) Carbon sequestration potential by global afforestation. In: Nakicenovic N *et al.* (eds) Long-term strategies for mitigating global warming. *Energy* 18: 104–112.

Nilsson S, Sallnas O, Hugosson M, Shvidenko A (1992) The forest resources of the former European USSR. Parthenon Publ. Group, Carnforth, UK, 407 pp.

Olson J, Watts JA, Alison LJ (1985) Major world ecosystem complexes ranked by carbon in live vegetation: A database. Carbon Dioxide Information Center, Oak Ridge National Laboratory, Oak Ridge, TN.

Pastor J, Post WM (1986) Influence of climate, soil moisture, and succession on forest carbon and nitrogen cycles. *Biogeochemistry* 2: 3–27.

Polgase PH, Wang YP (1992) Potential CO_2 enhanced carbon storage by the terrestrial biosphere. *Aust. J. Bot.* YO: 641–656.

Sampson RN (1992) Forestry opportunities in the United States to mitigate the effects of global warming. *Water, Air and Soil Pollut.* 64: 83–120.

Schlesinger WH (1977) Carbon balance in terrestrial detritus. *Ann. Rev. Ecol. Syst.* 8: 51–81.

Sedjo RA (1992) Temperate forest ecosystems in the global carbon cycle. *Ambio* 21: 274–277.

Semetchkin IV (1970) Dynamics of age structure and methods for its evaluation. In: Problems of Forestry, Vol 1 Institute of Forest and Wood, Russian Academy of Sciences, Krasnojarsk, Russia, pp 442–445. [In Russian].

Shanin SS (1965) Structure of pine and larch stands in Siberia. Forest Industry, Moscow, 104 pp. [In Russian].

Shaver GR, Hobbie JE, Houghton RA, Melillo GM, Moore B, Peterson BJ, Woodwell GM (1982) The role of terrestrial biota and soils in the global carbon budget. In: Brown S (ed) Global Dynamics of Biospheric Carbon, US Department of Energy, Conf - 8108131, pp 160–165.

Shvidenko A, Nilsson S, Rojkov V (1994) Status and possibilities for increased influences of terrestrial biota on the carbon budget in the territories of the former USSR. In: Mathai CV, Stenseand G (eds) Proceedings of the Air and Waste Mgmt. Assn., International Specialty Conference: Global Climate Change—Science, Policy and Mitigation Strategies, 5–8 April, 1994, Phoenix, AZ, pp 923–939.

Shvidenko A, Nilsson S, Rojkov V (1994a) Possibilities for increasing carbon sequestration by implementation of rational forest management in Russia. *Water, Air, Soil Pollut.* (Unpublished manuscript).

Shvidenko A, Nilsson S, Rojkov V (1994b) Possibilities for increased carbon sequestration by improvement of protection of Russian forests. *Water, Air, Soil Pollut.* (In press).

Smith TM, Weiskampel JF, Shugart HH, Bonan GB (1992) The response of terrestrial storage to climate change: Modelling C dynamics at varying temporal and spatial scales. In: Wisniewski J, Lugo AE (eds) Natural sinks of CO_2. Kluwer, Dordrecht, The Netherlands, pp 307–326.

Stocks BJ (1991) The extent and impact of forest fires in Northern Circumpolar countries. In: Levine JS (ed) Carbon Biomass Burning: Atmospheric, Climatic, and Biospheric Implications. Massachusetts Institute of Technology Publ., Cambridge, MA, pp 197–202.

Suchachev, VN (1964) Basic notions of forest biogeocenology. In: Suchachev VN, Dylis NB (eds) Foundations of forest biogeocenology. Nauka, Moscow, pp 5–49 [In Russian].
Tukey JW (1962) *Ann. Math. Stat.* 33: 13.
Trexler MC (1991) Minding the carbon store: Weighing US Forestry strategies to slow global warming. WRI, New York, 81 pp.
Woodwell GM, Whittaker RH, Reiners WA, Likens CE, Delwiche CC, Botkin DB (1978) The biota and the world carbon budget. *Science* 199: 141–146.
Zamolodchikov DG (1994) Carbon balance of tundra and forest tundra zones. *Nature (Priroda), Moscow* 7: 22–24. [In Russian].
Zavarzin GA (1992) International ecological convention. *Nature (Priroda), Moscow* 12: 3–7. [In Russian].

13 Conflicting objectives while maximising carbon sequestration by forests

Andreas Fischlin

Systems Ecology ETHZ, Institute of Terrestrial Ecology, Department of Environmental Sciences, Swiss Federal Institute of Technology Zürich (ETHZ), ETH-Zentrum, CH-8092 Zürich, Switzerland

INTRODUCTION

In the current debate on global climatic change, forests and forestry policies have gained much attention; in particular hope has been expressed that forestry policies have the potential to mitigate climatic change—at least partially (Sedjo and Solomon 1988; Sedjo 1989b; Johnson 1992; Kauppi 1992; Krapfenbauer 1992; Kurz et al. 1992; Marland and Marland 1992; Dewar 1993; King 1993; Turner et al. 1993; Wisniewski et al. 1993). Many issues however, are poorly understood and contradictory conclusions have been drawn by different authors (e.g., Harmon et al. 1990; Marland and Marland 1992). While some authors have demonstrated how much land would be needed, how great the economic costs of a full mitigation by such policies would be (Sedjo 1989a), and have even warned against costly premature actions (Sedjo 1989b), others have argued that any contribution is worthwhile (Marland and Marland 1992). Moreover, management options favouring carbon (C) sequestering are often discussed without considering potential conflicts with other objectives, such as wood production.

One study advocated a forestry policy which harvests wood at the maximum sustainable yield (MSY) while maximising a C flux from the forests to long-lived forest products. Marland and Marland (1992) found that a net gain in total C sequestered can be achieved with such a policy but admit that the needed growth rates and the assumed harvesting efficiencies may be difficult to put into practice. If such net gains could be obtained for average forests on the whole globe and not just in highly productive plantations, such a policy would be highly beneficial in many ways. Marland and Marland (1992) suggest that significant potential exists for sequestering C this way. Regardless of whether this is actually the case or not, such a policy relies heavily on a basic assumption—namely that it is possible to maximise C sequestration while still harvesting wood.

Other authors have derived differing conclusions and, in particular, have warned against turning mature forests into young, highly productive stands by asserting that no forestry policy could surpass the C sequestration potential of mature forests (Harmon et al. 1990). Similar findings were reported by other authors (e.g., Kurz et al. 1992; Fischlin and Bugmann 1994a, b), who found that an attempt to maximise C storage by a policy which maximises sustainable yield, and therefore optimally feeds the forest product sector, does increase total C-storage capacity, but not by large amounts. Again, regardless of whether or not this is actually the case, the main issue is: a policy which maximises C-storage risks running into conflict with harvesting of wood.

Thus the following questions arise: Is there really a basic conflict between harvesting of wood and C sequestration? Moreover, if this conflict exists, is it unrealistic to recommend

NATO ASI Series, Vol. I 40
Forest Ecosystems, Forest Management and the Global Carbon Cycle
Edited by Michael J. Apps and David T. Price

cessation of all harvesting of wood—for most nations an unacceptable proposition—and would such a policy be in conflict with today's call for a more sustainable use of natural resources (which should result in the replacement of fossil-fuel dependent production systems with their detrimental effects to the environment)? Therefore, the question remains, are there circumstances under which these conflicts disappear or where there are acceptable compromises? For instance, how defensible is a policy of harvesting the replacement yield at sub-optimal levels instead of the MSY, in face of the C sequestration benefits? Could such policies be put into practice without resorting to highly unlikely prerequisites such as high mean growth rates or extremely high harvesting efficiencies?

In this study we used a simple forest growth model, similar to that by Marland and Marland (1992), which we combined with several management options to explore possible answers to the above questions. We assumed that the model, despite its primitive structure, properly reflects the important basic characteristics of a growing forest capable of producing wood for the forest sector while also providing woody reservoirs for C. Parameters are taken for northern mid-latitude species and growth conditions which can be found within a small area—specifically along an altitudinal transect through the Alps. We suggest that the forest types studied are typical for hard- and softwood forests, i.e., for northern mixed-deciduous and boreal like forests.

MATERIAL AND METHODS

The model is basically a logistic growth model (Equation 1) of aboveground biomass Q_j (i.e., stem, branches, and foliage) combined with discrete-event harvesting (Equations 2–4).

$$\frac{dQ_j(\mathrm{t})}{dt} = r_j \cdot \frac{K_j - Q_j(\mathrm{t})}{K_j} \cdot Q_j(\mathrm{t}) \qquad [1]$$

where

- t = continuous time (yr)
- j = type of forest: A = Beech forest, B = Montane spruce, or C = Subalpine spruce
- Q_j = dry weight (DW) of aboveground biomass (including wood) (Mg ha^{-1})
- r_j = maximum relative growth rate (yr^{-1})
- K_j = carrying capacity (Mg ha^{-1})

By harvesting wood the biomass Q_j is reduced by the harvested amount H_j (Equation 2)

$$Q_j(\mathrm{t}) = Q_j(\mathrm{t}^-) - H_j(\mathrm{t}^-) \qquad [2]$$

Woody slash is assumed to decay quickly, and only the fraction μ of the harvested biomass H_j is transferred to durable wood products P_j (Equation 3).

$$P_j(\mathrm{t}) = P_j(\mathrm{t}^-) + \mu H_j(\mathrm{t}^-) \qquad [3]$$

where:

- P_j = biomass stored as durable wood products (Mg ha^{-1})
- μ = fraction of harvested biomass ending up in durable wood products (%)

The management options investigated represent a small spectrum of basic options (Table 13.1). To establish a reference point for C sequestration similar to that used by Harmon et al. (1990) one extreme option (u) was to harvest no wood at all. Other options are clear-cutting (cC) and selection cutting (or plenter) management (p) as currently practised in many European countries.

Table 13.1. Management options used in all simulations. The output variables of interest are C^* (mean total C sequestered in forests and forest products) and W^* (mean annual wood production).

MANAGEMENT	DESCRIPTION	EQUATION	OUTPUTS
u	Unused forests without any harvesting	1	C^*, –
cC	Clear cutting	1–3,4,5,6	C^*, W^*
p	Selection cutting (plenter management)	1–3,4',5',6	C, W^*

Conventional harvesting, i.e., clear-cutting, is modelled as a sequence of three cuts with typically 8 year separations, initiated as soon as Q_j reaches a specified fraction θ_c, say 90%, of K_j (Equation 4). For each type of forest the harvest H_j can be represented as a sequence of three state events occurring at the cutting times $t^- = t_h$, t_{h+8}, and t_{h+16}, where the level of harvest removal from the forest is h_i = 30, 50 and 70% (respectively) of the current Q_j (Equation 5):

$$H_j(t^-) = \begin{cases} h_i \cdot Q_j(t^-) & t^- = t_h + i \cdot 8 \quad i = 0,1,2 \\ 0 & \text{else} \end{cases} \qquad [4]$$

where

$$t_h = t\big|_{Q_j(t) = \theta_c \cdot K_j} \qquad [5]$$

and

H_j = harvested biomass (DW) (Mg ha^{-1})
h_i = fraction of harvested wood in percentages of currently present biomass Q_j (%)
t_h = harvesting time or time of first cut in a sequence of 3 cuts (yr)
t^- = continuous left-hand side of time before, and up to, the discrete event harvest (yr)

Alternatively, selective cutting takes place whenever Q_j exceeds the fraction θ of K_j by the tolerance ε_j. The amount harvested is assumed to be $2 \times \varepsilon_j$ (Equation 4', 5'):

$$H_j(t^-) = \begin{cases} 2 \cdot \varepsilon_j & t^- = t_h \\ 0 & t^- \neq t_h \end{cases} \qquad [4']$$

where

$$t_h = t\big|_{Q_j(t) = \theta \cdot K_j + \varepsilon_j} \qquad [5']$$

ε_j = harvested biomass (DW) (Mg ha^{-1})

A simple equation of exponential decay (Equation 6) is used to model the slow decomposition of durable wood products P_j :

$$\frac{dP_j(t)}{dt} = -d_j \cdot P_j(t) \qquad [6]$$

where

d_j = relative decay rate (yr^{-1})

The model parameters, including those used in the sensitivity analysis (see below), are listed in Table 13.2.

Table 13.2. Parameters used in Equations 1–6 to model forest growth and harvesting. p_0: mean or reference value used to calculate deviations, i.e., Δp', in the sensitivity analysis (parameters *r, K, d* vary depending on the forest type *j* = *A*, *B*, or *C*; forest types are characterised by the dominant species). The parameters represent average growing conditions of mid-northern latitudes and were determined from yield tables of central Europe. DW: dry weight. For management options see Table 13.1.

PARAMETER p	MEANING	REFERENCE VALUE, p_0			UNIT	MANAGEMENT OPTION
		($j = A$)[a]	($j = B$)	($j = C$)		
r	Growth rate	0.04	0.05	0.05	yr^{-1}	u, cC, p
K	Carrying capacity (DW)	550	600	170	Mg ha^{-1}	u, cC, p
d	Decay rate of wood	0.025	0.037	0.037	yr^{-1}	cC, p
μ	Harvesting efficiency	40	40	40	%	cC, p
θ	Fraction of *K* at which selection cutting starts	50	50	50	%	p
ε	Tolerance by which Q_j may exceed θK_j before selection cut occurs	40	80	25	Mg ha^{-1}	

a Parameter *j* is forest type: *A*: Beech forest; *B*: Montane fir-spruce; *C*: Subalpine spruce.

All basic management options can be modified by changing the parameters listed in Table 13.2. For instance the parameter θ determines the age at which harvesting starts, expressed in terms of the biomass *Q* held by a stand, while the parameter μ determines how efficiently C is transferred from the forest into long-lived wood products.

Background information on the model formalisms and details on the simulation technique can be found in Fischlin et al. (1994) and copies of the model can be obtained from the author. All simulations needed for this study were made with ModelWorks and the RAMSES software on Apple Macintosh IIfx and Quadra 950 computers (Fischlin 1991; Fischlin et al. 1994).

The model system provides two outputs of main interest in this study: First, the mean total C sequestered in forests and long-lived forest products, *C** (Figure 13.1a), and second, the mean annual wood production, *W** (Table 13.1). *C** is computed from *C*, i.e., the total C currently sequestered within the forest biomass and the forest products. *W** is computed from *W* which is at any time the annually harvested wood taken from the forest according to the currently employed management option. *W* is calculated from H_j assuming a linear relationship. Since harvesting has been formulated as a discrete event, both variables *C* and *W* tend to fluctuate strongly (Figure 13.1b). To represent large areas—the model just represents a single stand—we integrated *C* and *W* over time and divided by the elapsed simulation time to calculate *C** and *W**, respectively. The first 100 years of simulation time were always discarded (i.e., $t < t_o = 100$ years), since this period is a transient, primary successional mode (Figure 13.1a).

The sensitivity of C^* and W^* to changes in the model parameters (Table 13.2) was analysed by sampling normally distributed variates p according to a normal distribution with an expected value of p_o and a coefficient of variation of 10%, i.e., $p \sim N(p_o, 0.1p_o)$. Using the reference values p_o , the corresponding steady-state outputs C^*_o and W^*_o (at year 500, Figure 13.1) and the deviations $\Delta C^* = C^* - C^*_o$ and $\Delta W^* = W^* - W^*_o$, respectively, were computed and correlated with the parameter changes $\Delta p = p - p_o$ (Figure 13.2a, Table 13.3). Parameters were only changed independently from each other, i.e., while a variate for parameter p was sampled, all other parameter values were kept at their reference value p_o.

To assess sensitivities (where the correlation was significant) the slopes of linear regressions between ΔC^* and ΔW^* with Δp were calculated. The sample sizes were 30 per forest type (*A*, *B*, and *C*, Table 13.2) amounting to 90 simulation runs per regression (Table 13.3) or a total of 990 runs for the entire sensitivity analysis.

For the optimisation of C sequestration and wood production, a brute-force method was used: Parameters were simply sampled from normally distributed variates $p \sim N(p_o, 0.1p_o)$. However, in contrast to the sensitivity analysis, parameter values were changed simultaneously, which required further simulations to a total of 4320 runs.

RESULTS AND DISCUSSION

Figure 13.1 shows the reference simulation results obtained while using the reference parameter values p_o (Table 13.2). The differences between the three management options (Table 13.1) become obvious: in the case of unused forests (u) total C sequestered (C), as well as its mean over time, C^*, are solely determined by the amount sequestered within the forest or the standing biomass, Q_j. C^* reaches a maximum determined by the carrying capacity K. Under the clear cutting regime (cC), the forest biomass Q_j fluctuates strongly together with C. Long-lived forest products P_j are incremented step-wise as a result from the cuts. They decay subsequently till the next cut. C fluctuates strongly which is similar to the selection cutting regime (p) although in the latter case the amplitude is smaller. C is irregular and difficult to assess unless pooled and averaged over time which yields C^*; it represents the long-term benefit and is of greater use to evaluate a particular management option than the variable C.

The sensitivity analysis showed that for most parameters the output deviations ΔC^* and ΔW^* were significantly ($2\alpha = 0.1$) correlated with the parameter deviations Δp (Figures 2a–c; Table 13.3). Among the most sensitive parameters were the carrying capacity K and θ (note, both parameters are related, since the latter is the fraction of K at which selection cutting starts). However, due to the non-linear relationship between $\Delta\theta$ and the deviations in the mean annual wood production ΔW^* (Figure 13.2d), the correlation between these two variables were not always significant (Table 13.3, last row, last two columns).

The more sensitive ΔC^* or ΔW^* are to a particular parameter change, the easier it is to modify the system behaviour in a desired direction by changing this parameter. Any management policy which can increase a positively correlated parameter or which can decrease a negatively correlated parameter leads to a higher C sequestration.

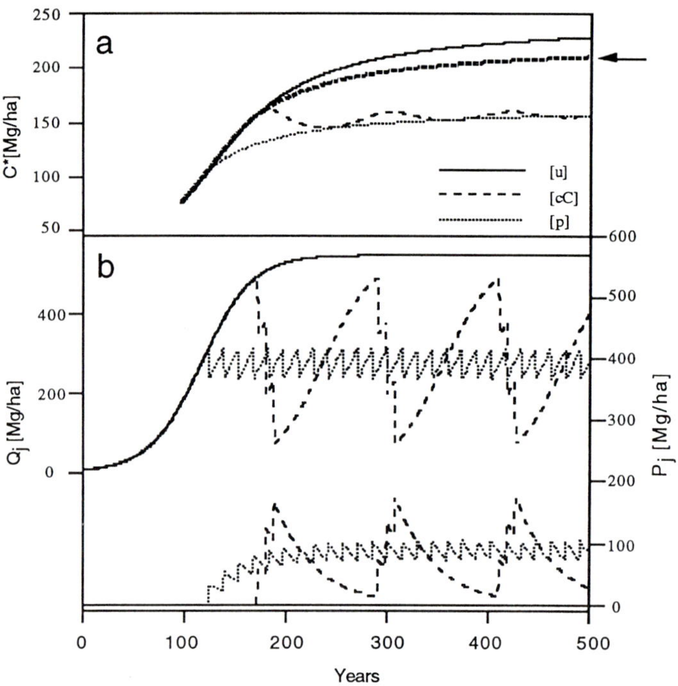

Figure 13.1. Typical simulation result obtained with the forest/forestry model used in this study. (a) Total C sequestered averaged over 400 years (C^*), for the standard parameters and three management options, respectively, for optimised $\theta = 0.8$ at arrow (see Figure 13.3, bar 8). (b) top: Forest biomass (Q_j) in beech forest ($j = A$); bottom: Beech wood in long-lived forest products (P_j). Management options applied are: unused forest (u); clear cutting (cC); selection cutting (p).

The most sensitive parameter is K (Table 13.3). However, K is a parameter which is largely determined by site conditions such as climate and soil characteristics and is therefore rather difficult or expensive to modify via a particular management. Of course, increasing the carrying capacity via fertilisation or manipulating age structure, preserving existing forests, are all management options which directly or indirectly modify K and lead to a higher C sequestration. If K is interpreted globally as the mean carrying capacity per ha, analogous arguments are even applicable for afforestation and reforestation.

θ was found to be the second most sensitive parameter. In contrast to K, θ is relatively easy to change in practice. ΔC^* is positively correlated, while ΔW^* is negatively correlated, with $\Delta\theta$ (Figure 13.2d). Thus, while optimising C^*, θ has to be increased as much as possible. However, this leads inevitably also to a decrease of mean annual wood production W^*.

Maximum annual wood production W^*, i.e., MSY, was found as expected with $\theta = \theta_o$ i.e., 0.5 (Figure 13.3, bar 4). The widely practised clear-cutting regime (cC) using only reference parameters is clearly sub-optimal, both in C sequestration as well as wood production (Figure 13.3, bar 1).

In terms of C sequestration, the best results can be obtained only under the management option of selective cutting (p). This is mainly because unused forests can store more C than

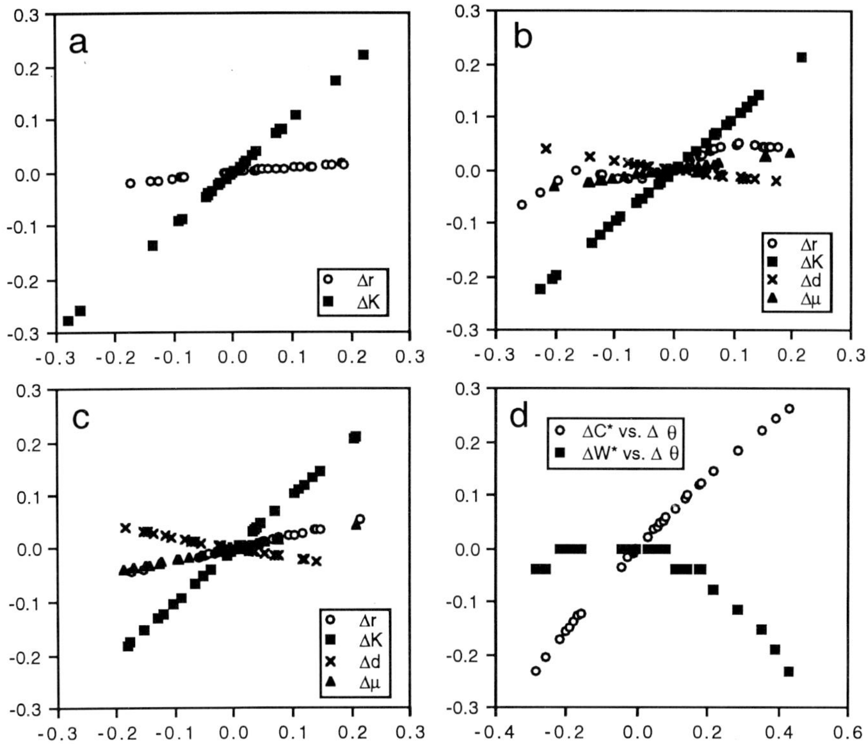

Figure 13.2. (a)–(c) Sensitivity analysis for beech forests using scatter grams drawn from a total of 990 simulation runs. Deviation of simulated $\Delta C^* = C^* - C_o^*$ (mean total C sequestered in forests and forest products) shown as a function of random variations in five selected model parameters $\Delta p = p - p_o$ ($p \sim N(p_o, 0.1p_o)$). (a) unused forest (u); (b) clear cutting (cC); (c) selection cutting (p). (d) Optimisation of beech forests. Deviation of simulated variables ΔC^* and ΔW^* (mean annual wood production) as a function of random deviations in parameter $\Delta\theta = \theta - \theta_o$ ($\theta \sim N(\theta_o, 0.1\theta_o)$) using data from 4320 simulation runs. ΔW^* decreases while ΔC^* is maximised, thus leading to a fundamental conflict between optimal wood production and maximum C sequestration.

can be transferred and kept in long-lived forest products. Note, there is a management option which gets very close to the potential of an unused forest (Figure 13.3, bar 9): Its sequestration potential (97.4%) is almost as high as that which can be achieved in the case of a management without any harvesting (u, 100%, Figure 13.3, bar 10). Its characteristics are selection cutting, little disturbance in the age structure, i.e., leaving the forest at a high mean age, but with a short rotation period. In addition, in this case (9) we assumed also first a very high harvesting efficiency (μ), similar to that assumed by Marland and Marland (1992), and second a maximum transfer into long-lived forest products.

However, we believe that the latter assumptions are unrealistic and require C flows into long-lived wood products, which can probably not be put into practice. More realistic is a lower harvesting efficiency ($\mu_O = 0.4$), similar to that shown in bar 8 of Figure 13.3, which reaches 91.8% (Figure 13.1a, see arrow) of the achievable, maximum C sequestration of 229.1 Mg ha^{-1}, which is remarkably high compared to only 68.4% obtained while harvesting at MSY ($\theta_o = 0.5$).

Table 13.3. Results of sensitivity analysis. Model parameters p were sampled from a normal distribution $N(p_o, 0.1p_o)$ and the sensitivity of the output variables C^* (mean total C sequestered in forests and forest products) respectively W^* (mean annual wood production) computed. Sensitivities were assessed by the slopes from linearly regressing deviations in the output variables, i.e., ΔC^* and ΔW^*, *vs.* the deviations in the respective parameter Δp. Note, ΔW^* is independent of Δd and $\Delta \mu$ (Equations 4 or 4'). p_o - mean model parameter of r, K, d, μ, or θ (Table 13.1); u, cC, p - management option (Table 13.1); L.R. - linear regression. All slopes based on 90 simulation runs (cf. Figure 13.2). r^2 - square of correlation coefficient. MIN(r^2), MAX(r^2) are minimum, maximum respectively of r^2 found in any of the three forest types.

OUTPUT VARIABLE	PARAMETER p	MANAGEMENT	MEAN SLOPE SENSITIVITY L.R.	MIN(r^2)	MAX(r^2)
ΔC^*	Δr	u	0.08	0.98	0.98
		cC	0.24	0.87	0.89
		p	0.23	1.00	1.00
	ΔK	u	1.00	1.00	1.00
		cC	0.99	1.00	1.00
		p	1.00	1.00	1.00
	Δd	cC	-0.15	0.99	0.99
		p	-0.19	0.99	0.99
	$\Delta \mu$	cC	0.15	1.00	1.00
		p	0.20	1.00	1.00
	$\Delta \theta$	p	0.76	1.00	1.00
ΔW^*	Δr	cC	1.20	0.74	0.85
		p	1.07	0.98	1.00
	ΔK	cC	1.05	1.00	1.00
		p	1.05	0.98	1.00
	$\Delta \theta$	p	-0.10	0.03[a]	0.46

a Correlation not significant, since estimated r^2 is less than $r^2_{2\alpha=0.1;\nu=28} = 0.0937$.

No management scenario yielded C sequestrations above that of an unused forest, in contrast to the findings by other authors (Marland and Marland 1992). This is not a true contradiction, since the differences can be explained by the different growth rates. Our growth rates (r in Table 13.2) were taken from European data (Fischlin and Bugmann 1994a; Fischlin and Bugmann 1994b) which represent average growing conditions, easily surpassed by intensively managed forests or tree plantations as assumed by Marland and Marland. Our rates are also very comparable to those reported by other authors (Sedjo 1989a, b) who listed for similar latitudes and conditions, i.e., for the U.S., average C accumulation rates of 0.82 Mg C ha^{-1} yr^{-1}; in our model these amounted to 0.83–0.84 Mg C ha^{-1} yr^{-1}. (Carbon accumulation rates in units of C ha^{-1} yr^{-1} is computed as $1.6 \times 0.5 \times e^r$, where 1.6 is the ratio between stemwood growth and other C pools such as branches, roots, debris etc., and 0.5 is the ratio between dry weight and C mass.)

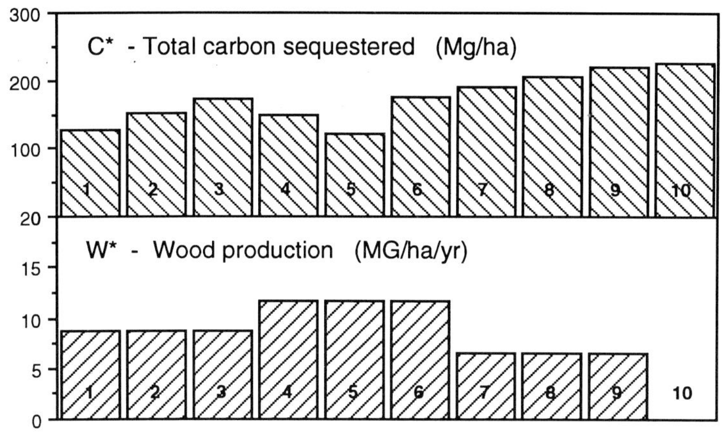

Figure 13.3. Comparison of total C-storage capacity C^* (forests plus long-lived forest products—top bars) and mean annual wood production W^* (bottom bars) for a set of management options and changed model parameters. Unless the value used is given together with the changed parameter in the following list, all other parameters were kept at their reference value p_0. For the management options see Table 13.1.: 1 - cC, d = 0.6 / 2 - cC / 3 - cC, $\mu = 0.75$ / 4 - *p* / 5 - p, d = 0.6 / 6 - p, $\mu = 0.75$ / 7 - p, $\theta = 0.8$, d = 0.6 / 8 - p, $\theta = 0.8$ / 9 - p, $\theta = 0.8$, $\mu = 0.75$ / 10 - u.

CONCLUSIONS

The following main conclusions were drawn from the presented findings:

- The greatest C sequestration is obtained in unused forests.
- The second greatest C sequestration is not with storing C in long-lived wood products while harvesting at MSY, but with augmenting standing crops while selectively cutting (plenter management). The period length should be relatively short and stands should be kept at a high mean age, i.e., with a biomass close to the carrying capacity.
- Objectives to maximise C sequestration conflict with wood production (e.g., MSY *vs.* maximum standing crop)
- Since the wood production is not independent of the state of a forest, strategies solely maximising harvest must be put in balance relative to optimal C storage strategies.

Obviously there are irresolvable conflicts between a management policy maximising wood production and that maximising C sequestration, which are too often overlooked. In its extremist form a policy of maximising C sequestration would even require protecting forests by fighting insects and fires, but completely C sequestration as a tool to mitigate climatic change, and may therefore inadvertently release C to the atmosphere. Since there is no single optimal solution (cf. Figure 13.3), the conflicting objectives have to be valued against each other. I believe, in a democratic society people have to make the decision how to weigh the conflicting objectives and accordingly choose management regimes between the two extremes. By doing so it may be helpful to recognise that solutions between these two extremes exist, solutions where wood production is not at its maximum, i.e., MSY, but harvesting is still possible while C sequestration is nearly optimal. Given the currently wide-spread, sub-optimal clear-cutting practices, such policies might also be economically more acceptable than appears to be the case at first glance.

ACKNOWLEDGEMENTS

I wish to thank Daniel Perruchoud and two anonymous referees for their most helpful critical comments. Many thanks also to Michael Ziege for his advice and efforts to redraw two figures.

REFERENCES

Dewar RC (1993) A mechanistic analysis of self-thinning in terms of the carbon balance of trees. *Ann. Bot.* 71: 147–159.

Fischlin A (1991) Interactive modeling and simulation of environmental systems on workstations. In: Möller DPF (ed) Analysis of Dynamic Systems in Medizinef Informatik-Fachberichte 275. Springer, Berlin, pp 131–145.

Fischlin A, Bugmann HK (1994a) Können forstliche Massnahmen einen Beitrag zur Verminderung der CO_2-Emissionen leisten? Ökologische Grundlagen und erste Abschätzungen. *Schweiz. Z. Forstwes.* 145: 275–292.

Fischlin A, Bugmann HK (1994b) Think globally, act locally! A small country case study in reducing net CO_2 emissions by carbon fixation policies. In: Kanninen M (ed) Carbon Balance of World's Forested Ecosystems: Towards a Global Assessment Academy of Finland, Helsinki, Finland, pp 256–266.

Fischlin A, Gyalistras D, Roth O, Ulrich M, Thoeny J, Nemecek T, Bugmann HK, Thommen F (1994) ModelWorks 2.2: An interactive simulation environment for personal computers and workstations. Systems Ecology Report No. 14, Institute of Terrestrial Ecology, Swiss Federal Institute of Technology ETH, Zurich, 324 pp.

Harmon ME, Ferrell WK, Franklin JF (1990) Effects on carbon storage of conversion of old-growth forests to young forests. *Science* 247: 699–702.

Johnson DW (1992) Effects of forest management on soil carbon storage. *Water, Air, Soil Pollut.* 64: 83–120.

Kauppi PE, Mielikäinen K, Kuusela K (1992) Biomass and carbon budget of European forests, 1971 to 1990. *Science* 256: 70–74.

King GA (1993) Conceptual approaches for incorporating climatic change into the development of forest management options for sequestering carbon. *Clim. Res.* 3: 61–78.

Krapfenbauer A (1992) Die Rolle des Waldes, der Holzbiomasse und der Solarenergie zur Stabilisierung des Glashauseffektes. *Centralblatt fuer das gesamte Forstwesen* 109: 85–103.

Kurz WA, Apps J, Webb TM, McNamee PJ (1992) The carbon budget of the Canadian forest sector: Phase I. Information report NOR-X-326, Forestry Canada, Northern Forestry Centre, Edmonton, AB, 93 pp.

Marland G, Marland S (1992) Should we store carbon in trees? *Water, Air, Soil Pollut.* 64: 181–195.

Sedjo R (1989a) Forests to offset the greenhouse effect. *J. For.* 87(7): 12–15.

Sedjo RA (1989b) Forests. A tool to moderate global warming? *Environment* 31: 14–20.

Sedjo RA, Solomon AM (1988) Climate and forests. In: Rosenberg NJ, Easterling III WE, Crosson PR, Darmstadter J (eds) Greenhouse warming: abatement and adaption Proceedings of a workshop held in Washington DC; 14–15 June, 1988, Resources for the Future, Washington DC, pp 105–119.

Turner DP, Lee JJ, Koerper GJ, Barker JR (1993) The forest sector carbon budget of the United States: carbon pools and flux under alternative policy options. 600/3-93/093, U.S. Environmental Protection Agency, Corvallis, OR, 202 pp.

Wisniewski J, Dixon RK, Kinsman JD, Sampson RN, Lugo A (1993) Carbon dioxide sequestration in terrestrial ecosystems. *Clim. Res.* 3: 1–5.

14 Retrospective assessment of carbon flows in Canadian boreal forests

Werner A Kurz and Michael J Apps[1]

ESSA Technologies Ltd., 1765 West 8th Avenue
Vancouver, BC, V6J 5C6, Canada

INTRODUCTION

The Canadian boreal forest is comprised largely of even-aged stands that for millennia have undergone cycles of disturbance (i.e., fires or insect-induced stand mortality) and regrowth. Previous studies of regional-scale carbon (C) budgets have assumed that, when averaged over a large area, forests not directly affected by human disturbance or land-use change are in steady state with regard to their net C exchange with the atmosphere (e.g., Houghton et al. 1983; Houghton et al. 1987). Underlying this assumption is the notion that, integrated over a large area, C uptake in forests regrowing after disturbance is balanced by C release from disturbances. The area of Canadian forests affected by disturbances varies between years and decades (Kurz et al. 1995) and therefore the assumption of a zero net balance between C uptake and release can, at best, be correct for some long-term average. Recent interest in the global C balance and the role of terrestrial ecosystems focuses on year-to-year variation and the record of the past decades (Sarmiento et al. 1992; Chapter 10, Houghton; Chapter 3, Townsend et al.). Analyses of forest C budgets that do not rely on *a priori* assumptions of steady-state conditions are thus required.

The objective of this paper is to assess changes in the net carbon exchange between Canadian boreal forests and the atmosphere as affected by natural disturbances and direct human impacts such as logging and forest protection. Indirect human impacts through, for example, atmospheric CO_2 increase, deposition of atmospheric nitrogen, or climatic change are not considered in this analysis. Some of these impacts at the global scale are, however, discussed in other Chapters in this book.

METHODS

The calculation of the carbon budget for a large forest area requires (1) a model of forest dynamics, (2) statistics on the major factors that affect forest dynamics during the time period of interest, and (3) an estimate of the initial forest conditions. Each of these three components is described in more detail below.

The Carbon Budget Model of the Canadian Forest Sector (CBM-CFS2)

The CBM-CFS2 is a dynamic simulation model that accounts for C pools and fluxes in forest ecosystems and in forest products, including landfills derived from harvested materials (Apps and Kurz 1993; Kurz et al. 1992; Kurz and Apps 1994; Kurz et al. 1995). It represents

1 Canadian Forest Service, Northwest Region,
5320-122 Street, Edmonton, Alberta T6H 3S5, Canada

NATO ASI Series, Vol. I 40
Forest Ecosystems, Forest Management and the Global Carbon Cycle
Edited by Michael J. Apps and David T. Price

Canadian boreal forests as an assemblage of several hundred forest ecosystem types, each represented by a characteristic growth curve (expressing forest biomass as a function of age) and a series of classifiers (e.g., site class and stocking). Each ecosystem type is further stratified into 27 age classes. The amount of C in biomass and soil pools is a function of the ecosystem type, the time since the last disturbance, and the disturbance history. Each record in the model's database represents a specific ecosystem type with a given age and contains information on the area and the C content of biomass and soil C pools.

Forest vegetation is simulated as 12 pools: four aboveground and two belowground (i.e., coarse and fine roots) biomass pools for both a hardwood and a softwood species group (Figure 14.1). All dead organic matter is represented by four soil C pools characterised by their decomposition rates. Decomposition rates are affected by soil C pool type, mean annual temperature, and stand conditions. Biomass C can be transferred directly to three of the soil C pools, released to the atmosphere by disturbances, or transferred to the forest product sector by harvesting. Decomposition releases C to the atmosphere and transfers residual organic matter to the fourth soil C pool, representing humified material with a slow decomposition rate.

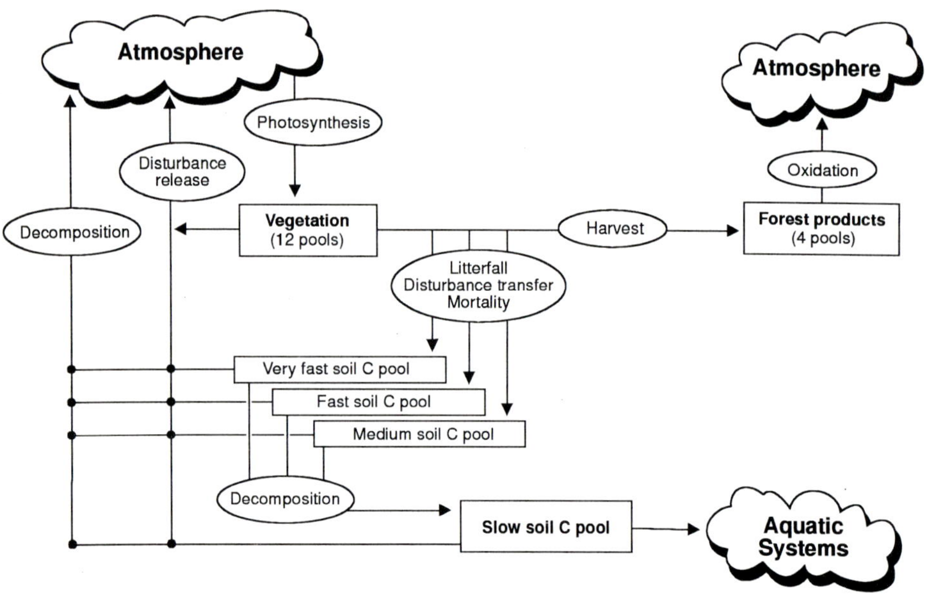

Figure 14.1. Overview of the major C pools and fluxes represented in the Carbon Budget Model of the Canadian Forest Sector (CBM-CFS2).

Continuous processes such as forest growth and decomposition affect all records in the model's database in every annual timestep. Growth removes C from the atmosphere and transfers it to vegetation C pools. Ecosystem dynamics (e.g., litterfall, stem mortality) result in C transfers between biomass and soil C pools, which include litter, coarse woody debris, and soil organic matter.

The growth curves in the model describe net biomass C dynamics for each ecosystem type. They are parameterized from the data contained in the national biomass inventory (Bonnor 1985). Unlike many traditional growth and yield models that do not define vegetation dynamics past mature stand conditions, the growth curves used here can specify a decline in biomass C pools in overmature stands, if the inventory data indicate such a decline. Biomass C associated with any decline is assumed to be detritus and is transferred to the appropriate soil C pool.

Harvested biomass C is transferred to the forest product sector where it is converted to products with different C retention times. Forest product C is released to the atmosphere through burning and landfill decomposition. The representation of biomass, soil, and forest product C pool dynamics is described in more detail elsewhere (Kurz et al. 1992; Apps and Kurz 1993; Kurz and Apps 1994).

Discontinuous processes, such as forest disturbances, affect only some of the records in any year. Rules specific to each ecoclimatic province define the area affected by each disturbance type, the ecosystem types and age classes selected, and how C is redistributed amongst different pools during disturbances (Kurz et al. 1992; Kurz et al. 1995). The model selects the eligible records for each disturbance type and transfers C between ecosystem C pools, releases it to the atmosphere (in the case of fire) or exports biomass C to the forest product sector (in the case of harvesting). Post-disturbance changes in the forest ecosystem are simulated by the growth and decomposition modules of the model; for example following wildfire, ecosystems can lose C for some time while decomposition losses from soil C pools exceed net biomass growth.

We used the CBM-CFS2 to conduct a retrospective analysis of the C budget of Canadian boreal forests over a period of 70 years (1920 to 1989). In this paper we consider only the boreal zone which includes three ecoclimatic provinces: boreal east, boreal west, and subarctic (for a map see Ecoregions Working Group 1989; Kurz et al. 1992; or Apps et al. 1993). The model represents this area (303 Mha) in 18 spatial units, defined as ecoprovinces within administrative Provinces and Territories. The CBM-CFS2 calculates and reports, for every time step, information on C pools and fluxes summed to the scale of spatial units. It can optionally report additional ecosystem-specific diagnostics that can be used for model validation.

Disturbances in Canadian Boreal Forests

Resource management agencies in Canada have been compiling statistics on forest disturbances and management activities since early this century. We have compiled, from a wide variety of federal and provincial sources, statistics on harvest levels, wildfires, insect activities, and planting for each of the spatial units represented in the model. This database covers the period 1920 to present for all areas of the country. A continuous record of disturbance statistics exists for most disturbance types in most provinces back to ~1920. Where necessary, we have augmented gaps in disturbance statistics by extrapolating available data over space or time using proxy data from adjacent spatial units or averages of time periods.

In the model, we represent salvage logging of stands previously disturbed by fire or insects based on data from government statistics. We do not, however, account for any increased likelihood of fires burning in insect-killed stands. Given the paucity of data from which to infer whether fires occurred in stands previously affected by insects, such a refinement of historic disturbance impacts is not possible at this time.

Disturbance statistics for the boreal and subarctic ecoprovinces in Canada are summarised as 5-year averages in Figure 14.2. Fires and insects are the dominant disturbance types. Forest harvesting levels have increased throughout the 70-year period. Total disturbances averaged 1.7 Mha yr^{-1} for the period 1920 to 1969 and increased to 3.9 Mha yr^{-1} in the period 1970 to 1989.

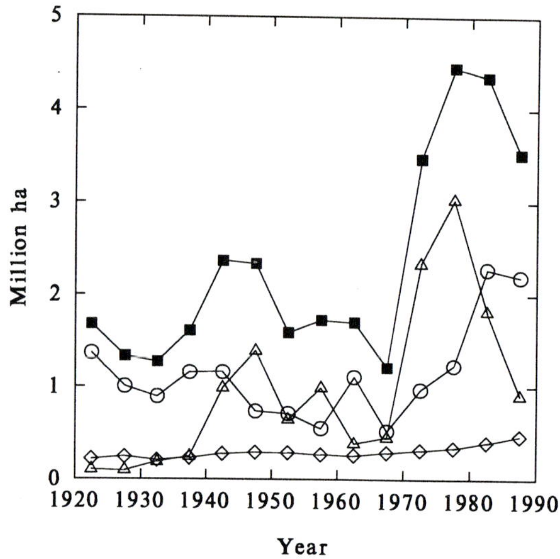

Figure 14.2. The five-year averages of the area annually disturbed by harvesting (◇), insects (△), and fire (○) in the Canadian boreal and subarctic ecoprovinces. The total (■) is less than the sum of the disturbance types because of salvage logging of insect- or fire-disturbed stands.

Determination of initial conditions

The model is initialised with forest conditions for 1920. These are derived by combining information from a number of sources. A more detailed description is provided in Kurz et al. (1995) and only a brief summary is presented here. The total area of boreal forest and its distribution by spatial units and ecosystem types is derived from the national forest inventory (Bonnor 1985). This inventory reports conditions of about 1970. The forest age-class structure in 1970 (derived from: Bonnor 1985; Forestry Canada 1988) is the result of stand-replacing disturbances during the preceding decades (Yarie 1981; van Wagner 1978). For each ecoclimatic province, we compiled the disturbance statistics for the period 1920 to 1989 and a set of rules with which to allocate disturbances to forest ecosystem types. We established the forest age-class structure in 1920 such that, when the 50-year disturbance statistics (1920 to 1969) have been simulated, the forest age-class structure in 1970 closely matches that observed in the inventory. We assumed that the total forest area and its distribution by ecosystem types in 1920 were similar to those in 1970. The distribution of the area by age classes differed between 1920 and 1970. Independent data do not exist to verify our inferred 1920 forest age-class structure for all of the Canadian boreal forest. The structure that we have obtained is, however, consistent with the observed 1970 age-class distribution once ageing and disturbances between 1920 and 1969 have been applied. Alterations to the rules by which disturbances are applied to ecosystem types and age classes could yield different initial conditions in 1920, but we have been unable to obtain significantly different initial conditions with rules that also meet ecological constraints.

The C content per hectare in the biomass and soil pools in each ecosystem type and age class differed between 1920 and 1970 (see Kurz et al. 1995). As in our earlier analyses (Kurz et al. 1992; Kurz and Apps 1994), we assumed that in all ecosystem types, two rotations of forest growth, each terminated by wildfire, preceded the natural regeneration of those stands used to define the initial conditions of 1920. We are conducting additional analyses to explore the effects of different assumptions about disturbance history, types, and frequency on the initial conditions of biomass and soil C pools.

RESULTS

Our reconstruction of forest conditions indicates that, in 1920, the average age of the Canadian boreal forest was 60.9 years (Figure 14.3). During the subsequent 50 years (1920 to 1969), the average forest age increased to 82.5 years. This age increase suggests that the rate of disturbances was lower during this period than it had been previously. A twofold increase in natural disturbances since 1970 decreased the average forest age to 76.4 years by 1989.

Biomass C dynamics were also affected by the changes in disturbance regimes over the 70-year period. Above- and belowground biomass C content of the forests increased from 5.6 Pg C in 1920 to 8.7 Pg C in 1970, and subsequently decreased to 7.1 Pg C in 1990 (Figure 14.4).

Total soil C, which includes all standing and downed coarse woody debris, litter, and soil organic matter (excluding peat) increased over the 70-year period (Figure 14.4). The change in C content differed between soil C pools (Figure 14.5) and reflects the different responses to changes in disturbance regimes and the resulting changes in biomass dynamics.

Over the 70-year period of the analysis, the Canadian boreal forest accumulated 8.3 Pg C, of which 1.4 Pg C and 6.8 Pg C accumulated in biomass and soil pools, respectively. In the period 1920 to 1979, net ecosystem productivity (NEP, i.e., the sum of net changes in all biomass and soil C pools) of the Canadian boreal forest was 147 Tg C yr^{-1} (Figure 14.6). After taking up C from the atmosphere for six decades, forest ecosystem C decreased by 57 Tg C yr^{-1} in the last decade of the analysis. Averaged over the 70-year period, the Canadian boreal forest was a C sink of 118 Tg C yr^{-1}. Periods of high disturbances (e.g., the 1940s and 1980s) result in reduced C accumulation in biomass C pools and a corresponding increase in soil C pools. The increase of dead organic matter pools associated with disturbances results in higher C losses from decomposition in the years following the periods of high disturbances. The recent 20-year period of increased disturbances will therefore affect soil C releases in the coming years (Kurz and Apps 1995).

The total area harvested during the 70-year period was 20.3 Mha (Figure 14.2) or 6.7% of the forest area. Total transfer of biomass C to the forest product sector was 0.78 Pg C, of which 0.30 Pg C (Figure 14.4) was retained in forest products and landfills in 1989. This increases the annual net C storage by 4 Tg C yr^{-1} on average over the period. We can include this storage in forest products pools in the overall C budget because we account for C removals from ecosystem pools even if this C is not immediately released to the atmosphere. Distributing this forest product C pool across the Canadian boreal forest would be equivalent to an average accumulation rate of 1.4 g C m^{-2} yr^{-1} over the 70-year period.

DISCUSSION

While it is not possible to obtain independent estimates of all of our results, some components of this analysis can be compared to estimates from the literature. Our estimates of the average aboveground biomass C content of the Canadian boreal forest range from 14.4 Mg C ha^{-1} in 1920, to 22.3 Mg C ha^{-1} in 1970, and 18.2 Mg C ha^{-1} in 1990. These estimates are within the range of recent biomass C measurements for the North American boreal forest of 18.8 ±4.5 Mg C ha^{-1} (Simpson et al. 1993).

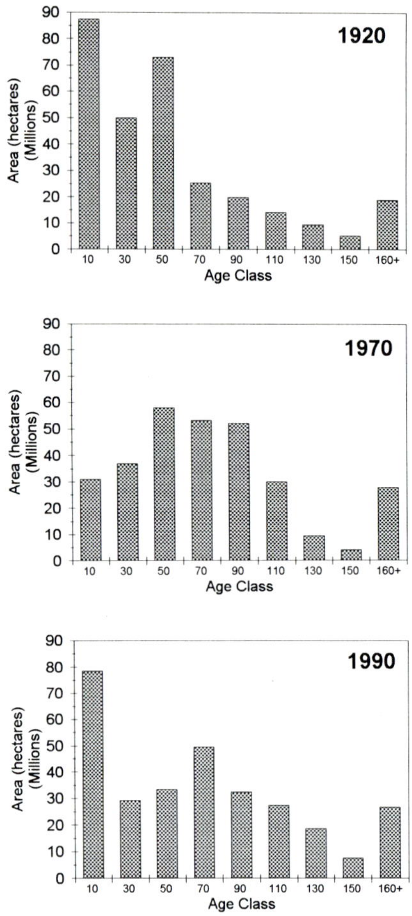

Figure 14.3. Forest age-class distribution of Canadian boreal forests in 1920, 1970, and 1990.

Our estimates of the net C accumulation over the 70-year period average to 6.8 g C m^{2} yr^{-1} for total biomass and 32 g C m^{-2} yr^{-1} for total soil C pools. The average NEP was 39 g C m^{-2} yr^{-1}. Bonan (1991) estimated NEP of 74 g C m^{-2} yr^{-1} for the boreal forest near Fairbanks, Alaska. In the absence of fire, net increase in dead organic matter and soil C pools in his analysis was 38 g C m^{-2} yr^{-1}. His analysis covered a single year with average conditions. Our estimates of NEP were not constant across the 70-year period. For example, in the period 1970 to 1989, biomass C pools decreased (Figure 14.4) and NEP was only 6.7 g C m^{-2} yr^{-1}. Thus, given the variability over time and the site differences, this agreement between Bonan's result for a specific region in Alaska and ours for the entire Canadian boreal forest is acceptable.

The largest uncertainties in our model are associated with the simulation of soil C pool dynamics. Over the 70-year period, the slow soil C pool (which receives input solely through decomposition of organic matter in the three faster soil pools) accumulated 3.4 Pg C. This represents an increase of 9% or 15.8 g C m^{-2} yr^{-1}. Schlesinger (1990) lists three studies that

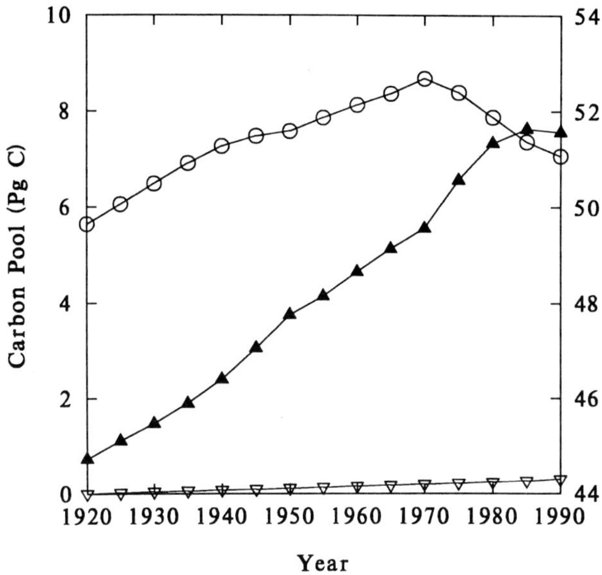

Figure 14.4. Biomass (○) (above- and below-ground), soil (▲), and forest product sector (▽) C in Canadian boreal forests 1920 to 1989. Note that soil C pool dynamics are plotted relative to the scale on the right.

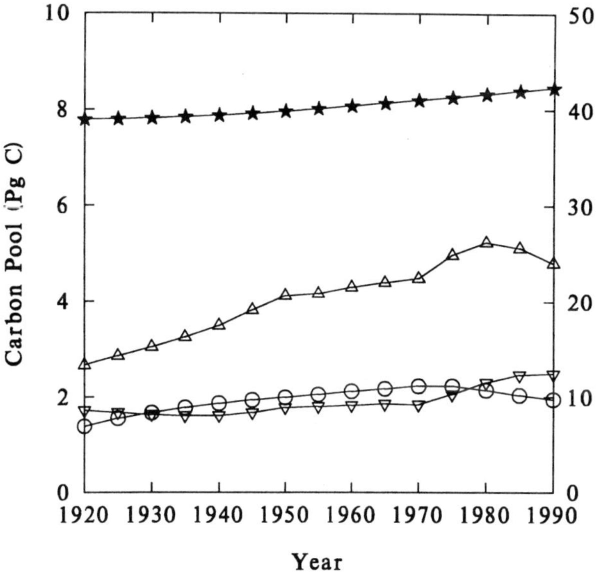

Figure 14.5. Carbon content of the four soil C pools in Canadian boreal forests 1920 to 1989. The very fast (○), fast (△), medium (▽), and slow (★) soil C pools are characterised by the biomass components entering each pool and by their decomposition rates. Note that the slow soil C pool dynamics are plotted relative to the scale on the right. See also Figure 14.1 for C transfer diagram.

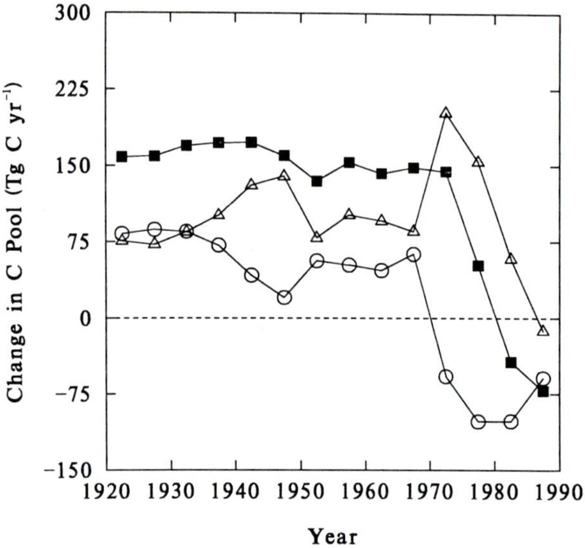

Figure 14.6. Annual net change in biomass (○), soil (△), and total ecosystem (■) C in Canadian boreal forests 1920 to 1989.

found long-term (i.e., 2700–5400 years) C accumulation rates in refractory humus substances in soils of boreal forests of 0.8, 2.2 and 11.7 g C m^{-2} yr^{-1}. Our higher estimate of 15.8 g C m^{-2} yr^{-1} is derived from the simulation of a 70-year period with significant changes in forest biomass and disturbance regimes. Soil humus accumulation rates during these transient periods could be higher than the long-term averages, but further research is needed to better quantify soil C pool sizes and dynamics in Canadian boreal forests (Trofymow et al. 1995).

The terrestrial C budget of the boreal forest region is affected by a number of processes that operate over a wide range of temporal scales. At the scale of centuries to millennia, climatic changes and glaciation affect forest distribution and C storage in biomass and soil pools (including peatlands) (Prentice et al. 1991; Harden et al. 1992). At the scale of decades to centuries, changes in disturbance regimes can modify age-class distributions and C dynamics (Kurz and Apps 1993; Kurz et al. 1995). At the scale of years to decades, changes in temperature and precipitation lead to variation in growth and decomposition rates (Dai and Fung 1993). Since these and other processes all contribute to the C exchange between forest ecosystems and the atmosphere, we suggest that C uptake by and release from forests that are not directly influenced by human activities and land use have a non-zero net contribution to atmospheric CO_2. Our simulation results for the 70-year period clearly show this for Canada. The assumption of zero net contribution was originally introduced for analyses that focused on the role of land-use change and in which little information was available on the area not directly affected by land use (e.g., Houghton et al. 1983, 1987). Our results suggest that it is inappropriate to make this assumption for forests where natural disturbance regimes are the dominant force driving vegetation dynamics over large areas. Our results also show that changes in these disturbance regimes—changes that may be significantly influenced by human activities, including modification of the climate—are critically important in determining the net atmospheric C exchange. In the circumpolar boreal forest, direct human intervention has in the past played a minor role in vegetation dynamics (Apps et al. 1993) and most of the area is influenced by other forces, particularly wildfire and insect damage. *A priori* assumptions of steady-state conditions are therefore not appropriate for such forest ecosystems.

Our sink estimate accounts for a component of the C currently unaccounted for in global C budgets (the so called "missing sink") because most of the C uptake occurred in forest ecosystems that were previously assumed to not contribute to the net C flux between vegetation and the atmosphere. Our result is based on model assumptions that do not include a number of processes such as nitrogen and CO_2-fertilisation and changes in temperature, that could influence the actual net C exchange. Although several attempts have been made to quantify these effects at the global scale (e.g., Schindler and Bayley 1993; Dai and Fung 1993), we feel that they are not easily quantified at the scale of the Canadian boreal forest and that neither the magnitude nor the direction of the impact of these changes are clearly understood.

We must emphasise that the C uptake that we estimate is primarily the result of fluctuations in natural disturbance regimes and that the estimated C uptake during the 70-yr period of this analysis should not be extrapolated outside this period. In the decades prior to 1920, forest disturbance regimes must have been extensive, as is indicated by the large areas in young age classes in 1920 (see Figure 14.3). Numerous independent studies have reported high disturbance impacts in the second half of the last century throughout North America (Blais 1983; Clark 1990; Cwynar 1977; Foster 1983; Tande 1979). The recent 20-year period of high disturbances will likely affect C dynamics in the coming decades; these disturbances have greatly increased the pools of decomposing organic matter, and the area of young regenerating forests whose C uptake rates are initially low. Scenario analyses of the Canadian forest sector C budget for the next 50 years suggest that, for many scenarios, the forest could remain a C source in the near future (Kurz and Apps 1995).

This analysis demonstrates that, in Canadian boreal forests, fluctuations in natural disturbance regimes at time scales of decades to centuries are influencing the C exchange with the atmosphere. We suggest that similar dynamics in the boreal forests of Alaska and Russia may also affect the global C budget but differences in climate and forest structure must be carefully considered. This could be of particular significance if the factors that influence fluctuations in natural disturbance regimes at the continental scale also synchronise these fluctuations at the circumpolar scale. For example, unusually cool periods were reported throughout the Northern Hemisphere in the second half of the 19th century (Jones and Bradley 1992).

Our results emphasise the need to re-examine some of the assumptions underlying the analysis that detect an imbalance in the global budget of C release and C uptake. Our results suggest that for the period 1920 to 1979 some of the missing C can be accounted for by fluctuations in natural disturbance regimes which have led to increasing C storage in boreal forests. The recent twofold increase in natural disturbances has resulted in a net release of C from these forests in Canada. These findings also challenge the notion that the processes that in the past contributed to the uptake of anthropogenic C releases will continue to operate in the future as assumed in the 1990 IPCC assessment (Houghton et al. 1990). To the contrary, fluctuations in natural disturbance regimes can at times contribute to the sources of atmospheric C.

ACKNOWLEDGEMENTS

This work was supported in part by the ENFOR (ENergy from the FORest) program of the Canadian Forest Service. We thank S Beukema for conducting and analysing model runs, T Lekstrum for compilation of historical disturbance data, and Dr J Greenough and two anonymous reviewers for comments on the draft manuscript.

REFERENCES

Apps MJ, Kurz WA, Price DT (1993) Estimating carbon budgets of Canadian forest ecosystems using a national scale model. In: Vinson TS, Kolchugina TP (eds) Carbon Cycling in Boreal Forest and Subarctic Ecosystems, U.S. EPA Report EPA/600R-93/084 Washington DC, pp 243–252.

Apps MJ, Kurz WA (1993) The role of Canadian forests in the global carbon balance. In: Carbon Balance of World's Forested Ecosystems: Towards A Global Assessment. Academy of Finland No. 3/1993 Helsinki, pp 14–28.

Blais JR (1983) Trends in the frequency, extent, and severity of spruce budworm outbreaks in eastern Canada. *Can. J. For. Res.* 13: 539–547.

Bonan G B (1991) Seasonal and annual carbon fluxes in a boreal forest landscape. *J. Geophys. Res. Atmos.* 96: 17329–17338.

Bonnor GM (1985) Inventory of Forest Biomass in Canada. Canadian Forestry Service, Petawawa National Forestry Institute. Chalk River, ON, 63 pp.

Clark SJ (1990) Fire and climate change during the last 750 yr in Northwestern Minnesota. *Ecol. Monogr.* 60: 135–159.

Cwynar LC (1977) The recent fire history of Barron Township, Algonquin Park. *Can. J. Bot.* 55: 1524–1538.

Dai AG, Fung IY (1993) Can climate variability contribute to the "missing" CO_2 sink? *Glob. Biogeochem. Cycles* 7: 599–609.

Ecoregions Working Group (1989) Ecoclimatic Regions of Canada, First Approximation. Ecoregions Working Group of Canada Committee on Ecological Land Classification. Ecological Land Classification Series, No. 23, Sustainable Development Branch, Canadian Wildlife Service, Conservation and Protection, Environment Canada, Ottawa. 119 pp. and map at 1:7 500 000.

Forestry Canada (1988) Canada's Forest Inventory 1986. Forestry Canada, Ottawa, ON. 60 pp.

Foster DR (1983) The history and pattern of fire in the boreal forest of southeastern Labrador. *Can. J. Bot.* 61: 2459–2471.

Harden JW, Sundquist ET, Stallard RF, Mark RK (1992) Dynamics of soil carbon during deglaciation of the Laurentide ice sheet, *Science* 258: 1921–1924.

Houghton JT, Jenkins GJ, Ephraums JJ (eds). (1990) Climate Change: The IPCC Scientific Assessment. Cambridge University Press, Cambridge, 365 pp.

Houghton RA, Boone RD, Fruci JR, Hobbie JE, Melillo JM, Palm CA, Peterson BJ, Shaver CR, Woodwell GM, Moore B, Skole DL, Myers N (1987) The flux of carbon from terrestrial ecosystems to the atmosphere in 1980 due to changes in land use: geographic distribution of the global flux. *Tellus* 39B: 122–139.

Houghton RA, Hobbie JE, Melillo JM, Moore B, Peterson BJ, Shaver CR, Woodwell GM (1983) Changes in the carbon content of terrestrial biota and soils between 1860 and 1980: a net release of CO_2 to the atmosphere. *Ecol. Monogr.* 53: 235–262.

Jones PD, Bradley RS (1992) Climatic variations over the last 500 years. In: Bradley RS, Jones PD (eds) Climate Since A.D. 1500. Routledge, London and New York, pp 649–665.

Kurz WA, Apps MJ (1993) Contribution of Northern forests to the global C cycle: Canada as a case study. *Water, Air, Soil Pollut.* 70: 163–176.

Kurz WA, Apps MJ (1994) The carbon budget of Canadian forests: A sensitivity analysis of changes in disturbance regimes, growth rates, and decomposition rates. *Environ. Pollut.* 83: 55–61.

Kurz WA, Apps MJ, Webb TM, McNamee PJ (1992) The Carbon Budget of the Canadian Forest Sector: Phase I. Forestry Canada, Northwest Region, Northern Forestry Centre, Information Report NOR-X-326, Edmonton, Alberta, 93 pp.

Kurz WA, Apps MJ (1995) An analysis of future carbon budgets of Canadian boreal forests. *Water, Air, Soil Pollut.* 82: 321–331.

Kurz WA, Apps MJ, Beukema SJ, Lekstrum T (1995) Twentieth century carbon budget of Canadian forests. *Tellus* 47B: 170–177.

Prentice IC, Bartlein PJ, Webb T III (1991) Vegetation and climate change in eastern north America since the last glacial maximum. *Ecology* 72: 2038–2056.

Sarmiento JL, Orr JC, Siegenthaler U (1992) A perturbation simulation of CO_2 uptake in an ocean general circulation model. *J. Geophys. Res.* 97: 3621–3645.

Schlesinger WH (1990) Evidence from chronosequence studies for a low carbon-storage potential of soils. *Nature* 348: 232–234.

Schindler DW, Bayley SE (1993) The biosphere as an increasing sink for atmospheric C: estimates from increased nitrogen deposition. *Glob. Biogeochem. Cycles* 7: 717–733.

Simpson LG, Botkin DB, Nisbet RA (1993) The potential aboveground carbon storage of North American forests. *Water, Air, Soil Pollut.* 70: 197–205.

Tande GF (1979) Fire history and vegetation pattern of coniferous forests in Jasper National Park, Alberta. *Can. J. Bot.* 57: 1912–1931.

Trofymow JA, Preston CM, Prescott CE (1995) Litter quality and its potential effect on decay rates of materials from Canadian forests. In: Apps MJ, Price DT, Wisniewski J (eds) Boreal Forests and Global Change, Kluwer, Dordrecht, pp 215–226.

Van Wagner CE (1978) Age-class distribution and the forest fire cycle. *Can. J. For. Res.* 8: 220–227.

Yarie J (1981) Forest fire cycles and life tables: a case study from interior Alaska. *Can. J. For. Res.* 11: 554–562.

15 The carbon pulse resulting from forest dieback related to insect outbreaks: case study of a forest district in the Sudety Mountains (southwest Poland)

Wojciech Galinski and Jan Witowski

Forest Research Institute, Division of Ecology and Environmental Protection
3, Bitwy Warszawskiej St, 00-973 Warsaw, Poland

INTRODUCTION

Increasing concern over the possible effect of air pollution on European forest ecosystems is very noticeable (for literature review see Schulze et al. 1989). The area of severe forest decline due to pollution was estimated to be less than 0.5% of the European forest area and was thus considered not to have had much impact on the forest resources of the continent (Kauppi et al. 1992). However, in the area of forest decline a decrease in growing stock and permanent deforestation is occurring. This decreases the carbon (C) storage in forest ecosystemsand causes CO_2 release into the atmosphere.

One of the most severe forest dieback events in Europe occurred in the Sudety Mountains region (SW Poland) during the 1970s and 1980s. Heavy industrial pollution promoted insect outbreaks and this was followed by intensive logging and partial deforestation. This paper attempts to estimate the C storage decrease and C pulse into the atmosphere during the period 1975–1989. The estimation procedure is based on forest inventory data, thus a part of the Szklarska Poreba Forest Inspectorate, i.e., Szklarska Poreba Forest District (FD) has been chosen as a study area. The estimate obtained is used to assess the pan-European C source from forest decline attributable to industrial pollution.

Description of Szklarska Poreba Forest District

Szklarska Poreba FD (6689 ha of forests) is partially (80%) located in the Izerskie Mountains and partially (20%) in the Karkonosze Mountains. The lowest elevation within the studied area is 520 m and the highest is 1140 m a.s.l. (i.e., below the tree line located at 1250 m). The annual average temperature and precipitation at 640 m a.s.l. are +5 °C and 1141 mm, respectively, while at an elevation of 1140 m a.s.l., the average temperature is +2.5 °C and the precipitation is higher than 1400 mm. Soil is mainly brown acidic soil with a pH of 4.0 to 5.5 before pollution began and a current pH of 2.5 to 4.0.

Monospecific Norway spruce (*Picea abies* L.) stands cover about 90% of the forested area in Szklarska Poreba, while Scots pine (*Pinus sylvestris* L.), beech (*Fagus sylvatica* L.) and birch (*Betula sp.* L.) in total cover the remaining 10%. Almost all the spruce was planted. The origin of seeds used to produce seedlings for regeneration is not precisely known but is generally thought to be non-local. However, the growth habit of spruce was evaluated as good before the industrial pollution began.

NATO ASI Series, Vol. I 40
Forest Ecosystems, Forest Management and the Global Carbon Cycle
Edited by Michael J. Apps and David T. Price

Industrial pollution

The Sudety Mountains are located in an area of predominantly western and south-western winds. Thus, beside Polish sources of industrial emission, more than 50% of pollutants come from abroad, mainly from the near-border industry of the Czech Republic and Germany (Mazurski 1986). Because of their proximity, the brown coal power stations in Turoszow (Poland), Hirschfelde and Bergsdorf (Germany) and Most (Czech Republic) have a severe impact on the pollution level in the region of Szklarska Poreba.

As early as 1974, the deposition of sulphur and nitrogen compounds in the Sudety Mountains region was estimated to be the highest in Europe (Swedish Ministry of Agriculture 1982). The regular monitoring of air pollution in the region began in 1985, so numerical data on pollution for the 1970s and early 1980s are scarce and dispersed. The average annual SO_2 concentration in the air (Figure 15.1) as measured in the Czech part of the Sudety Mountains increased ca. four times between 1960 and 1987 (for that period Polish data are not available). The values for the Polish part of the mountains for the period 1990–1992 were relatively low (Figure 15.1) which suggests a drop in the level of pollution in the region during the last few years. In agreement with this are pollution indicators measured regularly by passive methods since 1985 (deposits of SO_2, NO_x, Cd and Pb) which show a decreasing trend in the late 1980s (Wawrzoniak and Malachowska 1994).

Insect outbreaks

Until the year 1977, the larch leaf-roller *Zeiraphera diniana* Gn. had never had any economic importance in Polish forestry and had very rarely been found in Poland (Zukowski 1960). The insect feeds mainly on current needles of coniferous trees and prefers stands with wide spacing and good penetration of sunlight within the crown layer (Capecki et al. 1989). Such

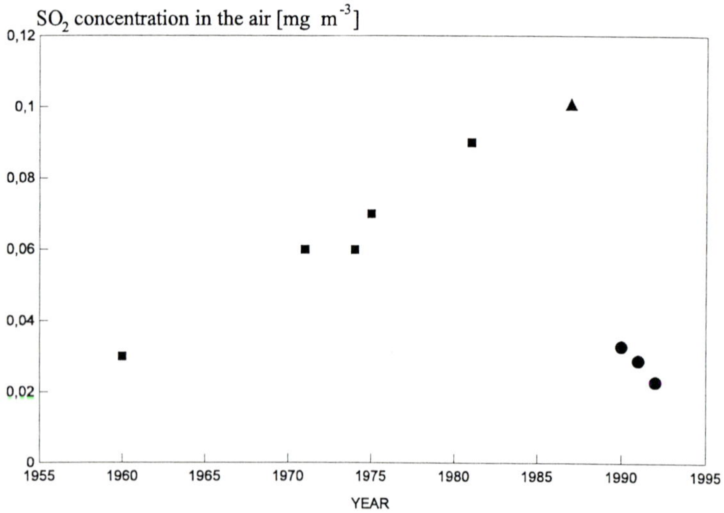

Figure 15.1. Average annual SO_2 concentration in the air as measured in Czech [■ Latocha (1985); ▲ Šima (1989)] and Polish [● Przybylski (1994)] parts of the Sudety Mountains.

conditions can hardly occur in healthy spruce forest. However, defoliation and the wind-breakage of trees weakened by pollution uncovered forest walls and promoted the outbreak of the larch leaf-roller. The outbreak began in 1977 and was highly dynamic (Figure 15.2). The sharp decline of the outbreak in 1982 cannot be attributed to an increase in the population of parasites or to chemical treatments and Capecki et al. (1989) suggest that it was the poor condition of the Sudety forests which caused the rapid decline. In their opinion it is most likely that repeated insect attacks decreased the production of current needles severely, causing the food resource of the larch leaf-roller to vanish and leading to a decline in the pest population due to a lack of food.

The outbreak of the larch leaf-roller in Szklarska Poreba FD was followed by outbreaks of *Ips typographus* L. and other secondary pests. Where populations of *Ips typographus* L. are at low levels, breeding is restricted to non-resistant material such as wind-thrown or cut stems of its host tree. If the population level is high enough trees can be killed (Weslien 1992). As the insects impair only sapwood, the trees are preferably used as saw timber if cut in time. The volume of wood harvested within the period 1975–1989 peaked in 1984, one year after the peak in the directly measured volume of standing trees infected by *Ips typographus* L. (Figure 15.2). The total volume harvested during that period resulted almost exclusively from sanitation cuts of the trees killed by *Ips typographus* L. Thus, the intensive logging in the 1980s can be attributed to pollution-induced forest dieback. The harvesting led ultimately to a large drop in the standing crop biomass and a substantial C pulse into the atmosphere due to the decay of non-timber parts of cut trees.

Estimation of carbon storage and pulse

In order to estimate the C storage change in forest biomass for Szklarska Poreba FD within the years 1975–1989, the C storage in the forest biomass was calculated for the initial year 1975 and repeated for 1989. The forest inventory data on merchantable wood provided by the local Regional Directorate of State Forests (RDSF 1989) were used to estimate the total tree biomass (including roots). The estimation requires the usage of arbitrary chosen coefficients, expressing the ratios of various components of tree biomass to merchantable wood (MW). Two independent sets of coefficients were used in two independent estimations of tree biomass to assess the range of possible results. In both cases the dry weight of the biomass was converted into C assuming a 50% C content of dry biomass (Larcher 1980). The estimation of changes in soil C and in the C already present in litter in 1975 were not taken into account due to lack of appropriate data.

The first estimation (Table 15.1) is based on the phytomass/MW ratios obtained from Isaev et al. (1993) for spruce of four age classes. The ratios resulted from an extensive survey of forest ecosystem biomass data from the former Soviet Union. Unfortunately, no Polish data were available.

The second estimation is based on the volume tables for the spruce forest in the polluted area of the Sudety Mountains (Zawada 1994) and the biomass estimation of various components of spruce trees according to Marklund (1987). The calculations were made for each ten-year age-class up to the class 111–120 years. Example calculations for the year 1975 and age class 81–90 are given in Table 15.2; data for other age classes and for the year 1989 were processed similarly.

Table 15.3 shows the change in the total C storage estimated for Szklarska Poreba FD. The two methods used gave very similar results, so an average value may be a fairly good estimation. The average change in storage was 31.6 Mg C ha^{-1} during the period 1975–1989.

Table 15.1. Estimation of carbon (C) storage in tree biomass for the Szklarska Poreba Forest District for the years 1975 and 1989 (method I).

AGE CLASS	RESOURCE[a] (VOL)		PHYTOMASS/VOL[b]	TOTAL C[c]	
	1975	1989		1975	1989
yr	m^3	m^3	Mg m^{-3}	Mg C	Mg C
0–20	1234	2438	1.19	734	1451
21–40	63 890	85 508	0.793	25 332	33 904
41–60	271 792	214 840	0.734	99 748	78 846
60+	957 970	469 425	0.759	363 550	178 147

a Merchantable wood
b Isaev et al. (1993).
c Assuming biomass C content of 50% (Larcher 1980).

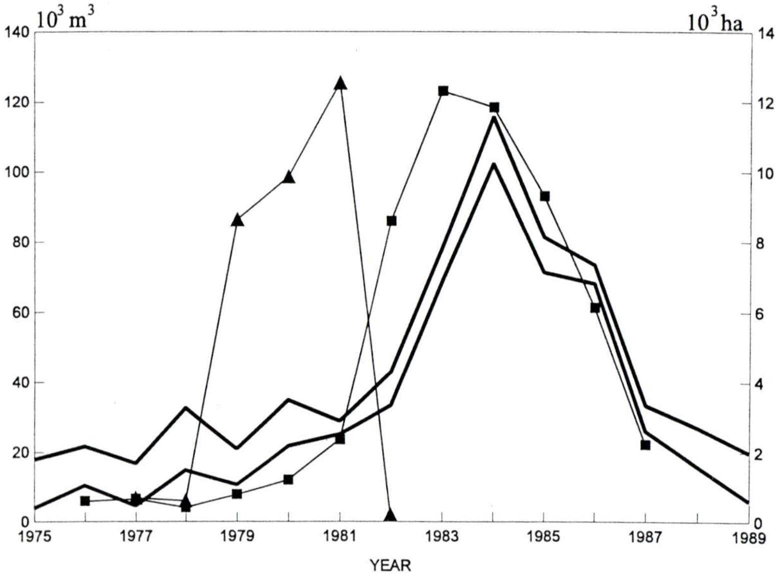

Figure 15.2. Insect outbreaks and logging cuts in Szklarska Poreba (FD) in the period 1975–1989. Upper thick line: volume of logging (including sanitary cuts); lower thick line: volume of sanitary cuts; ■ *Ips typographus* L. abundance, expressed as the volume of infected trees; ▲ *Zeiraphera diniana* Gn. abundance, expressed as the forest area covered by infected trees (data for *Zeiraphera diniana* abundance refer to Szklarska Poreba Forest Inspectorate).

For the purpose of estimating the C pulse released to the atmosphere resulting from logging, an estimation of the tree mass (including roots) decaying on-site after logging is needed. The estimate was obtained by multiplying the data on merchantable wood cut in the period 1975–1989 (627 134 m^3) (RDSF 1989) by the ratio of decayed tree mass to merchantable wood (1.36). The ratio was calculated using the biomass data resulting from method II for the year 1989. When converted to C the decaying tree mass yields a 221 755 Mg C pulse from Szklarska Poreba FD. The average pulse per hectare per year during the period 1975–1989 was 2.4 Mg C ha^{-1} yr^{-1}, thus more than three times higher than the average emission per hectare from Polish forest ecosystems for the year 1988 (0.7 Mg C ha^{-1} yr^{-1}) (Galinski and Küppers 1994).

Table 15.2. Estimation of carbon (C) storage in tree biomass for the Szklarska Poreba Forest District for age class 81–90 and year 1975 (method II). Numbers in parentheses refer to row numbers given in first column.

No.	Value	Unit	Description/reference
1	81–90	yr	Age class
2	567	ha	Area; (RDSF 1989)
3	209 930	m^3	Volume; (RDSF 1989)
4	370	$m^3\ ha^{-1}$	Volume per hectare; (3)/(2)
5	27	cm	DBH for age class (1); Zawada (1994)
6	602	trees ha^{-1}	Number of trees per ha; Zawada (1994), corrected for (4)
Dry weight (DW) of tree components as calculated from Marklund (1987) for (5)			
7	18.2	kg DW	Bark
8	83.3	kg DW	Living branches
9	29.3	kg DW	Needles
10	3.7	kg DW	Dead branches
11	134.5	kg DW	Total (7)+(8)+(9)+(10)
C content of biomass			
12	0.5	Mg C Mg^{-1}	Carbon/plant mass; Larcher (1980)
13	0.26	Mg C m^{-3}	Carbon/plant mass; Larcher (1980)
14	0.35	Mg C (Mg C)$^{-1}$	Root/shoot; Cannell (1982)
15	54 582	Mg C	C in stem; (3) × (13)
16	77 539	Mg C	Total C aboveground; (15)+(11) ×(6) ×(2) ×(12)/1000
17	27 139	Mg C	C in root; (14) ×(16)
18	104 677	Mg C	TOTAL C; (16)+(17)

Table 15.3. Comparison of the change in carbon storage in tree biomass for the Szklarska Poreba Forest District as estimated using methods I and II.

	Storage (Mg C)	
Year	Method I	Method II
1975	489 364	699 321
1989	292 348	474 070
Δ storage	197 016	225 251

DISCUSSION

The severe anthropogenic changes in the Sudety Mountains were initiated in the 14th century when landlords began to change forests into pastures and cultivated land. The forested area of the region had decreased by half at the end of the 18th century (Kirwald 1941; Köhler 1927). Radical changes in tree species composition took place in the 19th century, when spruce seedlings of non-local provenance were planted on large areas. Gradually, almost all forests were turned to spruce monocultures (Wilczkiewicz 1982). In the 20th century these forests appeared to be susceptible to windthrow and industrial pollution. Both factors resulted in decreasing stand density and impaired forest health, thus causing favourable conditions for insect outbreaks and large-scale forest dieback. Therefore, the ecological disaster in the

Sudety Mountains can be regarded as a human-triggered long-term sequence of events where biotic and abiotic factors combined together to destabilise the whole ecosystem.

The data used in the paper emerged from typical forest inventory procedures. Thus, they contain an error specific to the standard method of stand volume assessment. According to Szymkiewicz (the author of Volume Tables, pers. comm.) the error does not exceed 12%. The difference between the two estimations of change in C storage (12.5%) may approximate the error of the method applied. Therefore, the error of the method was similar to the error in the data.

Both sets of coefficients were age-sensitive, as the age structure changed dramatically within the period 1975–1989. Older age classes (by area) vanished almost completely and were replaced by young ones. This was caused by the higher susceptibility of older stands to pest-attack. The area of younger age classes was assumed according to foresters' reports which did not include the areas of regeneration failure due to severe frost damage, grazing and rodent damage.

The C pulse resulting from excessive logging was assumed to be equal to all C contained in non-timber tree parts. The C was not released immediately. However, we do not consider the time course of C emission but the total emitted amount only. The pulse is estimated to be within the range 32–45% of the C storage for the year 1975. Due to the high volume of cuttings, well beyond the wood haulage capacity of a single forest district, an unknown amount of cut wood was left at the site. The volume of this wood is not included in the preceding estimation which thus underestimates the 'real' C pulse.

The C pulse value (33.2 Mg C ha^{-1}) obtained within this paper may be multiplied by the area of severe forest decline caused by industrial pollution in Europe to obtain an estimate of the C pulse from all forests declining there during the period 1975–1989. If the pollution related C pulse (per hectare) from the declining European forests did not exceed the magnitude of the pulse from Szklarska Poreba FD during that period the obtained estimate is the upper one. Such an assumption is rational because forest decline in the Sudety region was among the most severe in Europe and locally resulted in permanent deforestation. Kauppi et al. (1992) estimated that the European area of severe forest decline caused by industrial pollution covered a maximum of 0.8 Mha. Thus, the upper estimate of the C pulse from all forests declining in Europe within the years 1975–1989 is 26.5 Tg C. This figure is approximately 3.8% of the C build-up in European forests during that period, assuming the annual C build-up in the forests is 50 Tg C yr^{-1} (Kauppi et al. 1992).

REFERENCES

Cannell MGR (1982) World forest biomass and primary production data. Academic Press, London, 391 pp.

Capecki Z, Grodzki W, Zwolinski A (1989) The outbreak of larch leaf-roller *Zeiphera griseana* Hb. (Lepidoptera, Tortricidae) in Poland within the years 1977–1983. *Prace IBL* 689: 95–152 [In Polish, with English abstract].

Galinski W, Küppers M (1994) Polish forest ecosystems: the influence of changes in the economic system on the carbon balance. *Climatic Change* 27: 103–119.

Isaev AS, Korovin GN, Utkin AI, Pryazhnikov AA, Zamolodchikov DG (1993) Estimation of carbon pool and its annual deposition in phytomass of forest ecosystems in Russia. *Lesovedenye* 5: 3–10 [In Russian, with English abstract].

Kauppi PE, Mielikäinen K, Kuusela K (1992) Biomass and carbon budget of European forests, 1971 to 1990. *Science* 256: 70–74.

Kirwald HS (1941) Forstgesichtlicher Überblick mit besonderer Berücksichtigung der Forsteinrichtung im Sudetenland. Tharandter Forstl. Jahrbuch. Bd. 92 Sonderh.

Köhler C (1927) Die Forstlichen Verhältnisse der Herrschaft Schaffgotsch im Reisen und Isergebirge. Jahrbuch d. Schlesischen Forstvereins. Breslau.

Larcher W (1980) Physiological plant ecology. Springer-Verlag, Berlin Heidelberg, 252 pp.

Latocha E (1985) The threat of industrial pollution to the mountain spruce stands in Central Europe. *Sylwan* 1: 17–28 [In Polish, with English abstract].

Marklund LG (1987) Biomass functions for Norway spruce (*Picea abies* (L.) Karst.) in Sweden. *Sveriges Lantbruksuniversitet Rapporter - Skog* Nr 43: 1–127.

Mazurski KR (1986) The destruction of forests in the Polish Sudetes Mountains by Industrial Emissions. *Forest Ecology and Management* 17 (4): 303–315.

Przybylski T (1994) The danger to spruce existence in Sudety Mountains due to environmental pollution. *Prace IBL* Ser. 'B' 21: 185–194 [In Polish, with English abstract].

RDSF (1989) Forest management plan for Szklarska Poreba Forest District for the period 1 Jan. 1989–31 Dec. 1998. Part I. Unpublished manuscript [In Polish].

Schulze E-D, Lange OL, Oren R (1989) Air pollution and forest decline: a study of spruce (*Picea abies*) on acid soils. *Ecological Studies* Vol. 77, Springer-Verlag, Berlin Heidelberg, 475 pp.

Šima A (1989) SO_2 concentrations in Krušné and Jizerské mountains as dependent on the meteorological conditions. *Zpr. Les. Vyzk.* 34 (3): 17–22 [In Czech].

Swedish Ministry of Agriculture (1982). Acidification Today and Tomorrow. Risbergs Tryckeri AB, Uddevalla, Sweden, 231 pp.

Wawrzoniak J, Malachowska J (1994) Technical and biological monitoring in West Sudety Mountains. *Prace IBL* Ser. 'B' 21: 9–24 [In Polish, with English abstract].

Weslien J (1992) Monitoring *Ips typographus* (L.) populations and forecasting damage. *Journal of Applied Entomology* 114: 338–340.

Wilczkiewicz M (1982) Historical background of forest economy in Sudety Mountains. *Sylwan* 6: 49–54 [In Polish, with English abstract].

Zawada J (1994) Site and stand analysis of spruce forests in lower mountain region of the Zywiec Beskid and the Sudety Mountains as well as directions of its further application. *Prace IBL* 759 [In Polish, with English abstract].

Zukowski R (1960) The observations on the occurrence and possible course of outbreaks, of some butterfly species within the years 1949–1959 in Pieninski National Park. *Sylwan* 10: 37–52 [In Polish, with English abstract].

16 Carbon budget of temperate zone forests during 1851–2050

Pekka E Kauppi

Finnish Forest Research Institute
Unioninkatu 40 A, FIN-00170 Helsinki, Finland

INTRODUCTION

The temperate zone covers lands in the mid-latitudes on five continents between the two other main biomes, the tropical and boreal zone (e.g., Heath et al. 1993). The human population has utilised forests for a long time in areas such as China and southern Europe. Only about half of the original forests have remained forested in the temperate zone, while the other half has been converted to other uses, mainly to agricultural and pastoral land. More recently and especially in the second half of this century, the emissions of CO_2, SO_2 , N-oxides and other gases from energy production and industry have changed the forest environment.

The carbon (C) reservoir of forests changes over different time horizons. The variation of disturbances from year to year, due to random events, can trigger oscillations in the time scale of decades to centuries (Kurz and Apps 1993). Climate variability, air pollution, human impact, development of soils, and the glacial cycle induce long term changes in reservoirs over various and overlapping time scales from decades to millennia. It is reasonable to postulate that the C reservoir of any given forest, at any time interval, is not in a steady state (Houghton et al. 1987; Lugo 1992; Dixon et al. 1994a). The problem is to learn about the C dynamics, in other words, to describe the development of C reservoirs of forest vegetation and soils over time, and to calculate the related fluxes of C between the ecosystem and the atmosphere.

Land management in the past has affected the current C reservoirs. The changes of gas concentrations in the atmosphere have had impacts, and can have increasing impacts in the future on climate and on forests. This paper focuses on historical, current and future trends of the C reservoirs in temperate zone forests as affected by changes in management and in the environment. In the past, the time interval from 1851 covers a period of many changes in land use in the different parts of the world. It also covers the rise of atmospheric CO_2 concentration from the pre-industrial level of 280–290 ppm to the current value of 355–360 ppm. The aim of this paper is to describe causal mechanisms which have affected the C reservoirs in temperate zone forests since 1851, to estimate the magnitude of changes in the reservoirs, and to discuss the possible development of the reservoirs by the year 2050. Special emphasis is given to mechanisms affecting the fluxes and reservoirs of C in the soil.

NATO ASI Series, Vol. I 40
Forest Ecosystems, Forest Management and the Global Carbon Cycle
Edited by Michael J. Apps and David T. Price

CRITERIA, INDICATORS, AND THE PRESENT RESERVOIRS

The two main reservoirs of interest are C in the soil, and C in living forest vegetation. The reservoirs in the temperate zones are smaller than those in the boreal or tropical regions (Table 16.1).

Table 16.1. Forested area, C storage in vegetation, and C storage in soil of the different world regions in about 1990 (Dixon et al. 1994a). The high latitude numbers refer only to Russia, Canada, and Alaska (Apps et al. 1993).

LATITUDE	BIOME	FORESTED AREA (10^6 km^2)	SOIL C STORAGE (Pg)	VEGETATION C STORAGE (Pg)
High	Boreal	13.7	471	88
Mid	Temperate	10.4	100	59
Low	Tropical	17.6	216	212

The processes responsible for changes in these reservoirs include forest increment, cutting removals, litter production and tree mortality, the transfer of dead organic matter from the forest, and fire (Figure 16.1). The reservoir in ground vegetation is omitted since it is quite small (e.g., Mälkönen 1987). The concepts *sink* and *source* are used in this paper to indicate changes in reservoirs where appropriate. An increasing reservoir is a sink, a decreasing reservoir is a source. Chapters 10, 19 and 21 (Houghton; Matthews; Maclaren) also discuss these terms in detail.

CARBON RESERVOIRS IN THE PAST

Impact of land management

Deforestation and afforestation

Deforestation accelerated and peaked in the temperate zone between 1851 and 1950 (Heath et al. 1993). In the United States, deforestation started in the Great Lakes region in 1870, and in the Pacific Northwest early this century (Stuiver 1978). Deforestation also peaked in 1851–1950 in New Zealand (Climate Change—The New Zealand Response 1994).

Stuiver (1978) estimated a net release of 120 Pg C from global terrestrial ecosystems into the atmosphere between 1851 and 1950. During that period of time, deforestation occurred mainly in the temperate zone. Assuming a share of 50–85%, the contribution from temperate zone forests would have been 60–100 Pg over this one hundred year period. According to this estimate, the C reservoir of temperate zone forests would have declined by 20–35% (see Table 16.1). The estimated average flux, 0.6–1.0 Pg C per annum, was about half of the estimated current releases from the reservoirs in tropical forests (Dixon et al. 1994a).

Stuiver's estimate was derived from C isotope ratios in tree ring records. The method does not distinguish soil from vegetation releases and, therefore, includes the C source contribution from the soil reservoir. Advantages of the method are that it is retrospective, the measurements can be repeated, and the data cover a long period of time 150–250 years ago.

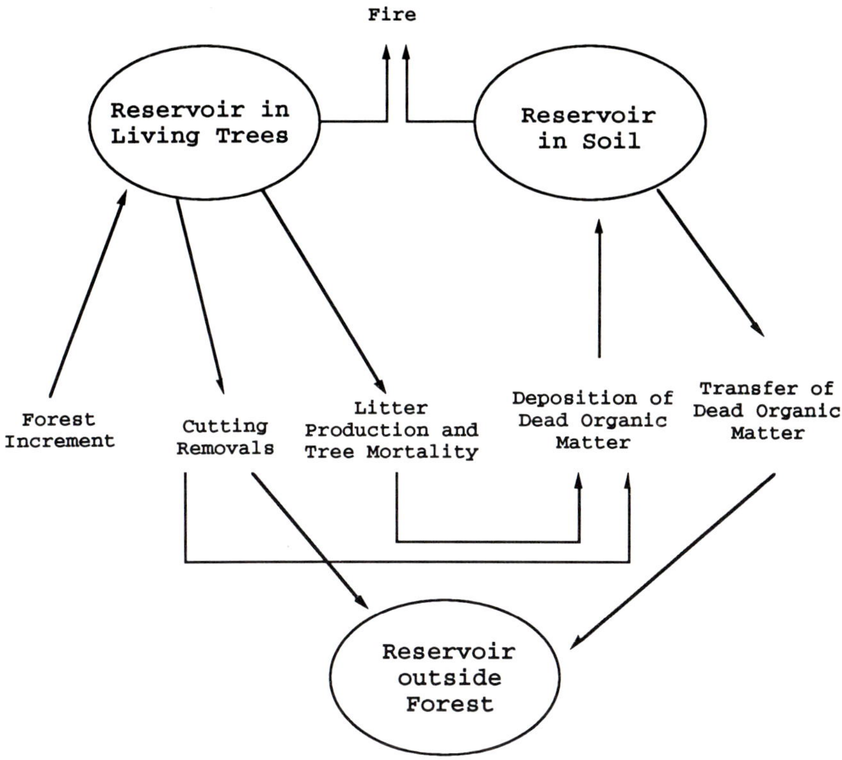

Figure 16.1. Biomass, C reservoirs and fluxes in forest ecosystems.

In fact, the investigation should be repeated and the uncertainties be described in detail, given the current interest in the history of deforestation, and in the mechanism of global C budget. In the meantime, Stuiver's (1978) estimate can be used as the best approximation for quantifying the history of C releases from forests.

Since 1951, although land has been reverting from non-forest to forest, the sink impact of afforestation has been small. In the temperate zone in Europe, excluding the area of the former USSR and the Mediterranean region, the area of forested land increased by 4% from 532 000 km^2 to 554 000 km^2 between ca. 1965 and 1990 (Kuusela 1994). The subsequent change of C reservoir has been much less than +4%, because of the insignificant C reservoir per unit of land area in young forest stands. The response is strongest 20–50 years after afforestation. The response is even slower in the soil (Dewar and Cannell 1992). Although the impact of early afforestation in 1951–1980 has generated a sink response in areas such as Europe, the southern U.S. and New Zealand (Kauppi et al. 1992; Heath and Birdsey 1993; Hollinger et al. 1994), the main sink effect of reverting lands to forests is only emerging (Dixon et al. 1994a; Climate Change—The New Zealand Response 1994).

Forest harvest rate

Sustainable logging, by definition, does not have a decreasing nor increasing impact on the C reservoirs. Wood removals are adapted to forest increment, and the growing stock deviates from a steady state only in cases of anomalies in the stand age structure. Logging on average has been *more than sustainable* in many regions in the temperate zone: Removals have been much less than forest increment (Kauppi et al. 1992; Heath and Birdsey 1993). The harvest rate (defined as the ratio of wood removals to forest increment) has declined to about 0.7 in many temperate region countries (Kuusela 1994). This has contributed as a strong sink term to the global C budget, especially in the past 20–30 years. Since 1965, the growing stock has increased by more than 40% in the temperate zone in Europe (Kuusela 1994). The low harvest rate on permanent forest land has clearly been the main cause of the sink effect of vegetation, in comparison to the rather small effect of afforestation (Sedjo 1992; Kuusela 1994).

Fire suppression, nature protection

Protective management has aimed at maintaining or increasing the growing stock. Fewer trees have been burned than would have burned if the forests had not been actively protected. Protective measures such as fire suppression have a secondary increasing effect on the C reservoir of the soils. (See also Chapter 8, Stocks et al., who explicitly discuss the role of wild fire and fire suppression.)

Trees in nature conservation areas are protected against cutting removals. According to IUCN statistics, the protected areas in Europe in 1994 covered about 500 000 km^2, or 10.2% of the land area (World Conservation Union (IUCN) 1994). The statistics cover the Nordic and Baltic countries and the Mediterranean region, but exclude Belarus, Russia and Ukraine. Complete cutting bans are applied only to part of the areas but all are subject to some restrictive policies. Protection has been established and implemented mainly in the 20th century. The impact of protection on C reservoirs has developed mainly in the latter part of the century. Although only a fraction of the protected areas are closed forests, protective land management has contributed to the sink effect.

Other changes in land management

Cattle grazing in forests and the collection of small sized wood for fuel were common practices in many areas until the mid-1900s. Those practices have virtually disappeared over large regions, especially in Europe. The C reservoir in small sized wood, in dead standing trees and in the soils has increased as a consequence. It is difficult to estimate the quantitative impact. Nevertheless, the changes have contributed to the sink effect.

Impact of changes in the atmospheric environment

Eutrophication

The atmospheric CO_2 concentration is increasing globally, while nitrogen deposition is also increasing, especially in the temperate zone. The productivity of ecosystems is increasing as a consequence (Kenk and Fischer 1988; Luxmoore et al. 1993; Johnson et al. 1995). The nitrogen deposition in areas such as The Netherlands and parts of Germany can be as high as 50 to 70 kg N ha^{-1} yr^{-1}.

The productivity response of CO_2 fertilisation is nutrient-limited (e.g., Houghton et al. 1987). Nitrogen deposition alleviates the limitation in the temperate zone. The increment rates of forest stands have increased in Europe especially since the 1960s (Kenk and Fischer 1988). As a response to eutrophication, the input of dead organic matter to forest soil has probably increased. Whether the decomposition rate has also increased is not known. Therefore, not even the signs of the impact of eutrophication on the C reservoir of forest soils can be evaluated at present. (Townsend and Rastetter discuss the issue of nutrient stimulation further in Chapter 3.)

Adverse effects of air pollutants

Forest damage due to air pollution has been investigated thoroughly in Europe and North America (e.g., Schulze 1989; Shriner et al. 1990). In Poland, the Czech Republic, and Germany, for example, complete forest stand damage has been observed in large areas (100–500 km^2). The C reservoir has declined in these areas, and the eventual recovery of the reservoir as a response to decreasing pollution levels and afforestation will take a long time (100 years or more). In Chapter 15, Galinski and Witowski discuss the interaction of air pollution and insect damage in a case study in south-west Poland.

Forest statistics do not indicate a general decline of the C reservoir in countries where air pollution damage has been a problem. Between 1980 and 1990, the growing stock increased by 8%, 100%, 7%, and 19% in eastern Germany (former GDR), western Germany, Czechoslovakia and Poland, respectively (Kuusela 1994). The reported change in western Germany is due mostly to a severe underestimation of the growing stock of 1980. A statistically representative forest inventory was carried out in western Germany at the end of the 1980s, and the estimate of 1990 was based on these new measurements. In spite of problems with the statistics, it is clear that there has not been a general decline of the vegetation C reservoir in these countries. The reported cases of severely damaged forests are exceptions, not the rule. Yet, the pollutants pose a future threat to forest ecosystems (Kauppi et al. 1992).

Carbon pump hypothesis

The annual production of foliage litter is a source of C into the forest soil. The canopy foliage of deciduous trees is recycled annually. In conifers, 10–30% of the needles are lost annually depending on species. Litter is also produced annually in the root system. Liski (1995) has reported more soil C near the trees than in small openings within a boreal forest stand. This indicates the impact of litter production from foliage and from the roots.

Litter deposition and turnover occur in every forest ecosystem world-wide; only the rate of deposition and turnover varies among systems. Depending on climate and soil, a small fraction of the annual deposition of litter can remain in the soil and be stored in deeper soil layers. This plant-to-soil sink mechanism is here referred to as the *carbon pump hypothesis*. It is justifiable to speak about a sink mechanism, if the C reservoir in the soil is accumulating as a consequence.

Because of the C pump mechanism, the C reservoir of soils can develop independently of the vegetation reservoir. For example, C may accumulate in the deep soil layers of unmanaged wilderness forests, even if fire and disturbance dynamics tend to maintain the vegetation near a steady state. The C pump would vary among ecosystems, even between neighbouring ecosystems, depending on vegetation, soil depth, hydrology, microclimate and the history of soil management. The depletion of mineral nutrients such as nitrogen, potassium and phosphorus from the soil can limit the CO_2 fertilisation effect (Johnson et al. 1995), and enrich soil biomass with C. Protective land management and increasing levels of atmospheric CO_2 may have strengthened the C pump.

Dixon et al. (1994b) evaluated the potential of various land-use practices in conserving and sequestering soil C. They concluded that increased C accumulation is associated with cool, wet, and more fertile soils as well as soils with balanced aeration. Increasing soil fertility may have contributed to an increase of C reservoir in forest soils. If measurements and monitoring show constant levels of C reservoir in deeper soil, the C pump hypothesis can be rejected.

Conclusions regarding 1851 to present

It has been stated that

> *"...the greatest changes in terrestrial carbon result from the transformation of forests to non-forests and from the transformation of non-forests to forests"* (Houghton et al. 1987).

This has been the case for temperate region forests in 1851–1950, and also more recently in New Zealand (Climate Change—The New Zealand Response 1994). More recently in the U.S. and in Europe, however, large changes in terrestrial living biomass have taken place on permanent forest land. The ratio of cuttings to forest increment has declined. The same has also been documented from northern Europe (Eriksson 1991). Although there are regions where the harvest rate has been high (Harmon et al. 1990; Kauppi et al. 1995), national forest statistics indicate that such regions are exceptions and not the rule (Sedjo 1992; Heath and Birdsey 1993; Kuusela 1994).

The C reservoir of vegetation (growing stock) can increase even in conditions of a decreasing trend of forest growth presuming that cutting removals remain at a lower level than forest increment, and mortality is insignificant (Figure 16.1). Also the opposite holds true. If the sum of cutting removals and mortality exceeds increment, the forest is a source of atmospheric C regardless of the rate of forest increment.

Increasing logging removals could eliminate an eventual increasing impact of eutrophication on the C reservoir in vegetation. The statistics indicate however, that this has not been the case. Cutting removals have not increased but been almost stable since 1965, while forest growth has increased substantially (Kuusela 1994). The positive development of increment has been due to improvements in silvicultural practices (including protective management) and, eventually, eutrophication. It is the high and increasing increment, rather than a decreasing trend of removals that has lowered the forest harvest rate. Therefore, this analysis supports the hypothesis that eutrophication has contributed to the vegetation sink.

Dixon et al. (1994a) estimated the sink of temperate region forests in 1990 as 0.26 ±0.09 Pg yr^{-1}. As the forest harvest rate has declined between 1950 and 1990, the sink must have been smaller in the 1950s and 1960s than in the late 1980s. The estimate of Dixon et al. (1994a) combines results from forest statistics based on primary measurements taken 5–15 years ago. As the sink has been increasing, the estimate of 0.26 ±0.09 Pg yr^{-1} must be too low to represent the sink in the early 1990s. For this period, a removal estimate of 0.3–0.4 Pg could be more justified. Assuming an average sink of 0.15 Pg in 1951–1990, the C reservoir of vegetation in the temperate zones would have increased by 6 Pg. These estimates however are tentative, and need to be re-examined in future research.

The results can contribute to detection of the sink in the global C budget which has remained unidentified (Detwiler et al. 1988; Tans et al. 1990). Assume that part of the remaining, unidentified sink is within the temperate zone: if it is in vegetation, there must have been a serious underestimation of the recent increase of the growing stock. Although this is not completely impossible, it is more likely that a sink has remained unidentified in the soil.

CHANGES OF THE CARBON RESERVOIR IN THE FUTURE

The idea of sustainable development is likely to prevent an outbreak of deforestation trends in the temperate zone. New policies are being implemented which tend to protect forests and enhance the sink effect. For example, the lowered harvests in old-growth forests in the U.S. have been estimated to bring about an additional sink of 10 Tg C yr^{-1} in the year 2000, compared to the sink of 1990 (U.S. Climate Action Report 1994). As a result of current trends in forest increment and restrictions in cuttings, the ratio of cutting removals to forest increment will continue falling in the near term in temperate zone forests. If that happens, it would strengthen the current sink impact.

In Europe, the impacts of afforestation will remain small at least until 2010. Conversely, in New Zealand, the C sink of planted forests is estimated to increase from 4.5 to 7.0 Tg C yr^{-1} between 1990 and 2000, and further to 10.0 Tg C yr^{-1} by 2010 (Climate Change–The New Zealand Response 1994). New Zealand has a large afforestation program, probably surpassed only by China. (See also Chapter 21, Maclaren, for a study of plantation forestry and the C sink in New Zealand.)

The socio-economic situation, affecting the sink/source development, varies among regions within the temperate zone. For example, economic development in China is faster than in most other regions of the world. Increasing wood demand and the adverse effects of air pollution can threaten vegetation and soils, and counteract the increasing effect of afforestation on the forest C reservoir.

There are many possibilities available for increasing the C reservoirs. A number of policies are sustainable also in the long term (Heath et al. 1993; Hoen and Solberg 1994). Protective land management is potentially important in the next few decades, in the temperate region just as in the tropics (Dixon et al. 1994a).

The low ratio of harvests to net increment has been the main reason for the observed sink effect in Europe since the 1950s. This low ratio is likely to persist for some more decades. In the long term, adverse impacts of air pollutants and climatic change, and land demand for the growing population can outweigh the sink impacts due to low harvest rate, land abandonment, and afforestation.

Temperate region forests will continue to be sink of atmospheric CO_2 for a few more decades, but the future thereafter is uncertain (Heath et al. 1993). By the year 2050 or so, the vegetation reservoir can approach saturation, unless policies such as large scale afforestation are developed to postpone the eventual saturation (Winjum et al. 1993; Dixon and Krankina 1993).

REFERENCES

Apps MJ, Kurz WA, Luxmoore RJ, Nilsson LO, Sedjo RA, Schmidt R, Simpson LG, Vinson TS (1993) Boreal forests and tundra. *Water, Air and Soil Pollut.*70: 39–53.

Climate Action Report–Submission of the United States of America Under the United Nations Framework Convention on Climate Change (1994) US Government Printing Office, Washington DC, 200 pp.

Climate Change–The New Zealand Response (1994) New Zealand's First National Communication under the Framework Convention on Climate Change. Ministry of the Environment Wellington, NZ, 69 pp.

Dewar CD, Cannell MGR (1992) Carbon sequestration in the trees, products and soils of forest plantations: an analysis using UK examples. *Tree Physiol.* 11: 49–71.

Detwiler RP, Hall CAS (1988) Tropical forests and the global carbon cycle. *Science* 239: 42–47.

Dixon RK, Krankina ON (1993) Forest fires in Russia: CO_2 emissions to the atmosphere. *Can. J. Forest Res.* 23: 700–705.

Dixon RK, Brown S, Houghton RA, Solomon AM, Trexler MC, Wisniewski J (1994a) Carbon pools and flux of global forest ecosystems. *Science* 263: 185–190.

Dixon RK, Winjum JK, Andrasko KJ, Lee JJ, Schroeder PE (1994b) Integrated land-use systems: Assessment of promising agroforest and alternative land-use practices to enhance carbon conservation and sequestration. *Climatic Change* 27: 71–92.

Eriksson H (1991) Sources and sinks of CO_2 in Sweden. *Ambio* 20: 146–150.

Harmon ME, Ferrell W, Franklin JF (1990) Effect on carbon storage of conversion of old-growth forest to young forests. *Science* 247: 699–702.

Heath LS, Kauppi PE, Burschel P, Gregor HD, Guderian R, Kohlmeier GH, Lorenz S, Overdieck D, Scholz F, Thomasius H, Weber M (1993) Contribution of temperate forests to the world's carbon budget. *Water, Air, Soil Pollut.* 70: 55–69.

Heath LS, Birdsey RA (1993) Carbon trends of productive temperate forests of the coterminous United States. *Water, Air, Soil Pollut.* 70: 279–293.

Houghton RA, Boone RD, Fruci JR, Hobbie JE, Melillo JM, Palm CA, Peterson BJ, Shaver GR, Woodwell GM, Moore B, Skole DL, Myers N (1987) The flux of carbon from terrestrial ecosystems to the atmosphere in 1980 due to changes in land use: geographic distribution of the global flux. *Tellus* B 39: 122–139.

Hoen HF, Solberg B (1994) Potential and economic efficiency of carbon sequestration in forest biomass through silvicultural management. *Forest Sci.* 40: 429–452.

Hollinger DY, Maclaren JP, Beets PN, Turland J (1993) Carbon sequestration by New Zealand's plantation forests. *NZ J. Forest Sci.* 23: 194–208.

World Conservation Union (IUCN) (1994) Parks for Life. Action for Protected Areas in Europe. The World Conservation Union. Gland, Switzerland, 150 pp.

Johnson DW, Ball T, Walker RF (1995) Effects of elevated CO_2 and nitrogen on nutrient uptake in ponderosa pine seedlings. *Plant Soil* 168–169: 535–545.

Kenk G, Fisher H (1988) Evidence from nitrogen fertilisation in the forests of Germany. *Environ. Pollut.* 54: 199–218.

Kauppi PE, Mielikäinen K, Kuusela K (1992) Biomass and carbon budget of European forests, 1971 to 1990. *Science* 256: 70–74.

Kauppi PE, Tomppo E, Ferm A (1995) C and N storage in living trees within Finland since the 1950s. *Plant and Soil* 168–169: 633–638.

Kurz WA, Apps MJ (1993) The carbon budget of Canadian forests: A sensitivity analysis of changes in disturbance regimes, growth rates, and decomposition rates. *Environ. Pollut.* 83: 55–61.

Kuusela K (1994) Forest Resources in Europe. Cambridge University Press, Cambridge, 154 pp.

Liski J (1995) Variation in the amount of organic carbon in soil within a forest stand: effect of trees and implications for sampling. The Finnish Forest Research Institute, Research Papers 537: 77–82.

Lugo AE (1992) The search for carbon sinks in the tropics. *Water, Air, Soil Pollut.* 64: 3–9.

Luxmoore RJ, Wullschleger SD, Hanson PJ (1993) Forest responses to CO_2 enrichment and climate warming. *Water, Air, Soil Pollut.* 70: 309–323.

Mälkönen E (1977) Annual primary production and nutrient cycle in a birch stand. *Communicationes Instituti Forestalis Fenniae* 91(5): 1–35.

Shriner DS, Hech WW, McLaughlin SB, Johnson DW, Irving PM, Joslin JD, Peterson CE (1990) Response of vegetation to atmospheric deposition and air pollution. NAPAP SOS/T Report 18, In: Acidic Deposition: State of Science and Technology, Volume III, National Acid Precipitation Assessment Program, Washington DC, 206 pp.

Sedjo RA (1992) Temperate forest ecosystems in the global carbon cycle. *Ambio* 21: 27–277.

Shulze ED (1989) Air pollution and forest decline in a spruce (*Picea abies*) forest. *Science* 244: 776–783.

Stuiver M (1978) Atmospheric CO_2 and carbon reservoir changes. *Science* 199: 253–258.

Tans PP, Fung IF, Takahashi T (1990) Observational constraints on the global atmospheric CO_2 budget. *Science* 247: 43–47.

UN Framework Convention on Climate Change (1992) United Nations, New York, 27 pp.

Winjum JK, Meganck RA, Dixon RK (1993) Expanding global forest management: an "easy first" proposal. *J. Forestry*, pp 38–42

17 WG2 Summary: Forests and the global carbon cycle: past, present, and future role

Sandra Brown[1,2], Anatoly Z Shvidenko[1,2], Wojciech Galinski[2], Richard A Houghton[2], Eric S Kasischke[2], Pekka Kauppi[2], Werner A Kurz[2], Ian A Nalder[2] and Vjacheslav A Rojkov[2]

[1] Co-Chairs of Working Group 2
[2] Addresses are provided in the participant list.

INTRODUCTION

Research efforts over the last decade or so on the interaction of human and natural disturbances on the carbon (C) dynamics of forests have increased our state of understanding about the role of forests in the global C cycle. However, at the same time new areas of uncertainty and new data gaps have emerged. In this chapter, we present areas of improved understanding, identify areas of uncertainty that require further research, identify new data needs, discuss the future role of forests in the global C cycle, and conclude with a discussion on the need for a system for tracking the C flux from changes in the cover, use, and management of forest lands. Material for this overview chapter is based for the most part on the chapters included in this Part of the book.

NEW AREAS OF UNDERSTANDING

State and change in cover, use, and management of forest land

In discussing the role of forests in the global C cycle, it is necessary to distinguish between land cover, land use, and land management. Lack of recognition of the differences in these terms has often led to confusion about the role of forests in the C cycle (Dixon et al. 1994). Land cover (LC) refers to the physical and biological appearance of the land surface (e.g., closed forests, open forests, grasslands, etc.) and change in LC converts land of one type of cover to another regardless of its use. Change in LC can be brought about by human (e.g., clearing of forests for agriculture) and natural disturbances (e.g., wildfires or tropical storms can convert mature forests to non-forests or successional forests). Land use (LU) refers to the purpose for which land is used by humans, e.g., nature preserves, forestry, plantations, agriculture. Change in LU may or may not cause a change in LC (e.g., conversion of logged forests to forest preserves causes little to no change in LC), but a human-induced change in LC is usually associated with a change in LU. Land management (LM), a relatively new term in this field, refers to how a given LU is managed by humans, e.g., selective versus clearcut harvesting, conventional versus whole tree removal, irrigated versus rain-fed agriculture. Each of these terms has different implications for monitoring change in area and impact on C storage and flux.

NATO ASI Series, Vol. I 40
Forest Ecosystems, Forest Management and the Global Carbon Cycle
Edited by Michael J. Apps and David T. Price

The status and rate of change in LC/LU/LM of the world's forests are among the major sources of uncertainty contributing to our inability to balance the terrestrial C budget. Major changes in forest to non-forest and *vice versa* can for the most part be monitored by both low and high resolution remote sensing platforms. However today's sensors, even high resolution ones, cannot always distinguish between cover changes caused by anthropogenic or natural factors. Furthermore, distinctions between changes in LU or LM (e.g., selectively harvested forests to protected forests, closed forests to degraded forests) cannot readily be detected by remote sensing techniques. In these situations, national inventories based on finer resolution data are required.

A more precise identification of the major processes causing change in the status of forest lands has significant implications for choice of appropriate monitoring methodologies. Major processes responsible for change in LC/LU/LM and subsequent changes in C budgets and fluxes have varied over time and by latitudinal region (Table 17.1). In general, changes in LC in the boreal zone have for the most part been and are likely to be in the future of little significance. The exception to this trend is the clearing of forests for activities associated with expansion of urban/industrial and mining/dam construction areas in the FSU (Former Soviet Union) and SCA (Scandinavian countries and European part of Russia's boreal forests). Natural disturbances have been and are projected to be the major causes of change in LC and subsequent C budgets of forests in the boreal zone (Table 17.1; see also Chapters 8, 12, and 14). Remote sensing techniques provide useful information for assessing the C budgets for boreal forests, particularly with respect to the area of forest cover, fire detection, and large insect outbreaks in these relatively remote areas (Cahoon et al. 1994; Kasischke and French 1995; Kasischke et al. 1995; and Chapters 7 and 8). More recently, changes in LU and LM are affecting forest C budgets in this zone. In this situation, national forest inventories coupled with statistical data on harvested forests are the appropriate monitoring methodology.

For temperate forests, changes in LU and LM associated with forest harvesting and changes in LC associated with conversion to and abandonment of agriculture and forestation appear to be the most significant processes affecting C budgets and flux (Table 17.1). In the future, changes in LU and LM are expected to become more important and changes in LC less important. As the major processes affecting forest C budgets are changes in LC/LU/LM, the judicious merging of the spatial and temporal capabilities of remote sensors with those from regular national forest inventories would be the method of choice.

The major processes affecting C budgets and flux in the tropics are mostly related to changes in forest cover and to a lesser degree to changes in use and management of forests (Table 17.1). For the very significant changes in LC since the 1970s, a combination of high resolution remote sensing imagery coupled with forest inventories would be the most appropriate methodology.

Initial conditions of terrestrial ecosystems

Most analyses of terrestrial C budgets have assumed that forests not directly converted to other uses since the pre-industrial era are in steady state with regard to net C exchange with the atmosphere (e.g., Houghton et al. 1983). This assumption was deemed appropriate in regions where the effects of anthropogenic disturbances are large relative to those of natural disturbances, and where data are lacking to support a more detailed analysis. The consequence of this assumption is partly responsible for the present inability to balance the terrestrial C budget.

The direct human influence on the C cycle of boreal forests of Russia and North America is small compared to the influences of natural disturbance regimes such as fire and insects (Table 17.1). Any deviation from the assumed steady state condition of forests not affected by human disturbances could therefore substantially alter the net C budget of these regions. Steady state conditions are only maintained if the periodic (annual, decadal) C uptake in forest ecosystems (i.e., the net balance of photosynthesis and respiration) is offset by the C release through forest fires and other disturbances over the same period. Fluctuations in natural disturbance regimes, however, can alter this balance, at time scales of years to decades (Chapters 9 and 14). Human influences, such as air and soil pollution, can also affect large areas and operate over time scales of years to decades (e.g., Chapters 13 and 15). Other processes, operating at even longer time scales may further cause a deviation from assumed steady-state conditions. For example, C storage in forest ecosystems, specifically in soil and detritus pools, could be changing in response to long-term temperature fluctuations. The immediate impacts

Table 17.1. Major processes causing change in land cover (LC), land use, (LU), and land management (LM), and subsequent change in the carbon (C) flux. The relative importance of the process of change during a given time period is designated by a score of 1 to 4, where 1 is very significant and 4 is insignificant. The impact of the change in LC/LU/LM on the C flux is given by the symbol in parenthesis: (+): C sink, (-): C source, (0): no effect, and (+/-): uncertain direction.

PROCESS		PAST 1850–1950	RECENT PAST 1950–1990	PRESENT 1990–2000	FUTURE 2000–2050
		BOREAL ZONE			
1. CHANGE IN LC					
• Forest to non-forest					
-permanent crops	FSU[a]	3(-)	4(0)	4(0)	4(0)
	NA[a]	4(0)	4(0)	4(0)	4(0)
	SCA[a]	1(-)	4(0)	4(0)	3(-)
-urban/industrial	FSU	2(-)	3(-)	3(-)	3(-)
	NA	4(0)	4(0)	4(0)	4(0)
	SCA	2(-)	2(-)	3(-)	3(-)
-mining/oil and gas/dams	FSU	2(-)	2(-)	2(-)	3(-)
	NA	4(0)	3(-)	3(-)	3(-)
	SCA	3(-)	2(-)	3(-)	2(-)
• Non-forest to forest					
-forestation	FSU	4(0)	3(+)	3(+)	2(+)
	NA	4(0)	4(0)	4(0)	3(+)
• Natural disturbances					
-wildfires	FSU/NA	2(+/-)	2(+/-)	2(+/-)	1(--)
	SCA	2(+/-)	2(+/-)	3(+/-)	2(--)
-insects/disease	All	3(+/-)	2(+/-)	2(+/-)	2/3(+/-)
• Pollution					
-air/soil	FSU	4(0)	2(-)	3(-)	3(-)
	NA	4(0)	4(0)	4(0)	3(?)
	SCA	4(0)	2(-)	2(-)	3(-)
2. CHANGE IN LU/LM					
-forest harvesting (including roads)	FSU	4(0)	2(-)	2(-)	2(-)
	NA	4(0)	3(-)	3(-)	2(-)
	SCA	4(0)	2(-)	2(-)	2(-)
-draining forested wetlands	All	4(0)	3/4(0)	3/4(0)	3(+/-)
-grazing	FSU	3(-)	3(-)	2(-)	3(-)
	NA	4(0)	4(0)	4(0)	4(0)

a FSU: Former Soviet Union.
NA: North America (Canada and Alaska).
SCA: Scandinavian countries and European part of Russia's boreal forests.

Table 17.1. (continued).

PROCESS	PRE-1900	PAST (1900–1990)	PRESENT (1990–2000)	FUTURE (2000–2050)
	TEMPERATE ZONE			
1. CHANGE IN LC				
• Forest to non-forest				
-permanent crops/pasture	1(-)	2(-)	3(-)	4(-)
-urban/industrial	4(-)	3(-)	3(-)	3(-)
• Non-forest to forest				
-forestation	4(0)	2(+)	3(+)	3(+)
-abandonment/succession	4(0)	3(+)	2(+)	3(+)
• Pollution	3(+/-)	1/2(+/-)	2(+/-)	2(+/-)
2. CHANGE IN LU/LM				
-1st harvesting of primary forests for fuelwood	1(-)	2(-)	3(-)	4(-)
-1st harvesting primary forests for wood products	2(-)	1(-)	2(-)	3(-)
-harvesting of regrowth for fuelwood	4(0)	3(+)	2(+)	1(+)
-harvesting of regrowth for wood products	4(0)	3(+)	2(+)	1(+)
-grazing	1(-)	2(+)	3(0)	3(0)
-protection	4(-)	2(-)	4(+)	4(+)
-fire suppression	4(0)	2(+)	2(?)	2(?)
-insect/disease suppression	4(0)	4(0)	3(+)	3(+)
	TROPICAL ZONE			
1. CHANGE IN LC				
• Forest to non-forest				
-shifting cultivation	2(+/-)	1(-)	2(-)	3(+/-)
-permanent crops	3(--)	2(--)	1(--)	1(--)
-permanent pasture[b]	4(-)	3(-)	2(-)	2(-)
-degraded lands	4(-)	3(-)	2(-)	3(-)
-urban/industrial	4(-)	4(-)	3(-)	3(-)
• Non-forest to forest				
-abandonment	4(+)	4(+)	3(+)	3(+)
-forestation	4(+)	4(+)	3(+)	3(+)
• Natural disturbances				
-wildfires	4(+1-)	4(+/-)	4(+/-)	3(?)
-insects/disease	4(+/-)	4(+/-)	4 (+/-)	3(?)
-tropical storms	4(+/-)	4(+/-)	4(+/-)	3(?)
• Pollution	4(+/-)	4(+/-)	4(+/-)	3(+/-)
2. CHANGE IN LU/LM				
-harvesting for timber[c]	4(+/-)	3(+/-)	2(-)	3(-)
-harvesting for fuelwood	4(-)	3(-)	3(-)	3(-)
-regrowth	4(+)	3(+)	2(+)	2(+)
-grazing	4(-)	4(-)	4(-)	4(-)
-protection	4(+)	4(+)	4(+)	3(+)

b Mostly in tropical America.
c Includes sanctioned and unsanctioned logging.

of land use and land management in these boreal regions must therefore be analysed against the background of medium and long-term fluctuations in the amount of C stored in biomass and soil C pools (e.g., Kasischke et al. 1995; Chapter 14).

In contrast to the boreal forests, tropical forests have been subject to human disturbance and undergone many changes in cover and use over the centuries (Brown et al. 1993). Humans have lived in tropical forest areas for millennia and their populations have fluctuated widely as civilisations have risen and fallen (e.g., Mayans and Incas of tropical America and the past civilisations of Cambodia and Sri Lanka). On top of this has been the gradual overall increase in human populations over time and a parallel increase in forest disturbance. With this long history of human habitation, it is unlikely that the tropical forest landscape has been in C steady state since the pre-industrial era. Carbon losses are likely to occur in areas of increased human activity, but C gains are just as likely to occur in previously disturbed areas that are now abandoned (e.g., Brown et al. 1993, 1994; Lugo and Brown 1992). The magnitude of the C gains and losses will be different across different geographic regions, and their balance in time and space is largely unknown. Recent analysis of the C budget for the forest sector of Brazil demonstrated that the C gains in the vast areas of regrowing secondary forests was almost enough to offset the C losses from deforestation (Schroeder and Winjum 1995).

The above discussion emphasises that the analytical methods selected to estimate the C budget of a region should be based on an understanding of the major processes that are driving the C cycle (cf. Table 17.1). Furthermore, assumptions related to steady state conditions, adopted for the most part by convenience and lack of data at the time, need to be relaxed. Complete forest sector C budgets for the important natural and anthropogenic processes causing change in LC/LU/LM and associated gains and losses of C need to be developed.

Biomass-C densities of forests

The other major source of uncertainty in estimating forest sector C budgets is the C density of live and dead forest organic matter. In particular, estimates of forest biomass-C densities vary widely across and within biomes and contribute to the large uncertainty around the C-flux estimates caused by changes in forest cover, use and management. Biomass-C densities vary naturally across the forest landscape as a result of many factors, including climatic, edaphic, natural disturbance regimes, and age-class distributions (Chapters 11 and 12). In addition to natural factors, human activities, particularly in the tropics, also affect biomass-C densities causing either decreases or increases depending upon the intensity of human pressure and accessibility to forests. Capturing this variability in biomass-C density estimates in the world's forests remains a challenge for scientists modelling and estimating C budgets.

Several important advances regarding biomass-C densities have emerged over the last decade or so. These are:

- Biomass-C densities of forests are a "moving target", that is, they are increasing or decreasing over time depending on frequency and intensity of natural and human disturbances.
- Forest degradation is widespread in many regions of the tropics due to such activities as illicit harvesting of wood for timber of fuel, shortening forest fallows, fragmentation, and grazing (FAO 1993; Chapter 11). The result in these cases is a net loss of biomass C in the forest, a loss additional to those brought about by outright deforestation (Houghton and Hackler 1994).
- The best source of data for making estimates of biomass-C densities is forest inventories (Chapter 11) because the data are generally collected from the population of

interest (regional to national) and collected in a statistically sound manner. Ecological data are of limited value because plot selection tends to be biased toward "mature" stands which generally do not represent the larger region. More effort should be given to improving methods for converting inventory data to biomass-C densities.

- Improvements in C-flux estimates from changes in LC/LU/LM can be made if location of change in area is matched to appropriate biomass-C densities, in other words, spatial distributions of C densities are needed. Recent work has demonstrated that producing maps of C densities using a modelling approach, coupled with large-scale inventory data, in a geographic information system (GIS) is feasible (Brown et al. 1993).
- The C density of other forest components, particularly belowground biomass (roots) and woody debris, and the allocation of C density into slow or fast decomposing pools after disturbance are still poorly known.

Methodologies and data availability

Two approaches have generally been used to estimate the net exchange of C between forests and the atmosphere (Chapter 10). These two approaches use different data and include different processes and assumptions. Differences in their results may be helpful in quantifying the changes in C resulting from different processes, but differences have also led to considerable confusion regarding the global C imbalance.

One approach developed earlier is concerned with changes in land use (Houghton et al. 1983). It is based on two types of data: rates of land-use change and changes in C density as a result of disturbance. Ecosystems not deliberately affected by clearing, logging, abandonment, or forestation are assumed to be in steady state with respect to C. The approach keeps track of the C in declining and recovering forest vegetation and soil as well as in the slash and wood products generated as a result of land-use change.

The second approach is based on data from complete forest inventories, including both growing stocks and wood products (Chapter 14). Changes in these pools determine the net flux of C between forest and the atmosphere. This approach has typically been used for temperate and boreal zone forests only because national-level inventory data are generally lacking for the tropics. However, even within the temperate and boreal regions, different inventory techniques are used for different regions which can produce results with varying accuracy and precision.

The approach most useful for a particular region is determined by the availability of data and by the processes important in changing forest ecosystems. In the tropics, for example, the major process affecting C storage is conversion of forest to agricultural land (Table 17.1). Fire, insects, disease, and logging are all generally followed by regrowth, and the net change in C over time is small relative to permanent conversion of forest to cleared land. Outside the tropics, the conversion of forests to cleared lands is no longer of major importance (Table 17.1). The net change is determined largely by the balance of disturbance and recovery; i.e., logging and regrowth, fires, insects, and disease. Further, in most countries in the temperate and boreal zones, forest inventories provide the data necessary for reconstructing emissions and accumulations of C.

The processes included in the two approaches are, thus, not identical, and one might expect the calculated fluxes of C to be different. Changes observed between successive inventories, for example, should include the losses and gains in forest area as well as the changes resulting from logging, fire, and other disturbances. The second approach, therefore, yields a calculated flux that should approach the total net flux of C between forests and the atmosphere. The first approach, in contrast, includes only a part of this total flux. Because the data used in the first approach pertain only to deliberate changes in land use (e.g., clearing, logging, etc.), Houghton (Chapter 10) has distinguished between deliberate and inadvertent changes in forest ecosystems. This distinction is not always clear, for example, with fire and insects which may be deliberately suppressed in some cases and not in others. Nevertheless, the distinction is useful in interpreting the results of the two approaches, and it is useful in addressing the imbalance in the global C cycle.

Because the approach based on changes in LU is the one that can be used globally, it has been used to define the global imbalance (Chapter 10). To the extent that the approach based on forest inventories includes land-use change as well as other, inadvertent, processes, the difference between the two estimates of flux may explain some fraction of the imbalance. Regrowing forest may or may not explain the imbalance. If the regrowth is a result of logging, it has been included in the land-use term, and it does not help explain the imbalance. On the other hand, forest growth resulting from pre-industrial human disturbance, fire, or other natural processes is not included in the land-use flux and helps explain the imbalance. Nevertheless, it is clear from this discussion that uncertainties in resolving the imbalance still exist.

UNCERTAINTIES AND GAPS IN KNOWLEDGE

Uncertainties in forest C budgets exist for many reasons. These reasons can be ranked into three groups according to their impact on the final results as follows:

- incomplete data of sufficient accuracy and overall lack of data in some areas;
- lack of knowledge of some basic processes; and
- uncertainties caused by the present suite of models and methods.

Uncertainties due to lack of data arise mostly from the spatial and temporal scales. For basic parameters, such as C densities of forest vegetation and soil and NEP, spatial and temporal variation is so high and the data sets are so limited that statistical analyses are of limited usefulness. Furthermore, the procedures used to take direct measurements over large territories are inadequate for estimating the associated errors and data important for constructing C budgets are often omitted. For example, few inventories in the more forested countries record whether stands are even-aged or uneven-aged.

Uncertainties and errors in primary data are introduced in a multitude of ways, including use of different methods of data collection, different goals and sampling design methods, use of different equipment, use of unclear definitions, and uncoordinated classification schemes. Short-term measurements can be another source of uncertainty for processes that occur over longer time scales. For example, the time for the commercial wood C pool to decompose in northern taiga can reach 150–200 years, but historically there are no records which would be sufficient for estimating of quantity of commercial wood produced many decades ago.

Other gaps in knowledge that create an inability to parameterize basic processes are caused by incomplete knowledge of the processes themselves, especially their mechanism and/or structure. This applies to both natural and anthropogenic processes. For example major uncertainties are observed with regard to soil processes, post-disturbance C dynamics, etc. While litter and woody debris pools can be modelled with some certainty, there are large uncertainties in the role of the slow soil C pools, including humus and peat. Problems of leaching of C to aquatic systems or into deep soil horizons are also poorly known. In the realm of human activities, our inability to project future LC/LU/LM change is caused by our lack of understanding of how changes in population and economic activities drive these processes.

Uncertainties generated by the present models and methods deal mostly with:

- simplified approaches/models (there are trade-offs, though simpler does not imply always more uncertain);
- difficulties with model validation and consequently with uncontrolled biases;
- application of deterministic models for stochastic and/or fuzzy processes;
- uncertainties generated by use of different forecasts and projections from GCMs.

For example, analysis of the latest versions of GCMs outputs for large regions of the boreal zone shows that the projected levels of climatic change by year 2050 is of the same order of magnitude as the differences between the models.

Uncertainties of models are also directly tied to the lack of knowledge and data. This results in the scientific debate about the effects of CO_2 fertilisation on life-cycles and productivity of forests over large territories and the effects of forest harvesting and forest fires on nutrient cycling. Finally many uncertainties are related to the typical fuzzy character of a majority of processes and phenomena.

FUTURE ROLE OF FORESTS

Forests are presently and will continue to be critical to the global C balance and in the coming decades will be looked to increasingly by policy makers as means of mitigating C emissions from fossil fuels. Through various management options and land-use decisions, forests can be used to sequester part of these emissions.

It is unrealistic, however, to expect forests to be managed solely for C sequestration. Rather C sequestration must be carried out in the context of a sustainable development, industrial and fuelwood production, traditional uses, protection of natural resources, recreation, and the like. This integrated approach to forest management provides opportunities for C sequestration by three main management strategies: (1) conservation management—conserve existing C pools by controlling deforestation, increase forest protection, change harvest regimes, and control disturbances such as fires and pest outbreaks; (2) storage management—increase the area of forests through forestation (afforestation, reforestation, agroforestry) activities, increase the C density of existing forests, and increase the wood products pool; and (3) substitution management—increase the transfer of forest biomass C into products and fuels rather than using fossil-fuel based products (Brown et al. 1995).

The most significant potential for C conservation is in the tropics where the highest rates of deforestation currently occur (Brown et al. 1995). Reducing deforestation, however, is a difficult policy to implement because deforestation is mainly caused by conversion to agricultural land which in turn is driven by population growth and economic development. Protection of forests for conservation of biodiversity and sustainable timber production is likely to be more successful because of the Global Forest Principles and Agenda 21 promulgated at the 1992 UNCED and the International Tropical Timber Agreement (agreed in January 1994).

Many techniques will be applied to increase the area of forest cover. Forestation of all suitable and available areas will be an important option throughout the globe. In the tropics, forest rehabilitation and agroforestry will also be important. In temperate regions, reclamation of abandoned or marginal agricultural land will be a significant option for increasing the area devoted to forests.

As many of the world's forests are not at their potential or maximum C density, there is considerable scope for increasing the C density of existing forests up to their potential. It is particularly important to integrate the effects of C density improvement schemes over long time scales, typically over several rotations or disturbance intervals. Effects on soil C pools as well as tree biomass should be considered. Forest fertilisation, increased rotation ages, changing to new, higher C density species, suppression of fire and insects, and reduction of slash and burn are just a few of the techniques that may be employed to increase C density. Choice of the appropriate techniques will depend largely on economic factors. Analyses indicate that the costs per Mg C are far lower than C taxes used or proposed in some countries (see Chapter 26).

Forest management for C sequestration must also extend to management of the end fate of products taken from the forest. The decomposition curve of wood products can in some cases be extended by making wood more resistant to decay, e.g., through the use of preservatives, and by diverting a greater proportion of the wood into more stable pools such as houses. More significant gains are possible by substituting wood for fossil fuels. Finally, construction materials with a high energy input in their production, such as concrete, steel, and aluminium, can be replaced by wood in many applications.

Management options such as those described above have the potential to conserve and sequester almost 90 Pg C over the next 50 years, an amount which could offset up to 15% of the total fossil-fuel emissions over this same period (Brown et al. 1995). About 80% of this quantity could potentially occur in the tropics. Such a reduction in emissions to the atmosphere would reduce the greenhouse warming potential and thus could be expected to mitigate climate change. Furthermore, the C conserved and sequestered has the potential to balance the C emissions from changes in cover and use of the land.

The estimated C conservation and sequestration potential presented here assumes no change in climate or human demographics. Including these factors in the analysis is likely to result in a reduction in the amount of C sequestered and conserved in forests globally because of the diminished area of tropical forests (Brown et al. 1995). Not only is the amount of land available in the tropics likely to decrease substantially due to increased need for agricultural production, but a changed climate and atmospheric composition is likely to result in an overall decrease in net C uptake. In the temperate and boreal forest zones, land area suitable and available for C sequestration is projected to remain constant or increase, but it is still unclear as to how rates of net C uptake will change under a new warmer climate, with a potential increased fire frequency, and a changed atmospheric composition.

FOREST CARBON BUDGETS AND NATIONAL EMISSION INVENTORIES

To improve scientific ability to balance the terrestrial C budget, to provide policy makers with a sound scientific basis for negotiations on the UN Framework Convention for Climate Change (UNFCCC), to allow for the development of national actions plans, and to verify compliance of actions taken to mitigate greenhouse gas emissions, a system for tracking the C flux from changes in terrestrial ecosystems is needed. This system must be able to track net national C gains and losses brought about by past, present, and near-term projected changes in LC/LU/LM, natural disturbances, climate, and atmospheric composition.

Forests are dynamic ecosystems that contribute both to the storage of C and to the annual exchange of C with the atmosphere. Continuous processes such as photosynthesis and respiration contribute to the daily fluxes of C. Episodic events such as fire and deforestation occur only periodically but play an important role both in terms of their contribution to C fluxes and transfers within the ecosystem and because they alter the post-disturbance continuous processes such as regrowth and slash decomposition. Any system designed to estimate the net exchange of C between forest ecosystems and the atmosphere must be designed both to quantify and integrate the small fluxes that occur continuously over large areas and at small rates and those that occur episodically over smaller areas but at much higher rates. Moreover, the system must be designed to establish the net exchanges of C.

Direct measurement of the net C exchange over relatively large areas are possible, but are logistically constrained. Alternative approaches exist, which may consist of one or more methods, many of which are referred to above and in the next chapters. The appropriate methodology depends on the primary processes driving the C exchange as well as the available data for the region. The available methods include:

- Successive inventories of forest biomass and soil C pools.
- Monitoring of areas undergoing disturbance (addresses primarily cover changes) and intensity of disturbances, e.g., through remote sensing and aerial photography.
- Collection of national-level statistical data on land use and land management, including areas and intensity of timber and fuelwood harvesting, plantation establishment, rotation lengths, etc.
- Modelling tools to update inventories, to estimate impacts of natural and human disturbances on C flux, to extrapolate point measurements in space and time, and to integrate change estimates in space and time

The basic need is to develop a tracking system that accounts for all important fluxes (continuous and episodic) both in and out of terrestrial ecosystems. The system must track *net* C exchange, not merely emissions, associated with fires, harvesting, or changes in the cover, use, and management of the land. To achieve the main objective of the UNFCCC to "stabilise greenhouse gas concentrations in the atmosphere at a level that would prevent dangerous anthropogenic interference with the climate system" will require the development of a C flux tracking system along the lines outlined here for implementation by the ratifying countries.

REFERENCES

Brown S, Sathaye J, Cannell M, Kauppi P, (1995) Management of forests for mitigation of greenhouse gas emissions. Ch. 24, Working Group II, IPCC 1995 Assessment, Cambridge University Press. (In press).

Brown S, Iverson L, Lugo AE (1994) Land-use and biomass changes of forests in Peninsular Malaysia during 1972–82: use of GIS analysis. In: Dale VH (ed) Effects of land-use change on atmospheric CO_2 concentrations: Southeast Asia as a case study. Springer-Verlag, New York, pp 117–144.

Brown S, Iverson L, Prasad A, Liu D (1993) Geographical distribution of carbon in biomass and soils of tropical Asian forests. *Geocarto International* 4: 45–59.

Cahoon Jr DR, Stocks BJ, Levine JS, Cofer III WR, Pierson JM (1994) Satellite analysis of the severe 1987 forest fires in northern China and southeastern Siberia. *J. Geophys. Res.* 99(D9): 18627–18638.

Dixon RK, Brown S, Houghton RA, Solomon AM, Trexler MC, Wisniewski J (1994) Carbon pools and flux of global forest ecosystems. *Science* 263: 185–190

Food and Agriculture Organization (1993) Forest resources assessment 1990. Tropical countries. FAO Forestry Paper 112. Rome.

Houghton RA, Heckler JL (1994) The net flux of carbon from deforestation and degradation in south and southeast Asia. In: Dale VH (ed) Effects of land-use change on atmospheric CO_2 concentrations: Southeast Asia as a case study. Springer-Verlag, New York, pp 301–328.

Houghton RA, Hobbie JE, Melillo JM, Moore B, Peterson BJ, Shaver GA, Woodwell GM (1983) Changes in the carbon content of terrestrial biota and soils between 1860 and 1980: a net release of CO_2 to the atmosphere. *Ecological Monographs* 53: 235–262.

Kasischke ES, French NHF (1995) Locating and estimating the areal extent of wildfires in Alaskan boreal forests using multiple season AVHRR NDVI composite data. *Remote Sens. Environ.* 50: 263–275.

Kasischke ES, Christensen Jr NL, Stocks BJ (1995) Fire, global warming and the mass balance of carbon in boreal forests, *Ecological Applications* 5(2): 437–451.

Lugo AE, Brown S (1992) Tropical forests as sinks for atmospheric carbon. *Forest Ecol. Manage.* 54: 239–256.

Schroeder P, Winjum JK (1995) Balancing Brazil's carbon budget: the potential importance of secondary forests. *Interciencia* (In press).

Part III

The global carbon cycle, forest products and forest biomass

III The global carbon cycle, forest products and forest biomass

Introduction

David T Price and Michael J Apps

Whereas the Chapters in Part II presented a "top-down" view of the role of forests in the context of global C cycling, the articles in Part III look at the global consequences from the viewpoint of forest management and harvesting. Here, the contribution made by forest products to carbon storage is considered explicitly, and the consequences of forest management for these products on ecosystem C storage are examined. Several models and simulation results are reported that provide estimates of the carbon stored in regional forests and wood products today and potential changes in the foreseeable future. The importance of considering fossil energy displacement (through both direct and indirect substitution) when assessing the C offset benefits of using forests and wood products to mitigate fossil-fuel emissions was a recurrent theme in the working group discussions and appears in a number of the Chapters. Another key issue addressed by Part III is the need for comprehensive methods to assess forest C budgets and to subject them to rigorous comparison.

Which forests sequester the most carbon?

Nabuurs (Chapter 20) reports a broad comparison of the potential for sequestering C in ten representative world forest ecosystems, using the model CO2FIX to quantify the C budget for each (biomass, soils and products). The model allows for recycling of products including use as bioenergy, and considers reductions in emissions achieved through direct and indirect substitution. To evaluate the C sequestering potential, C stocks were assessed on four criteria (long term average C in biomass plus products, in products alone, in soil organic matter, and total C in biomass and products at the end of the first rotation), while fluxes were compared as average net annual flux over the first rotation. He concludes that a given forest type may show high potential for C sequestration under one criterion, but low potential under another. Thus, for example, if maximum sequestration in the short-term is required, then average net annual C flux in the first rotation is the key criterion, but if maximum long-term storage in forest is the objective, then a different criterion is needed.

Nabuurs finds that different forest types vary greatly in C stocks at the end of the rotation. Of the forest types considered, selectively-logged evergreen rainforest contains the largest stock in biomass and products—207 Mg C ha^{-1}—while Norway spruce in central Europe contains 137 Mg C ha^{-1}. Soil organic matter typically stores a high proportion of total ecosystem C, but only in boreal spruce does it greatly exceed biomass C. Products typically contain

NATO ASI Series, Vol. I 40
Forest Ecosystems, Forest Management and the Global Carbon Cycle
Edited by Michael J. Apps and David T. Price

small amounts of C but range from 10% in selectively logged rainforest to 66% in slash pine in Brazil. Short rotation forests achieve the highest rates of average net annual fixation, with *Pinus radiata* in Brazil the highest at 4.5 Mg C ha yr^{-1}.

As an introduction to the principles of forestry and their implications for managing forests for C storage, Maclaren (Chapter 21) uses the (mostly planted) New Zealand (NZ) forest estate as a case study. For many commercial forest operations, the management objective is to maintain steady flows of wood to mills and industry. This requires equal annual allocations of resources to forest operations such as planting, stand tending and harvesting, and leads to the conceptual ideal of the "normal forest". A normal forest (which never exists in reality) contains equal areas of land within each age-class, with the oldest age-class equal to rotation age. In comparison, natural forests typically have a higher proportion of "mature" stands that are older than the commercial rotation age and hence have higher standing volume than a normal forest. Greater C storage will be therefore be achieved in a natural forest, unless the products harvested from a managed forest displace fossil fuels (directly or indirectly). Based on the detailed forest inventory data available for NZ, Maclaren calculates that national forests (growing *P. radiata*) can sequester about 112 Mg C ha^{-1} in perpetuity, ignoring the potential for energy substitution and assuming that storage in products is insignificant. He also implicitly assumes that there will be no change in the growing conditions, for example, due to climate change.

If long-term sequestration is the goal, and plantable land is limited, Maclaren argues it is often better to establish species with a long rotation age, even if their mean annual increment (MAI) is lower than that of other species which culminate more quickly. Nabuurs notes that fast growing short rotation crops usually exhibit high net annual C fluxes over a short period, but culminate in a relatively low equilibrium biomass (see also Schlamadinger and Marland, Chapter 18), but also explains that this is true only when the energy substitution potential is not being considered. Real forests (of natural or anthropogenic origin) typically possess uneven age-class distributions (particularly when the planting history is driven by changes in government policies and short-term economic criteria), so even if environmental conditions are constant, the carbon stored will also vary among individual age-classes. Maclaren shows that when age-classes of greatly varying size are felled and regenerated on schedule, the net annual C flux can change dramatically, with an "echo effect" being passed on to the next rotation. In practice, however, forest managers will delay or advance felling dates to smooth wood flows and avoid market gluts, which tends to attenuate this effect over the long-term. For this reason, the managed forest estate can shift from being a net sink to a net source. Maclaren points out that countries reliant on a large area of actively growing forests as C sinks cannot necessarily use this to advantage in offsetting national GHG emissions, because these forests will become sources at some later date, a phenomenon also reported by Kurz and Apps (Chapter 14) for Canada and by Kauppi (Chapter 16) for Europe.

Which management regimes sequester the most carbon?

In Chapter 18, Schlamadinger and Marland review the carbon implications of different forest management strategies. They point out that it is possible to influence atmospheric CO_2 concentration by: (1) sequestering C in biomass, soils and litter; (2) displacing fossil fuel by renewable biofuels; (3) changing the amounts of fossil fuel consumed in forest management, harvesting and wood product manufacture; (4) storing C in wood products, and manipulating their average life-spans; and (5) changing the land area used for bioenergy or timber production. Referring to previous work, they conclude:

> *"The relative impacts on the C balance depend very much on what is done with any harvest and the efficiency with which it is done."*

They then compare several forestry management scenarios over a time horizon of 100 years using a simple but comprehensive model. In the short-term (first 20 yr) the greatest gains are obtained by establishing trees on former agricultural land. For conventional and fast-growing (short rotation) plantations assessed over longer time-frames however, productivity becomes more important in comparison to the previous land use (see also Nabuurs, Chapter 20). They find that with longer rotations, net C storage benefits are increasingly dependent upon the extent to which renewable wood products, including biofuels, can substitute for fossil fuels (direct substitution) and energy-intensive materials (indirect substitution). Although there may be significant short-term net storage of C in trees, soils and products, all achieve steady state at some level, and then provide no further sequestration. Substitution for fossil fuels, on the other hand, results in continued reductions in emissions, although this is strongly dependent upon the efficiency with which biofuels are used. Afforestation is a useful means of sequestering additional C in forests, but without harvesting—for timber or bioenergy—there is no opportunity for fossil energy substitution, and hence an inevitable decline in net annual C uptake towards zero, a point also made by Maclaren.

For the USA, comparing a "business-as-usual scenario" with several alternatives, Heath et al. (Chapter 22) show that a halt to forest harvesting would reduce emissions and transfers to landfills, but provide a smaller offset to fossil fuels (58 $Tg\,C\,yr^{-1}$) and no additional C sequestration. Direct reduction in fossil energy consumption is still a more effective means of reducing emissions.

Price et al. (Chapter 23) use a boreal forest management unit in western Canada as a case study to demonstrate how forest products may act as a transient net sink. The sink must eventually disappear, however, when net additions from new products balance the decay of existing products. Using the CBM-CFS model (see also Kurz and Apps, Chapter 14), they compare carbon dynamics of the "natural forest" to that of the "managed forest" using data and current yield projections obtained from the local forest management plan. These simulations suggest that management and harvesting may actually *increase* C sequestration relative to the natural forest ecosystem under some circumstances. This conclusion appears to be in stark contrast to the results reported by Kellomäki and Karjalainen (Chapter 5), but the difference is in the disturbance regimes of the two forest systems under consideration. Where the natural life-span is typically terminated by disturbances, such as fire, with a mean return interval between disturbances *shorter* than the optimal rotation age, the net effect is an increase in the spatially-averaged C density. Taking advantage of this for C sequestration purposes would require effective protection from fires and other disturbances (which may be impracticable as discussed in Stocks et al., Chapter 8). The potential benefits include increased average forest age (therefore higher average C density); transfer of wood into long-lived forest products (that would otherwise be lost to disturbance or decomposition), and reduced losses of forest litter (with greater transfers to soil C pools).

Matthews et al. (Chapter 24) summarise the potential of forest management for increasing C sequestration. Given the wide diversity of forest ecosystems and available management practises, many options are very site-specific. The most appropriate management regime will depend on site productivity, potential harvest efficiency, and current land use: "*There are no blanket prescriptions for forest management and forest policy to enhance forest C sequestration.*" They remind us that there are fossil fuel costs associated with managing the forest and caution that the resulting emissions should not negate the C benefits of increased productivity.

Briefly, the useful options at the manager's disposal are:

1. preservation and protection of existing forests.

2. afforestation.
3. better utilisation of harvested wood and processing residues.
4. longer rotations for managed stands (see also Fischlin, Chapter 13).
5. site amelioration.
6. fertilisation.
7. tree breeding for increased productivity and carbon density.
8. silvicultural treatments aimed at increasing the proportion of harvested wood in high-value, longer-lived, product categories (e.g., thinning, pruning).
9. using wood wastes as fuel in manufacturing.
10. urban forestry.

How can forest products affect C balance?

Heath et al. (Chapter 22) examine the dynamics of forest C pools and fluxes in the USA by using a forest products accounting model, HARVCARB, to estimate the contribution of past, current and future wood harvested and removed since 1900. The fate of harvested wood C is tracked from trees to roundwood, via lumber and finished fibre, to end uses including products, landfills, fuelwood and emissions. They also projected the C in wood products from 1990–2040 assuming a business-as-usual scenario, and considered four alternative projections where harvest production was increased or decreased, or emissions were decreased. They estimate that during the period 1900–1990, 10.7 Pg C were removed from U.S. forests. At the end of this period, 3.7 Pg C remained in products and landfills, 3.7 Pg C had been burned for energy (particularly by wood products industries), and the remainder was lost through decomposition. Consumption of wood carbon is expected to increase at least until 2040. The net current flux into products and landfills is about 37 Tg C yr^{-1}, or about 40% of the flux into U.S. forest vegetation. A further 50 Tg C yr^{-1} are released from burning of fuelwood. Compared to a 76 Tg C yr^{-1} reduction target for U.S. emissions under the UN FCCC agreement, these are not insignificant quantities.

In contrast, for NZ, Maclaren finds that only about 12% of the C in a harvested stand has a life-span (as products) greater than 5 years, a proportion he considers insignificant. He recognises that net storage in products really depends on how they are treated—do they decay on average slower or faster than dead wood in the forest? He argues that a complete life-cycle analysis is needed to determine the net benefits of plantation forestry for mitigating GHG emissions, including an accounting of all energy sources used in tending, harvesting and processing operations.

Price et al. find the net C accumulation in products may be only 5–10% of ecosystem carbon (including soils), but believe this to be a potentially important contribution. They also suggest that long-term accumulations of products (including pulp) disposed in landfills may become an important component of C storage, especially in cool wet conditions because decay rates are much slower in anaerobic conditions, although as Maclaren points out, this may also lead to increased emissions of methane from landfills.

For recycling and substitution, Nabuurs compares three forest types: long rotation mixed deciduous (mainly oak) producing large volumes of sawtimber that could be used as fuelwood; short rotation hardwoods producing pulpwood and packing timber; and medium rotation conifer from which many products can be obtained. He finds that the absolute C storage benefits of recycling are reduced if the associated energy costs are considered. The relative increase in products C stock, however, can be very high: 29% for oak, 167% for poplar and 73% for Norway spruce. Generally, forest types yielding the widest variety of products have the greatest potential for C storage. When the displacement of fossil energy by wood is also considered however, short rotation high productivity forest types have the greatest potential to reduce C emissions.

Resolving conflicts between C storage and other forest management objectives

Matthews et al. (Chapter 24) address the wider impacts of forest management to enhance C sequestration in attempting to formulate recommendations to policy-makers, forest managers and the public. There is a risk that other values from forests may be sacrificed for the sake of C storage, although both negative and positive side-effects are possible. As an example, biodiversity may be reduced by intensive management of productive sites, but enhanced when forests are established on marginal lands. In general, the authors believe, it should be possible to manage forests so that traditional forest benefits such as timber, fuel and food, are maintained without greatly compromising C sequestration objectives. The balancing of C storage benefits with other social, cultural and economic values is also discussed from a socio-economic perspective in Part IV.

In fact, several socio-economic issues bear directly on the use of forests, wood products and biofuels for mitigating atmospheric CO_2 increase. Mitigation options are frequently equated with increasing the extent and carbon density of forests. Matthews et al. suggest better results can often be achieved by more careful utilisation of existing forest resources and more efficient harvesting practices. It may also be necessary, in some circumstances, to accept early net losses of C from forest stands to obtain net gains in the longer term. Expansion of forest cover (desirable for many reasons) could be stimulated by offering incentives to encourage the displacement of energy-intensive materials by wood products; or to generate markets for wood in the energy sector; or by allowing tax breaks or other financial incentives to store carbon in forest lands.

Cost-benefit analysis of the potential damage from climate change may show financial attractiveness in using forest land to mitigate C emissions, leading to policies to achieve this. Nabuurs argues strongly that a comprehensive cost–benefit analysis should consider the value of environmental damage *avoided* by investing in forest operations (a point taken up in more depth by Sedjo and Sathaye, Chapters 25 and 26). Marginal benefits could also be compared, such as the C benefits of management *versus* no management (see also Price et al., Chapter 23). Nabuurs also states the need for attention to other forest management objectives—including sustainability, biodiversity and soil conservation:

> *"...government policies should stimulate use of wood products from sustainably managed forests in preference to energy-consuming alternatives, and should stimulate recycling of wood products to help achieve a favourable C balance."*

Need for standardised procedures for C budget accounting

Matthews (Chapter 19) provides an extensive review of published methods of forest C budget assessment. He draws attention to the importance of proper accounting for all factors when determining whether management has positive or negative benefits for C storage. (As an aside, Heath et al. also warn about "double accounting"—emissions from burned wood products are sometimes reported with fossil emissions.) Matthews demonstrates that more optimistic expectations are obtained with more comprehensive assessments. That is, there is a higher expectation for forest management to result in enhanced C sequestration when more factors are considered in the C budget assessments. In particular, he finds that accounting for direct and indirect energy substitution makes a critical difference.

This theme is reiterated by the Working Group (Matthews et al.) which finds that wood products are an important factor, especially when direct and indirect energy substitution are considered (e.g., Schlamadinger and Marland; Nabuurs). They use this to emphasise the need for comprehensive standardised procedures to evaluate and compare management impacts on C in forest ecosystems, forest products, wood processing technology and the supply and consumption of energy. There are several possible approaches to performing assessments, including direct measurements, 'book-keeping' models and process-based models. The only published 'standard' however, is the IPCC/OECD (1994) procedure for the 'forestry sector' which estimates changes in carbon stocks in forest biomass and soil pools, but is incomplete in its consideration of wood products and takes no account of energy substitution. The procedure for the 'energy sector' tracks fuel emissions but does not distinguish between fossil and renewable sources. Matthews et al. therefore conclude that the IPCC/OECD procedures are inadequate for properly quantifying the impacts of forestry schemes on the global C budget.

Nabuurs provides some guidelines, in addition to which the Working Group presents five criteria. To summarise, a standard C budget procedure should:

1. meet with scientific approval.
2. account for all components of the forest ecosystem (i.e., above- and below- ground vegetation, dead organic material, including soil C and peat, products, fossil energy consumption and substitution).
3. consider alternative possible management scenarios.
4. consider long-term consequences (i.e., 100 years or more).
5. recognise that not all data may be available in all cases.

Looking up to the global perspective

Heath et al., extrapolating their results for the USA to the globe, estimate the annual transfer to products (not including landfills) is about 100 Tg C yr^{-1}, with an additional 360 Tg C yr^{-1} released through burning of wood fuels in the tropics alone. They conclude that harvested wood products do have significant impact on C storage within USA, but may not be so significant globally, while wood burning in particular helps mitigate fossil C emissions, even though C is not physically sequestered, both within the USA and in tropical developing countries.

Matthews et al. cite Dixon et al. (1994) who estimated global storage of 1146 Pg C in forests, and 1990 net emissions of 0.9 ±0.2 Pg C, primarily from deforestation. To improve on such global estimates, they propose that more research should be directed to quantifying stocks and fluxes of C in forest ecosystems (at regional scales). They suggest the data needed to perform comprehensive global C budgets, must include:

1. areas and C densities of forest ecosystems, and changes in areas and C densities associated with forest growth, succession and management.
2. data to permit tracking harvested wood and impacts of wood utilisation on the products C pool.
3. data with which to assess impacts of forest management, wood processing and wood use in terms of fossil fuel consumption.

REFERENCES

Dixon RK, Brown S, Houghton RA, Solomon AM, Trexler MC, Wisniewski J (1994) Carbon pools and flux of global forest ecosystems. *Science* 263: 185–190.

IPCC/OECD (Intergovernmental Panel on Climate Change and Organisation for Economic Cooperation and Development Joint Programme) (1994) Greenhouse Gas Inventory. IPCC draft guidelines for national greenhouse gas inventories. 3 vol. IPCC/US-EPA, Bracknell, UK, and Washington, DC.

18 Carbon implications of forest management strategies

Bernhard Schlamadinger[1] and Gregg Marland

Environmental Sciences Division, Oak Ridge National Laboratory
Oak Ridge, TN 37831-6335, USA.

INTRODUCTION

Forest management is frequently cited (e.g., Winjum et al. 1993; Kolchugina and Vinson 1993; U.S. Office of Technology Assessment 1991) as a strategy for attempting to ameliorate the build-up of the greenhouse gas CO_2 in the Earth's atmosphere. Combustion of fossil fuels is the primary cause of the atmospheric increase of CO_2, with additional emissions coming from the clearing of forests to accommodate other land uses. The management of forests and forest resources can affect the global cycling of C in several ways. The amount of CO_2 in the atmosphere can be altered by changing:

- the amount of C stored in living biomass, soils and forest litter
- the amount of C in fossil fuel displaced by biomass fuel
- the amount of fossil fuel used for forest management activities, for biomass conversion processes and to make wood products
- the amount of C stored in wood products and their lifetime
- the specific land area chosen for production of bioenergy or wood products and its history of land use.

One of the first studies comparing strategies of fossil fuel substitution and C-sequestration in trees was done by Hall et al. (1990). For many circumstances (efficient use of biomass and substitution of coal) fossil fuel substitution was seen as the better alternative. Houghton (1990) calculated the net C balance for tropical forests under (1) continued deforestation, (2) reforestation without further deforestation, and (3) harvest of wood fuel without any further deforestation. He concluded that only the latter strategy, where wood fuels displace fossil fuels, could provide a potentially permanent solution to stabilising the CO_2 concentration of the atmosphere. In a subsequent modelling exercise, Marland and Marland (1992) examined the relative merits of growing trees to sequester C versus growing trees to supply fuels to replace fossil fuels. They showed that the rate of forest production, the efficiency with which wood is used to produce energy, the prior use of land converted to forest management, and the time scale under consideration were important variables. The advantage, in terms of net CO_2 emissions to the atmosphere, of using trees to replace fossil fuels rather than simply storing C in standing trees, increased as the productivity increased, the time interval of consideration increased, and the efficiency with which wood fuels substituted for fossil fuels increased. In this analysis we have developed a somewhat more complex model to permit consideration of wood products, energy embodied in products, and soil C; and to allow

1 Permanent address: Joanneum Research, Elisabethstrasse 11, A-8010 Graz, Austria

NATO ASI Series, Vol. I 40
Forest Ecosystems, Forest Management and the Global Carbon Cycle
Edited by Michael J. Apps and David T. Price

comparison of different forest management strategies over time. Similar studies have been carried out by Nabuurs and Mohren (1993), Dewar (1991), and Dewar and Cannell (1992) with detailed analyses of the C pools and fluxes in forests and, in addition, taking into account the C temporarily stored in wood products. Burschel et al. (1993a, b) have also included material substitution and energy substitution in their model. Schlamadinger et al. (1995) show the net C benefit of using logging residues for energy.

THE MODEL

The model which we used for this analysis provides a simplified description of C flows associated with management of forests. We tried to use realistic values for model parameters but emphasis is on representing the critical components and capturing the form of the relationships over time. The model calculates C accumulation in trees, in long and short-lived products, in fossil fuels not burnt because generation of products that are made from wood instead of other materials requires less energy, and in fossil fuels substituted by bioenergy. The resulting diagrams also contain other C pools, i.e., soil organic matter, roots, and litter. The C gain/loss in soils and forest litter is generated exogenously by making simple assumptions about the quantity and rate of C gain/loss based on the type of change in management regime. In this version of the model roots and litter are treated as a single pool. The model calculates the C balance in a unit of area. The structure of the model is shown in Figure 18.1. All input parameters are listed in Appendix 18.1 while Appendix 18.2 shows the main functions used in the model to calculate the 18 C fluxes depicted in Figure 18.1.

Starting from a specified initial standing stock of aboveground C (indicating if there is an old growth forest, a forest regrown from a previous harvest, or agricultural land) an initial harvest (in case there is one) can be defined by parameters describing:

- share of the initial harvest used for long-lived products, short-lived products, very short-lived products and biomass fuels,
- C emissions from making products from wood in comparison to producing them from other materials,
- efficiency of the bioenergy system in comparison to the displaced fossil energy system, and C emission rate of the displaced fossil fuel,
- fossil fuel input for maintaining the plantation (e.g., fertilisers), harvesting, processing biofuels and for conversion of biofuels into heat and/or electricity,
- and similarly for the substituted fossil energy system.

All of the parameters can be changed for subsequent harvests. The model uses a very simple growth function for trees (Marland and Marland 1992). We assumed a maximum supportable biomass of 160 Mg C ha^{-1} and linear growth up to 80 Mg C ha^{-1} (see Appendix 18.2). The standing stock at subsequent harvest is defined by the rotation length and the above mentioned growth rate. Six parameters are necessary to describe soil, roots and litter processes and for product decay: net soil C uptake (or loss), time for soil C uptake (or loss), net roots and litter C uptake (or loss), time for roots and litter C uptake (or loss), average lifetime of long-lived products and average lifetime of short-lived products. We assumed that for very short-lived products C storage can be neglected.

Four of the factors used in this model are characterised as *displacement factors* and need additional explanation. If a biofuel containing one unit of C is burned, the quantity of fossil fuel displaced will depend on the heat content and chemistry of the biofuel and of the fossil fuel it displaces and on the efficiency with which the two fuels can be burned. Our

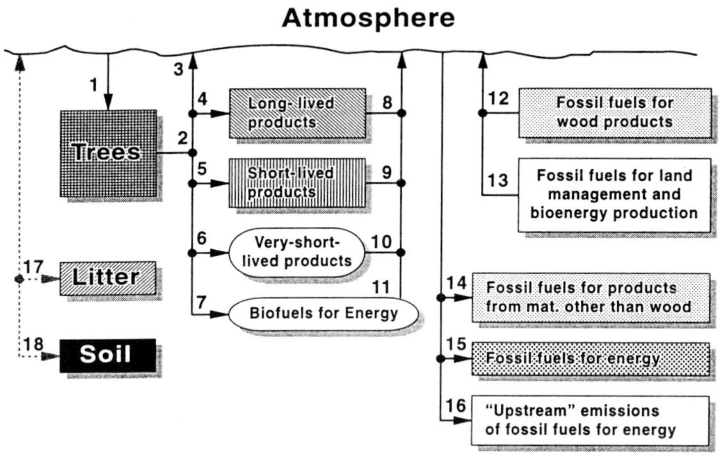

Figure 18.1. Carbon pools and fluxes as described by the model. Net forest growth corresponds to arrow 1. Arrows 2 to 7 describe C fluxes at harvest: 4, 5, and 6 are C fluxes into wood product pools with different lifetimes, arrow 7 is the amount of C used for energy (e.g., logging residues) and arrow 3 represents C left on the site that is released to the atmosphere. Arrows 8 to 11 are C flows to the atmosphere due to decay or burning of products. Arrows 12 and 13 represent fossil C emissions from land management, bioenergy production, and making wood products. Fluxes 14 to 16 are "saved" fossil C emissions (and thus represented by arrows away from the atmosphere). At this stage of model development soil and litter C dynamics are not calculated by the model but net soil and litter uptake/loss are assumed exogenously. Thus arrows 17 and 18 are net C fluxes to or from the litter and soil pools. Shading of the various boxes corresponds to the shading in Figures 18.2 to 18.7. For example the sum of "fossil fuels for wood products" and "fossil fuels for products from materials other than wood" corresponds to "energy for products" in Figures 18.2 to 18.7. Having only indirect C implications (for the various fossil C pools) without storing C, "very short-lived products" and "biofuels for energy" have no equivalents in Figures 18.2 to 18.7.

displacement factor fuel describes the units of C in fossil fuel displaced by 1 unit of C in a biomass fuel and is the ratio of fluxes 15 and 7 in Figure 18.1. Thus a displacement factor of 0.6 implies that using biomass fuels containing 1 Mg C will displace fossil fuels containing only 0.6 Mg C. Similarly for products, a *displacement factor products* of 0.6 would imply that when 1 Mg C is used for wood products, 0.6 Mg C in fossil fuels are not burned. This is because the wood products require less energy input than do alternate products which provide the same service. The *displacement factor products* is the ratio of fluxes 4, 5, or 6 to the difference of the corresponding shares of fluxes 14 and 12.

The model output is presented in graphs with time on the horizontal axis and net cumulative C flows on the vertical axis. Net changes in the respective C pools are represented so that the total increase in net C storage is the sum of the net increases in storage in the individual pools. In many of the plots the bottom line of the plot drops below zero to represent gross C losses due to initial harvest of standing biomass (for example the bottom line drops to -100 Mg C ha^{-1} if there is an initial harvest of 100 Mg C ha^{-1}) and/or net losses of soil or litter C. In these cases, regrowth of trees following initial harvest is then represented above this base line as a net increase in C in trees. If the rotation length is 60 years the diagram shows the parcel harvested at time = 0 and again in the 60th year. The plots show, successively from the bottom, net C uptake in soils, net C uptake in litter (as for soil, net decreases are represented by a drop in the baseline of the plot), net C increase in trees, net C storage in long-lived products, net C storage in short-lived products, net C retained in fossil fuels due to substitution of wood-based materials for other materials like steel, and net C retained in fossil fuels displaced by biofuels. It is important to recognise that whereas the top line of each plot represents the total of C which is sequestered in either the forest, wood products, or displaced fossil fuels, it does not always represent the net savings in C emissions for the scenario under consideration. There is an input of fossil fuels required for land management, for processing of biofuels, etc., which is in most cases larger than the comparable upstream energy requirements for processing the fossil fuels that are displaced. This difference in fossil fuel requirements, flux 13 minus flux 16 in Figure 18.1, can result in a negative effect on the total C balance represented in the diagrams, so we have in these cases subtracted the appropriate amount of C emissions and represent the final net effective C sequestration for each scenario with a black line, indicated by an arrow in each diagram.

RESULTS

We have defined 8 scenarios (for parameter assumptions see Appendix 18.1) in order to illustrate the impact of various biofuels strategies on the global C cycle and to draw comparisons among alternative forest management strategies. We caution that these scenarios are largely illustrative. There are many parameters that are widely variable depending on the specifics of geography, timing, land management strategy, land use history, and resource allocation that play an important role and this range cannot be represented in a small number of scenarios. Also, many complex processes are represented very simply in this model. While the scenarios will not necessarily portray any given area or project, our hope is that they capture the essence of the forest management alternatives and focus attention on some of the critical features of the C flows. Further refinements of the C flows must come from better and/or more site-specific values on forest and forest-product dynamics.

As a reference case we consider the simplest circumstance, an existing second growth forest that is protected and allowed to continue growing. For comparison with later scenarios we assume that the forest has a standing stock of 100 Mg C ha^{-1} and that the mass of C asymptotically approaches the maximum supportable standing stock for the site, 160 Mg C ha^{-1}. One hundred and sixty Mg C ha^{-1} might represent a very good site in the south-eastern U.S. or in western Europe (Birdsey 1992; Turner et al. 1993; Nabuurs and Mohren 1993). The decline in growth rate with time is seen in Figure 18.2 and after 100 years the forest has stored about 50 Mg C ha^{-1} in addition to the original 100 Mg C ha^{-1}. Under these circumstances one might reasonably expect a slow increase in soil and litter as well but we do not try to capture that here.

To maintain the generality of the figures, the legend to Figure 18.2 includes all of the elements of the subsequent figures even though the only change over 100 years is a gradual increase in the C stored in the trees. In a sense the C uptake on this site provides a pattern against which to evaluate alternative strategies and the scenarios below will demonstrate that the relative impacts on the C balance depend very much on what is done with any harvest and the efficiency with which it is done.

Another simple way to store additional C in forests is to afforest areas which are not now forest, e.g., agriculture or pasture-land. Marland and Marland (1992) recommend this strategy—from a C balance perspective—to be applied especially for sites with low productivity. The scenario in Figure 18.3 is based on a relatively high early growth rate of

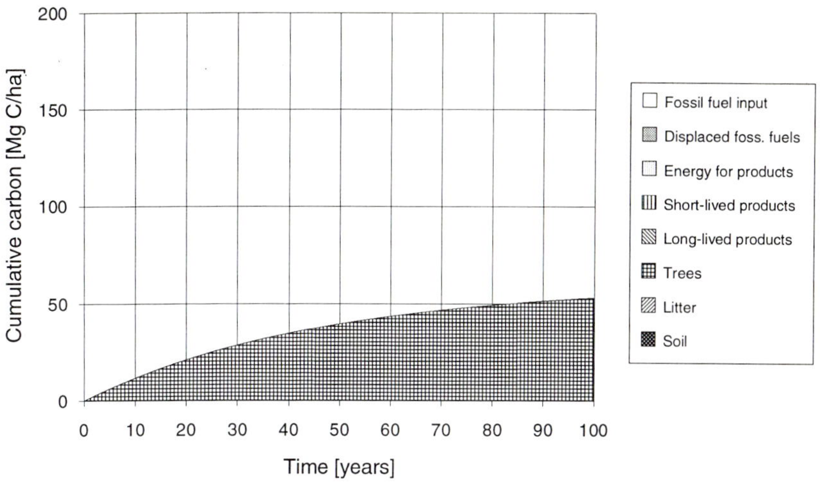

Figure 18.2. Net cumulative reduction in flows of C to the atmosphere—in this case net cumulative flows of C from the atmosphere to the biosphere—for a forest of 100 Mg C ha^{-1}, that is not harvested but left to grow further. The flows are for 1 ha of forest and numbers on the y-axis represent C storage above the initially standing 100 Mg C ha^{-1}. All figures showing model results have the same legend even if not all components show.

1.7 Mg C ha^{-1} yr^{-1}, a value within the range of U.S. commercial forests but well above the mean (the growth rate of the forest in Figure 18.2 was the same in its early lifetime, prior to time = 0 of our scenario). Such an afforestation project is expected to result in an increase of C in litter and soil, and it certainly results in an increase of aboveground C, as shown in Figure 18.3. For the parameter assumptions chosen here (given in Appendix 18.1), the C uptake is about 175 Mg C ha^{-1} during the first 100 years. During the initial 60 years the rate is higher, around 2.3 Mg C ha^{-1} yr^{-1}, and during years 60 to 100 it is only 1 Mg C ha^{-1} yr^{-1}. The reason for this is that the forest growth rate declines. This effect is, over longer time periods, an important disadvantage of afforestation without harvest compared to those scenarios where the forest is harvested either for wood products or bioenergy, because in the latter case fossil fuels continue to be substituted for and there is no decline in the net annual C flow.

"Conventional" forestry is represented by a clear-cut harvest at time = 0 of one hectare of forest which contained 100 Mg C ha^{-1}. The forest is harvested again in the 60th year after growing at an average rate of 1.66 Mg C ha^{-1} yr^{-1}. Figure 18.4 shows C flows on the hectare harvested, with the harvested C distributed among product pools and credit given for fossil-fuel displacement, both by the portion of the harvest used as a fuel and because some of the wood products displaced more energy-intensive products. Carbon is shown to be lost from soils and litter in the years immediately following the clearcut but the model does not show that soil and litter C would likely re-accumulate during the latter years of the cycle. There is a slight surplus of fossil fuels for forest management and bioenergy production as compared to upstream emissions of fossil fuels that are displaced by biofuels. This surplus is subtracted from the top line in Figure 18.4 and the net savings in C emissions are represented with the line indicated by the arrow. Perhaps the weakest parameter used in our model runs is the amount of energy displaced as a result of using wood in place of alternative materials. The basic assumption here is that long-lived wood products displace structural steel and concrete

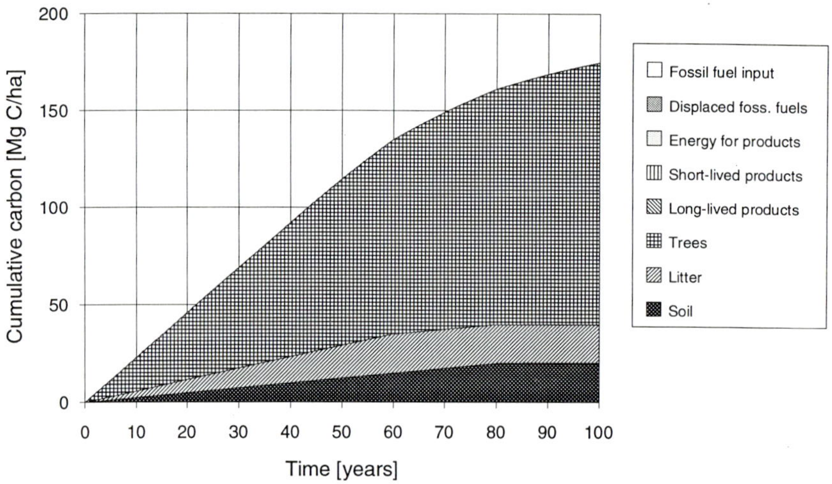

Figure 18.3. Net cumulative reduction in flows of C to the atmosphere—in this case net cumulative flows of C from the atmosphere to the biosphere—for afforestation of previously agricultural land.

and that there is subsequently a net savings in fossil fuels burned. According to Kürsten and Burschel (1993), citing Burschel et al. (1993b), the reduction of C emissions by substituting wood for materials such as aluminium, steel or concrete in construction (e.g., buildings) is estimated at 0.28 Mg C m^{-3} of wood. With a C content of 0.25 Mg C m^{-3} of wood, the *displacement factor products* is roughly 1. We used this value for the optimistic scenario in Figure 18.5, whereas for the scenario in Figure 18.4 we used 0.5. According to Burschel et al., 0.28 Mg C m^{-3} is valid only for trunk wood used in construction. Thus we used 1 and 0.5 for long-lived products, whereas for short and very short-lived products (the energy input to produce paper products or chipboards for furniture is higher than to produce sawed timber) we used smaller values, 0.5 and 0.25 respectively. Acknowledging that the numeric values are poorly known and certainly widely variable (depending on the details of the materials and applications for which products are substituted) the prominence of this component in some of our model output makes a strong case that we need better understanding of appropriate values to use.

Harmon et al. (1990) argued that harvest of old-growth forests resulted in a net flux of C to the atmosphere and that the sum of net C sequestration did not return to its original value for hundreds of years. In this analysis (although not an old-growth forest) we show similarly that the net C flow is negative following harvest; but when we give credit for fuel displacement this scenario produces net positive C sequestration after about the 40th year. This 40 years to break-even is, of course, dependent on many variables, not the least of which is the high forest productivity assumed here (see, for example, Marland and Marland 1992).

There is an equilibrium between the oxidation of wood products and the production of new products, which is reached sooner for short-lived products. All further C offsets depend on the extents to which biofuels displace fossil fuels, and to which fossil fuels used to make products can be saved. The model does not recognise that some fraction of wood products ends up in modern landfills and is not oxidised over very long times. If scrap wood were used as a fuel after its primary use, this could be represented by a transfer of C from the product pool

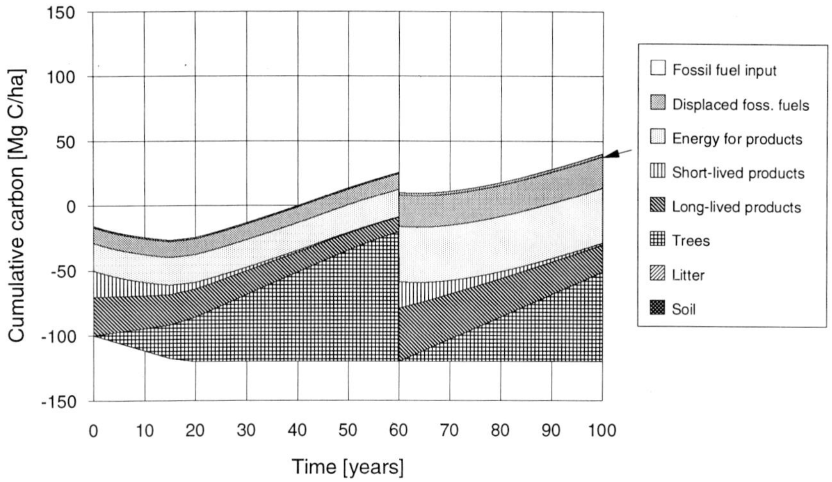

Figure 18.4. Net cumulative reduction in flows of C to the atmosphere for conventional forestry producing both long-lived and short-lived products and some biofuels on a 60-year rotation. Increases in soil and litter C would be shown according to the hatching in the legend, but a decrease, as in this scenario, is represented by a drop in the bottom line. The line marked with the arrow represents the net C balance and is slightly lower than the top line, because here the C emissions from auxiliary fossil fuels for bioenergy are somewhat higher than the upstream emissions of the fossil fuel displaced by the biofuel.

to the displaced fossil-fuel pool with the magnitude of the transfer dictated by the share of scrap wood burned and the efficiency with which it displaced fossil fuels.

To put an upper bound on the potential for conventional forestry to mitigate net emissions of C in the future, we examined a scenario with the same parameters for forest harvest and resource allocation but more efficient utilisation of the harvest in terms of product life and fossil fuel displacement (Figure 18.5). We have, for example, increased the fossil fuel displacement factor from 0.6 to 1.0. That 1 unit of C in wood could displace 1 unit of C in coal for generation of heat or power is not now achievable (except for substitution of very inefficient coal boilers/stoves or for combined heat and power production with biomass), but prospects for pre-drying or gasifying the wood or co-firing wood chips with coal suggest that this might be approached (Williams and Larson 1993; Demeter and Piscitello 1993). We did not change the harvest efficiency—in both scenarios 23% of the aboveground, logged biomass is left on site where it decays immediately to CO_2.

As it can be seen, more efficient use of the harvest would improve the C balance significantly. For the set of parameters chosen for Figure 18.5, the net C balance is never negative. A key is again that use of forest products displaces use of material which would require greater inputs of fossil fuels.

Lest Figures 18.4 and 18.5 be taken as a license for unlimited harvest, we draw a comparison with the scenario represented in Figure 18.2. The 3 scenarios all start with 100 Mg C ha^{-1} on site in the standing forest and assume average forest growth rates of 1.66 Mg C ha^{-1} yr^{-1}. While all three scenarios have a positive net C balance at the end of 50 years, it takes over 100 years for the conventional forestry scenario in Figure 18.4 to achieve the same net C benefit as the forest protection scenario, and the highly efficient conventional forestry scenario in Figure 18.5 is at about the same point as the forest protection scenario

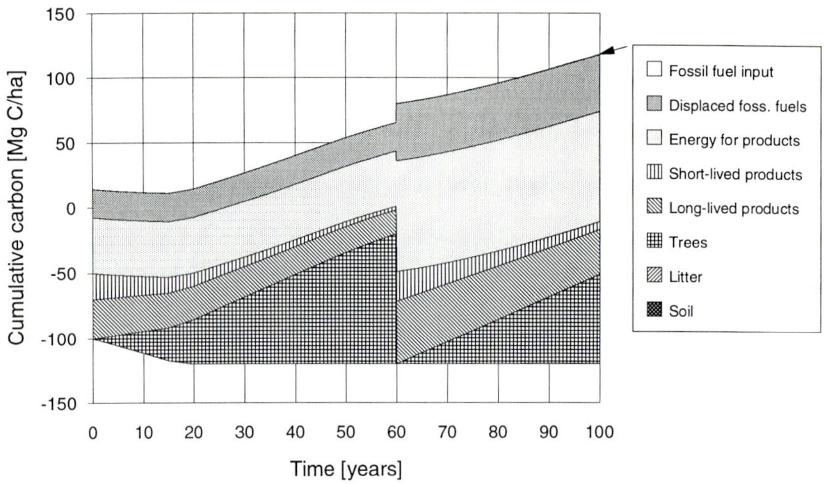

Figure 18.5. Net cumulative reduction in flows of C to the atmosphere for more efficient conventional forestry producing both long-lived and short-lived products and some biofuels on a 60-year rotation.

after about 40 years. Again, although these numeric results are very scenario dependent they provide an interesting comparison. Also, we do not suggest that net C balance is, or should be, the ultimate objective of forest management, but we try to illustrate the impact of forest management alternatives on the C balance.

An excess of agricultural and grazing land in many places raises the possibility of using fast-growing trees on short rotation as a dedicated energy crop. In short-rotation forestry, trees are mechanically harvested with a cycle of 4 to 12 years (7 years in our scenarios) and regrow from the same root. Where short-rotation forestry is implemented on agricultural land (Figure 18.6) we assume that there will be a slight build-up of soil and litter C (Ranney et al. 1991). In this scenario we assume again that one unit of C in wood substitutes for one unit of C in fossil fuels, i.e., for example, wood fuel is used with virtually the same net thermal efficiency as the coal it displaces, a state not now achieved in wood combustion systems. Because of a variety of factors involving economics, stage of development of the technologies, scale factors, fuel moisture content, etc., conversion efficiencies for wood tend to be notably less than for coal in current systems.

Some care is required in comparing the figures for our short-rotation scenarios (Figures 18.6 and 18.7) with those for the conventional forestry scenarios (Figures 18.4 and 18.5). In Marland and Marland (1992) comparisons were drawn between different forest management strategies on the same parcel of land and with the same tree species so that the expected rates of forest growth were the same regardless of the intended fate of the standing trees. In these scenarios we have envisioned conventional forestry on good forest land while the short rotation forests are envisioned to be fast-growing species with an intensive management regime. The difference in C balance is thus not simply a matter of the management strategy but also reflects a difference in assumed species selection and rates of forest productivity. The basic growth rate used for our short-rotation scenarios is 6 Mg C ha^{-1} yr^{-1}, within the range observed on experimental plots in the U.S. short-rotation-woody-crops program (Wright 1994; Graham et al. 1992). Turhollow and Perlack (1991) assumed the annual harvest to be an average of 7 Mg C ha^{-1} yr^{-1} (14.0 dry tons of biomass ha^{-1} yr^{-1} for hybrid poplar). This

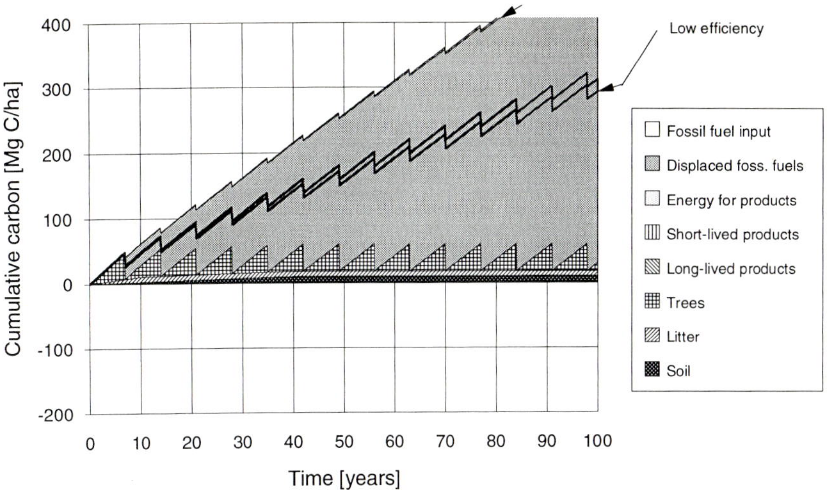

Figure 18.6. Net cumulative reduction in flows of C to the atmosphere for short rotation forestry on previously agricultural land, with efficient conversion of the harvest to replace coal combustion (top lines) and if the conversion efficiency is 60 % as high ("low efficiency" lines).

6 Mg C ha^{-1} yr^{-1} is 3.6 times the growth rate used in the conventional forestry scenarios above and suggests that these are generally not competing scenarios to be considered on the same lands. Short-rotation forestry is most likely to compete for cropland or potential cropland (Graham et al. 1992), land where optimal growing conditions can be combined with careful species selection and intensive management to produce very high yields. A large positive C balance is due to the commitment of the full harvest to fossil-fuel displacement, especially when fuel substitution is very efficient. As can be seen in Figure 18.6, C storage effects for short-rotation forestry occur both on the site (trees, eventually litter and soil) and in the displaced fossil fuel pool. If we consider a similar scenario for short-rotation forestry but with less efficient use of the wood harvest, the net C balance looks more like the line designated "low efficiency" in Figures 18.6 and 18.7. Less efficient use may indicate simply lower conversion efficiencies in the process of conversion into heat and/or power. It might also suggest that biofuels are substituted for a less C intensive fuel. The curves labelled low efficiency in Figures 18.6 and 18.7 are based on values typical for power generation from wood by current practice whereas the high efficiency curves represent advanced technology.

Figure 18.7 depicts model results when an area initially stocked with a forest of 100 Mg C ha^{-1} is harvested and then planted with highly-productive, short-rotation woody crops. For the initial harvest we assume the same parameters as for the conventional forestry scenario in Figure 18.4 and thus have a negative C balance at the beginning as in the scenario in Figure 18.4. After the initial harvest, harvest from the short-rotation plantation is used with a fuel displacement factor of 1. As a consequence, in this scenario with very high forest growth rate and very high efficiency in fuel use, the C balance turns positive within the first decade. A comparison with the scenarios in Figure 18.6 shows that even with high forest productivity and high fuel-use efficiency, the net C benefit is notably better when implemented on agricultural or pasture land than when land initially in forest is cleared to create a short-rotation plantation. This distinction is even more striking when fuels are used with the "low efficiency" typical today.

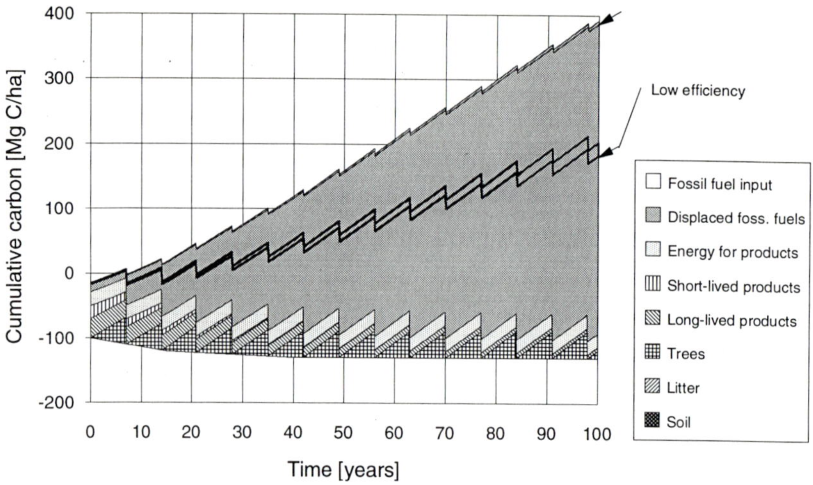

Figure 18.7. Net cumulative reduction in flows of C to the atmosphere for short-rotation forestry with efficient conversion of the harvest to replace coal combustion (top lines) and if the conversion efficiency is 60% as high ("low efficiency" lines). The land was previously forested and the initial harvest allocated as for conventional forestry in the scenario in Figure 18.4.

Note that in the scenarios depicted in our Figures 18.4, 18.5 and 18.7, when forest is cleared it is immediately replaced by newly growing forest or forest plantation and the net C balance eventually reaches its initial point, sooner when the growth rate is high and products are produced and used efficiently. In a scenario frequently played out, forest is cleared and the land is put into other uses or recovers slowly. Given the starting condition of our scenarios, this would lead to long-standing net losses of C exceeding 100 Mg C ha^{-1} as soil and litter C continued to oxidise as did any durable products produced. In other words, if the land were not regrown with forest, the net C balance would remain below the bottom line in Figures 18.4, 18.5, or 18.7.

SUMMARY AND CONCLUSION

We have tried to imagine the C implications of forestry projects over a 100-year time period. To do so, we have represented in a simple model many of the important components of the C balance, even if we were not able to do so with the detail and accuracy we would prefer. Because C flows are very site- or project-specific and because some flows are poorly understood, we have chosen 8 scenarios to try to illustrate the important features of various forest management strategies.

Figure 18.8 summarises the output from the 8 scenarios analysed in this study. The columns represent the net reduction of C emissions from 0 to 20, from 0 to 50, and from 0 to 100 years. The 4 groups of columns shown to the left are for conventional forestry scenarios and those to the right for fast growing plantations (short rotation forestry). It can be seen that for a time frame of 20 years the scenarios on agricultural land (initial standing stock = 0 Mg C ha^{-1}) have the highest C benefit. The longer the time perspective, the more important is the growth rate in comparison to the initial land use. It should be emphasised that the four

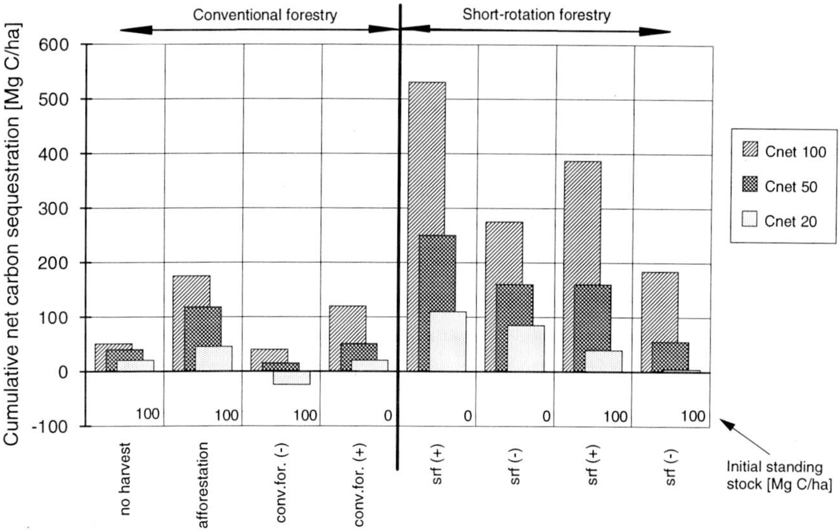

Figure 18.8. Net reduction of C emissions for 8 land use scenarios with a time period of 20 (C_{net} 20), 50 (C_{net} 50) and 100 (C_{net} 100) years. "srf": Short-rotation forestry, "+": Efficient use of the harvest to replace fossil fuel combustion and store C in products, "-": Less efficient use of the harvest to replace fossil fuels and store C in products.

conventional forestry scenarios shown here are based on highly productive commercial forest land and all have the same productivity. The four short-rotation-forestry scenarios assume a very high productivity which would require intensive management on optimal lands.

The scenarios suggest that, over long time periods, an important feature of the C flows is the extent to which renewable biofuels displace fossil-fuel use and fossil fuels used to make products from energy-intensive materials are saved by using wood instead. Although there can be significant short term, net storage of C in trees, soils, forest litter, and wood products, all of these C pools achieve equilibrium at some level and then provide no further sequestration over time. On the other hand, displacing fossil-fuel use results in a reduction in the net flux of C to the atmosphere as long as fossil fuels are still available and in use and thus can be displaced. In all cases, the efficiency with which biofuels are able to substitute for fossil fuels plays a major role in the C balance.

An important additional point raised by this study has to do with the amount of fossil-fuel-based energy displaced when wood products are substituted for products from more energy-intensive materials. Acknowledging that the numerical values are poorly known and certainly widely variable (depending on the details of the materials and applications for which products are substituted) the prominence of this component in some of our model output makes a strong case for better understanding of this factor.

This analysis points out the importance of C flows as a function of time. In some of our scenarios, biofuels strategies result in long-term benefits in net C fluxes despite small or negative benefits in the early stages. In this same vein, our scenarios remind us that the current land use may be an important consideration affecting the net C balance. A strategy to be implemented on land which is currently in pasture or agriculture may look far less attractive if implemented on land which is now in forest. The potential for forest management to contribute to mitigating the growth of atmospheric CO_2 will depend on the amount of appropriate land which is available.

Net C flows are not, and should not be, the sole criteria on which land-use decisions are based. Nevertheless, it appears that C flows may become an increasingly important factor in some land management decisions. To the extent that this is true, it is important that we have a realistic view of the full implications of the decisions and of the options available.

ACKNOWLEDGEMENTS

The work for this study was partly carried out within the Environment Research Program of the European Union with funds from the Austrian Federal Ministry of Science and Research and Joanneum Research. Gregg Marland was supported by the U.S. Department of Energy, under contract DE-AC05-84OR21400 with Martin Marietta Energy Systems, Inc.

REFERENCES

Bernow SS, Gurney K, Prince G, Cyr M (1992) The potential for biomass to mitigate greenhouse gas emissions in the northeastern U.S., a report to CONEG Policy Research Center. Tellus Institute, Boston, MA, 137 pp.

Birdsey RA (1992) Carbon storage in trees and forests. In: Sampson RN, Hair D (eds) Forests and global change, Volume 1: Opportunities for increasing forest cover. American Forests, Washington DC, 285 pp.

Burschel P, Kürsten E, Larson BC, Weber M (1993a) Present role of German forests and forestry in the national carbon budget and options to its increase. *Water, Air, Soil Pollut.* 70: 325–340.

Burschel P, Kürsten E, Larson BC (1993b) Die Rolle von Wald und Forstwirtschaft im Kohlenstoff-haushalt—Eine Betrachtung für die Bundesrepublik Deutschland, Forstliche Forschungsberichte Nr. 126, Universität München, D-8050 Freising, Germany, 135 pp.

Demeter CP, and Piscitello ES (1993) Biomass co-firing: chipping away at the power market? *Biologue* 12: 9–11.

Dewar RC (1991) Analytical model of carbon storage in the trees, soils and wood products of managed forests. *Tree Physiology* 8: 239–258.

Dewar RC, Cannell MGR (1992) Carbon sequestration in the trees, products and soils of forest plantations: an analysis using UK examples. *Tree Physiology* 11: 49–71.

Federer CA (1984) Organic matter and nitrogen content of the forest floor in even-aged northern hardwoods. *Can. J. For. Res.* 14: 763–767.

Graham RL, Wright LL, Turhollow AF (1992) The potential for short-rotation woody crops to reduce U.S. CO_2 emissions. *Climatic Change* 22: 223–238.

Hall DO, Mynick HE, Williams RH (1990) Carbon sequestration versus fossil fuel substitution—alternative roles for biomass in coping with greenhouse warming. Princeton University CEES Report 255, Princeton, NJ, 27 pp.

Harmon ME, Ferrell WK, Franklin JF (1990) Effects on carbon storage of conversion of old-growth forests to young forests. *Science*: 247: 699–701.

Houghton RA (1990) The future of tropical forests in affecting the CO_2 concentration of the atmosphere. *Ambio* 19: 204–209.

Johnson DW (1992) Effects of forest management on soil carbon storage. *Water, Air, Soil Pollut.* 64: 83–120.

Kolchugina TP, Vinson TS (1993) Carbon sources and sinks in forest biomes of the former Soviet Union. *Glob. Biogeochem. Cycles* 7: 291–304.

Kürsten E, Burschel P (1993) CO_2-mitigation by agroforestry. *Water, Air, Soil Pollut.* 70: 533–544.

Marland G (1983) Carbon dioxide emission rates for conventional and synthetic fuels. *Energy* 8, 981–992.

Marland G, Marland S (1992) Should we store carbon in trees? *Water, Air, Soil Pollut.* 64: 181–195.

Nabuurs GJ, Mohren GMJ (1993) Carbon fixation through forestation activities: a study of the carbon sequestering potential of selected forest types, commissioned by the Foundation FACE. IBN Research Report 93/4, Face/Institute for Forestry and Nature Research, Arnhem/Wageningen, The Netherlands, 205 pp.

Pastor J, Post WM (1986), Influence of climate, soil moisture, and succession on forest carbon and nitrogen cycles, *Biogeochem.* 2: 3–27.

Ranney JW, Wright LL, Mitchell CP (1991) Carbon storage and recycling in short-rotation energy crops. In: Mitchell CP (ed) Bioenergy and the greenhouse effect. Proceedings of a seminar of the International Energy Agency, Bioenergy Agreement, Stockholm, April 15–16, 1991, pp 39–51.

Schlamadinger B, Marland G (1995) The role of forest bioenergy in the global carbon cycle. *Biomass and Bioenergy* (In press).

Schlamadinger B, Spitzer J, Kohlmaier GH, Lüdeke M (1995) Carbon balance of bioenergy from logging residues. *Biomass and Bioenergy* (In press).

Turhollow AF, Perlack RD (1991) Emissions of CO_2 from energy crop production. *Biomass and Bioenergy* 1: 129–135.

Turner DP, Lee JJ, Koerper GJ, Barker JR (eds) (1993) The forest sector carbon budget of the United States: carbon pools and flux under alternative policy options. EPA/600/3-93/093, Environmental Protection Agency, Environmental Research Laboratory, Corvallis, OR, 201 pp.

U.S. Office of Technology Assessment, U.S. Congress (1991) Changing by degrees: steps to reduce greenhouse gases. OTA-O-482, U.S. Government Printing Office, Washington DC, 354 pp.

Williams RH, Larson ED (1993) Advanced gasification-based biomass power generation. In: Johansson TB, Kelly H, Reddy AKN, Williams RH (eds) Renewable energy: sources for fuels and electricity. Island Press, Washington DC, pp 729–785.

Winjum JK, Dixon RK, Schroeder PE (1993) Forest management and carbon storage: an analysis of 12 key forest nations. *Water, Air, Soil Pollut.* 70: 239–258.

Wright LL (1994) Production technology status of woody and herbaceous crops. *Biomass and Bioenergy* 6: 191–209.

APPENDIX 18.1: INPUT PARAMETERS OF THE MODEL: PARAMETER ASSUMPTIONS FOR 8 LAND MANAGEMENT SCENARIOS

No.	Parameter		Scenario*							
			No harvest (For) (Figure 2)	Afforestation (Agr) (Figure 3)	Conventional forestry (For, –) (Figure 4)	Conventional forestry (For, +) (Figure 5)	Short rotn. forestry (Agr, +) (Figure 6)	Short rotn. forestry (Agr, –) (Figure 6)	Short rotn. forestry (For, +) (Figure 7)	Short rotn. forestry (For, –) (Figure 7)
Initial harvest										
1	Initial aboveground C	$Mg\ C\ ha^{-1}$	100	0	100	100	0	0	100	100
2	Fossil C-emissions from harvesting	$Mg\ C\ (Mg\ C)^{-1}$	-	-	0.01	0.01	-	-	0.01	0.01
3	Share of harvest to biofuel	$Mg\ C\ (Mg\ C)^{-1}$	-	-	0.22[a]	0.22[a]	-	-	0.22[a]	0.22[a]
4	Share of harvest to long-lived products	$Mg\ C\ (Mg\ C)^{-1}$	-	-	0.30[a]	0.30[a]	-	-	0.30[a]	0.30[a]
5	Share of harvest to short-lived products	$Mg\ C\ (Mg\ C)^{-1}$	-	-	0.20[a]	0.20[a]	-	-	0.20[a]	0.20[a]
6	Share of harvest to very short-lived products	$Mg\ C\ (Mg\ C)^{-1}$	-	-	0.05[a]	0.05[a]	-	-	0.05[a]	0.05[a]
7	Share of harvest not used	$Mg\ C\ (Mg\ C)^{-1}$	-	-	0.23[a]	0.23[a]	-	-	0.23[a]	0.23[a]
8	Displacement factor fuel #	$Mg\ C\ (Mg\ C)^{-1}$	-	-	0.6	1	-	-	0.6	0.6
9	Displacement factor—long-lived products	$Mg\ C\ (Mg\ C)^{-1}$	-	-	0.5[b]	1[b]	-	-	0.5[b]	0.5[b]
10	Displacement factor—short-lived products	$Mg\ C\ (Mg\ C)^{-1}$	-	-	0.25[b]	0.5[b]	-	-	0.25[b]	0.25[b]
11	Displacement factor—very short-lived products	$Mg\ C\ (Mg\ C)^{-1}$	-	-	0.25[b]	0.5[b]	-	-	0.25[b]	0.25[b]
12	Fossil C-emissions from bioenergy production	$Mg\ C\ (Mg\ C)^{-1}$	-	-	0.05	0.03	-	-	0.05	0.05
13	Upstream C-emissions of displaced fossil fuel	$Mg\ C\ (Mg\ C)^{-1}$	-	-	0.08[c]	0.08[c]	-	-	0.08[c]	0.08[c]
Subsequent harvests										
14	Early growth rate	$Mg\ C\ ha^{-1}\ yr^{-1}$	1.72	1.72	1.72	1.72	6[g]	6[g]	6[g]	6[g]
15	Rotation length	yr	-	-	60	60	7[g]	7[g]	7[g]	7[g]
15a	Aboveground C affected by harvest‡	$Mg\ C\ ha^{-1}$	-	-	100	100	42	42	42	42
16	Annual fossil C-emissions from cultivation	$Mg\ C\ ha^{-1}\ yr^{-1}$	-	-	0	0	0.1	0.1	0.1	0.1
17	Fossil C-emissions from harvesting	$Mg\ C\ (Mg\ C)^{-1}$	-	-	0.01	0.01	0.03	0.03	0.03	0.03
18	Share of harvest to biofuel	$Mg\ C\ (Mg\ C)^{-1}$	-	-	0.22[a]	0.22[a]	0.80[g]	0.80[g]	0.80[g]	0.80[g]
19	Share of harvest to long-lived products	$Mg\ C\ (Mg\ C)^{-1}$	-	-	0.30[a]	0.30[a]	0	0	0	0
20	Share of harvest to short-lived products	$Mg\ C\ (Mg\ C)^{-1}$	-	-	0.20[a]	0.20[a]	0	0	0	0
21	Share of harvest to very short-lived products	$Mg\ C\ (Mg\ C)^{-1}$	-	-	0.05[a]	0.05[a]	0	0	0	0
22	Share of harvest not used	$Mg\ C\ (Mg\ C)^{-1}$	-	-	0.23[a]	0.23[a]	0.20[g]	0.20[g]	0.20[g]	0.20[g]

No.	Parameter	Unit								
23	Displacement factor fuel #	Mg C (Mg C)$^{-1}$	-	-	0.6	1	1	0.6	1	0.6
24	Displacement factor—long-lived products	Mg C (Mg C)$^{-1}$	-	-	0.5[b]	1[b]	-	-	-	-
25	Displacement factor—short-lived products	Mg C (Mg C)$^{-1}$	-	-	0.25[b]	0.5[b]	-	-	-	-
26	Displacement factor—very short-lived products	Mg C (Mg C)$^{-1}$	-	-	0.25[b]	0.5[b]	-	-	-	-
27	Fossil C emissions from bioenergy production	Mg C (Mg C)$^{-1}$	-	-	0.05	0.03	0.03	0.03	0.03	0.03
28	Upstream C emission of displaced fossil fuel	Mg C (Mg C)$^{-1}$	-	-	0.08[c]	0.08[c]	0.08[c]	0.08[c]	0.08[c]	0.08[c]
Other parameters										
29	Soil C uptake	Mg C ha^{-1}	-	20	-10[d]	-10[d]	10[e]	10[e]	-15	-15
30	Soil C uptake time	yr	-	80	20[d]	20[d]	40[e]	40[e]	40	40
31	Roots and litter C uptake	Mg C ha^{-1}	-	20	-10[d]	-10[d]	8[e]	8[e]	-15	-15
32	Roots and litter C uptake time	yr	-	60	15[d]	15[d]	10[e]	10[e]	15	15
33	Average lifetime long-lived products	yr	-	-	60[f]	120[f]	-	-	60[f]	60[f]
34	Average lifetime short-lived products	yr	-	-	15[f]	30[f]	-	-	15[f]	15[f]

Symbols and notes

- * The general differences between the scenarios are explained in the text.
- \+ Efficient use of the harvest to replace fossil fuel combustion and store C in products
- – Less efficient use of the harvest to replace fossil fuel combustion and store C in products
- Agr Previously agricultural land
- For Previously forested land
- \# The displacement factor fuel (α) is calculated as follows:

  α = (efficiency of bioenergy system/efficiency of fossil energy system) × (C emission rate of fossil fuel / C emission rate of biofuel).

  The C emission rate of biofuels is slightly higher than of the current fossil fuel mix, and the efficiency is on average slightly lower. $\alpha = 0.6$ reflects this and $\alpha = 1$ is more optimistic.
- ‡ Not an input parameter, but calculated by the model (integral of annual growth rate over one rotation, see Appendix 18.2)

Letter superscripts refer to the references cited below. Additional details on the parameter assumptions are provided in Schlamadinger and Marland (1995).

a Barnow et al. (1992)
b Kürsten and Burschel (1993), Burschel et al. (1993b)
c Marland (1983)
d Johnson (1992), Federer (1984), Pastor and Post (1986)
e Ranney et al. (1991)
f Burschel et al. (1993b)
g Turhollow and Perlack (1991)

APPENDIX 18.2: EQUATIONS AND FUNCTIONS USED IN THE MODEL

All numbers in parenthesis represent the parameters in Appendix 18.1 (see numbers in left column).

Table 18.A2.1. Carbon fluxes at the time of harvest.

FLUXES ILLUSTRATED IN FIGURE 18.1	EQUATION FOR A POSSIBLE INITIAL HARVEST	EQUATION FOR SUBSEQUENT HARVESTS
Harvested C, flux 2	= (1)	= (15a)
Harvested C to atmosphere, flux 3	= (1) × (7)	= (15a) × (22)
Harvested C to long-lived products, flux 4	= (1) × (4)	= (15a) × (19)
Harvested C to short-lived products, flux 5	= (1) × (5)	= (15a) × (20)
Harvested C to very-short-lived products, flux 6	= (1) × (6)	= (15a) × (21)
Harvested C to biofuels, flux 7	= (1) × (3)	= (15a) × (18)
Very-short-lived products to atmosphere, flux 10	= (1) × (6)	= (15a) × (21)
Biofuels to atmosphere, flux 11	= (1) × (3)	= (15a) × (18)
Atmosphere to fossil fuels for products from material other than wood, minus fossil fuels for wood products to atmosphere, flux 14 – flux 12	= (1) × (4) × (9) + (1) × (5) × (10) + (1) × (6) × (11)	= (15a) × (19) × (24) + (15a) × (20) × (25) + (15a) × (21) × (26)
Atmosphere to fossil fuels for energy, flux 15	= (1) × (3) × (8)	= (15a) × (18) × (23)
Atmosphere to upstream emissions of fossil fuels for energy minus fossil fuels for land management and bioenergy production to atmosphere, flux 16 – flux 13	= (1) × (3) × (8) × (13) – (1) × (3) × (12) – (1) × (2)	= (15a) × (18) × (23) × (28) – (15a) × (18) × (27) – (15a) × (17) – (15) × (16)

Table 18.A2.2 Annual C fluxes at any time t, not only time of harvest.

FLUXES ILLUSTRATED IN FIGURE 18.1	EQUATION
Atmosphere to forest, flux 1	= (14), if standing stock < 80 Mg C ha^{-1} or = 2 × (14) × [1 – Sn / 160], if standing stock > 80 Mg C ha^{-1} (Sn is the standing stock at time t. See Marland and Marland 1992)
Long-lived products to atmosphere, flux 8	= Plong / exp [1 / (33)] (Plong is the C stored in long-lived products at time t – 1, that has remained from previous harvests and not yet decayed)
Short-lived products to atmosphere, flux 9	= Pshort / exp [1 / (34)] (Pshort is the C stored in short-lived products at time t – 1, that has remained from previous harvests and not yet decayed)
Atmosphere to soil, flux 17	= (31) / (32) for t < (32) or = 0 for t > (32)
Atmosphere to litter, flux 18	= (29) / (30) for t < (30) or = 0 for t > (30)

19 The influence of carbon budget methodology on assessments of the impacts of forest management on the carbon balance

Robert Matthews

Forestry Commission, Research Division, Alice Holt Research Station, Wrecclesham, Farnham, Surrey GU10 4LH, UK

INTRODUCTION

There is now a substantial body of literature in scientific journals on the carbon (C) balance of forest ecosystems at all geographical scales, and some authors have proposed changes in the way forests are managed in order to enhance the offsetting or reduction of emissions of C to the atmosphere. As yet, however, there is no widely accepted and applied method for estimating the C budget of forests and the impacts of forest management on C sequestration. This paper describes how the results of C budgeting exercises depend strongly on the methodologies employed, and briefly considers the scope for developing conventions for the calculation and presentation of C budgets for forest ecosystems and forestry projects. These points are illustrated using examples of model simulations of the effects of alternative methods of forest management in Britain, and by comparison of published results of studies in other countries based on different C budgeting methodologies.

OBJECTIVES OF CARBON BUDGETING AND IMPLICATIONS FOR METHODOLOGIES

Carbon budgets of forest ecosystems are required for two primary purposes: firstly, they are an essential contribution to global statistics on C fluxes to and from the atmosphere, necessary for modelling and predicting climate change; secondly, they form the factual basis for negotiations and policy formulation on the management of forests for offsetting or reducing global C emissions. For technical and logistical reasons, these two tasks cannot be easily accomplished using one set of measurement, analysis and modelling techniques.

In compilations of regional and global statistics on C fluxes for national and international reports, forest ecosystems comprise just one form of land use out of many. Data on C in terrestrial plants are just one element in a complete C budget, which also includes soils, aquatic systems and the burning of fossil fuels. The linkages between these elements, in particular linkages between forest management, wood utilisation and energy consumption, are of key importance to the evaluation of the impact of forest policy and management on the direction and rate of flow of C to or from the atmosphere. Among others Hall et al. (1991), Koch (1992), Marland and Marland (1992) and Matthews (1992a, b, 1994) have emphasised the potential of harvested wood to substitute for fossil fuels as a source of energy (*direct substitution*), or for more energy-expensive materials (*indirect substitution*).

NATO ASI Series, Vol. I 40
Forest Ecosystems, Forest Management and the Global Carbon Cycle
Edited by Michael J. Apps and David T. Price

COMPARISON OF ALTERNATIVE CARBON BUDGETS FOR BRITISH FORESTS

Matthews (1992b, 1994) presented C budgets for a theoretical plantation forest in Britain, and showed how excluding and including direct or indirect substitution in budget calculations could lead to diametrically opposed conclusions about future forest management to achieve maximum C sequestration. When the substitution role of wood was not accounted for in any form, comparison of C budgets for alternative future management options led to the conclusion that preservation of the forest with minimal intervention and little or no harvesting of trees was the best option for achieving maximum C sequestration. On the other hand, allowance for either direct or indirect substitution in the calculations gave results which suggested that active management of the forest for sustainable timber production was more effective as a means of sequestering C than management based on non-intervention.

Table 19.1 shows the results obtained when two different budgeting methods are applied at the national level in order to evaluate the C sequestration potential of a series of alternative forest policy options in Britain. Brief descriptions of the changes to forest management assumed in the various scenarios are shown in the first column of Table 19.1. The next pair of columns in Table 19.1 shows results for 50 year and 500 year simulation periods obtained using a C budget methodology (referred to below as the *ecosystem methodology*) which accounts for exchanges of C between the atmosphere and forest biomass, forest necromass, forest soils and harvested wood products. The final pair of columns in Table 19.1 shows results for 50 year and 500 year simulation periods obtained using an alternative C budget methodology (referred to below as the *forestry sector methodology*), which is identical to the *ecosystem methodology*, but also accounts for effective C sequestration attributable to direct and indirect substitution by harvested wood. All results were produced using C sequestration models developed by the Research Division of the Forestry Commission (Thompson and Matthews 1989; Matthews 1992a, 1994), used in combination with national forest inventory data for Britain. The values shown in Table 19.1 represent, for a series of possible changes in forest management, the resultant change in the rate of C sequestration by British forests in Tg C yr^{-1}, relative to that obtained by continuing to manage forests according to current practice.

In order to estimate changes in national C sequestration, it was necessary to compute C budgets for British forests based on a *business-as-usual* scenario covering many decades from the distant past through to the distant future. The estimates of national C sequestration for the business-as-usual scenario are shown at the foot of Table 19.1. The construction of the business-as-usual scenario, and the estimation of future C sequestration for all scenarios, required many assumptions to be made about past, present and future forest cover, C densities, forest management, wood utilisation and energy consumption. An exhaustive description of these assumptions is beyond the scope of this paper. The following discussion highlights and contrasts certain details of the two methodologies as well as those used in previous case studies of C sequestration by British forests.

Inspection of the estimated business-as-usual rates of C sequestration in Table 19.1 reveals that negative rates of C sequestration are estimated using the *ecosystem methodology*. This is because 'C sequestration' is defined here as the net rate of removal of carbon dioxide from the atmosphere (in terms of fixed C), a negative rate therefore indicating net release of C to the atmosphere. It is interesting to contrast these results with previous reports on the C balance of British forests (Adger et al. 1992; Matthews 1992a; Cannell and Dewar 1995), all of which suggest positive rates of C sequestration. The apparent disagreement can be explained in terms of the system boundaries and methodological details adopted in the different studies. The *ecosystem methodology* used in this study and those of Matthews

(1992a) and Cannell and Dewar (1995) are in fact similar, except that in this study C sequestration by vegetation and soils of non-forested land in Britain is taken into account, whereas the other studies considered only land that is forested. The negative rates of C sequestration estimated using the *ecosystem methodology* in this study can be explained mainly by the release of C from soils with initial high C content including peatlands which have been disturbed by conversion to forestry and, more importantly, conversion to intensive agriculture.

Table 19.1. Evaluation of forest policy/management options for enhancing carbon (C) sequestration potential of British forests[a], including comparison of two alternative C budget methodologies.

	Change in annual C sequestration by British forests[a] relative to business-as-usual scenario (Tg C yr^{-1})			
	Ecosystem methodology (not including direct or indirect substitution)		Forestry sector methodology (including direct and indirect substitution)	
Example management option	Over 50 yr	Over 500 yr	Over 50 yr	Over 500 yr
Shorter rotations[b] (-20 yr)	-1.2	-0.2	-1.3	-0.8
Longer rotations[b] (+20 yr)	0.6	0.1	-0.6	-0.8
Harvest whole trees	0	0	0	0
Utilise unmanaged forests	-0.3	0	0.1	0.3
Improve timber quality	0	0.1	-0.1	0.1
Stop all harvesting and felling	5.4	0.7	-1.5	-6.4
Increase forest area by 20%, (0.5 Mha) over 50 years				
(with conifers)	0.7	0.1	1.1	2.2
(with broadleaves)	0.5	0.1	1.1	2.2
Business-as-usual[c] absolute rates	-0.2	-0.3	7.7	7.6

a 'British forests' refers to high forest and coppice on rotations longer than 5 years but not short rotation wood-fuel coppice crops.
b Conifer plantation forests are typically grown on a rotation between 40 and 70 years. Broadleaf plantation forests are typically grown on rotations between 50 and 200 years.
c Does not include wood imported from other countries.

Non-forested lands were included in the estimation of the business-as-usual C budget in recognition of the fact that certain changes in forest policy, specifically those involving land-use change such as afforestation, would inevitably have impacts on the C budgets for other sectors of land use. In order to correctly evaluate the impact of afforestation policies on C sequestration, it is essential to account for the initial C balance of all lands given the current pattern of land use, any changes in land use, and the changes in C fluxes likely to arise from afforestation. In this study, the business-as-usual scenario assumes that the existing forest area in Britain is maintained but not expanded in the future. Alternative scenarios involving afforestation assume that new forests are established on land previously under intensive agriculture, with relatively low initial C stocks in both biomass and soil.

Adger et al. (1992) have already attempted to estimate the overall C balance of all land use types in Britain, and the changes in the C balance attributable to changes in land use.

Superficially, therefore, the *ecosystem methodology* in this study resembles quite closely that of Adger et al. (1992). The apparent disagreement in the estimates of C sequestration made in the two studies reflects merely differences in the time periods to which the estimates are applicable. The estimate of 1.3 Tg C yr^{-1} for the total change in soil and biomass C in Britain for the year 1980 presented by Adger et al. (1992), in fact agrees well with an equivalent estimate of 1.4 Tg C yr^{-1} calculated using the *ecosystem methodology*. This agreement, however, masks many differences in the detail of the methodologies of the two studies, notably with respect to assumptions about the magnitudes and time-courses of forest biomass accumulation, the types and C contents of soils under forest, and the fate of wood products particularly from thinnings.

The results produced by the *ecosystem methodology* i.e., not accounting for direct or indirect substitution, may be summarised thus:

- Reducing rotation age by 20 years could cause a significant reduction in C sequestration, while increasing rotation age by 20 years could cause a significant increase in C sequestration.
- Harvesting whole trees appears to have a negligible effect on C sequestration.
- Improving timber quality appears to have a negligible effect on C sequestration.
- Utilising unmanaged forests could cause a reduction in C sequestration.
- Stopping all harvesting and felling of trees could cause a significant increase in C sequestration.
- Afforestation with broadleaves and/or conifers could cause a significant increase in C sequestration.

The policy/management option identified by the *ecosystem methodology* as most effective at enhancing C sequestration is to stop all harvesting and felling of trees. The results also suggest that extended rotation ages and afforestation would be significant but less effective options. Comparison of the results for 50-year and 500-year simulation periods indicates that the significant effects identified above are not sustained in the long term.

The above results contrast markedly with those produced by the *forestry sector methodology* which accounts for direct and indirect substitution. These alternative results are summarised below:

- Reducing or increasing rotation age by 20 years could cause a significant reduction in C sequestration.
- Harvesting whole trees appears to have a negligible effect on C sequestration.
- Improving timber quality appears to have a negligible effect on C sequestration.
- Utilising unmanaged forests could cause an increase in C sequestration.
- Stopping all harvesting and felling of trees could cause a significant decrease in C sequestration.
- Afforestation with broadleaves and/or conifers could cause a significant increase in C sequestration.

The policy/management option identified by the *forestry sector methodology* as most effective at enhancing C sequestration is afforestation with either broadleaves or conifers. The results for the 500 year simulation period suggest that utilisation of unmanaged forests could be a significant but less effective option. Comparison of the results for 50-year and 500-year simulation periods indicates that the significant effects identified above are sustained in the long term, and in some cases (e.g., stopping harvesting and felling, or afforestation) the effects are only fully realised in the long term.

It is evident from a comparison of the results produced by the *ecosystem* and *forestry sector* methodologies that exclusion or inclusion of direct and indirect substitution in the calculation of C budgets produces results from which diametrically opposed conclusions would be drawn about optimal strategies for forest policy and management for maintaining or enhancing C sequestration. The question therefore arises: which budgeting methodology is the "correct" one to apply? Unfortunately, it is difficult to give a straightforward answer, because the relevance of a given C budget methodology depends on the specific objectives of the study in question, and may also depend to some extent on any economic, social and environmental objectives, costs and benefits associated with the forests under consideration. It is clear in this specific example, however, that ignoring direct and indirect substitution in C budgeting would result in the misrepresentation of the C sequestration potential of timber production from industrial plantation forests such as those currently forming 90% of the forested land area in Britain.

ANALYSIS OF PUBLISHED CARBON BUDGETS

Although restricted to British forests, the above results suggest that many of the apparent disagreements between different authors over optimal forest policy and management for maintaining or enhancing C sequestration could arise simply from differences in the C budget methodologies adopted in different studies. Table 19.2 presents a review and analysis of papers on forest management and potential C sequestration, in which the papers have been classified according to the type of C budget methodology used and the conclusions and recommendations made on optimal forest management.

Published studies differ widely in terms of the range of processes considered relevant to the estimation of the C budget for forests and their management. To a large extent, though not entirely, this reflects differences in the objectives and subject matter covered by different authors. In Table 19.2 a series of six ticks are used to indicate the comprehensiveness of the budgeting methodologies used in particular reports. Comprehensiveness of the C budget methodology adopted in each study is summarised in Table 19.2 in terms of the number of ticks, using a score ranging from 1 to 4. Higher scores indicate more comprehensive budget methodologies (see Table 19.2 for key).

Conclusions about the potential for offsetting emissions of carbon dioxide to the atmosphere through active forest management range from virtual dismissal (e.g., Freedman et al. 1992) to very positive (e.g., Sampson 1992). Table 19.2 states briefly any conclusions and recommendations made in each paper. A score ranging from 1 to 4 is assigned to each paper, indicating the impacts, in terms of C sequestration, that the authors consider forest policy or management are likely to have. Papers with a score of 1 conclude that any impact will be insignificant, while papers with a score of 4 report significant short- and long-term impacts, in response to a combination of changes in forest management and wood utilisation (see Table 19.2 for key).

Table 19.2. Analytical review of published results on carbon (C) sequestration potential of forest projects. The budget components considered in column 3 are: B: biomass; S: soil; P: wood products; F: fuel costs; D: direct substitution; and I: indirect substitution. The scores in column 5 are allocated as follows: *First digit:* assesses the comprehensiveness of C budget methodology with an entry of: 1 if only one box is ticked (either biomass or soil); 2 if several boxes are ticked, but the study does not account for direct and indirect substitution; 3 if several boxes are ticked, and the study accounts for direct and/or indirect substitution; and 4 if all boxes are ticked. *Second digit:* Paper concludes that, in terms of C sequestration, changes in forest policy or management are likely to have: 1: insignificant impact; 2: significant impact, but only in the short- to medium-term; 3: significant, long-term impact; 4: significant short- and long-term impacts, but a combination of changes in forest management and wood utilisation is required. (Note: An attempt has been made to eliminate duplicate references.)

		Budget components considered[a]							
Year	Source	B	S	P	F	D	I	Summary of conclusions	Scores[b] allocated
1977	Dyson	✓	✓			✓		Short term, CO_2 released by burning fossil fuels can be controlled by growing trees. Long term, CO_2 emissions could be cut by increased use of renewable energy, e.g., wood fuel.	3,3
1979	Breuer	✓						Worldwide reforestation could, within 15/20 years, sequester C at rates comparable to current global emissions. This effect would be limited to several decades.	1,2
1979	Dyson and Marland	✓						Short/medium term, C can be transferred from atmosphere to biosphere by growing long-lived trees that are left unharvested. There are logistical/political constraints to implementation.	1,2
1981	Sharpe and Johnson	✓		✓				Reforestation and changes in wood utilization patterns could be implemented to increase C storage in biomass. Harvesting mature forests and conversion to wood products and young plantations results in net emission of CO_2.	2,2
1983	Cooper	✓						C sequestration by forested lands depends on rotation age. Forests managed for timber production contain typically one-third of the C of equivalent undisturbed forests.	1,2
1983	Cropper and Ewel	✓	✓					Short term, afforestation and changes in forest management can have a significant effect on C sequestration.	2,2
1984	Armentano	✓						Increased harvesting of forests for biofuel would result in less C storage in forest biomass, compared to current quantities.	1,2
1987	Cropper and Ewel	✓	✓					Conversion of mature forest to short rotation wood fuel crops would result in a 30–40% reduction in C storage.	2,2
1988	Marland	✓		✓		✓		Forestry alone cannot offset emissions of CO_2. Short term, it could play a significant role. There is a conflict between reforestation to limit atmospheric CO_2 and attempting to increase wood production, unless we can store wood such that it is not oxidised, or can replace fossil fuel consumption with wood consumption.	3,2
1989	Jarvis	✓		✓				It is not practical to carry out afforestation to offset the rise in atmospheric CO_2. Harvesting of wood results in a net release of CO_2. Any C sequestration due to afforestation would be short/medium term.	2,1
1989	Thompson and Matthews	✓		✓				Effect of forest management on C sequestration depends on relative rates of biomass accumulation and wood products decay. For any management option, C storage approaches an asymptotic quantity.	2,2
1990	Dewar	✓		✓				Forest management for maximal C storage depends on relative rates of productivity of forest, and longevity of wood products. For a given management option, C storage approaches an asymptotic quantity.	2,2
1990	Hall et al.	✓				✓		Displacing fossil fuel with biomass grown sustainably and efficiently converted into useful energy would be more effective in decreasing atmospheric CO_2 than sequestering C in trees.	3,3
1990	Harmon et al.	✓	✓	✓				Conversion of 'old-growth' forests to secondary forests causes net release of CO_2.	2,2

Table 19.2. (continued).

YEAR	SOURCE	BUDGET COMPONENTS CONSIDERED[a] B	S	P	F	D	I	SUMMARY OF CONCLUSIONS	SCORES[b] ALLOCATED
1990	Hendrickson	✓	✓	✓	✓	✓		Effect of forest management on C sequestration depends on relative rates of biomass accumulation and wood products decay, and on impacts on soil C. Biofuel production identified as promising option.	3,4
1990	Kayes et al.	✓	✓	✓				Afforestation to achieve C sequestration is only effective in short/medium term. C sequestration by afforestation would be small compared to global emissions of C.	2,2
1991	Adger et al.	✓	✓	✓				Doubt cast on effectiveness of forests for sequestering C. Growing of plantation forests and clear-cutting not optimal in terms of C sequestration.	2,1
1991	Dewar	✓	✓	✓				Harvesting mature forests for products is only more effective at C sequestration than leaving the forest undisturbed if products are long-lived and the rotation age is equal to the rotation for maximum sustained yield.	2,2
1991	Graudal	✓		✓				Short/medium term, afforestation and encouraging the manufacture of long-lived wood products would significantly enhance C sequestration.	2,2
1991	Johnson and Kern		✓					High priority should be given to maintaining C stocks in soil pool. Each soil type has a maximum C holding capacity, which cannot be exceeded.	1,2
1991	Maclaren and Wakelin	✓		✓				Relative to CO_2 emissions, uptake by afforested land in New Zealand is highly significant in short/medium term.	2,2
1991	Vitousek	✓	✓	✓		✓		Conversion of old growth forests into products and young plantations causes net emission of CO_2. Afforestation for C sequestration is only effective short/medium term. Forest plantations for bioenergy could achieve a longer term reduction in CO_2 emissions.	3,4
1992	Alban and Perala	✓	✓	✓			✓	Changes in ecosystem C reflect changes in biomass. If wood can be used for durable wood products, or to substitute for fossil fuels, short rotation plantation forests sequester more C than long rotations.	3,3
1992	Bekkering	✓						Afforestation and forest conservation would sequester only moderate amounts of C.	1,1
1992	Brown et al.	✓	✓					Discouraging human disturbance of forest ecosystems increases C sequestration.	2,2
1992	Dewar and Cannell	✓	✓	✓				Species choice/site selection can strongly influence effectiveness of afforestation for C sequestration. Thinning reduces total C storage. Wood products constitute a minor C store. Systems based on short rotations do not achieve high C storage.	2,2
1992	Freedman et al[c]	✓	✓	✓		✓		Intensively managed, unlogged plantation forests attain maximal C storage sooner than naturally regenerated forests. Afforestation to achieve significant C sequestration is unrealistic.	3,1
1992	Gregor	✓	✓	✓			✓	Afforestation and protection of forests from damage can make a significant contribution to offsetting emissions of CO_2. On-site C sequestration will be significant in short term, but impacts of indirect substitution are potentially long term.	3,3
1992	Koch	✓			✓		✓	Measures to prevent logging and preserve natural forest ecosystems could lead to massive increases in fossil fuel energy use.	3,3
1992	Marland and Marland	✓	✓	✓		✓		The best management option for C sequestration depends on forest productivity. If productivity is low it is best to leave forest undisturbed to maximise C storage on site. If productivity is high, best options include harvesting wood as biofuel and long lived products.	3,4
1992a	Matthews	✓	✓	✓	✓	✓		Maximising C storage in biomass or wood products would be effective in the short/medium term. Long term, sustainable use of wood as biofuel or as an energy-efficient material could reduce net emissions of CO_2.	3,3

Table 19.2. (continued).

YEAR	SOURCE	BUDGET COMPONENTS CONSIDERED[a] B	S	P	F	D	I	SUMMARY OF CONCLUSIONS	SCORES[b] ALLOCATED
1992	Winjum et al.	✓						Global afforestation could sequester 50–100 Gt C.	1,2
1993	Birdsey et al.	✓	✓	✓				C sequestration by forests has significantly offset emissions of CO_2 in the past. This should continue in the short/medium term, followed by a drop in rates of C sequestration. These changes reflect changes in forest age structure.	2,2
1993	Nabuurs and Mohren	✓	✓	✓				Afforestation significantly enhances the forest C sink but only in the short/ medium term. C sequestration is higher in areas of high forest productivity, and is lower in logged forests compared to unlogged forests.	2,2
1993	Turner et al.	✓	✓					Reductions in harvests and increased recycling of wood products are likely to lead to increased C sequestration. Short/medium term, afforestation enhances C sequestration.	2,2
1994	Fischlin and Bugmann	✓		✓		✓		Reforestation the most effective option for offsetting CO_2 emissions in the short/medium term. Encouraging manufacture of durable wood products and increased use of wood as bioenergy could reduce net CO_2 emissions.	3,4
1994	Heil *et al.*	✓	✓	✓	✓			Although afforestation could offset emissions of CO_2, the quantities are small compared to the potential for reduction in emissions through a general improvement in energy efficiency.	2,2
1994	Hoen and Solberg	✓		✓	✓			Fertilisation of forests enhances C sequestration. Very long rotations combined with late, heavy fellings and regeneration is identified as an effective management option for C sequestration.	2,2
1994	Matthews	✓	✓	✓	✓	✓	✓	Forest management affects C storage in the forest, in wood products and in fossil fuels through direct and indirect substitution. Results suggest the latter effects are the most significant.	4,4
1994	This study	✓	✓	✓	✓	✓	✓	A combination of changes in forest management and wood utilisation could have long term, significant impact on C sequestration.	4,4

a Budget components: B: biomass; S: soil; P: wood products; F: fuel costs; D: direct substitution; I: indirect substitution.

b Allocated scores: *First digit.* Comprehensiveness of C budget methodology evaluated as follows: 1: only one box ticked (either biomass or soil); 2: several boxes ticked, but not accounting for direct and indirect substitution; 3: several boxes ticked, with accounting for direct and/or indirect substitution; 4: all boxes ticked. *Second digit.* Paper concludes that, in terms of C sequestration, changes in forest policy or management are likely to have: 1. an insignificant impact; 2. a significant impact, but only in the short- to medium-term; 3. a significant, long-term impact; 4. significant short- and long-term impacts, but a combination of changes in forest management and wood utilisation is required.

c Although Freedman et al. consider biofuel plantations, they do not fully account for direct substitution.

Table 19.3. Classification of publications on carbon (C) sequestration potential of forests by comprehensiveness of C budget and stated conclusions.

CONCLUSIONS SCORE[a]	COMPREHENSIVENESS OF C BUDGET METHODOLOGY[b]			
	SCORE = 1: only one box ticked.	*SCORE = 2:* several boxes ticked, but without accounting for direct or indirect substitution.	*SCORE = 3:* several boxes ticked, with accounting for direct and indirect substitution.	*SCORE = 4:* all boxes ticked.
SCORE = 4: significant short and long-term impacts, but requiring changes to forest management and wood use.	-	-	5	3
SCORE = 3: significant, long-term impact.	-	-	6	-
SCORE = 2: significant impact, but only in the short- to medium- term.	6	17	1	-
SCORE = 1: insignificant impact.	2	2	1	-

a In terms of scores assigned in Table 2.
b In terms of scores assigned based on boxes ticked in Table 2.

Each paper in Table 19.2 is characterised by the combination of the score attributed for comprehensiveness of the C budget methodology and the score attributed for the conclusions and recommendations, as recorded in the last column of Table 19.2. Table 19.3 shows the numbers of papers for each combination of the two scores. There is an evident positive correlation between the attributed scores, in particular, a *conclusions score* of 3 or 4 is always associated with a *budget comprehensiveness score* of 3 or 4, while a *budget comprehensiveness score* of 1 or 2 is always associated with a *conclusions score* of 1 or 2. As was observed in the above analysis of British forests, inclusion of direct and/or indirect substitution as a component of the C budget methodology appears to have a strong influence on conclusions about the C sequestration potential of forests and forestry. There is little evidence in Table 19.2 to suggest that C budgets are becoming progressively more comprehensive with time.

CONCLUSIONS

Sharp inconsistencies and diametric viewpoints are revealed by an analysis of published conclusions and recommendations about the potential for enhancing C sequestration by forests through changes in forest management. Much but not all of this apparent disagreement is attributable to differences in the C budgeting methodologies adopted in different studies. Results from studies which have accounted for direct and indirect substitution by wood products suggest that changes in wood utilisation are potentially more important than changes in forest management in terms of C sequestration. The sensitivity of results to changes in underlying budgeting methodology places considerable onus upon modellers, forest managers and forest policy-makers to ensure that methodologies are adopted that are consistent with stated objectives, and to evaluate published C budget statistics critically and objectively. A clear and correct picture of the extent to which forest management can be modified to enhance C sequestration by forests will only emerge when conventions are formulated for the modelling and reporting of C budgets for forests and their management. These conventions will need to be agreed and accepted by scientists, foresters and policy-makers. Although much work still needs to be done, it is already apparent that models of the impact of forest policy and management on C fluxes need to consider impacts not only on the C pool in the forest itself, but also on other C pools very distant from the forest. The diversity of forest types, their many economic, social and environmental roles, and the complexity of the system of C flows present obstacles to the development of a single standard methodology for C budgeting.

ACKNOWLEDGEMENTS

The author wishes to acknowledge the advice of two anonymous referees, the editorial comments of Mike Apps and Caroline Johns, and the assistance of June Bell with the formatting of tables.

REFERENCES

Adger WN, Brown K, Shiel RS, Whitby MC (1991) Dynamics of land use change and the carbon balance. ESRC Countryside Change Initiative Working Paper 15. University of Newcastle-on-Tyne, Newcastle, UK, 42 pp.

Adger WN, Brown K, Shiel RS, Whitby MC (1992) Carbon dynamics of land use in Great Britain. *J. Environ. Manage.* 36: 117–133.

Alban DH, Perala DA (1992) Carbon storage in Lake States aspen ecosystems. *Can. J. For. Res.* 22: 1107–1110.

Armentano TV (1984) Effects of increased wood energy consumption on carbon storage in forests of the United States. *Environ. Manage.* 8: 529–538.

Bekkering TD (1992) Using tropical forests to fix atmospheric carbon: the potential in theory and practice. *Ambio* 21: 414–419.

Birdsey RA, Plantinga AJ, Heath LS (1993) Past and prospective carbon storage in United States forests. *For. Ecol. Manage.* 58: 33–40.

Breuer G (1979) Can forest policy contribute to solving the CO_2 problem? *Environment International*, 2: 449–451.

Brown S, Lugo AE, Iverson LR (1992) Processes and lands for sequestering carbon in the tropical forest landscape. *Water, Air, Soil Pollut.* 64: 139–155.

Cannell MGR, Dewar RC (1995) The carbon sink provided by plantation forests and their products in Britain. *Forestry* 68: 35–48.

Cooper C (1983) Carbon storage in managed forests. *Can. J. For. Res.* 13: 155–166.

Cropper WP, Ewel KC (1983) Computer simulation of long-term carbon storage patterns in Florida slash pine plantations. *For. Ecol. Manage.* 6: 101–114.

Cropper WP, Ewel KC (1987) A regional carbon storage simulation for large-scale biomass plantations. *Ecol. Model.* 36: 171–180.

Dewar RC (1990) A model of carbon storage in forests and forest products. *Tree Physiol.* 6: 417–428.

Dewar RC (1991) Analytical model of carbon storage in trees, soils and wood products of managed forests. *Tree Physiol.* 8: 239–258.

Dewar RC, Cannell MGR (1992) Carbon sequestration in the trees, products and soils of forest plantations: an analysis using UK examples. *Tree Physiol.* 11: 49–71.

Dyson FJ (1977) Can we control the carbon dioxide in the atmosphere? *Energy* 2: 287–291.

Dyson FJ, Marland G (1979) Technical fixes for the climatic effects of CO_2. In: Elliott WP, Machta L (eds) Workshop on the global effects of carbon dioxide from fossil fuels, Miami Beach, Florida, March 7–11, 1977, US Department of Energy, CONF-770385, US Department of Energy, Washington DC, pp 111–118.

Fischlin A, Bugmann HKM (1994) Think globally, act locally! A small country case study in reducing net CO_2 emissions by carbon fixation policies. In: Kanninen M (ed) Carbon balance of the world's forested ecosystems: towards a global assessment. Proceedings of a workshop held by the Intergovernmental Panel on Climate Change AFOS, Joensuu, Finland, 11–15 May 1992, Painatuskeskus, Helsinki, pp 256–266.

Freedman B, Meth F, Hickman C (1992) Temperate forest as a carbon-storage reservoir for carbon dioxide emitted by coal-fired generating stations. A case study for New Brunswick, Canada. *For. Ecol. Manage.* 55: 15–29.

Graudal L (1991) The greenhouse effect and carbon storage in forests and forest products with special reference to Danish forests. *Dansk Skovbrugs Tidsskrift* 76: 41–62.

Gregor HD (1992) The potential of temperate forests as sinks for CO_2—examples from the German environmental policy against global warming. Should we store carbon in trees? *Water, Air, Soil Pollut.* 64: 197–212.

Hall DO, Mynick HE, Williams RH (1990) Carbon sequestration versus fossil fuel substitution. Alternative roles for biomass in coping with greenhouse warming. Princeton University Center for Energy and Environmental Studies Report 255, Princeton University, Princeton, NJ, 44 pp.

Hall DO, Mynick HE, Williams RH (1991) Cooling the greenhouse with bioenergy. *Nature* 353: 11–12.

Harmon ME, Ferrell WK, Franklin JF (1990) Effects on carbon storage of conversion of old-growth forests to young forests. *Science* 247: 699–701.

Heil GW, Gerrits M, Janssen LHJM, Weenink JB (1994) A simulation model to evaluate CO_2 sink enhancement—modelling the carbon storage and the carbon balance of forested ecosystems in the Netherlands. In: Kanninen M (ed) Carbon balance of the world's forested ecosystems: towards a global assessment. Proc. Workshop Intergovernmental Panel on Climate Change AFOS, Joensuu, Finland, 11–15 May 1992, Painatuskeskus, Helsinki, pp 215–239.

Hendrickson OQ (1990) How does forestry influence atmospheric carbon? *Forestry Chron.* 66: 469–472.

Hoen HF, Solberg B (1994) Potential and economic efficiency of carbon sequestration in forest biomass through silvicultural management. *For. Sci.* 40: 429–451.

Jarvis PG (1989) Atmospheric carbon dioxide and forests. *Phil. Trans. Roy. Soc. Lon.* B 324: 369–392.

Johnson MG, Kern JS (1991) Sequestering carbon in soils: a workshop to explore the potential of mitigating global climate change, 26–29 February 1990, Corvallis, Oregon, USA. USEPA Report EPA/600/3-91/031. USEPA, Corvallis, OR, 85 pp.

Kayes RJ, Arden-Clarke C, Taylor PJ (1990) An assessment of the feasibility of large-scale afforestation in Britain to offset carbon dioxide emissions. Political Ecology Research Group Research Report RR-19. Political Ecology Research Group, Oxford, 98 pp.

Koch P (1992) Wood versus nonwood materials in U.S. residential construction: some energy-related global implications. *For. Prods. J.* 42: 31–42.

Maclaren JP, Wakelin SJ (1991) Forestry and forest products as a carbon sink in New Zealand. Forest Research Institute Bulletin 162. FRI, Rotorua, New Zealand, 360 pp.

Marland G (1988) The prospects of solving the CO_2 problem through global afforestations. Oak Ridge National Laboratory Report DOE/NBB-0082. National Technical Information Service, Springfield, TN, 66 pp.

Marland G, Marland S (1992) Should we store carbon in trees? *Water, Air, Soil Pollut.* 64: 181–195.

Matthews RW (1992a) Forests and arable energy crops in Britain: can they help stop global warming? In: Richards GE (ed) Wood: Fuel for Thought. Proc. Conf. Energy Technology Support Unit, Bristol, England, 23–25 October 1991, Harwell, UK, pp 39–62.

Matthews RW (1992b) Role of carbon budgeting in assessing environmental costs and benefits of forestry schemes. In: Richards GE (ed) Wood: Energy and the Environment. Proc. Conf. Energy Technology Support Unit, Harrogate, England, 9–10 September 1992, Harwell, UK, pp 21–32.

Matthews RW (1994) Towards a methodology for the evaluation of the carbon budget of forests. In: Kanninen M (ed) Carbon balance of the world's forested ecosystems: towards a global assessment. Proc. Workshop Intergovernmental Panel on Climate Change AFOS, Joensuu, Finland, 11–15 May 1992, Painatuskeskus, Helsinki, pp 105–114.

Nabuurs GJ, Mohren GMJ (1993) Carbon fixation through forestation activities. Institute for Forestry and Nature Research Report 93/4. FACE/Institute for Forestry and Nature Research, Wageningen, The Netherlands, 205 pp.

Sampson RN (1992) Forestry opportunities in the United States to mitigate effects of global warming. *Water, Air, Soil Pollut.* 64: 157–180.

Schroeder P (1992) Carbon storage potential of short rotation tropical tree plantations. *For. Ecol. Manage.* 50: 31–41.

Swisher JN (1992) Cost and performance of CO_2 storage in forestry projects. *Biomass and Bioenergy,* 1: 317–328.

Thompson DA, Matthews, RW (1989) The storage of carbon in trees and timber. Forestry Commission Research Information Note 160. Forestry Commission, Edinburgh, 5 pp.

Turner DP, Lee JJ, Koerper GJ, Barker JR (1993) The forest sector carbon budget of the United States: carbon pools and flux under alternative policy options. US Environmental Protection Agency Report EPA/600/3-93/093. USEPA, Washington DC, 202 pp.

Vitousek PM (1991) Can planted forests counteract increasing atmospheric carbon dioxide? *J. Environ. Qual.* 20: 348–354.

Winjum JK, Dixon RK, Schroeder PE (1992) Estimating the global potential of forest and agroforest management practices to sequester carbon. *Water, Air, Soil Pollut.* 64: 213–227.

20 Significance of wood products in forest sector carbon balances

Gert-Jan Nabuurs

Institute for Forestry and Nature Research
Bosrandweg 20, PO Box 23, NL-6700 AA Wageningen, The Netherlands

INTRODUCTION

Global forests are thought to be important in mitigating the increase in atmospheric carbon dioxide (Dixon et al. 1991; Kauppi et al. 1992; Sampson et al. 1993). The gross global fluxes of CO_2 between the atmosphere, biosphere and oceans are huge compared to the net annual increase in atmospheric CO_2. This suggests that relatively small changes in these gross vegetation-atmosphere fluxes might change the net increase in the atmospheric CO_2 concentration. To elucidate, very many studies have focused on the role of the biosphere in the global carbon (C) cycle (Apps et al. 1993; Armentano and Ralston 1980; Auclair and Bedford 1993; Houghton 1991; Lugo and Brown 1992; Sedjo 1992; Tans et al. 1990) and on the C sequestering potential of different forest types (Brown et al. 1985; Dewar and Cannell 1992; Dixon et al. 1991; Mohren and Klein-Goldewijk 1990a; Nabuurs and Mohren 1993; Schroeder and Ladd 1991; Winjum et al. 1992).

Only a few studies go into detail concerning the role of wood products in forest sector C balances (Karjalainen et al. 1994), and the results of the previously mentioned studies are difficult to compare because of discrepancies in the methods used, compartments of the forest ecosystem considered, and criteria for evaluating the C storing capacity. Most of the studies tend to concentrate on intensive, short rotation silvicultural systems with mean annual increments that are only realistic on the best sites.

This paper describes a study that attempted to give an overview of the C sequestering potential of representative and realistic world forest ecosystems. In addition, the significance of the evaluation criteria and the significance of the wood products are examined.

METHODS

The C sequestering potentials of ten representative world forest ecosystems were calculated using the CO2FIX model, developed by Mohren and Klein-Goldewijk (1990b) to quantify the C budget of a given forest type at the stand level, i.e., per hectare per year. The model comprises all tree organs, the forest floor, stable humus and wood products (the latter divided into five groups). It produces an annual output of stocks and fluxes of C for the various biomass, wood product and soil compartments.

The model was parameterised with data on stem growth from yield tables, forest biomass and forest soil C obtained from sources including Cannell (1982), Kimmins et al. (1985), Millemann and Boden (1985, 1986) and de Vries et al. (1990).

NATO ASI Series, Vol. I 40
Forest Ecosystems, Forest Management and the Global Carbon Cycle
Edited by Michael J. Apps and David T. Price

CO2FIX describes the C cycle from annual growth and loss rates of the main biomass compartments of the forest ecosystem, in combination with the accumulation and turnover of soil organic matter. Harvested wood is divided into five product groups in proportions depending on diameter and tapering. An average life span for each product group can be specified. If the products are used once only, an exponential decay function is assumed. If recycling is stipulated, the total amount of products is transferred to a product group of a lower quality at the end of their hypothetical life time. For example, it is assumed that the total amount of wood in spruce sawn timber (saw losses have already been deducted and are used as fuelwood) can be re-used as particle board and can be re-used again as fuelwood. In this study the average residence times of C in the once-only use option were assumed to be 1 year for fuelwood, 2 years for paper, 3 years for packing wood, 20 years for particle board and 35 years for sawn timber. In the recycling scenario, the life spans of these products were assumed to be 0.5, 0.5, 2 and 5 years respectively. The life span of sawn timber depends on the tree species; it was set at 18 years for spruce and poplar wood and at 45 years for oak and beech wood. A carbon content of 50% was used for living biomass and litter (Ajtay et al. 1977) and of 58% for stable humus (Buringh 1984).

The reductions in C emissions achieved through material substitution and using wood to produce energy were calculated for three forest types by comparing the amount of energy necessary to produce the wooden materials with the amount necessary to manufacture their plastic, steel or concrete alternatives. Table 20.1 gives the wood products, their alternatives and the reduction in C emissions per m^3 of wood product. For details on the data used, see Sikkema and Nabuurs (1994). The volume required, material density, energy required for manufacturing and the C emission factor were taken into account when comparing the products. Table 20.2 gives the reduction in C emissions per m^3 wood (on the basis of energy content) that were used to calculate the reduction in emissions obtained by using wood products as a source of energy instead of using coal.

Table 20.1. Comparison of wood products, their alternatives and the reduction in carbon (C) emission per m^3 product when wood is used instead of the alternative (Burschel et al. 1993; Sikkema and Nabuurs 1994).

Wood product group	Product	Alternative	Reduction in C emission (Mg C m^{-3})[a]
Sawn timber	Oak sleeper	Concrete	0.30
	Spruce window frame	Steel	0.31
	Oak furniture	Steel	0.24
	Oak parquet	Carpet (polyamide)	0.50
	Spruce construction wood	Steel	0.29
Packing wood	Spruce, poplar palette	Plastic (polyethylene)	0.15
Particle board	Particle board from spruce and beech	No alternative	note b
Paper	Paper from spruce and poplar	No alternative	note b

a For details on harvested volumes, allocation to wood product groups, energy required to manufacture the wood product, energy required to manufacture the alternative and C emission per Gigajoule, see Sikkema and Nabuurs (1994).

b Since there are no substitutes for these two materials, the required energy in their manufacture has been subtracted from the energy released in their combustion (See Table 20.2).

Criteria for evaluating the carbon sequestering potential of forest types

When evaluating a forest type for its potential to sequester C, both the fluxes and the stocks should be considered. Only then is it possible to evaluate the efficacy of silvicultural systems for sequestering atmospheric C. In this study, C stocks (represented in Mg C ha^{-1}) were subdivided into 4 criteria. Only one criterion was used for net C fluxes (in Mg C ha^{-1} yr^{-1}). The criteria are:

1. Long-term average stock of C in biomass and forest products
2. Long-term average stock of C in forest products;
3. Long-term average stock of C in soil organic matter;
4. Total stock of C at the end of the rotation in biomass and products;
5. Average net annual C storage flux over the first rotation.

Forest types and model input

The forest types studied were:

1. Regrowing, heavily logged evergreen tropical lowland Dipterocarp rainforest (*Shorea* spp.) in Indonesia in three site classes in rotations of 70 years with mean annual increment (MAI) values over the regrowing period of 11.7, 14.1 and 18.7 m^3 ha^{-1} yr^{-1}.
2. Regrowing, selectively logged evergreen tropical lowland Dipterocarp rainforest (*Shorea* spp.) in Indonesia in three site classes in rotation of 70 years with MAI values over the regrowing period of 12.4, 15.8 and 18.1 m^3 ha^{-1} yr^{-1}.
3. Afforestation of tropical wastelands in Indonesia with *Albizia falcataria* as a pioneer and underplanting with Dipterocarps (*Shorea* spp.) in a rotation of 70 years and an MAI value in the *Shorea* stand of 7.0 m^3 ha^{-1} yr^{-1}.
4. Norway spruce (*Picea abies*) forests in montane central Europe in rotations of 120 years with MAI values of 5.8, 9.3 and 14.2 m^3 ha^{-1} yr^{-1} (Schober 1975).
5. Mixed deciduous forests of pedunculate oak (*Quercus robur*) and beech (*Fagus sylvatica*) in central Europe in rotations of 150 years with MAI values of 3.5, 5.4 and 8.1 m^3 ha^{-1} yr^{-1} (Wiedemann 1942; Schober 1975).
6. Douglas-fir (*Pseudotsuga menziesii*) forests in the Pacific Northwest of the USA in rotations of 100 years with MAI values of 11.5, 14.9 and 18.6 m^3 ha^{-1} yr^{-1} (Curtis et al. 1982).
7. Norway spruce (*Picea abies*) forests in the boreo-nemoral zone of Russia in rotations of 100 years with MAI values of 3.2, 4.6 and 6.0 m^3 ha^{-1} yr^{-1} (Kazimirov and Morozova, cited in Cannell 1982).
8. Poplar (*Populus* × *euramericana*) on former agricultural land in western Europe in rotations of 45 years with MAI values of 5.5, 10.4 and 13.2 (Sevenster 1991).
9. Radiata pine (*Pinus radiata*) in New Zealand and Australia in rotations of 40 years with MAI values of 17.5, 22.2 and 27.0 m^3 ha^{-1} yr^{-1} (Lewis 1954, cited in Scott 1960).
10. Slash pine (*Pinus elliottii*) in Brazil in rotations of 30 years with MAI values of 17.2, 26.0 and 34.8 m^3 ha^{-1} yr^{-1} (Schneider 1984).

For more details on the input data, see Nabuurs and Mohren (1993).

Table 20.2. Combustion energy of various materials (GJ Mg^{-1}, assuming 10% moisture content) and the reduction in carbon emission achieved by using them instead of coal to produce the equivalent amount of energy (Sikkema and Nabuurs 1994).

MATERIAL	COMBUSTION ENERGY ($GJ\ Mg^{-1}$)	C EMISSION REDUCTION ($Mg\ CO_2\ m^{-3}$)
Wood waste	16.0	0.86
Oak/beech fire wood	14.8	0.96
Spruce fire wood	16.4	0.70
Poplar fire wood	15.8	0.66
Paper	14.0	0.46
Particle board	16.1	0.69
Coal	29.3	–

RESULTS

The ten forest types considered differ greatly in the C stocks contained in the biomass and soil organic matter at the end of a rotation (Figure 20.1) and in long-term average stocks of C (column 1, Table 20.3). Selectively logged evergreen rainforest seems to contain the largest long-term average stock of C in biomass and products: 207 Mg C ha^{-1}. Using selective rather than conventional logging techniques may result in an additional storage of 63 Mg C ha^{-1}. A widespread silvicultural system, for example the Norway spruce forests in central Europe, may contain a long-term average stock of C in biomass and products of 137 Mg C ha^{-1}. The amount of C in forest biomass on sites of limited fertility (as chosen in this study) is approximately 60–65% of the amount in forests on very fertile sites. The soil organic matter compartments of litter, stable humus and dead wood under each forest type contain considerable stocks of C.

The long-term average stocks in wood products (column 2, Table 20.3) are small compared to the stocks in forest biomass and soil organic matter. The wood products pool is relatively important in the forest types that deliver a wide range of products, e.g., Norway spruce, slash pine and radiata pine plantations. The stocks of C in the wood products as a percentage of the stocks in the forest biomass range from 10% in the selectively logged tropical rainforest, to 66% in the slash pine forest type.

The short rotation forest types perform well according to the criterion of net annual fixation during the first rotation. The highest net annual C fixation in the first rotation (a storage rate of 4.54 Mg C ha^{-1} yr^{-1}) is achieved with *Pinus radiata* in Brazil. However, this value is only valid for the first rotation, because the net C flux tends to zero in the long term.

The general conclusion from these results is that a forest type may show a high potential for C sequestration according to one criterion, but may show the opposite effect according to another criterion. It is therefore very important to always show the sequestering potential according to a set of criteria. Which criterion is deemed to have priority over another is a political issue. It also depends on the goals that have been defined. If results have to be achieved within a short period, the average net annual C flux over the first rotation is a very important criterion.

The effects of product recycling (Table 20.4) and of the reduction in C emissions achieved through material substitution and production of fuelwood (Table 20.5) were evaluated for mixed deciduous, poplar and Norway spruce forests.

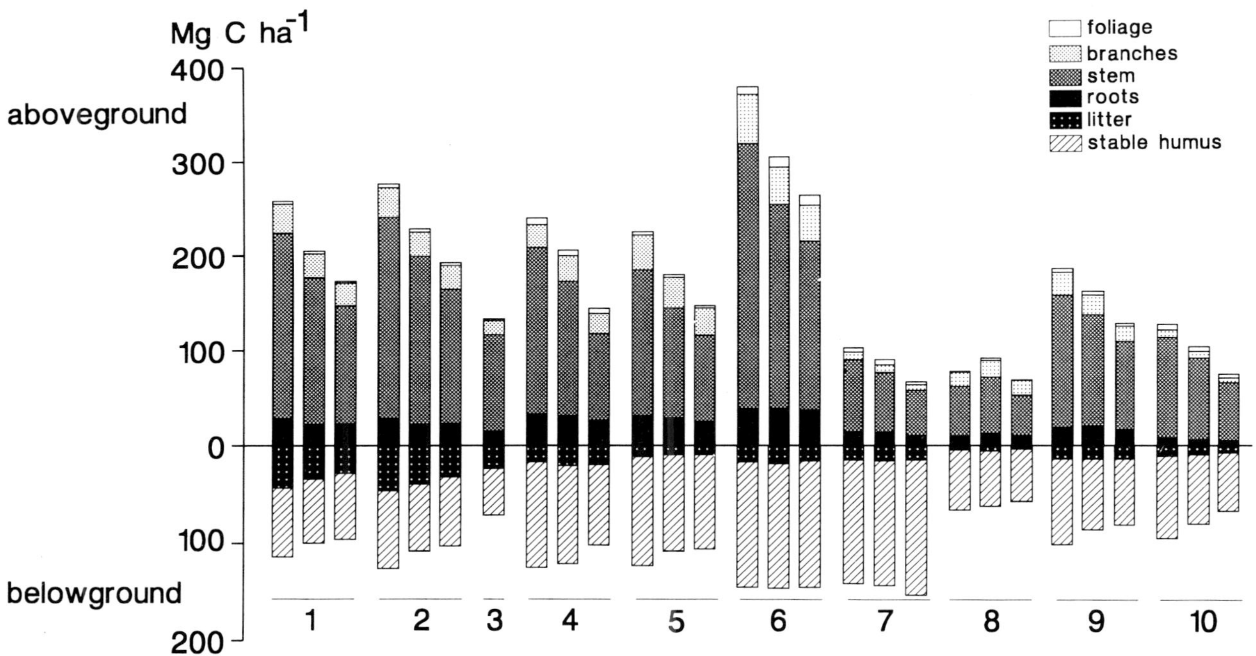

Figure 20.1. Bar diagram showing the stocks of carbon in each compartment at the end of a rotation. The left-hand bar of each forest type represents the fertile site, the bar in the middle the sites of moderate fertility and the right hand bar the sites of limited fertility. The forest types are: 1. Heavily logged evergreen tropical rainforest in Indonesia; 2. Selectively logged evergreen tropical rainforest in Indonesia; 3. Afforestation of tropical wastelands in Indonesia with *Albizia falcataria* as a pioneer; 4. Norway spruce in montane central Europe; 5. Pedunculate oak and beech in montane central Europe; 6. Douglas-fir in the Pacific Northwest of the USA; 7. Norway spruce in the boreo-nemoral zone of Russia; 8. Poplar on former agricultural land in western Europe; 9. Radiata pine in New Zealand and Australia; 10. Slash pine in Brazil.

The effect of the full recycling scenario varies with forest type. The three forest types of Table 20.4 represent: a long rotation forest type from which considerable amounts of sawnwood can be produced, but which can produce large amounts of firewood; a short rotation forest type which mainly produces paper and packing wood and a forest type from which a wide range of products can be produced, with a large proportion of sawn wood. The absolute stocks in the wood product pool remain small and the effect of recycling would have been less if the energy necessary for recycling had been taken into account. However, the relative increase in the C stock in wood products can be very high: 29% for the deciduous forest type, 167% for poplar and 73% for Norway spruce (Table 20.4). The allocation of oak wood to various products is limited and only a small amount of the harvested wood of the final cut (23%) is allocated to the sawnwood product group. Thus recycling can only be applied to a small amount of wood, resulting in a small increase in the long-term average stocks sequestered in products. In contrast, in the Norway spruce forest type almost all parts of the harvested wood can be used for some kind of product (usually with smaller sawing losses). This results in a larger quantity of wood products and thus in better results in the case of recycling. The same is true for the poplar forest type.

So far, only the results of the C storing capacity of the forest types have been presented. The conclusion is that forest types that yield the widest variety of wood products store the largest stocks of C in products, and that long-rotation forest types contain the largest long-term average stocks of C in the living biomass. The wood products compartment usually contains one third of the C in the forest biomass and is relatively important in the intensive pine plantations. The contribution of the wood products can be increased by recycling wood products. However, the results change drastically when the reduction of C emission is also taken into account (Figures 20.2 and 20.3, and Table 20.5).

Table 20.3. Carbon (C) sequestering potential of the ten forest ecosystems as indicated by the five criteria. Carbon stocks are represented in Mg C ha^{-1} and C fluxes in Mg C ha^{-1} yr^{-1}. The results are valid for sites of moderate fertility.

		CRITERIA[a]				
FOREST TYPE	MAI[b]	1	2	3	4	5
Heavily logged evergreen rainforest	14.1	144	16	92	206	2.40
Selectively logged evergreen rainforest	15.8	207	19	102	229	2.85
Afforested waste land	9.3	121	12	67	133	1.69
Norway spruce in central Europe	9.3	137	35	117	225	2.02
Mixed deciduous central Europe	5.4	110	20	105	199	1.35
Douglas-fir NW USA	14.9	196	46	143	327	3.43
Boreal spruce	4.6	53	13	139	95	1.03
Poplar	10.5	62	15	75	93	2.23
Radiata pine	22.2	126	35	97	171	4.34
Slash pine	26.0	111	44	80	112	3.88

a Column headings represent the following criteria:
1 = Long-term average stock of C in forest biomass and forest products.
2 = Long-term average stock of C in forest products alone.
3 = Long-term average stock of C in soil organic matter.
4 = Total stock of C at the end of the rotation in forest biomass and products.
5 = Average net annual C storage flux over the first rotation.

b MAI is mean annual increment for sites of moderate fertility (m^3 ha^{-1} yr^{-1}).

Table 20.4. Comparison of the long-term average stocks of carbon (C) in wood products under a once-only use scenario versus an optimal recycling scenario (Mg C ha^{-1}).

FOREST TYPE	ROTATION LENGTH [a] (yr)	MAI (m^3 ha^{-1} yr^{-1})	PRODUCT C STOCKS (ONCE–ONLY USE)	PRODUCT C STOCKS (OPTIMAL RECYCLING)
Mixed deciduous				
Oak/beech	100	6	7	9
Poplar	15	15	3	8
Norway spruce	75	12	19	33

a These rotation lengths are somewhat shorter than the ones used in Table 20.3, resulting in smaller C stocks under the scenario of once–only use in comparison to the results of Table 20.3.

Table 20.5. Potential contribution to carbon (C) storage and emission reduction (Mg C ha^{-1} 300 yr^{-1}) of four forest types (Sikkema and Nabuurs 1994).

		CONTRIBUTION CLASS[a]			
FOREST TYPE	ROTATION (yr)	1	2	3	4
Mixed deciduous oak/beech	100	110	50	263	423
Norway spruce	75	97	214	352	663
Poplar	15	28	178	425	631
Poplar	5	-29[b]	0	1579	1550

a The column headers represent:
1 Long-term average stocks of C in forest biomass and wood products.
2 Reduction of C emission achieved through material substitution.
3 Reduction of C emission achieved through production of fuelwood.
4 Total C storage and emission reduction in 300 years.

b Manufacture of fertiliser used in this forest type accounts for 6.0 GJ ha^{-1} yr^{-1}.

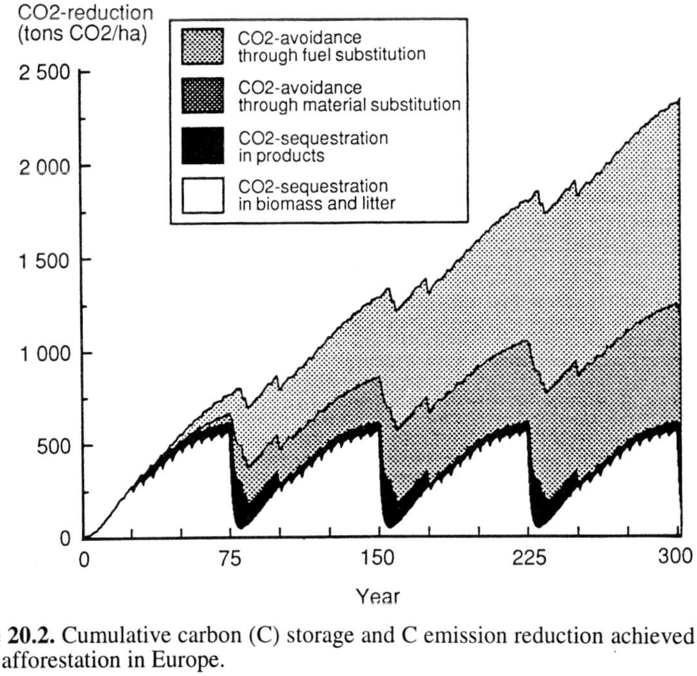

Figure 20.2. Cumulative carbon (C) storage and C emission reduction achieved by a Norway spruce afforestation in Europe.

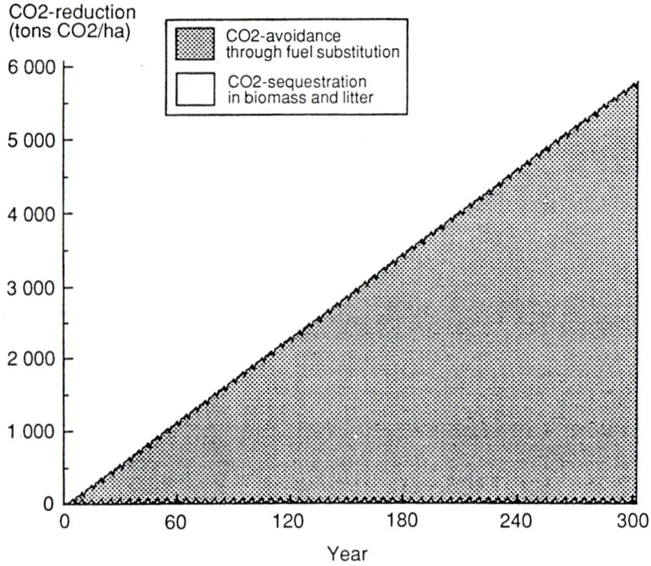

Figure 20.3. Cumulative carbon (C) storage and the C emission reduction achieved by a short-rotation poplar fuelwood plantation in Europe.

Including the C emission reduction factor changes the C sequestering potential of the forest types drastically. The total effect including the emission reduction gives a positive correlation with the total production of biomass in the period of 300 years. The short rotation forest type can reach the highest MAI in a period as short as 5 years. The cumulative effect over 300 years is therefore the largest in this forest type. The significance of the wood products also changes drastically. In all forest types, whether long or short rotation, the influence of wood products eventually becomes predominant mainly because of the production of wood for energy. This underlines the importance of the criteria chosen to evaluate forest types for their C storage capacity.

DISCUSSION

Reliability of results

The possible sources of uncertainty in the outcome of the calculations as presented here are: (1) the input parameters (a lack of sufficiently detailed data) and (2) lack of knowledge of forest growth and development under future site and climatic conditions. Uncertainty analysis (van der Voet and Mohren 1994) can be used to deal with (1). Monte Carlo sampling of the parameter space, specified from information on the variation in parameter estimates, was applied to the CO2FIX model and the outcome will be dealt with in detail in another article (for preliminary results on one forest type seen Nabuurs and Mohren 1993). The second uncertainty is directly associated with the use of a dynamic model in which a limited number of assumptions are assembled in order to be able to make predictions. The predictions about

C storage in the forest ecosystem rely heavily on assumptions about the stability of site conditions for growth. The simulation runs for a period of 300 years and assumes that the site conditions remain the same as in the initial situation, i.e., that site productivity is maintained, that there is no climate change or soil acidification and no major catastrophes occur. The validity of these assumptions can be disputed, however, since several recent publications indicate that the growth rates of large areas in Europe are higher than previously expected (Kauppi et al. 1992). These phenomena have not been taken into account in the analysis presented here.

Importance of evaluation criteria

Column 1 in Table 20.3 is an important criterion for evaluating the C sequestering potential of forest types, because this is the long-term (300 years) average amount which is withdrawn from the atmosphere and stored in biomass and products as a result of afforestation. Since this is a running average, it is virtually insensitive to the point in time from which it is presented. It quickly rises to a level close to its equilibrium.

The criterion "net annual C flux" (Table 20.3, column 5) is only valid for the first rotation, because net storage stops when the long-term average amount of C in biomass, products and soil reaches equilibrium. Although this criterion is often used in evaluations of C storage potential, it is somewhat misleading, because it will always tend to zero in the long-term. It is even more misleading when it is presented for short rotation, intensive silvicultural systems, because then this criterion scores very high.

Table 20.3 implies that certain forest types can perform very well according to the criterion "net annual C flux", but perform in a limited way according to the criterion "average amount of C in the biomass and products". It is important to distinguish between these two criteria. Fast growing short rotation plantations usually show high net annual C fluxes during a short period, but they soon reach their relatively low equilibrium biomass and the high flux quickly declines to zero (if we disregard the effect of the C emission reduction).

The importance of a generally accepted set of evaluation criteria is clearly shown by the effect of taking the emission reduction factor into account as well. Doing so results in the preference for certain forest types changing completely from long rotation types to short rotation forest types.

The set of evaluation criteria presented here is not yet complete. It does not include the cost efficiency versus C efficiency, i.e., the amount of C that can be stored for a certain sum of money. If a forest type has a credit balance, C storage might even be lucrative. However, efficacy of a forest type can be calculated in many ways and results are often confusing and difficult to compare (Richards and Stokes 1995). A truly comprehensive cost-benefit analysis should even take into account the rate of return of the prevented damage due to the afforestations that were carried out. The evaluation criteria could be further expanded by taking into account only the marginal effects of the afforestation, i.e., comparing the C storage results with the anticipated natural regeneration.

The other important criteria that should always be taken into account when forest types are evaluated are sustainability, biological diversity and degree of soil protection. These are often neglected when forest types are evaluated for their C sequestering capacity. The analysis as presented here is still not complete. Although all major C pools were taken into account, the energy necessary for planting, management and transport was assumed to be negligible in the forest part of the balance. However, the energy required to manufacture wood products and their alternatives has indeed been taken into account when comparing them for their potential to reduce C emission.

A proper comparison of these results with results presented by others remains difficult because of differences in model parameters and in the criteria used to evaluate the C storing capacity. For example, the model presented by Dewar (1990) which is the most similar to CO2FIX, nevertheless contains very different parameters. CO2FIX uses a specified lifetime per wood product group, whereas Dewar's model uses one lifetime for wood products, i.e., the time taken to reach the maximum mean annual increment in the stand (MAI_{max}). In CO2FIX, the C content in organs other than the stem is set by a combination of relative allocation with turnover rates, whereas Dewar's model uses asymptotic amounts for these organs. CO2FIX uses a litter residence time, a litter humification factor and a stable humus residence time for the soil compartment. Dewar's model uses a fractional decomposition of certain groups of material in the litter layer, together with a fractional decomposition of soil organic matter. For a more detailed comparison between the results presented here and results from others, see Nabuurs and Mohren (1995).

OVERALL CONCLUSIONS AND RECOMMENDATIONS

This study indicates that the wood products compartment is of moderate interest when only its C stocks are considered in comparison to the stocks in the forest biomass. The wood products compartment usually contains one third of the C in the forest biomass. This study also showed that the largest long-term average stocks in forest biomass and products can be achieved in the long rotation forest types. However, the outcome changes drastically when the reduction in C emission by material substitution and by production of fuelwood is taken into account. The best results are achieved in short rotation fuelwood plantations, the forest type with the highest mean annual increment. Recycling of wood products can also increase the significance of wood products as a C stock. Optimal recycling may increase the C stock in products by as much as 160%. However, the absolute stock in products may remain small. It is doubtful whether full recycling can be achieved, but the results confirm the worth of a recycling strategy.

When evaluating C sequestering possibilities in forests, there is a tendency in the international literature to concentrate on highly productive forest types on fertile sites. Results may then be extrapolated to large areas even though their applicability to these large areas is dubious.

Evaluation for C storage has tended to ignore criteria such as socio-economic and ecological sustainability. Including these criteria would certainly improve the set of criteria used in this study. In any case there is a clear need to standardise the criteria and methods used to assess the C sequestering potential of forest types, to allow comparison of these assessments. Whatever the result of using a standard set of criteria, ultimately the decision about which forest to plant will always be a political issue in which the original criteria are reassessed in the light of their political importance. As this study showed, the wood products part of the C balance became very important when reduction in C emission is also taken into account. Therefore, government policies ought to stimulate the use of wood products from sustainably managed forests in preference to energy-consuming alternatives. Furthermore, they should stimulate the recycling of wood products to help achieve a favourable C balance.

ACKNOWLEDGEMENTS

This chapter is based on a study of the carbon sequestering potentials of 13 global forest types commissioned by the FACE Foundation (Forests Absorbing Carbondioxyde [sic] Emission) and on a study of carbon emission reduction potential of Dutch forest types carried out in co-operation with the Foundation for Forest and Wood (SBH) in Wageningen, the Netherlands. The author wishes to thank Richard Sikkema for his work in the study of carbon emission reduction and Frits Mohren for supervision during the various projects.

REFERENCES

Ajtay GL, Ketner P, Duvigneaud P (1977) Terrestrial primary production and phytomass. In: Bolin B, Degens ET, Kempe S, Ketner P (eds) The Global Carbon Cycle. SCOPE 13, pp 129–181.

Apps MJ, Kurz WA, Luxmoore RJ, Nilsson LO, Sedjo RA, Schmidt R, Simpson LG, Vinson TS (1993) Boreal forests and tundra. *Water, Air, Soil Pollut.* 70: 39–53.

Armentano TV, Ralston CW (1980) The role of temperate zone forests in the global carbon cycle. *Can. J. For. Res.* 10: 53–60.

Auclair A, Bedford J (1993) New perspectives on the terrestrial carbon flux: importance to the policy debate. ORD EPA Technical Briefing. April 28, (1993) Environmental Protection Agency, Washington DC, 4 pp.

Brown S, Lugo AE, Chapman J (1985) Biomass of tropical tree plantations and its implications for the global carbon budget. *Can. J. For. Res.* 16: 390–394.

Buringh P (1984) Organic carbon in soils of the world. In: Woodwell GM (ed) The Role of Terrestrial Vegetation in the Global Carbon Cycle: Measurement by Remote Sensing. SCOPE 23, pp 91–110.

Burschel P, Kürsten E, Larson BC (1993) Die Rolle von Wald und Forstwirtschaft im Kohlenstoffhaushalt. Eine Betrachtung für die Bundesrepublik Deutschland. (Role of Forests and forestry in the carbon cycle; a case-study for Germany.) Forstliche Forschungsberichte München. Schriftenreihe der Forstwissenschaftlichen fakultät der Universität München und der Bayer. Forstlichen Versuchs und Forschungsanstalt. 126 (1993) 135 pp.

Cannell MGR (ed) (1982) World forest biomass and primary production data. Natural Environment Research Council. Institute of Terrestrial Ecology. Academic Press, London-New York, 391 pp.

Curtis RO, Clendenen GW, Reukema DL, De Mars DJ (1982) Yield tables for managed stands of coast Douglas-fir. U.S. Dept. Agric. For. Ser., Gen. Tech. Rep. PNW-135, Pacific Northwest Forest and Range Experiment Station, Portland, OR, 182 pp.

Dewar RC (1990) A model of carbon storage in forests and forest products. *Tree Physiol.* 6: 417–428.

Dewar RC, Cannell MGR (1992) Carbon sequestration in the trees, products and soils of forest plantations: an analysis using UK examples. *Tree Physiol.* 11: 49–71.

Dixon RK, Schroeder PE, Winjum JK (1991) Assessment of promising forest management practices and technologies for enhancing the conservation and sequestration of atmospheric carbon and their costs at the site level. EPA/600/3-91/067. United States Environmental Protection Agency, Office of Research and Development, Washington DC, 138 pp.

Houghton RA (1991) Tropical deforestation and atmospheric carbon dioxide. *Climatic Change* 19: 99–118.

Karjalainen T, Kellomäki S, Pussinen A (1994) Role of wood-based products in absorbing atmospheric carbon. *Silva Fennica* 28(2): 67–80.

Kauppi PE, Mielikainen K, Kuusela K (1992) Biomass and carbon budget of European forests, 1971 to 1990. *Science* 256: 70–74.

Kimmins JP, Binkley D, Chatarpaul L, de Catanzaro J (1985) Biogeochemistry of temperate forest ecosystems: Literature on inventories and dynamics of biomass and nutrients. Canadian Forestry Service Information Report PI-X-47E/F, Petawawa National Forestry Institute, Chalk River, ON, 227 pp.

Lewis ER (1954) Yield of unthinned Pinus radiata in New Zealand. NZ For. Serv., Forest Research Note 1(10), 19 pp.

Lugo AE, Brown S (1992) Tropical forests as sinks of atmospheric carbon. *Forest Ecol. Manage.* 54: 239–255.

Millemann RE Boden TA (1985) Major world ecosystems complexes ranked by carbon in live vegetation: A database. Carbon Dioxide Information Centre, Oak Ridge National Laboratory, Oak Ridge, TN, 164 pp.

Millemann RE, Boden TA (1986) Worldwide organic soil carbon and nitrogen data. Carbon Dioxide Information Centre. Oak Ridge National Laboratory, Oak Ridge, TN, 136 pp.

Mohren GMJ, Klein Goldewijk CGM (1990a) Forestry in the Netherlands as a measure in the framework of the climate change carbon dioxide problems. Part 1. Carbon storage in forests. De Dorschkamp, Institute for Forestry and Urban Ecology, Ministry of Agriculture, Nature Management and Fisheries, Wageningen, The Netherlands, 86 pp [In Dutch].

Mohren GMJ, Klein Goldewijk CGM (1990b) CO2FIX: a dynamic model of the carbon dioxide fixation in forest stands. Institute for Forestry and Urban Ecology, Report No 624. Wageningen, The Netherlands, 35 pp.

Nabuurs GJ, Mohren GMJ (1993) Carbon fixation through forestation activities. Institute for Forestry and Nature Research. IBN Research report 93/4. Wageningen, The Netherlands, 205 pp.

Nabuurs GJ, Mohren GMJ (1995) Modelling analysis of potential carbon sequestration in selected forest types. *Can. J. For. Res.* (In press).

Richards KR, Stokes C (1995) Regional studies of carbon sequestration: A review and critique. Contribution to IPCC Working Group 3, Costs of Climate Change Mitigation (Unpublished manuscript).

Sampson RN, Apps M, Brown S, Cole V, Downing J, Kauppi P, Ojima DS, Smith TM, Solomon AM, Twilley RR, Wisniewski J (1993) Workshop Statement: Terrestrial biospheric carbon fluxes: Quantification of sinks and sources of CO_2. *Water, Air, Soil Pollut.* 70: 3–15.

Schneider PR (1984) Betriebswirtschaftliche und Ertragskundliche Grundlagen der Forsteinrichtung in Südbrasilien am Beispiel von *Pinus elliottii*. Dissertation, Faculty of Forestry, Albert Ludwigs University, Freiburg, Germany.

Schober R (ed) (1975) Ertragstafeln wichtiger Baumarten. Sauerländers Verlag, Frankfurt, Germany, 154 pp.

Schroeder P, Ladd L (1991) Slowing the increase of atmospheric carbon dioxide: a biological approach. *Climatic Change* 19: 283–290.

Scott CW (1960) *Pinus radiata*. FAO Forests and Forest Products Studies. No. 14, 328 pp.

Sedjo RA (1992) Temperate forest ecosystems in the global carbon cycle. *Ambio* 21: 274–277.

Sevenster J (ed) (1991) Yield tables for important tree species in the Netherlands. Commission yield tables. Draft report. 188 pp.

Sikkema R, Nabuurs GJ (1994) Forest and forest products on the carbon balance; study of the CO_2 emission reduction possibilities of four Dutch forest types. Dutch Foundation for Energy and Environment, NOVEM Report 9416. Foundation for Forest and Wood in co-operation with the Institute for Forestry and Nature Research, Wageningen, The Netherlands. 46 pp + app. [In Dutch, with English summary].

Tans P, Fung IY, Takahashi T (1990) Observational constraints on the global atmospheric CO_2 budget. *Science* 247: 1431–1438.

van der Voet H, Mohren GMJ (1994) An uncertainty analysis of the process-based model FORGRO. *Forest Ecol. Manage.* 69: 157–166.

de Vries W, Hol A, Tjalma S (1990) Literature study of stocks and flows of minerals in forest ecosystems. Winand Staring Centre-DLO. Report 94. Wageningen, The Netherlands, 205 pp.

Wiedemann E (1942) Der Eichenbestand mit Buchen Unterwuchs. [Oak stands with an undergrowth of beech]. *Zeitschr. Forst. Jagdw.* 74: 305–335.

Winjum JK, Dixon RK, Schroeder P (1992) Estimating the global potential of forest and agroforest management practices to sequester carbon. *Water, Air, Soil Pollut.* 64: 312–227.

21 Plantation forestry—its role as a carbon sink: conclusions from calculations based on New Zealand's planted forest estate

J Piers Maclaren

NZ Forest Research Institute, PO Box 465, Rangiora, New Zealand

INTRODUCTION

It is common to encounter misunderstanding with regard to the role of forests as a means of mitigating the postulated enhanced greenhouse effect (Houghton et al. 1990, 1992). Some of the confusion appears to lie in loose terminology. In particular, the words *carbon sink*, and *carbon reservoir* need to be clearly distinguished (see Glossary). Only if a reservoir increases in capacity does it constitute a net sink, and then only for the period during which it is expanding in size. The word *store*, used as a noun, may be synonymous with *reservoir*, but as a verb it means *to create a reservoir* or *to act as a sink*; hence use of *store* should be avoided.

Confusion also surrounds the word *forest*, which is frequently used synonymously with *stand*. Forestry terms used throughout this chapter will be as defined by the Society of American Foresters (1917) and by Ford-Robertson (1971); more important terms are italicized upon first use and definitions provided in the Glossary. Whereas a stand comprises trees of similar composition, age-structure and management regime, a forest is a larger unit that may consist of hundreds or thousands of stands. The dynamics of a forest are considerably more complex than those of a stand, as shall be explained later.

It is readily acknowledged that *all* forests are C–reservoirs, in that they contain biomass which comprises about 50% C by oven-dry mass (Rowell 1984; Kinerson et al. 1977) but it is not obvious whether any given forest is a net C sink, a C source or is C-neutral. Living trees are usually net C absorbers. In the case of a forest, net C fixation occurs in the live vegetation, i.e., excess of C obtained by photosynthesis over C lost by plant respiration, but C is also lost through decomposition or harvesting. The difference between the C entering and the C leaving the forest will determine if the forest is a net sink or a net source.

In a production forest, C leaves the forest in the form of logs or other vegetation. Some writers assume that durable wood products are a permanent C-sink, and that a steady-state production forest therefore acts as a *C pump* by adding continuously to the reservoir of C in wood products (Sedjo 1989; Hendrickson 1990; Schroeder 1992). But wood products, however long-lasting, have a finite life. The decay or combustion of such products release CO_2 in a way similar to dead wood on the forest floor. Just as with the forest itself, the difference between the C entering and leaving the forest products C pool determines the effectiveness of the sink. It will be shown later that the size of the products pool is small compared to the forest pool, and it can safely be assumed that the *change* in size of the products pool is also small compared to the actual and potential *changes* of the forest pool.

Even if the products pool is not expanding in size, in which case it does not constitute a net C-sink *per se*, it may act as a surrogate sink, as wood products substitute for non-wood products that require fossil fuel energy to process. Wood production, therefore, indirectly

NATO ASI Series, Vol. I 40
Forest Ecosystems, Forest Management and the Global Carbon Cycle
Edited by Michael J. Apps and David T. Price

reduces fossil fuel consumption (Matthews 1994). This is elaborated later in this Chapter and by others in Part III.

In this Chapter, plantation forestry in New Zealand (NZ) is used to illustrate the general concepts of C sequestration by forests. The concepts and methodologies are developed for a simple—but real-world case—but may be applied to more complex situations. The land area of NZ comprises 23.0% natural forest (Ministry of Forestry (MOF) 1993) which is mainly unavailable for extraction of wood. It is not clear whether this C reservoir constitutes a net source or a net sink. A further 4.8% of land area is vegetated with exotic plantations managed specifically for wood production. Ninety percent of this plantation area consists of a single species, radiata pine (*Pinus radiata* D. Don), which is grown in large even-aged stands. Considerable data are available to enable both wood volumes and total biomass to be calculated with a precision of ±10%. Furthermore, computer models exist which enable the composition of the estate to be simulated for future decades. The main uncertainty in such simulations is the rate of forest expansion: it is not possible to predict the extent to which private growers will choose to convert pasture to forest for commercial reasons. Lesser uncertainties lie in the productivity of new plantations and in future management decisions: in particular, at what age managers will choose to clearfell stands.

THE CONCEPTUAL FRAMEWORK

The normal forest

In a *normal forest*, the removal of forest products from the oldest stand exactly counterbalances the growth of those products in all other stands (Figure 21.1). Thus there is no change in biomass from year to year and therefore no change in C. Once established, the forest will continue to be a reservoir, but will not be a C sink or C source. However, the act of establishment of a normal forest on a site previously containing little C in the form of vegetation will constitute a net C sink. This establishment phase will last one rotation with the greatest sequestration occurring in the year before the first harvest.

Figure 21.2 shows idealised C accumulation (from a STANDPAK run, West 1993) for successive 30-year rotations. In order to calculate the mean C content (Mg C ha^{-1}) of a normal forest, the total C content of all stands is divided by the total area. Alternatively, the C content of every age class can be summed and divided by the rotation length. In other words, the same result is obtained by averaging over time as over space. As normal forests do not exist in reality the first method is preferred in practical situations.

If biomass growth is linear, a stand with mean C content will also be a stand of mean age, which in the case of a normal forest is half the rotation age. In practice, however, the growth of a stand is not linear (Figure 21.2) and the stand with mean C content will be slightly older than the average age.

Harvest residues add to the C accumulated by the second and subsequent rotations. If the initial stand contains no residues from a previous crop, C levels in the first rotation have a different profile to subsequent rotations. The long-term level of accumulated C will therefore depend on the time period under consideration. Work is underway to refine the decay profile of this forest component but it is anticipated that the C in the forest debris pool will not continue to accumulate indefinitely but will fluctuate within distinct bounds.

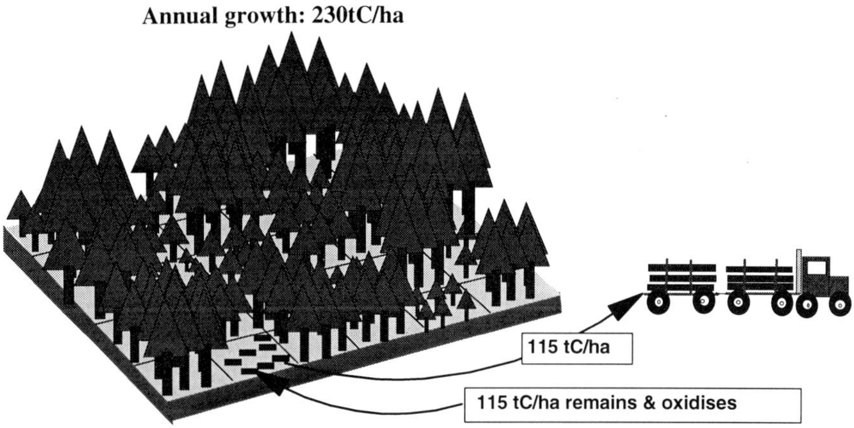

Figure 21.1. A normal forest is a C reservoir, but not a net sink or a net source (each square represents one hectare).

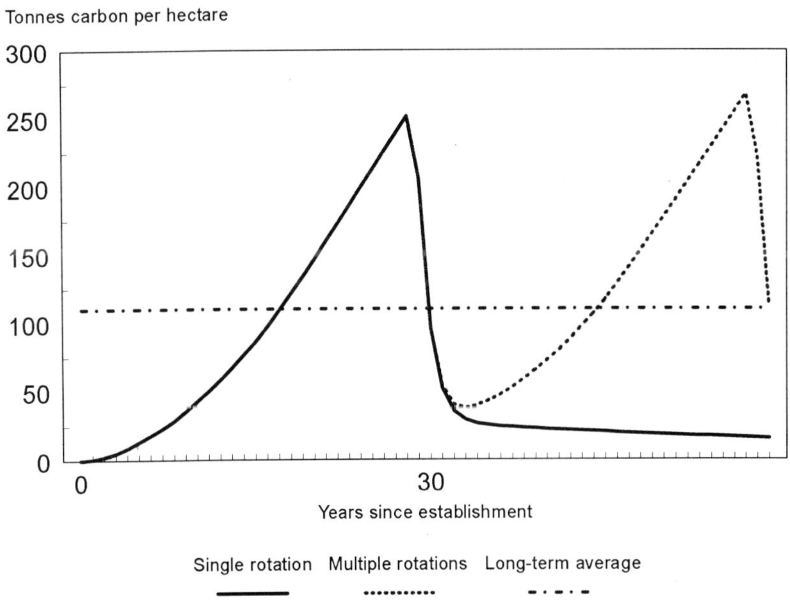

Figure 21.2. Cumulative C sequestration in a stand through two successive rotations.

The harvested wood

In NZ, approximately half the C content of a typical stand is removed at harvest. The balance is contained in stumps and roots, branches, needles, unmerchantable wood, understorey vegetation and litter, and remains on the site to decay exponentially over a period of years.

Conceptually, it makes no difference if the harvested wood is left to rot in the forest or is removed to decay elsewhere—except that the slope of the decay curves differs. Some wood products are consumed and burnt in a shorter period than the wood that decays naturally in the forest. Other products are protected, for example in houses, and preserved from decay for longer than would occur naturally.

Of all wood harvested in NZ, approximately 40% is classified as pulpwood, in that it is too juvenile, too small, or too defective to be considered for higher-value solid wood uses (MOF 1993). Some pulpwood is destined for long lasting products (particleboard, fibreboard, books, etc.) while some will enter landfills and persist for decades, but we can assume that most pulpwood will be oxidised within a very short term (< 5 years). Of the sawlog component, there is considerable loss (ca. 45%) at sawmills in the form of sawdust and offcuts. Waste continues to occur through each subsequent process, so that, for example, 15% of sawn timber delivered to a building site does not appear in the final construction. In summary, only 12% of the C present in a harvested stand has a life expectancy of longer than 5 years (Maclaren and Wakelin 1991).

A natural forest

An unmanaged normal forest is no different in principle from a managed normal forest, except that dead wood decays on the *forest floor* rather than in the form of wood products outside the forest. Also, the stands are not so clearly defined. The boundaries of stands are merged, and may even be superimposed, so that a younger age class may exist as an understorey of an older age class.

It is a common growth habit of trees to persist for a long period in the mature form, but with very little volume growth. A natural normal forest would contain many stands with a high standing volume and a low proportion of stands in the phase of rapid growth. The mean C content of a natural forest of a given type would therefore tend to be greater than that of a managed forest (Tate et al. 1993). While net ecosystem productivity is still zero, net primary productivity (NPP) is lower, because the average stand has less vigour and fixes less C annually.

Clearing a natural forest in order to replace it with a managed forest will therefore substitute a reservoir of large capacity with a smaller one. While a managed forest of young fast-growing trees has a high NPP, this is not of central interest in determining C budgets. Net ecosystem productivity (i.e., NPP less removals, both anthropogenic or by natural mortality and decay) in both systems is zero, for a normal forest or for an uneven-aged forest averaged over a very long term. To reduce atmospheric concentrations of C, it is clearly undesirable to make the substitution (Harmon et al. 1990) unless it is argued that wood products displace the use of fossil fuels directly or indirectly.

Growth rate

It is often stated (e.g., Read 1994; Bekkering 1992; Sedjo 1989) that *growth rate* is important for C sequestration. Faster growing species, it is said, will sequester more C than slow growing ones. But once again, there is considerable ambiguity in the English language. Growth rate has at least two distinct meanings:

1. *Growth rate* is often used as a synonym for rotation age. Radiata pine on a 30 year rotation is thereby deemed to be faster growing than Douglas-fir on a 60 year rotation.
2. *Growth rate* is also used by foresters to indicate the quantity of wood (net or gross) that is created every year on a given hectare. Radiata pine in NZ averages a total recoverable volume, inside bark, of 575 $m^3 ha^{-1}$ at age 30 (Neumann and Perley 1992), whereas Douglas-fir averages 959 $m^3 ha^{-1}$ at age 60. Thus radiata pine at 19.2 $m^3 ha^{-1} yr^{-1}$ is deemed to be faster growing than Douglas-fir at 16.0 $m^3 ha^{-1} yr^{-1}$.

Douglas-fir is therefore slower growing than radiata pine in both senses of the expression. These two meanings have conflicting implications on the value of a species for C sequestration. Figure 21.3 shows modelled growth of both a radiata pine and a Douglas-fir forest. Note that these results refer to a normal *forest* not a *stand*, so that at age 30 in radiata pine and at age 60 in Douglas-fir there is an even distribution of all age classes and no subsequent net C sequestration. It can be seen that for the first 48 years radiata pine provides greater greenhouse gas mitigation benefit than Douglas-fir—more C has been sequestered per hectare. After that age, however, the Douglas-fir forest continues to accumulate C whereas the radiata pine forest does not. The mature Douglas-fir forest has a higher standing volume of wood than the radiata pine in just the same way that an unmanaged forest tends to have a higher volume than a managed one.

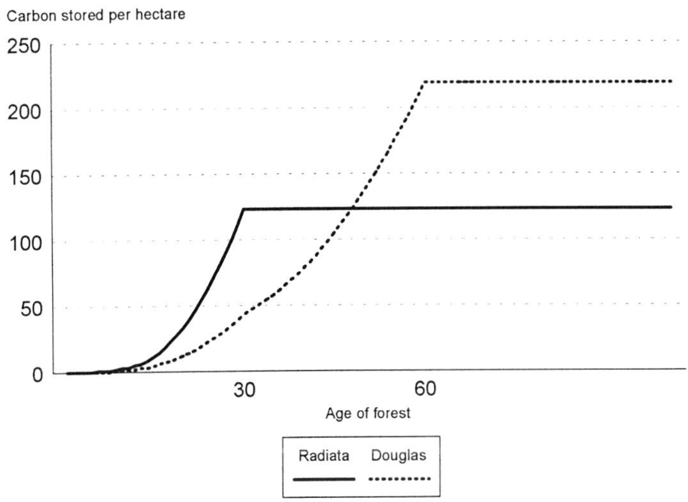

Figure 21.3. The influence of rotation age on the capacity of C reservoirs—the establishment and maintenance of forests (i.e., aggregates of stands) of radiata pine and Douglas-fir.

The principle demonstrated here is that, if long-term C sequestration in the living forest is the goal, and if plantable land is limited, it is advantageous to select species with a long rotation age, despite slower growth in terms of annual biomass production, provided that those species can generate a standing volume commensurate with their long rotation ages. At the opposite extreme, a forest of very short rotation, coppicing eucalypts or willows, would contain very little C per hectare.

Non-normal forests

The *normal forest* is an abstract notion used for heuristic and planning purposes. Real forests, both natural and man-made, are considerably more complex. A real forest contains stands of varying areas and ages. In NZ, for example, there was a planting boom from 1926 to 1936, followed by very little planting until 1964 (Figure 21.4). From 1964 until 1988 there was a second planting boom, followed by several years of reduced planting. Latest figures (MOF 1993) indicate that NZ is embarking on a third boom. The result of this uneven rate of planting is a forest that consists of a disproportionate area in certain age classes. As these age classes reach maturity they are clear-felled and replanted thus perpetuating the uneven distribution.

As well as being uneven in age-class distribution, a real forest is uneven in the distribution of species, growth rates and management policies. For example, radiata pine now comprises 90% of the NZ forest estate, whereas in 1920 it constituted only 60% of state plantings (Burdon and Miller 1992). Early stands were unthinned and unpruned, whereas over 50% of the existing resource is now thinned and pruned to at least 4.0 m. Many early plantings were restricted to steep infertile sites as better land was reserved for agriculture. Most new planting is now believed to take place on ex-pasture sites where tree growth may be substantially higher (West and Dean 1988). To model a real forest therefore requires an intimate knowledge of its component stands and the ability to predict growth for each situation.

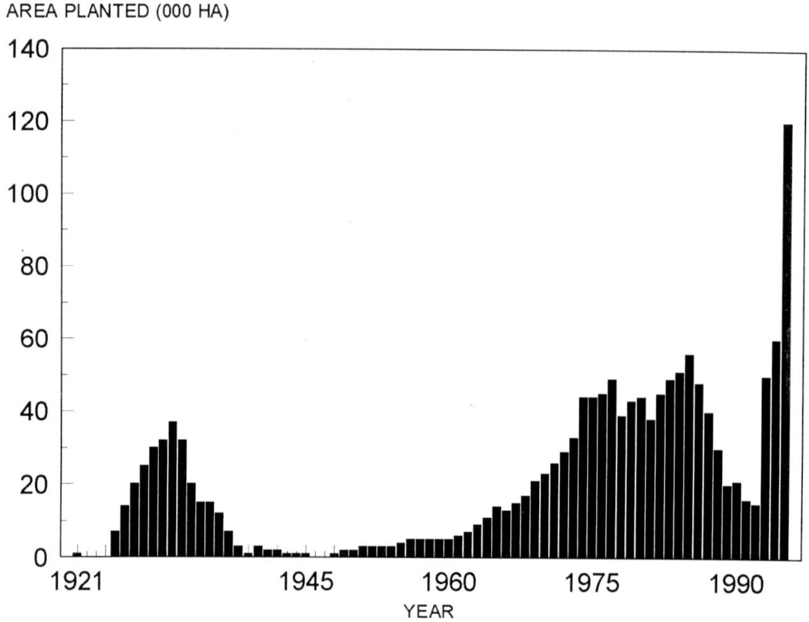

Figure 21.4. Rates of new land planting in New Zealand, 1921–1994.

Figure 21.5 shows the C content of NZ forests since 1921. There is a general increase in C contained in the estate commensurate with the increased area. Nevertheless, there are some anomalies. It can be seen that the C content *declined* from 1972 to 1979 despite an increasing forest area. The reason for this was that the sizeable age-cohort of 1930s plantings reached maturity and was felled within a short interval to be replaced by young seedlings. The average age, and therefore the average biomass per hectare, declined during this period. This effect counteracted the increased forest area. In 1972–1979, therefore, a plantation estate of fast-growing species and increasing area became a C *source*.

Again, the age cohort that replaced the 1930s plantings sequestered a disproportionate quantity of C in the period around 1990 and the forest estate was then a powerful *sink*. The slope of the line was not, however, typical of the longer term. About the year 2000, this age cohort will be felled over a short interval, resulting in below normal net sequestration for the national estate. This echo effect repeats itself every rotation but becomes diluted in time as a result of an attempt by forest managers to smooth wood flows.

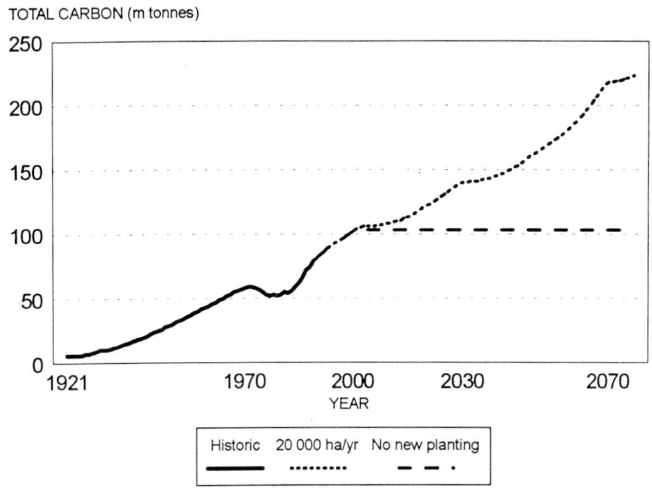

Figure 21.5. Total stock of carbon (in Tg C) in New Zealand's plantation estate since 1921 and under two forest expansion scenarios to the year 2070. (Source: Maclaren JP, Wakelin SJ (1992). Contract report for the Ministry for the Environment, June 1992).

CALCULATIONS

Four types of calculation are discussed here:

1. Calculations that describe the sequestration by a typical stand of radiata pine planted on a site devoid of biomass and replanted after clear-felling.
2. Calculations that give the net C sequestration by NZ's plantation estate at or around the year 1990.
3. Calculations that extrapolate C sequestration by the existing plantation estate in future decades, using various scenarios of forest estate expansion and assumptions about future management intentions.
4. The fate of wood products, and the fossil fuel energy that is associated with them.

The typical stand

The biomass growth of any radiata pine stand varies with the site productivity and the management regime. *Stemwood* volume is simulated for most common situations using the modelling system STANDPAK (West 1993). Based on equations derived from an extensive data set of permanent sample plots (Pilaar and Dunlop 1990), it is widely used throughout the NZ forestry sector. Using STANDPAK, forest managers derive yield tables for each category (*crop-type*) of the forest. Categorisation is subjective, but crop-types are distinguished by variation in site quality, location, or management regime. By weighting local *yield tables* with the area in each crop-type, a national yield table has been obtained for total *merchantable volume* (Newmann and Perley 1992). This can be converted to total stem volume by multiplying by 1.15 (Goulding 1986).

Total stemwood volume is converted to total stand biomass using regression equations that link wood density to tree age and location (Cown et al. 1990). Total biomass (i.e., stem, needles, branches, live roots, understorey vegetation and litter) has been derived empirically

(Beets and Brownlie 1987) and formulated as a computer model DRYMAT. The ratio of total biomass to total stemwood mass varies with age and with management regime (Maclaren et al. 1993). Combining DRYMAT with stemwood predictions from STANDPAK, gives the total biomass per hectare for a given site and management regime. (The STANDPAK predictions of stem volume are preferred to those of DRYMAT, because the former modelling system is based on a larger data set.)

Having derived the total stand biomass, a conversion factor of 0.5 can be used to express the biomass in terms of the mass of elemental C. DRYMAT was run for four first-rotation management regimes representing the four major regime types in the country (Newmann and Perley 1992):

1. Minimum tending without production thinning 38%
2. Minimum tending with production thinning 9%
3. Intensive tending with production thinning 17%
4. Intensive tending without production thinning 38%

The DRYMAT results were weighted by the areas in each regime to obtain the ratios of stem volume to total C per hectare at each age, for a weighted-average regime. Multiplying by the total stem volume per hectare from the national yield tables, provides the mass of C per hectare for each age of this regime. The results (Table 21.1) show a steady accumulation of C in the stand—with some anomalies in earlier years, especially at ages 7–9. During this period, thinning and pruning take place, and it is common for stand volume to decline. At age 30, a typical harvesting age, there are 231 Mg C ha^{-1}, of which 115 Mg (50%) is in harvested stemwood (575 m^3 of merchantable wood with a mean density of 0.4 Mg m^{-3} and a C content of 50%). The remainder decays exponentially; work is in progress to refine the decay curve for this component. Table 21.2 shows how long-term or *forest-average* C levels vary with rotation age.

Table 21.1. Growth and C content of a weighted-average hectare of radiata pine.

AGE	TOTAL STEM VOLUME (m^3)	RATIO[a]	C CONTENT (Mg ha^{-1})	AGE	TOTAL STEM VOLUME (m^3)	RATIO[a]	C CONTENT (Mg ha^{-1})
1	0	0.86	0	26	537	0.37	199
5	36	0.77	27	30	661	0.35	231
10	124	0.59	73	35	799	0.35	280
15	225	0.46	104	40	919	0.34	312
20	360	0.40	144	45	955	0.34	325
25	507	0.37	188	50	957	0.34	325

a Ratio of total elemental C (Mg ha^{-1}) to total stem volume (m^3 ha^{-1}) obtained from DRYMAT.

Table 21.2. Long-term or *forest-average* C storage by rotation age.

ROTATION AGE (yr)	20	25	30	35	40	45	50
FOREST AVERAGE (Mg C ha^{-1})	72	91	112	133	154	172	187

Baseline estimates of C sequestration

Several attempts have been made to calculate a base-line of C sequestration for NZ so that subsequent performance can be monitored. The approach has been to compare C inventories at two separate dates. The NZ Ministry of Forestry maintains a computerised database of the national forest estate (Turland and Newmann 1992) which is divided into 820 crop-types each with a corresponding yield table. The stemwood volume derived from this database is converted to oven-dry weight, and then to total standing biomass and C, as described previously. Details are given by Maclaren et al. (1993).

It would be possible to repeat the process each year and the differences in inventories would yield the statistic of interest: the C sequestration in the interval. This is not a satisfactory method because a small difference between two large numbers is computed. Each year there are slight changes in crop-type definitions, yield tables, and the percentage of the estate that is sampled, and these differences mask the real changes that have occurred. A more satisfactory method is to use the same yield tables that were used to generate the standing volumes of wood in the first period to *time-shift* the forest estate forward one year, and to subtract C leaving the system in the form of harvested logs and decay from harvest residues. Table 21.3 describes five baseline estimates, using this general approach, but differing in methodological detail, and in the details of forest statistics used.

Table 21.3. Baseline estimates for the standing C and the annual sequestration in NZ's plantation forest estate. Reference 3 gives a lower estimate of standing C than the other four references. This calculation was for an earlier year, and no account was taken of harvest residues, which can comprise a substantial part of the standing C in a forest.

REFERENCE (SEE NOTES)	STANDING CARBON (Tg C) (YEAR ENDING 31 MARCH)	SEQUESTRATION (Tg C) (YEAR ENDING 31 MARCH)
a	113 (1990)	3.5 (1991)
b	121 (1990)	3.2 (1991)
c	88 (1988)	4.5 (1989)
d	117 (1990)	3.3 (1991)
e	122 (1990)	5.1 (1991)

a Maclaren JP, Wakelin SJ (1991)
b Maclaren JP, Wakelin SJ (1992)
c Maclaren JP, Hollinger DY, Beets PN, Turland J (1993)
d Maclaren JP, Wakelin SJ, Beets PN (1993)
e Maclaren JP, Te Morenga L, Beets PN (1994)

Why was the NZ plantation estate in 1990 acting as a sink? The answer lies in the age-class distribution. Figure 21.6 gives the age-class distribution of the estate as at 1 April 1990 (Turland and Neumann 1992). The area sampled was believed to be 92.5% of the total area, with 66% of the resource under 15 years of age. If 30 years is a typical rotation age, then it is clear that the forest estate is immature. Even if there was no new planting, so that the forest area remained constant, the forest would continue to act as a sink until the annual harvest volume equalled annual forest growth. It is estimated that this would take 26 years, assuming a likely scenario of management policies. (Kurz and Apps report similar changes in Canadian forests in Chapter 14.)

Simulation for future decades

A widely-used forest estate planning model, FOLPI (Forestry Oriented Linear Programming Interpreter, García (1984)) was used to simulate C sequestration under various scenarios of new planting rates (see references in Table 21.3). It was assumed that managers would harvest the oldest stands of trees, and that they would not allow the yield to decline in

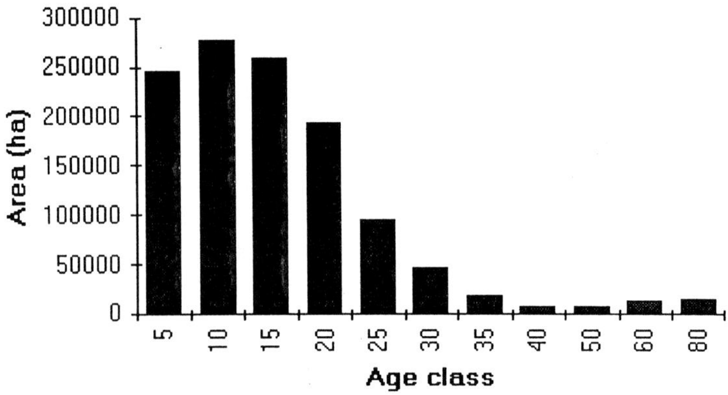

Figure 21.6. New Zealand forest plantation area by age class as at 1 April 1990.

subsequent years, nor to increase by more than 10%. It was also assumed that when a stand was harvested, it was replanted immediately into the same crop-type, and that any new land planting would consist of a crop-type that was the area-weighted mean of existing crop types. In reality, new land planting is taking place on improved land with regimes and genetic stock that differ from previous practice, and harvesting policies will depend on fluctuating international prices and other unpredictable criteria. Despite these reservations, the results in Table 21.4 provide a clear lesson: future C sequestration depends on an active programme of new land planting. The land must be currently in pasture or short vegetation, because, as discussed previously, no advantage is gained by replacing high forest with plantations.

Table 21.4. Carbon sequestration (Tg C yr^{-1}) in 2015 for three scenarios of new land planting. (Numbers in parenthesis give the rate of forest expansion, or new-land planting, for which the C sequestration was calculated).

Reference (see notes)	Low rate of forest expansion (10^3 ha yr^{-1})	High rate of forest expansion (10^3 ha yr^{-1})
a	5.3 (46)	13.1 (90)
b	6.0 (40)	18.0 (100)
c	8.7 (60)	14.2 (100)

a Maclaren JP and Wakelin SJ (1991)
b Maclaren JP and Wakelin SJ (1992)
c Maclaren JP, Te Morenga L, Beets PN (1994)

Given the country's estimated 7.6 Tg C emissions from fossil fuel combustion in 1990 (Wratt et al. 1991), it can be seen that for NZ, plantation establishment (at the current rate of about 100 000 ha yr^{-1}) is a powerful offset to emissions. It is debatable whether this rate of forest expansion can be maintained for a long period.

The fate of wood products

Maclaren and Wakelin (1991) estimate that in 1990, the NZ forest products C pool contained approximately 22 Tg C. This is small compared to the 113–122 Tg C present in the plantation estate (Table 21.3). Whereas the area in plantation estates is increasing at about 7% yr^{-1}, the forest products pool is probably expanding at a rate similar to the increase in population: 0.76% yr^{-1} (Statistics NZ 1994). It is unlikely, therefore, that forest products comprise a significant C sink relative to the living forest. (This may not be the case for all countries as discussed in Chapters 5, 14, 19, 23 and 24).

Processing of forest products consumes energy, some of which is derived from fossil fuels. Secondly, forest products can be seen as substituting for fossil fuels, either directly or indirectly. Thirdly, the decay of wood products may not result in the release of CO_2 alone. Other greenhouse gases, especially methane, may be produced. Since methane has 11 times the Global Warming Potential of CO_2 over a 100-year time-span (Isaksen et al. 1992), this factor merits consideration.

A complete life-cycle analysis should be used to determine the benefits or otherwise of plantation forestry for mitigation of greenhouse gases. Ford-Robertson (1994) has calculated that fossil-fuel emissions resulting from the cultivation, harvesting, transport and processing of NZ's production forests amount to 27 Mg C per harvested hectare with the current mix of electricity sources (mainly hydroelectric) and 49 Mg C if the electricity used was generated entirely from fossil fuels. Given that a hectare of forest with a 30 year rotation will sequester 112 Mg C in perpetuity (Table 21.2), it would take more than two rotations before an afforestation project returned a negative C balance. The pulp and paper industry consumes almost 90% of the total energy used by the NZ forest industry, and therefore energy use is very sensitive to the composition of the processing industry, and to energy efficiency in the pulp and paper sector. Fossil fuels, as opposed to wood-derived fuels, are currently used to process wood because they are cheaper, but there are no technical limitations to the development of a forest sector that is entirely self-sufficient in energy and therefore does not erode the initial benefits of afforestation.

Wood products generally use less energy to manufacture than substitutes (Table 21.5). Care should be used in interpreting such tables, because a complete life-cycle and end-use analysis is necessary to establish the ratio of product by quantity for a given use, and the relative duration in service.

It is also noteworthy that 40% of the C content of a typical NZ municipal waste landfill is derived from paper or wood (Royds Consulting Ltd. 1994). After a delay of 3–5 years, methane production reaches a peak and averages 55% of landfill gas.

DISCUSSION AND IMPLICATIONS

The significance of mean rotation age

Maximum mean annual increment of stemwood volume (see Table 21.1) for an average stand of radiata pine is 23 m^3 ha^{-1} yr^{-1}, attained with a 38–39 year rotation. Because of the effect of discount rates in reducing the value of longer rotations, current practice is to harvest at younger ages. Most stands are harvested between ages 25 and 35 (Maclaren 1994), but there is a plateau of profitability where little is lost by delaying the rotation (Maclaren 1990). Although the growing cost has to be borne for longer, older trees are more valuable.

Table 21.2 shows the substantial gains that can be made in the long-term reservoir capacity of a forest by increasing the average rotation lengths of component stands. It would be technically feasible, for example, to double the C capacity of existing plantation forests. The cost of this would be prohibitive but more modest improvements in C-reservoir capacity could be made for little cost.

The significance of age-class distribution

In NZ, stand age has considerable influence on stand volume (Table 21.1), wood density (Cown et al. 1991), and the proportion of total biomass to stemwood volume (Table 21.1). Both plantation and natural forests consist of stands with an uneven distribution of age classes. In natural forests, the distribution may be the result of past perturbations (fire, wind, disease, etc.—see Chapter 14). As each stand reaches maturity, and a new stand grows, the distribution pattern is carried over to the next generation. This echo effect can persist for generations until a new perturbation occurs although it may become blurred with time. Over a long period, a forest can be assumed to be C-neutral, but in the shorter-term C capacity will fluctuate: when there is a high proportion of older age classes the forest will contain a high level of C, and *vice versa*.

It is possible, for example, that Northern Hemisphere forests could be a C sink as a result of perturbations created by excessive harvesting and inadequate planting during World War II —although alternative explanations are given in Chapters 14 and 22. When the age-cohort that supplanted those harvested sites reaches maturity and is replaced, the same forests can become a C source. Countries with a non-expanding forest area that claim credits for the fact that their forests have a young average age class should be conscious of the fact that there will come a time when the reverse is the case.

The significance of growth rate

The two distinct meanings of growth rate previously discussed must be clearly distinguished because they have separate implications. An increase in productivity (biomass fixed per hectare per year) provides a more immediate C sequestration benefit from a planting programme but a modest lengthening of rotation age provides a greater long-term benefit (Figure 21.3). Presumably, delayed sequestration is of lesser value than more immediate sequestration but the appropriate discount factor is beyond the scope of this paper.

If a higher concentration of atmospheric CO_2 fertilises global forests, it may increase growth rate in the former sense (Strain and Cure 1985; Wittwer 1986). It may also have the effect of reducing average rotation age or wood density and thereby lowering, or at least not increasing, the average capacity of forest reservoirs (Phillips and Gentry 1994).

The significance of wood products

Wood products have an impact on fossil-fuel consumption in the wider economy. This can be positive or negative. Wood products can also influence the emission rate of greenhouse gases other than CO_2, especially methane. A sophisticated input–output table would be required to account for all the ramifications of a major increase in wood production resulting from an expanded programme of plantation forestry managed for production.

Table 21.5. Energy required to manufacture some construction materials. Source: Honey and Buchanan (1992).

PRODUCT	GJ Mg^{-1}	GJ m^3
Treated timber	2.4	1.2
Glue-laminated timber	9	4.5
Structural steel	59	448
Reinforced concrete	3.1	7.3
Aluminium	145	362

CONCLUSIONS

The establishment, and replanting upon harvest, of one hectare of radiata pine on New Zealand pastureland, using a regime typical of present-day practices, will sequester approximately 112 Mg of elemental C in perpetuity. This can be viewed as a one-time transfer of C from the air to the land surface. Some of this benefit will be lost if fossil fuels are used to process the wood or if the wood products decay anaerobically in landfills to release methane, but there may be benefits that result from the substitution of wood for less energy-intensive products.

Mitigation of C emissions by afforestation provides a temporary palliative because the on-site benefits depend on a vigorous programme of new planting on unforested land, but suitable land is limited in extent. New Zealand has perhaps 5 Mha of pasture that could profitably be converted to forest which equates to 50 years of planting at an annual rate of 100 000 ha yr^{-1}.

Forests are considerably more complex than stands and it is essential to understand the age-class structure of a forest in order to be able to quantify the C present and to predict future levels of sequestration or emissions. Forests that are currently sinks can become sources as the age-class composition changes.

Fast-growing plantations can provide a lower long-term benefit in terms of sequestration than forests comprising stands of long rotations and high biomass carrying capacity. Carbon dioxide fertilisation, for example, may not result in an increase in net sequestration by global forests.

REFERENCES

Bekkering TD (1992) Using tropical forests to fix atmospheric carbon: the potential in theory and practice. *Ambio* 21(6): 414–419.

Beets PN, Brownlie RK (1987) Puruki experimental catchment: site, climate, forest management, and research. *NZ J. Forestry Sci.* 17: 137–60.

Burdon RD, Miller JT (1992) Introduced forest trees in New Zealand: recognition, role, and seed source. 12. Radiata pine (*Pinus radiata* D. Don). NZ Forest Research Institute Ltd, FRI Bulletin No. 124, 59 pp.

Cown DJ, McConchie DL, Young GD (1991) Radiata pine wood properties survey. NZ Ministry of Forestry, FRI Bulletin No. 50, 50 pp.

Ford-Robertson FC (ed) (1971) Terminology of forest science, technology practice and products. The multilingual forestry terminology series, No. 1. Society of American Foresters, Washington DC.

Ford-Robertson JB (1994) Carbon balance of the plantation forest industry in New Zealand. (Unpublished manuscript)

Goulding CJ (1986) Measurement of tree crops. In: Levack H, Forestry Handbook. NZ Institute of Foresters, Wellington, New Zealand, pp 80–81.

García O (1984) FOLPI, a forestry-oriented linear programming interpreter. In: Ngumo H et al. (eds) Proceedings IUFRO Symposium on Forest Management, Planning and Managerial Economics, University of Tokyo, pp 293–305.

Harmon MF, Ferrel WK, and Franklin JF (1990) Effects on carbon storage of conversion of old-growth forests to young forests. *Science* 247: 699–702.

Hendrickson OQ (1990) How does forestry influence atmospheric carbon? *For. Chron.* 66: 469–472.

Honey BG, Buchanan AH (1992) Environmental impacts of the New Zealand building industry. Dept. of Civil Engineering, University of Canterbury, Christchurch, NZ.

Houghton JT, Jenkins GJ, Ephraums JJ (eds) (1990) Climate Change. The IPCC Scientific Assessment. Intergovernmental Panel on Climate Change, Cambridge University Press, Cambridge, 364 pp.

Houghton JT, Callander BA, Varney SK (eds) (1992) Climate Change 1992. The Supplementary Report to the IPCC Scientific Assessment. Intergovernmental Panel on Climate Change, Cambridge University Press, Cambridge, 200 pp.

Isaksen ISA, Ramaswamy V, Rodhe H, Wigley TML (1992) Radiative forcing of climate, In: Houghton JT, Callander BA, Varney SK (eds) Climate Change 1992. The Supplementary Report to the IPCC Scientific Assessment, pp 47–67.

Kinerson RS, Ralston CW, Wells CG (1977) Carbon cycling in a loblolly pine plantation. *Oecologia (Berl.)* 29: 1–10.

Maclaren JP (1994) Radiata pine growers' manual. NZ Forest Research Institute, FRI Bulletin No. 184, 140 pp.

Maclaren JP (1990) Optimisation of final crop stockings for clearwood production in radiata pine. In: James RN, Tarlton GL (eds) New Approaches to Spacing and Thinning in Plantation Forestry. Proc. IUFRO Symp, Forest Research Institute, Rotorua, NZ, 10–14 April 1989. pp 171–183.
Maclaren JP, Wakelin SJ (1991) Forestry and forest products as carbon sink in New Zealand. Ministry of Forestry, Forest Research Institute Bulletin No. 162, FRI, Rotorua, New Zealand, 360 pp.
Maclaren JP, Wakelin SJ (1992). Contract report for the Ministry for the Environment. June 1992, 40 pp. (Unpublished manuscript.)
Maclaren JP, Hollinger DY, Beets PN, Turland J (1993) Carbon sequestration by New Zealand's plantation forests. In: Kanninen M (ed) Carbon balance of the World's Forested Ecosystems. Proc. of the IPCC AFOS Workshop held in Joensuu, Finland, 11–15 May 1992, pp 201–214.
Maclaren JP, Wakelin SJ and Beets PN (1993). Contract report for the Ministry of Forestry. June 1993. (Unpublished manuscript.)
Maclaren JP, Te Morenga L and Beets PN (1994). Contract report for the Ministries of Forestry, the Environment, Foreign Affairs and Trade and Commerce. April 1994 (Unpublished manuscript.)
Matthews RW (1994) Towards a methodology for the evaluation of the carbon budget of forests. In: Kanninen M (ed) Carbon balance of the world's forested ecosystems: towards a global assessment. Proc. Workshop Intergovernmental Panel on Climate Change AFOS, Joensuu, Finland, 11–15 May 1992, Painatuskeskus, Helsinki, pp 105–114.
Ministry of Forestry (1993) New Zealand Forestry Statistics 1993. Ministry of Forestry, Wellington, New Zealand, 126 pp.
Newmann A, Perley C (1992) A national exotic forest description yield tables as at 1 April 1991. Ministry of Forestry, Wellington, New Zealand, 270 pp.
Pilaar CH, Dunlop JD (1990) The permanent sample plot system of the New Zealand Ministry of Forestry. In: Adlard P and Rondeux J (eds) Forest Growth Data: Retrieval and Dissemination. Proc. joint IUFRO workshop S4.02.03 and S4.02.04, April 1989, Bulletin des Recherches Agronomiques de Gembloux, Gembloux, Belgium.
Phillips OL, Gentry AH (1994) Increasing turnover through time in tropical forests. *Science* 263: 954–958.
Read P (1994) Responding to Global Warming—the Technology, Economics and Politics of Sustainable Energy. Zed Books Ltd, London, New Jersey, 304 pp.
Royds Consulting Ltd (1994) Estimate of methane production from New Zealand landfills. Ministry for the Environment contract report. (Unpublished manuscript).
Rowell R (1984) The chemistry of solid wood. *Advances in Chemistry*, p 207.
Schroeder P (1992) Carbon storage potential of short rotation tropical tree plantations. *Forest Ecol. Manage*. 50: 31–41.
Sedjo RA (1989) Forests—a tool to moderate global warming? *Environment* 31(1): 14–20.
Sedjo RA (1989) Forests to offset the Greenhouse Effect. *J. Forestry* (July 1989): 12–15.
Society of American Foresters (1917) Forest terminology. *J. Forestry* (Jan 1917): 71–101.
Statistics NZ (1994) The New Zealand Official Yearbook '94. 97th edn, Statistics NZ.
Strain BR, Cure JD (1985) Direct effects of atmospheric CO_2 enrichment on plants and ecosystems: a bibliography with abstracts ORNL/CDIC-13. Oak Ridge Natl. Lab., Oak Ridge, TN.
Turland J, Neumann A (1992) A national exotic forest description as at 1 April 1990, 7th edn. Ministry of Forestry, February 1992, Wellington, New Zealand.
United Nations Conference of Environment and Development (1992) Earth Summit—convention on climate change. Rio de Janeiro, June 1992.
West GG (1993) A review of the development and use of the New Zealand modelling system: STANDPAK. Intl. Symp. System Analysis and Management Decisions in Forestry. Valdivia, Chile, 299 pp.
West GG, Dean, MG (ed) Maclaren JP (1988) Use of existing tree growth models on farm sites. In: 1988. Proceedings of the agroforestry symposium, Forest Research Institute, Rotorua, New Zealand, 24–27 November 1986. FRI Bulletin No. 139, pp 121–129.
Wittwer SH (1986) Worldwide status and history of CO_2 enrichment—an overview. In: Enoch HZ, Kimball BA (eds) Carbon dioxide enrichment of greenhouse crops. Vol. 1: Status and CO_2 sources. CRC Press, Boca Raton, Florida, pp 3–15.
Wratt DS, Mullan AB, Clarkson TS, Salinger MJ, von Dadelszen J, Plume H (1991) Climate change—the consensus and the debate. NZ Climate Change Programme, Ministry for the Environment and NZ Meteorological Service. May 1991, 48 pp.

22 Carbon pools and fluxes in U.S. forest products

Linda S Heath, Richard A Birdsey[1], Clark Row[2] and Andrew J Plantinga[3]

U.S. Department of Agriculture, Forest Service
Northeastern Forest Experiment Station
Located at:
Pacific Northwest Research Station
1221 SW Yamhill
P.O. Box 3890
Portland, OR 97208 USA

INTRODUCTION

Increasing recognition that anthropogenic CO_2 and other greenhouse gas emissions may effect climate change has prompted research studies on global carbon (C) budgets and international agreements for action. At the United Nations Conference on Environment and Development in 1992, world leaders and citizens gathered and initiated the Framework Convention for Climate Change (FCCC), an agreement to address global climate change concerns. Over 160 nations have signed the FCCC, whose ultimate goal is to stabilise the atmospheric concentration of greenhouse gases to prevent significantly negative effects on the climate system. To reach this goal, some nations, including the United States, have committed to reduce greenhouse gas emissions to 1990 levels by the year 2000. Knowledge of the magnitude and processes in C cycles is essential in developing effective strategies to mitigate anthropogenic emissions.

The forest sector C budget is of special interest because it is a large active pool with many components that are difficult to enumerate, and it offers a variety of mitigation options. One pool that is often not included is the amount of C in wood harvested from forests because its size is estimated to be small relative to other pools (Dixon et al. 1994). However, Dixon et al. (1994) also state that the wood products pool contributes to much of the uncertainty in the forest C budget estimates for mid- and high-latitude forests.

Carbon in wood products has been accounted for in generalised frameworks, such as that of Dewar (1991), who presented an analytical model for forest trees, soil, and wood products, and that of Marland and Marland (1992), who examined the possibility of using harvested trees as fuel to replace fossil fuels. Row and Phelps (1991) calculated percentages of C remaining in various categories of wood products in the U.S. for up to 100 years following harvest under current utilisation standards but did not present absolute estimates. In their national C budget for Canada, Kurz et al. (1992) developed a forest product submodel using historical harvest rates. Plantinga and Birdsey (1993) and Heath and Birdsey (1993) presented estimates of C in harvested wood from projected future harvests through the year 2040 in the

1 USDA Forest Service, Northern Stations Global Change Research Program, PO Box 6775, Radnor, PA 19087, USA
2 Institute for Forest Analysis, Planning, and Policy, Baltimore, MD, USA
3 Dept. of Agricultural and Resource Economics, University of California, Berkeley, CA 94720, USA

NATO ASI Series, Vol. I 40
Forest Ecosystems, Forest Management and the Global Carbon Cycle
Edited by Michael J. Apps and David T. Price

U.S. Neither of these studies includes estimates of C in historical removals. In this paper, we estimate historic (post-1900) and current C pools and fluxes in wood harvested and removed from forests (removals) in the U.S. We also project C in removals through 2040, and explore how these may be influenced by different assumptions about the future.

METHODS

We compiled historical and projected removals and calculated C in wood products by employing multiplier estimates from the HARVCARB model (Row and Phelps 1991). The definition of *removals* is wood that is cut on-site and removed for use. We do not consider the fate of logging residue, which is usually modelled as part of the forest ecosystem, not part of the products pool.

HARVCARB

HARVCARB (Version 1) is a model that transforms wood C in several phases: (1) from harvested trees to roundwood, (2) from roundwood to solid wood and finished fibre, and (3) from solid wood and fibre to end uses including products such as construction lumber, paper and newsprint. There are four end-use disposition categories: products, landfills, wood burned for energy, and emissions. The products category includes wood in buildings, repairs and improvements, pallets, and furniture and fixtures. Wood burned to produce energy includes wood used in municipal solid waste incinerators, home heating, wood-burning power plants, and power facilities associated with forest product industries. Changes between end-use categories are included. For example, C may be stored in cardboard five years after harvest, in a landfill ten years later, and decomposed as emissions after twenty additional years. Figure 22.1 illustrates the possible fates of C in the model.

HARVCARB estimates were summarised as percentages of harvested C that remained in each disposition category for five-year time increments following harvest. Separate sets of percentages were developed for each region, hardwood or softwood type, and pulpwood or sawtimber size. Figure 22.2 shows one example of disposition patterns. Note that summing the percentages across all disposition categories at a given time yields 100%.

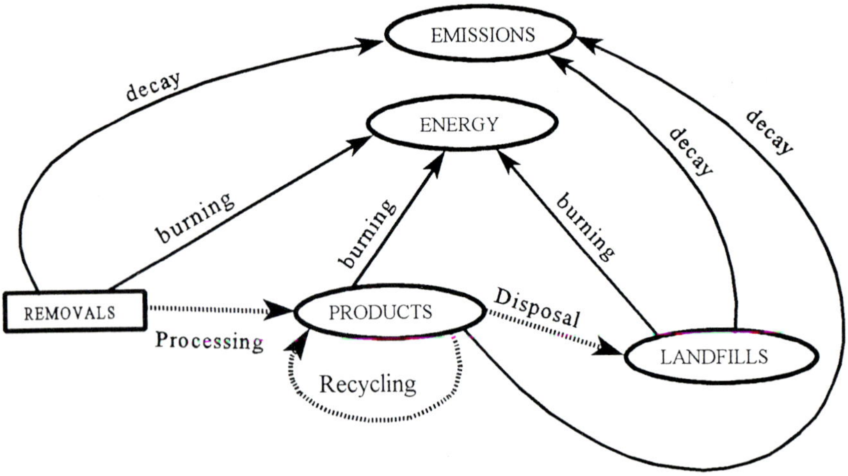

Figure 22.1. Simulated C pools and flows from removals.

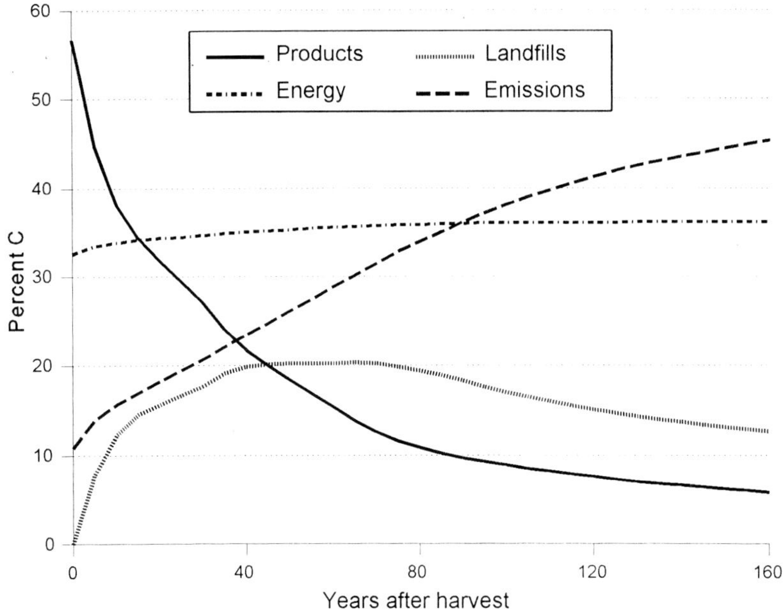

Figure 22.2. Carbon disposition curves for softwood sawtimber in the southern U.S.

Carbon was tracked over time by multiplying the amount of wood harvested in a given year by the percentages in each disposition category, and accumulating amounts remaining from previous years. HARVCARB estimates were based on current utilisation standards, but we assumed that they adequately represented historical standards and projected changes in disposition patterns.

Sources of data for removals

Historical estimates were gathered from a variety of United States Department of Agriculture (USDA) Forest Service publications of summary forest statistics, the latest from Powell et al. (1992). There is usually only one annual harvest estimate per decade, so we assumed that this annual estimate was equivalent to the average annual estimate for the period. Projected removals were obtained from the Timber Assessment Market Model (TAMM) (Adams and Haynes 1980) and the Aggregate Timberland Assessment System (ATLAS) (Mills and Kincaid 1992). The projections are annual estimates. TAMM projects price, consumption, and production trends in both stumpage and product markets in the U.S. forest sector in conjunction with ATLAS, which projects timber inventories on private lands. Because we calculate C in wood products based on amount of wood removed from U.S. forests only, we do not count C in imports. However, we do include C in exports, even if the logs are processed in other countries. Overall, about 12% of the timber consumed in the U.S. is imported; U.S. exports of roundwood and primary products are about 5% of harvest (Haynes 1990).

Scenarios

We projected C in wood products from 1990 to 2040 for a base scenario. The base scenario represents current expectations of what will happen in the future; it is a business-as-usual scenario. For specific information about the base, see Haynes et al. (1995). We adopted alternate scenarios to investigate the extent to which changes in assumptions affect the results. We looked at four scenarios that affected projections only: increase percent in harvested C in the products category by 50%, decrease percent in harvested C in the products category by 50%, decrease the percent of C in the emissions category by 50%, and stop harvesting in 1990. Because the percentages for the four disposition categories must sum to 100, increasing one category by 50% means the other categories collectively must be reduced by the amount of the increase. We assumed the percent reduction applied equally to the other three categories. These four scenarios were considered to provide an idea of how the magnitudes of C fluxes would vary if large changes occurred in wood use. We ran two other scenarios for a sensitivity analysis, one in which historical harvests (1900–1989) were assumed 25% greater than the base, and a second in which historical harvests were assumed 25% less than the base.

RESULTS AND DISCUSSION

Carbon pools

The cumulative fate of C removed from U.S. forests since 1900 is illustrated in Figure 22.3. By 1990, about 3.7 Pg of the 10.7 Pg of C removed in harvests during the period 1900–1990 remained sequestered in products and landfills, while another 3.7 Pg was burned for energy. By comparison, in 1987 an estimated 13.8 Pg C was stored in trees in conterminous U.S. forest ecosystems, and 24.7 Pg in soils, forest floor, and understory (Birdsey 1992). Thus, the

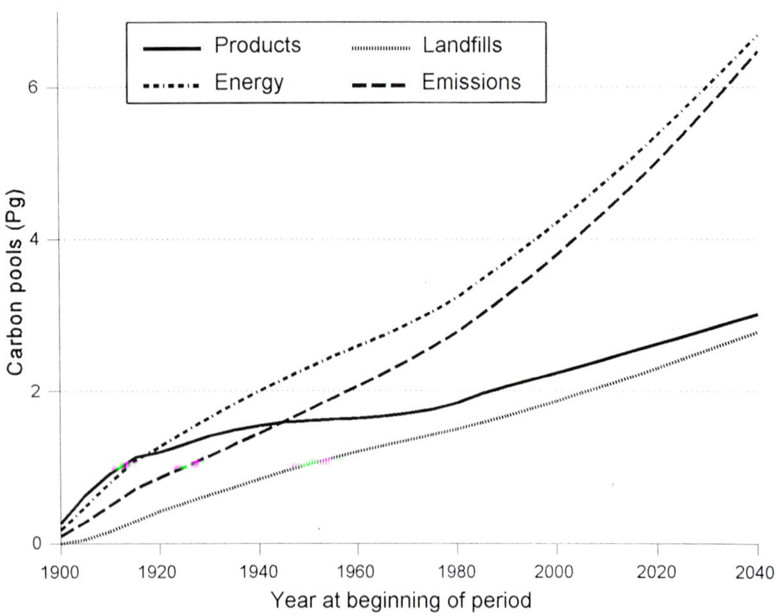

Figure 22.3. Carbon pool sizes (Pg) for the base scenario.

amount of C removed was significant when compared to the amount currently stored in trees in forests. The upturn in the estimates after 1990 indicates that consumption of wood products is expected to increase at least through 2040. Emissions and energy pools are greater than the products and landfills pools at the end of the projection period because over time C in products and landfills moves into emissions and energy pools, but C never transfers out.

On a percentage basis, by 1990 approximately 35% of the total C removed is stored in products and landfills, 30% has returned to the atmosphere through decay or burning without energy production, and 35% has been burned for energy, partially offsetting fossil fuel use. Forest products industries contribute to the last end use by burning virtually all residues used during processing of logs, and wood is burned in power plants, waste incinerators, and the home. This category is often a source of confusion in reporting C in wood products. Because such C is not sequestered in a solid state, it is considered an emission in some accounting frameworks. However, it has helped mitigate increasing atmospheric CO_2 concentration by offsetting fossil fuel use, and in this respect it is not equivalent to emissions from decay. We take the view that it is important to list the amount of C burned for energy, but that it should be listed separately.

Carbon fluxes

The net flux of C into each disposition category is displayed in Figure 22.4. The net current average annual flux into products and landfills is about 37 Tg C yr^{-1}, which is almost 40% of the estimated C flux into the tree component of U.S. forests (Birdsey and Heath 1995). About 50 Tg C yr^{-1} is emitted when wood is burned for energy. All estimates are positive, indicating that the size of the C pool in each category increases in each successive period. The sizes of these C flux estimates are quite significant considering the United States must reduce its annual emissions by 76 Tg yr^{-1} by the year 2000 to fulfil the FCCC agreement (U.S. Department of Energy 1994).

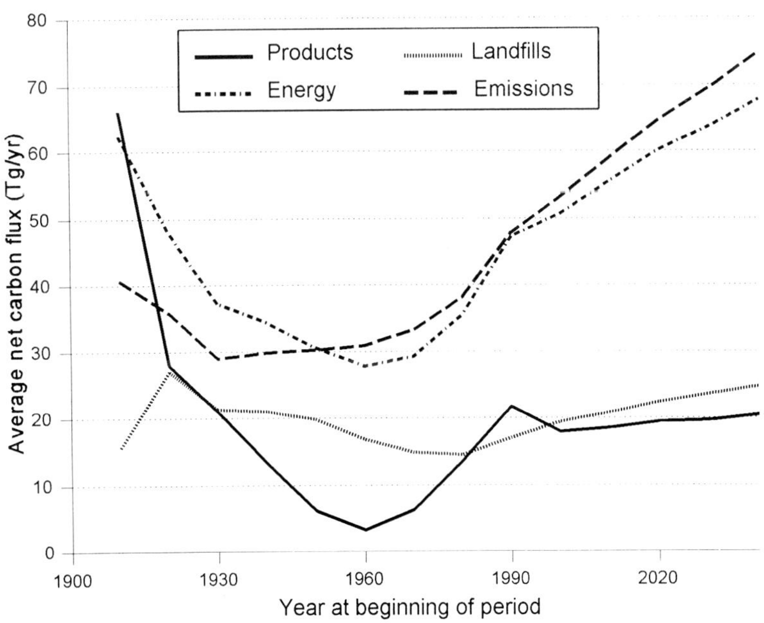

Figure 22.4. Average annual net carbon fluxes (Tg C yr^{-1}) into disposition categories for U.S. timber removals.

Because removals estimates are increasingly unreliable for earlier years, we also performed a sensitivity analysis on the historical estimates (Table 22.1). Results indicate that even if historical harvest estimates are 25% off from actual harvests, current and future products fluxes would change by an average of only 5 Tg C yr^{-1}, landfill flux by 0.5 Tg yr^{-1}, wood burned for energy by 0.4 Tg yr^{-1}, and emissions by 5 Tg C yr^{-1}. Thus historical removals contribute most to current and future products and emissions categories, and relatively little to the other two categories. Note that greater harvests in the past produce a smaller future net flux into products and landfills. This occurs by definition; change between pool size (flux) becomes smaller if initial pool size is increased while the next pool size is held constant.

Table 22.1. Projected net C fluxes by disposition category for variations in historical (1900–1989) harvest estimates of +25% and -25%.

	Net C flux by disposition category (Tg C yr^{-1})							
	Products		Landfills		Energy		Emissions	
Period	-25%	+25%	-25%	+25%	-25%	+25%	-25%	+25%
1990–1999	27	9	17	21	50	51	47	59
2000–2009	24	13	21	20	55	56	54	65
2010–2019	24	15	23	21	60	61	60	70
2020–2029	23	16	25	22	64	64	65	75
2030–2039	23	18	27	23	68	68	70	79

Scenarios

We examined how change in disposition rates may affect flux magnitudes under alternative scenarios (Table 22.2). The scenario that increased C sequestration most as compared to the base was to decrease percent in emissions by 50% (assuming that percent in the other categories increase proportionally). The scenario to stop harvesting in 1990 is presented only for comparison. If harvesting ceased, emissions and amount in landfills would decline significantly. However, less wood (average 58 Tg yr^{-1}) would offset fossil fuels for energy production, and 40 Tg yr^{-1} on average would not be sequestered in products as compared to the base.

Comparison to carbon in global harvesting

We can make inferences about C in wood products on a global scale through comparison with global wood products statistics. Compared to the rest of the world, in 1990, the U.S. produced 20% of the sawnwood, 25% of wood-based panels, 30% of paper and paperboards, and only about 5% of the fuelwood (Brooks 1993). Based on these percentages, we can roughly estimate that globally C in wood products (excluding landfills) is currently increasing on the order of 100 Tg C yr^{-1}. For global analyses, it may be justifiable to omit the wood products C flux component, as this amount is fairly small compared to an estimated global total net C flux to the atmosphere of 900 Tg C yr^{-1} (Dixon et al. 1994). However, we found that wood products should be included in C budgets for the U.S. because this component is significant compared to U.S. forest ecosystem flux estimates.

The other pool that is important to examine globally is wood burned for energy. Historically, there has been a trade-off from the use of wood to fossil fuel as nations become industrialised. As fossil fuel use grows, so do forests as they are harvested less for fuel. The potential for this pattern exists today. In 1989, harvests from tropical zone forests, located mostly in developing countries, were about 36% greater than harvests in temperate zone

Table 22.2. Estimated difference between scenario and base case in average annual net C flux into each disposition category during 1990–2040.

	NET C FLUX BY DISPOSITION CATEGORY (Tg C yr^{-1})			
	PRODUCTS	LANDFILLS	ENERGY	EMISSIONS
Scenario: Increase percentage in products by 50%				
1990–1999	8.8	-2.8	-2.6	-3.4
2000–2009	9.1	-2.7	-3.1	-3.5
2010–2019	9.6	-2.8	-3.3	-3.5
2020–2029	9.7	-2.9	-3.4	-3.5
2030–2039	10.2	-3.0	-3.6	-3.7
Scenario: Decrease percentage in products by 50%				
1990–1999	-7.2	3.9	35.2	37.9
2000–2009	-7.9	4.5	39.8	43.6
2010–2019	-8.5	5.4	43.7	48.6
2020–2029	-8.8	6.6	47.1	53.4
2030–2039	-9.4	7.1	50.6	58.0
Scenario: Decrease percentage in emissions by 50%				
1990–1999	9.0	8.6	9.0	-26.6
2000–2009	10.0	9.6	10.0	-29.5
2010–2019	10.9	10.6	10.9	-32.3
2020–2029	11.7	11.4	11.7	-34.8
2030–2039	12.6	12.3	12.6	-37.4
Scenario: Stop harvesting in 1990				
1990–1999	-52.8	-11.4	-48.2	-28.7
2000–2009	-40.7	-22.1	-54.0	-37.1
2010–2019	-37.6	-26.3	-58.9	-43.9
2020–2029	-33.7	-29.8	-62.7	-50.3
2030–2039	-31.6	-32.1	-67.0	-57.2

forests (Brooks 1993), and about 80% of tropical zone harvests were burned for fuel. With 80% of the total roundwood production of 1.922×10^9 m^3 burned for fuel in developing countries, and assuming 234 kg C m^{-3} (Birdsey 1992), then roughly 360 Tg C yr^{-1} are released to the atmosphere in the tropics alone. If developing nations begin to use more fossil fuels instead of wood for energy, examining only forest ecosystem C budgets would likely show that tropical forests were sequestering more C, although emissions for the entire system might be greater.

CONCLUSIONS

Carbon pools and fluxes from the disposition of C in harvested wood are significant relative to the U.S. forest ecosystem C budget. It is important to include this C when tracking C budgets to compare forest management and utilisation policies. Although wood burned for energy is not a pool of C that is physically sequestered, it helps mitigate increasing atmospheric CO_2 concentration by offsetting fossil fuel use. The wood for energy flux is larger than the uptake of C in products and landfills in the United States, and we estimate it is larger in proportion in developing countries, particularly those which feature tropical zone forests. Tracking the end-use C pools, including wood burned for energy, along with forest ecosystems C pools would more accurately reflect overall C budget effects.

REFERENCES

Adams DM, Haynes RW (1980) The Timber Assessment Market Model: Structure, projections, and policy simulations. *Forest Sci. Monogr.* 22, Society of American Foresters, Bethesda, MD, 64 pp.
Birdsey RA (1992) Carbon storage and accumulation in United States forest ecosystems. USDA For. Serv., Gen. Tech. Rep. WO-59, Washington DC, 51 pp.
Birdsey RA, Heath LS (1995) Carbon changes in U.S. forests. In: Joyce LA (ed) Climate change and the productivity of America's forests. USDA For. Serv., Gen. Tech. Rep., Rocky Mtn. Forest and Range Expt. Stn., Fort Collins, CO, 146 pp. (In press).
Brooks DJ (1993) U.S. forests in a global context. USDA For. Serv., Gen. Tech. Rep. RM-228, Rocky Mtn. Forest and Range Expt. Stn, Fort Collins, CO, 24 pp.
Dewar RC (1991) Analytical model of carbon storage in the trees, soils, and wood products of managed forests. *Tree Physiology* 8: 239–258.
Dixon RK, Brown S, Houghton RA, Solomon AM, Trexler MC, Wisniewski J (1994) Carbon pools and flux of global forest ecosystems. *Science* 263: 185–190.
Haynes RW, Adams DM, Mills JM (1995) The 1993 RPA Timber Assessment Update. USDA For. Serv., Gen. Tech. Rep. RM-259, Rocky Mtn. Forest and Range Expt. Stn, Fort Collins, CO, 66 pp.
Haynes RW (coord) (1990) An Analysis of the Timber Situation in the United States: 1989–2040. USDA For. Serv., Gen. Tech. Rep. RM-199, Rocky Mtn. Forest and Range Expt. Stn., Fort Collins, CO, 268 pp.
Heath LS, Birdsey RA (1993) Carbon trends of productive temperate forests of the coterminous United States. *Water, Air, Soil Pollut.* 70: 279–293.
Kurz WA, Apps MJ, Webb TM, McNamee TJ (1992) The carbon budget of the Canadian forest sector: Phase I. Forestry Canada Information Report NOR-X-236, Northwest Region, Edmonton, Alberta, 93 pp.
Marland G, Marland S (1992) Should we store carbon in trees? *Water, Air, Soil Pollut.* 64: 181–195.
Mill JR, Kincaid JC (1992) The Aggregate Timberland Assessment System—ATLAS: A comprehensive timber projection model, USDA For. Serv., Gen. Tech. Rep. PNW-281, Pacific Northwest Research Stn, Portland, OR, 160 pp.
Plantinga AJ, Birdsey RA (1993) Carbon fluxes from U.S. private timberland management. *Climatic Change* 23: 37–53.
Powell DS, Faulkner JL, Darr DR, Zhu Z, MacCleery DW (1993) Forest resources of the United States, 1992. USDA For. Serv., Gen. Tech. Rep. RM-234, Rocky Mtn. Forest and Range Expt. Stn, Fort Collins, CO, 132 pp.
Row C, Phelps RB (1991) Carbon cycle impacts of future forest products utilization and recycling trends. In: Agriculture in a World of Change, Proceedings of Outlook '91, U.S. Department of Agriculture, Washington DC, 786 pp.
US Department of Energy, coord. (1994) The Climate Change Action Plan: Technical Supplement. U.S. Dept. of Energy, Office of Policy, Planning, and Program Evaluation. Washington DC, 148 pp.

23 Effects of forest management, harvesting and wood processing on ecosystem carbon dynamics: a boreal case study

David T Price, Ralph M Mair, Werner A Kurz[1] and Michael J Apps

Natural Resources Canada, Canadian Forest Service, Northwest Region, Northern Forestry Centre, 5320 - 122 Street, Edmonton, AB T6H 3S5, Canada

INTRODUCTION

Many northern forests are managed with the objective of achieving and maintaining sustained yield. Sustained yield management implies continuous production so planned as to achieve at the earliest practical time, a balance between annual growth and harvesting. It is often argued that with silvicultural prescriptions, and protection from natural disturbances, responsible management will allow wood harvests and other benefits to be elevated in the long term, and thereafter maintained at increased levels indefinitely (when compared to unmanaged forest growing in a similar environment).

Taking this optimistic perspective, what are the likely impacts on the global carbon (C) balance of practising maximum sustained yield harvesting on a global scale? To answer this question it is necessary to account for the steady annual processing of harvested materials into wood products (i.e., manufactured goods ultimately derived from harvested trees), and their subsequent incineration or disposal in landfills. A working hypothesis would be that in the long term, assuming no changes in average climate, the annual global releases of C from decay and consumption of old wood products must eventually balance the annual additions of C from newly manufactured forest products (assuming the yield can be sustained), i.e., the net rate of accumulation of C in the forest products pool must tend asymptotically to zero. When this happens, the global product pool reaches a maximum: it may be said to be saturated.

The net accumulation of C in forest products that will take place until saturation occurs represents a finite potential sink for atmospheric CO_2. What is the capacity of this sink and how long will it last? Placing an upper limit on the value of this sink could be useful for developing global policy on the use of forests for mitigation of greenhouse gas (GHG) emissions. Additionally, wood could be used as a renewable energy source (which does not prevent other uses before burning it), in which case the energy value could also be used as an offset against emissions from fossil fuel energy production. What is the global maximum possible magnitude for this offset?

It is not currently possible to assess fully the dynamics of the global forest products pool, and how these dynamics relate to the conceptual management objective of maximum sustained yield applied to the world's forests. An alternative approach is to use individual forests as case studies, and possible models for the global picture. This is the approach adopted here, using a single forest management unit in central Alberta.

---

1 ESSA Technologies Ltd., 3rd Floor, 1765 West 8th Avenue, Vancouver, B.C., V6J 5C6, Canada

NATO ASI Series, Vol. I 40
Forest Ecosystems, Forest Management and the Global Carbon Cycle
Edited by Michael J. Apps and David T. Price

This paper examines some aspects of boreal forest C dynamics under a small range of defined future scenarios, and reports on the expected changes in C storage that would result under each of these assumed scenarios. The focus is on the roles of forest management and the forest product C pool in altering net C storage compared to the unmanaged and unharvested ecosystem. Specifically, it will address the following questions:

1. How does operational harvesting affect C sequestration in a managed forest, as compared to leaving it in an unexploited state?
2. What is the approximate time required to saturate the local forest products C pool, assuming the management objective of maximum sustained yield harvesting can be achieved?
3. What limits the size and duration of the local forest products sink?
4. What is the magnitude of the C emission offset if all products are also used for energy production?
5. Can insights obtained from the local analysis be extended to the global scale?

METHODS

C budget model

A simulation experiment was carried out for the forest management agreement (FMA) area of the Foothills Forest located in west central Alberta (Price et al. 1994) using the Carbon Budget Model of the Canadian Forest Sector (CBM-CFS). The model, described extensively elsewhere (e.g., Kurz et al. 1992; Apps and Kurz 1994; Kurz and Apps 1994), uses data derived from forest inventories, ecosystem classification, soil surveys and forestry statistics from government and industry, to estimate the C content of three distinct pools: biomass, soils and products. It uses empirical equations to simulate annual forest growth and soil decomposition for representative stand-types, while disturbance effects (including harvesting) are assessed over longer timesteps (typically five years). In previously reported assessments, CBM-CFS has been used to estimate forest sector C budgets for Canada as a whole (Kurz et al. 1992; Apps and Kurz 1994; Kurz and Apps 1994), and for specific ecoregions of Canada (e.g., Chapter 14; Apps et al. 1993; Kurz and Apps 1993).

This paper reports three simulation experiments using the dynamic version of the model (CBM-CFS2) to perform comparative C budget analyses of the Foothills FMA area. In the first simulation (Experiment I), only the effects of natural ecosystem processes on the forest sector C budget were examined for a 230-year period beginning in 1958 (taken to be the start of active management operations). It was assumed that there was no human intervention, that a natural 50-year disturbance interval persisted for the period prior to 1958, and that no changes in average environmental conditions occurred since then. In the second experiment (Experiment II), the effects of harvesting and management as practised by the current FMA licensees, Weldwood of Canada, on forest C dynamics were examined for the same 230-year period. This scenario assumed that yield projections would be met and that natural disturbances could be completely suppressed. Annual harvesting rates were assumed to be those for the known past (based on areas regenerated since 1958) and for the projected future from 1988 onwards (based on planned harvests). Management assumptions included increased site productivity following harvesting, (anticipated from Weldwood's silvicultural prescriptions), and no change in environmental conditions. A third experiment (Experiment III) examined the consequences of failing to achieve these projected yield increases, but again assuming complete suppression of natural disturbances.

Site description

Covering approximately 1.2 Mha, Foothills Forest is one of ten "model forests" established within Canada to promote sustainable development and integrated management of Canada's forest resources. The dominant species are lodgepole pine (*Pinus contorta* Dougl.), black spruce (*Picea mariana* (Mill.) B.S.P.), white spruce (*Picea glauca* (Moench) Voss.) and aspen (*Populus tremuloides* Michx.). In addition to pest control and suppression of wildfires carried out over the entire area, approximately 800 000 ha are considered operable, where harvesting and silviculture are practised. Silvicultural treatments include: stand tending of white spruce stands following removal of aspen overstorey; selective logging on sensitive sites; scarification to promote natural seeding or prior to artificial seeding or planting; brush and weed control; and juvenile spacing of lodgepole pine stands (Weldwood 1992). A systematic coverage of about 3000 permanent growth sample plots (PSPs), dating back to the mid-1950s, is maintained. The data from these plots are used to develop local yield models for calculating and updating projections of annual harvest volumes. Virtually all harvested timber is supplied to local pulp and sawmills, which produce only kraft pulp, construction lumber and particle board.

Information on local ecosystem C was derived from previously estimated parameters for the Upper and Lower Boreal–Cordilleran ecoregions in which almost all of the productive portion of the Foothills FMA area is located (Alberta Forestry, Lands and Wildlife 1992). Soil C density in the FMA area was estimated to be 84.3 Mg C ha^{-1}, based on the average for the Cordilleran ecoregion contained in the Canadian Forest Service soil C data base (Siltanen et al. 1995).

Disturbances

The age-class structure of the Foothills Forest, as observed in 1988, is shown in Figure 23.1. No significant harvesting occurred before management began in the mid-1950s, and since then, there have been few losses to fires or insect attack. Hence, the areas in the 0 to 30-year age classes represent, almost exclusively, those regenerated following harvesting, averaging about 4500 ha yr^{-1}. It was therefore possible to estimate the age-class structure of the forest as it existed in 1958, assuming that the older age classes were harvested in proportion to their area in the forest at that time.

The CBM-CFS2 was then initialised using this reconstructed age-class structure, and run forward in time from 1958 for 230 years. Van Wagner (1978) provides estimates of the mean disturbance interval derived for stands in the area of the Foothills FMA (referred to in his paper as "the forest of Northwestern Pulp and Power Ltd."). He found that prior to ca. 1915, the mean disturbance interval was approximately 50 years, but during the period 1915–1960, it increased to approximately 65 years, presumably in response to fire suppression practices in the area.

Biomass carbon pool

Kurz et al. (1992) used aboveground biomass inventory data collected by Bonnor (1982, 1985), to develop relationships between stand age and dry biomass (referred to as "biomass over-age" curves), for 457 forest stand-types distributed across Canada (Kurz and Apps 1994). These curves differentiate four distinct phases of forest biomass accumulation: regeneration delay, immature (growth), mature and overmature (where "stand-break-up" may occur).

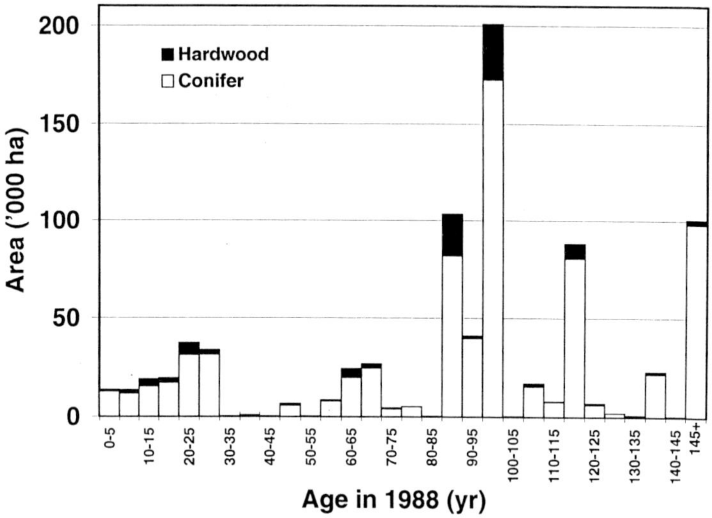

Figure 23.1. Age-class structure of the Foothills Forest determined from 1988 forest inventory data (Weldwood 1992).

For the analysis reported here, the Weldwood PSP data were used to parameterize the CBM-CFS2 biomass-over-age curves for the Foothills FMA area. Weldwood's 1991 Detailed Forest Management Plan explains how growth measurements obtained from the PSPs are used to derive approximately 70 volume yield curves (YCs), considered to represent the range of growing stock conditions within the FMA area (Weldwood 1992). Each stand identified in the forest inventory is matched to one YC, of which 56 are applicable to operable areas. The Weldwood YCs provide separate data for the softwood and hardwood components of each stand-type, reported as merchantable timber volumes (m^3 ha^{-1}) at 10-year intervals ranging from 10 to 150 years, after which they are assumed to remain constant. The approach adopted was therefore, firstly to transform the data for each YC into an equivalent biomass-over-age curve for use by CBM-CFS2; and secondly to estimate the C dynamics of forest biomass in the Foothills Forest using Weldwood's stand inventory data. The inventory data, contained in a GIS data base of over 300 000 records, were aggregated into 56 stand-types, corresponding to each YC. The Weldwood GIS inventory covers only about 50% of the FMA area, but was considered extensive enough to be representative of the entire area-weighted age-class distribution (Figure 23.1).

The data were used to estimate the dry biomass for each 10-year age class from its volume equivalent, using species-specific merchantable volume-to-biomass conversion equations for central Canadian species (Singh 1984). The CBM-CFS2 biomass-over-age curves were then fitted to the biomass yield data. A fixed regeneration delay of five years was assumed, after which the immature growth phase was fitted to a logistic growth function. For YCs where maximum biomass was reached after 100 years (typically conifers), the mature phase was generally considered to last from 150 to 200 years, with the biomass at 150 years being maintained until 200 years. For stands reaching maximum volume before 100 years (typically aspen), the mature phase was considered to last until 160 years. Following the mature phase, stand break-up was simulated using a negative exponential with a decay rate of 0.01 yr^{-1} for conifers and 0.02 yr^{-1} for hardwoods. With natural disturbance and harvesting intervals of order 50–75 years, the representation of the stand break-up phase is expected to have little impact on estimated C dynamics. Figure 23.2 shows a typical example of a CBM-CFS2 biomass-over-age curve fitted to a Weldwood YC.

The productivity of many sites is expected to increase significantly following harvesting of older stands. This occurs because lodgepole pine stands of fire-origin often achieve such high stem densities (sometimes exceeding 100 000 stem ha^{-1}) that volume growth stagnates within a few years (Pearson et al. 1984; Keane and Weetman 1987; Tait et al. 1988). The density of stands regenerated following harvesting can be controlled, leading to significantly improved merchantable volume production (often as much as 50% increase). Relatively few studies however, report corresponding increases in aboveground biomass productivity. With decreasing stand density, Pearson et al. (1984) observed lower root:shoot biomass ratios, and greater aboveground biomass; Keane and Weetman (1987) reported greater leaf area index, and increasing sapwood area to leaf area ratio (both of which would support greater biomass productivity). The hypothesis has been advanced that very densely stocked stands invest a very high proportion of photosynthate in fine root turnover in order to exploit limited soil nutrients and water, with the result that very little aboveground biomass accumulation occurs (JP Kimmins, University of British Columbia, 1988, pers. comm.).

Weldwood represents increased volume growth in lodgepole pine stands regenerated following harvesting by allocating new YCs. Since there is some question about the impacts of density control on biomass productivity, it may be incorrect to assume that the volume-to-biomass ratios determined for fire-origin stands will also be appropriate for those where silvicultural treatments have been applied. The significance of this will be discussed later.

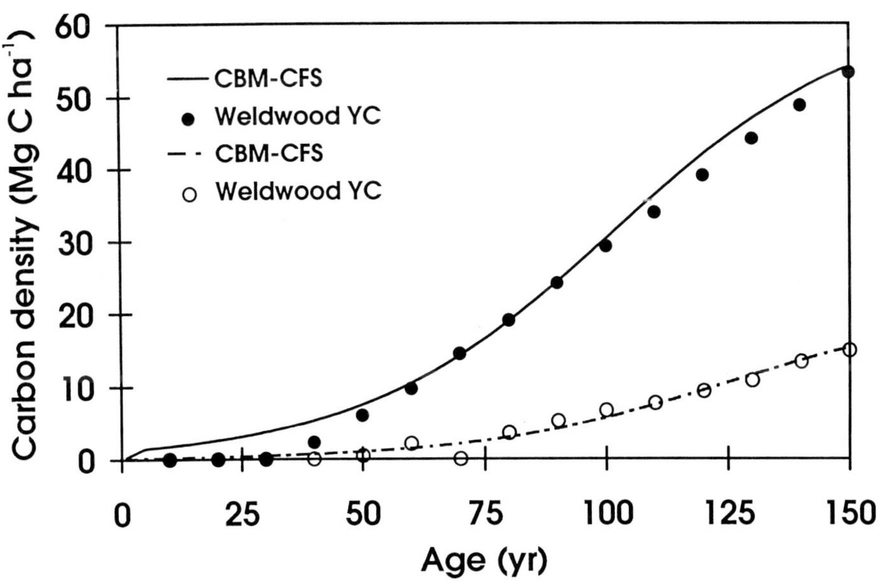

Figure 23.2. Example of a Carbon Budget Model of the Canadian Forest Sector (CBM-CFS) biomass-over-age curve fitted to a Weldwood Yield Curve (YC) for the Foothills Forest. Volume data for softwoods (●) and hardwoods (○) are shown as C density equivalents.

Soil carbon pool

As forest stands grow, decay or are destroyed by disturbances, CBM-CFS2 transfers litter and coarse woody debris from the biomass pool to one of three soil pools. These materials are then allowed to decay using appropriate decomposition constants, releasing much of their C as CO_2, and transferring the residues to a fourth "slow decay" pool (Kurz and Apps 1994).

For the simulations, the C content of the slow soil C pool for each age class of each stand-type was initialised using CBM-CFS2, beginning with the average soil C density reported above, and assuming a 50-year disturbance interval. When a smooth negative exponential age-class structure (Van Wagner 1978) had been obtained (and the total soil C pool equilibrated), the soil C densities obtained for each simulated age class were then applied to the same stand types in the 1958 age-class structure.

Harvesting

Data on current and projected annual harvesting rates, and on the wood consumed annually by company mills, were estimated from the management plan (Weldwood 1992). Harvesting is planned to occur at approximately 2 million m^3 yr^{-1} for the 20 year period 1988–2007, of which 1.9 million m^3 yr^{-1} are expected from coniferous stands, and the balance from hardwoods. Weldwood believes this rate of harvesting to be sustainable indefinitely, assuming that current silvicultural practices are maintained, that environmental conditions do not change appreciably, and that timber losses due to fire and insect pests do not increase from their current low incidence.

Operational harvesting is subject to certain constraints which needed to be considered by CBM-CFS2. Firstly, operable stands must carry a minimum of 47.5 m^3 ha^{-1} of merchantable conifer timber. Secondly, the annual harvest is determined by the need for a sustained supply of 2 million m^3 yr^{-1}; hence the area harvested and regenerated annually may vary. Thirdly, current harvesting is concentrated on stands which are usually overmature, with highest volume areas harvested first (although it will shift to younger stands harvested closer to economic rotation age as the supply of older timber is exhausted). The latter strategy is, however, subject to access constraints, which means that some lower volume stands may be harvested each year. Fourthly, for reasons discussed above, the production from many sites is expected to increase significantly due to silvicultural treatments following harvesting of the older age classes.

Of these four constraints, the first two were easily simulated by CBM-CFS2. The third was less easy because the second constraint required the model to harvest the highest volume (or biomass) stands first, but since the model lacks explicit spatial representation, it was not possible to substitute lower volume stands on the basis of accessibility. This problem was overcome to some extent by arbitrarily limiting the annual harvest from the highest volume stands to 20% of the total. The fourth constraint required a slight modification to CBM-CFS2 to track changes in the areas allocated to each stand-type, as older age classes were harvested and then "converted" to higher productivity sites.

Wood processing

All hardwood material is exported to an oriented strand board (OSB) plant, where the bark is used as an energy source for processing. Conifer roundwood is used for producing lumber and bleached kraft pulp. Sawn timber production is approximately 470 000 m^3 yr^{-1}, while pulp production currently averages about 0.385 Tg yr^{-1} (air-dry) requiring approximately 5 m^3 of raw material per Mg for a total annual consumption of 1.925 million m^3 yr^{-1}. The kraft pulpmill accepts both roundwood and sawmill residues, of which over 60% originates from within the FMA area, the balance coming from other suppliers. From these statistics, annual balance sheets were constructed showing the amount of forest biomass C transferred to the soil pool during harvesting (Table 23.1), and the amounts harvested within the FMA area entering the forest products C pool (classified as pulp products, construction lumber and bioenergy), or dumped as waste, or lost to the atmosphere (Table 23.2).

Using data on the energy contents of various fuels (Hall and Scurlock 1993) and of wood (Bonnor 1985), the energy equivalents of each component of the balance sheet were estimated. Approximately 74% of mill energy requirements are derived from pulp by-products (hogfuel and black liquor), releasing 0.032 Tg C yr^{-1} from locally produced wood for 1.28 PJ yr^{-1} energy. Other energy needs are met from natural gas (25%) and diesel fuel (1%), yielding approximately 0.45 PJ yr^{-1}, which would release an additional 6200 Mg C yr^{-1} assuming comparable conversion efficiencies. Fossil fuel consumed in forestry operations (vehicles, harvesting equipment, etc.) is not included here.

Forest product decay

Forest product decay functions developed for the national-scale CBM-CFS2 forest product sector sub-model (Kurz et al. 1992), were used to estimate releases of C from different forest products as they decay in buildings and landfills, or are burned as waste or for energy. These decay functions were applied to Weldwood's products, assuming that for any particular type of product manufactured within Canada, the geographic origin does not have any significant impact on its average lifespan. In addition, the Weldwood mills produce a certain amount of waste material, the quantity of which needed to be known in order to close the C balance presented in Table 23.2. To do this, it was estimated that 0.38 Mg C is released as CO_2 from the raw material for every Mg of pulp product, following data from the Canadian Pulp and Paper Association (1991, cited in Wellisch 1992). It was further assumed that the C content of wood is 50% of oven-dry weight. Air-dry bleached kraft pulp was assumed to contain 10% moisture, and a solid fraction of pure cellulose with 45% C, while the solid component of black liquor was assumed to be pure lignin containing 70% C (Salisbury and Ross 1978; K. Hunt, Pulp and Paper Research Institute of Canada, Vancouver, 1994, pers. comm.). These values led to an estimate of 66% for the average C content of the mill waste dumped in the landfill, when calculated as the ratio of residual C to residual wood product mass.

Table 23.1. Balance sheet showing estimated contributions of annual Foothills Forest harvest to soils and net transfers of roundwood to mills, in terms of wood volume, biomass, C content and energy equivalents. (o.d. indicates oven-dry mass.)

	NET TRANSFERS TO ROUNDWOOD AND SOIL				
RAW MATERIALS CATEGORIES	(10^3 m^3)	(Gg o.d.)	(Gg C)	(% C)	(PJ)
Net roundwood extracted					
Conifer (to stud- and pulp- mills)	1843	718	359.0	79.2	14.36
Hardwood (to OSB plant)	97	39	19.5	4.3	0.78
Total net roundwood extracted (to mills)	1940	757	378.5	83.5	15.14
Harvesting residues left on-site					
Stumps, tops, broken stems	60	23	11.5	2.6	0.47
Foliage and branches	N/A	126	63.0	13.9	2.04
Total residues (transferred to soil C pool)[a]	60	149	74.5	16.5	2.51
Total harvest, including losses	2000	906	453.0	100.0	18.12

a Volume of residues is from merchantable timber only, but dry mass includes foliage and branches.

Table 23.2. Balance sheet showing estimated contributions of annual net Foothills Forest round-wood harvest extracted in Table 23.1, to forest products, by-products and landfills, in terms of biomass, C contents and energy equivalents. (o.d. indicates oven-dry mass.)

END PRODUCT CATEGORIES	END PRODUCTS (Gg o.d.)	(Gg C)	(% C)	(PJ)
Construction lumber				
Oriented strand board (OSB)	33.1	16.6	4.4	0.66
Dimension lumber	174.6	87.3	23.1	3.49
Total construction lumber	207.7	103.9	27.5	4.15
Pulp				
Kraft pulp produced from local chips	229.3[a]	114.7	30.3	4.59
Bioenergy				
Bark waste from OSB production (used in process at OSB plant)	5.8	2.9	0.8	0.12
By-products from local chips (used at mill)	59.5	29.2	7.7	1.17
Total bioenergy	65.4	32.1	8.5	1.28
CO_2 emissions from pulpmill and OSB plant (excludes fuel emissions)	153.5	61.4	16.2	–
Pulpmill wastes (transferred to landfill)	100.9	66.4	17.5	2.66
Total consumption/production	756.8	378.4	100.0	12.68

a Converted from air-dry mass assuming 10% moisture content.

RESULTS

Figure 23.3a shows estimated changes in C stored in forest biomass, soils, and the total of these, for the Foothills Forest FMA area since 1958, for Experiment I (no human intervention). Only natural disturbances with an average return interval of 50 years are assumed. It can be seen that the C stored in forest biomass decreases steadily during the first 100 years, as the older stands are lost to wildfires and the sites regenerate naturally, after which it stabilises at about 20 Tg C. Soil C shows a slight increase and then slowly decreases to a stable 110 Tg. Projected total ecosystem storage therefore decreases quite significantly over current-day values before stabilising at about 130 Tg C. These results are explainable by the initial age-class structure (Figure 23.1) which does not follow a smooth negative exponential, and which developed during a period of reduced disturbances. Similar results are reported by Kurz et al. (1994) and Kurz and Apps (Chapter 14), who suggest that reduced disturbances in the early part of the 20th century have contributed, until recently, to a net C sink in the Canadian boreal forest.

The consequences of management, including complete disturbance suppression, and sustained harvesting at 2 million m^3 yr^{-1} were investigated in Experiment III, shown in Figure 23.3c. No change in stand productivity following harvesting is assumed (i.e., regenerated stands follow the same biomass/age curve as the harvested stands they replace). Soil C increases to a stable 160 Tg C, after 170 years (i.e., almost 50% greater than in the unmanaged scenario) while biomass C remains at 50–60 Tg C for about 100 years before decreasing, as harvesting shifts from old-growth to younger stands. Total ecosystem C increases to a maximum of about 220 Tg C by about 130 years, before decreasing with the reduction in biomass C, though it appears to be stabilising around 200 Tg C at 230 years.

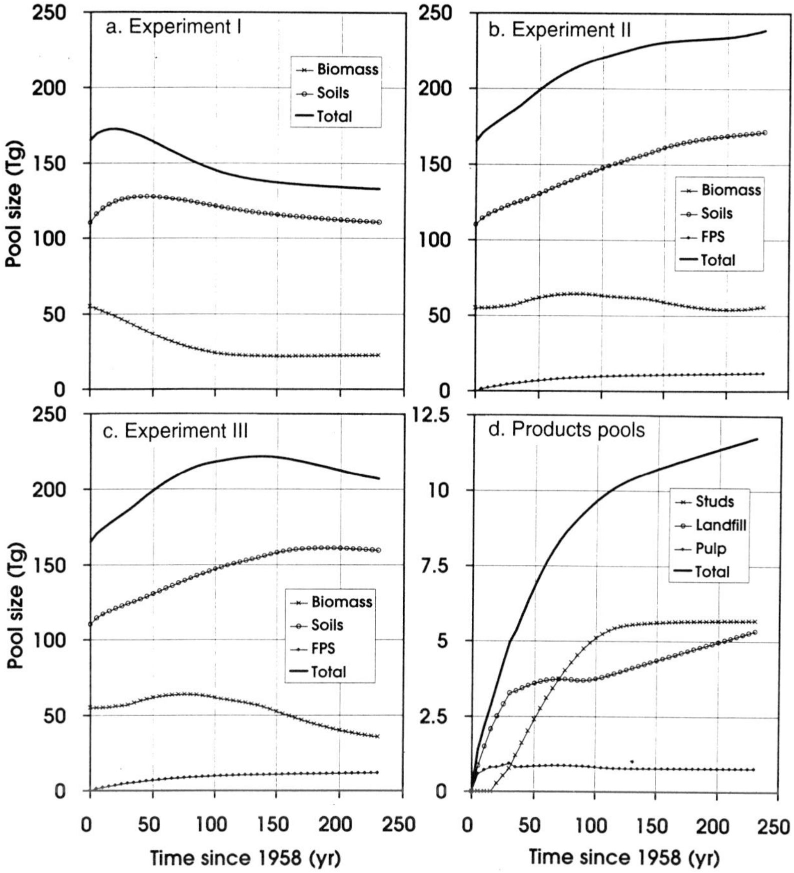

Figure 23.3. 230-year projections of C budget for Foothills Forest for three different management/harvesting scenarios. (a) no management or harvesting, but with natural disturbances occurring at an average return interval of 50 years (estimated from the 1958 age-class structure derived from Figure 23.1). (b) intensive management and sustained yield harvesting, assuming that productivity of lodgepole pine stands increases following harvesting, based on the results of volume yield studies. No natural disturbances assumed to occur since 1958. (c) as (b), but no changes in site productivity following harvesting are assumed. (d) forest products C pool broken down by product categories, for the scenarios simulated in (b) and (c).

Over the same period, Experiment II (Figure 23.3b) shows the additional effect of silviculture, if expected improvements in merchantable volume production are achieved, and if the volume-to-biomass ratios established for fire-origin stands of lodgepole pine are not seriously altered for stands regenerated following harvesting. Silviculture mainly avoids the decline in biomass C after the older age classes are removed, and increases transfers to the soil because of greater stand productivity. This result indicates that planned harvesting rates can be sustained if anticipated improvements in merchantable volume productivity are realised, and will even allow an increase in total C storage compared to the "no silviculture" management scenario

(Experiment III). Consequently, total ecosystem C storage continues to increase throughout the simulation, exceeding values in Figure 23.3c at about 100 years, and appearing to stabilise at about 240 Tg C after 200 years, compared to 135 Tg C for the unmanaged scenario (Experiment I), and 210 Tg C when only fire suppression and harvesting were simulated (Experiment III).

In both managed scenarios, storage of C in soils increases steadily, ultimately reaching significantly higher levels than in the unmanaged scenario, because suppression of wildfire reduces transfers of soil C to the atmosphere, and allows greater transfers from biomass in litterfall. It must be emphasised however, that soil C storage is very sensitive to the assumed decomposition rate. This implies that the estimates of future ecosystem C storage are to be treated with caution, and that the actual soil C storage could be very sensitive to future changes in climate. Nevertheless, these caveats must apply to all scenarios, indicating that within about 100 years, management will likely result in significantly greater ecosystem C storage.

Role of forest products

Figure 23.3d expands the vertical scale of Figures 23.3a–c, in order to examine more closely the long term trends in C storage within the products pool. Note that the harvesting data for Figures 23.3b and 23.3c are identical, since CBM-CFS2 harvests the required timber volume in both Experiments II and III, and differences in tree size are not considered. Initially, pulp and landfills contribute most of total products storage, because lumber production did not start immediately in 1958, and because a high proportion of annual pulp production is immediately transferred to landfills. After production of construction lumber begins, however, its contribution to total product C storage increases rapidly, stabilising at about 5.5 Tg C after 130 years (maximum harvesting rate begins only 30 years into the simulation, and 95% of products are assumed to decay within 100 years). From then on, the C stored in lumber products remains constant, as annual inputs balance losses (25-year half-life). By comparison, pulp products have a much shorter average lifespan, causing production to balance annual losses and hence achieve an approximately steady level under 1 Tg C within 30 years of the start of harvesting.

Role of landfills

The rate of C storage in landfills is directly proportional to the value selected for the recalcitrant fraction. Construction lumber initially contributes only 20% of total products C (when landfill contents are included). After 130 years, however, C storage in lumber products contributes six times the storage in pulp products, even though only 40% of harvested roundwood volume is used in their manufacture (Table 23.2). This apparent benefit of long-lived products is, however, small compared to the steady accumulation of C in landfills. Since 90% of previously manufactured pulp products (and all residues not used for bioenergy) are ultimately discarded in landfills, where it is assumed that 20% will persist indefinitely, there is a rapid accumulation of undecayed pulp material, which contributes significantly to long-term products C storage. Even if the assumption of zero decomposition for this 20% were replaced by one of very slow turnover rates (i.e., thousands of years), the inferences made for the time-horizon of this study would not be substantially affected.

DISCUSSION

Can forest management increase carbon storage?

The results of the simulations reported above indicate that harvesting combined with silviculture are very probably beneficial for long term C sequestration in the specific case of the boreal forest ecosystem considered here. In common with earlier studies (e.g., Chapter 21; Vitousek 1991; Harmon et al. 1990), these results demonstrate that management does affect forest sector C pools, although there are important differences. In general, a matrix of managed stands will store much less total C than the old-growth forest they replace, because the natural disturbance interval governing the average age of the old growth is typically much longer than the economic age of rotation. This is particularly true of the coastal forests of the Pacific north-west considered by Harmon et al. (1990) which exhibit typical natural lifespans of 200 years or more, but which have been replaced by forests managed on a 70–80 year rotation. In our study of a boreal ecosystem, the natural disturbance interval (50 years), is significantly shorter than the typical harvest age (80 years). Hence, management practices aimed at suppressing natural disturbances would also increase C storage.

Secondly, projected increases in lodgepole pine volume production following harvesting are assumed to result in proportionately greater rates of C uptake than would occur in natural (fire-origin) stands of the same age (compare Figures 23.3b and 23.3c). If this hypothesis is correct, then anticipated increases in stand productivity due to silviculture following harvesting will contribute to extra C storage in the lodgepole pine ecosystem.

Contribution of land-filled residues to ecosystem total C storage

At 30 years, total storage in products is about 5 Tg C or 2.6% of the ecosystem total of 190 Tg C. This value is consistent with proportions reported in Matthews et al. (Chapter 24) for global forests (1.3%) and UK forests (1.8%), considering that the Foothills FMA area is managed primarily for wood production and that current availability of large size trees is very high.

The assumption that 20% of wood products discarded in landfills will never decay has a strong cumulative effect on total C storage in the products pool. Anecdotal evidence derived from excavations of old landfill sites suggests that very long decay times are realistic. Not only are anaerobic conditions common in many landfills, but also the by-products of the kraft pulp process may be particularly recalcitrant. Figure 23.3c indicates that if 20% or more of the input material does persist for centuries, its contribution to total ecosystem C storage is also strongly dependent on its C content. The C content was estimated only as a residual term to close the C and mass balances of Table 23.2; hence the true value is currently unknown. Since mass conservation must apply, correct allocation of C among the various end-products could have a significant impact on the contribution of the landfill term. Analysis of land-filled residues would help resolve this problem.

If the assumptions used in the simulations are valid, by the end of the 230-year simulation period, pulp products, and wood products in landfills, would contribute 50% of total products C storage, totalling 12 Tg C, or approximately 5% of total ecosystem C.

Other factors affecting future forest C budgets

The projected long-term increases in C storage due to management and harvesting are based on several assumptions: (1) environmental conditions will not change in the future; (2) complete suppression of natural disturbances (fires and pathogen attacks) can be achieved; (3) projected increases in aboveground productivity resulting from silvicultural treatments are realistic and sustainable; (4) below-ground C dynamics (i.e., fine root turnover) are not significantly changed by management practices; (5) these protection and management procedures require no major increases in fossil energy consumption.

Changes in climatic conditions are expected to lead to changes in forest productivity, as well as in disturbance frequency, resulting from direct effects on growth physiology, seed production and germination success, and indirect effects on the population behaviour of endemic insect pests and disease organisms (e.g., see Chapter 7; Flannigan and Van Wagner 1991; Kurz and Apps 1994; Sirois 1992; Bergeron and Flannigan 1995; Fleming and Volney 1995). Such changes would generally influence the C dynamics of both managed and unmanaged ecosystems, and most likely in the same direction, but detailed analysis of the differential effects are beyond the scope of this paper. Briefly, even if the simulated absolute gains in C storage resulting from management cannot be achieved in reality, it seems plausible that a net increase will occur relative to the natural ecosystem in the area of this case study, assuming identical changes in climate. Conversely, it is difficult to conceive of a situation where the managed ecosystem would lose more C than the unmanaged under a warming climate, when management objectives are aimed at maximising wood production, because the unmanaged/unprotected/unharvested forest is likely to incur more widespread losses from disturbances.

Offsetting fossil energy C emissions

Approximately 7% of harvested roundwood is used for energy (Table 23.2), which represents a net reduction in fossil energy releases of approximately 0.018 Tg C yr^{-1} during processing if compared to a mill entirely dependent upon fossil energy. The total energy content of mill products is approximately 12.7 PJ yr^{-1}, of which approximately 2.7 PJ yr^{-1} is in pulp wastes. After subtracting the 0.45 PJ yr^{-1} currently obtained from fossil sources and 1.3 PJ yr^{-1} derived from bioenergy (Table 23.2), the managed forest could be viewed as a net producer of about 11 PJ yr^{-1}, which if fully utilised would correspond to a reduction in annual fossil C emissions of 0.15 Tg (natural gas) to 0.31 Tg (coal).

A refined analysis should account for all fossil fuel consumed in forest management and harvesting operations. It is expected, however, that the consequent emissions have little impact on the overall net C balance within the FMA area. More importantly, such an accounting should be balanced against the reduced GHG emissions obtained by substituting wood products for equivalents requiring much greater energy inputs to manufacture (e.g., galvanised steel studs). A complete C budget should therefore include the C cost of providing alternatives to wood products to consumers (see also Chapters 18 and 24).

Global implications

At the global scale, the time taken to saturate the forest products pool may be long indeed, because as the CBM-CFS2 simulations show very clearly, it will depend very significantly on the long-term fates of some products. If we accept the assumption that 20% of material deposited in landfills persists indefinitely, and that long-term management of landfills does not alter (e.g., due to changes in waste management policies), then the landfill contribution could significantly increase the effective long-term C storage attributable to global forest products.

In reality, the global products pool is unlikely ever to be fully saturated. Not all forests are exploited commercially, while for many of those which are, sustained yield management objectives are lacking. Furthermore, for some regions, there are inaccurate records of past and current production, and it is likely that current practices will change dramatically in the future as commercial demands and public expectations change. Such considerations will inevitably make estimating the maximum size of the global products sink a difficult proposition, but it may yet be possible to derive a conservative value for its upper limit. Carbon budget analyses similar to the one presented here could be carried out for representative forest management units in other regions of the world. Where maximum sustained yield harvesting is both a stated objective and demonstrably achievable, it should be possible to make estimates of regional saturation times (following first entry to the forest resource) and given estimates of sustained yield forest areas, the approximate magnitudes of the regional products sinks. The total of these regional estimates could then be used in global C accounting as a known offset for current and future GHG emissions.

The magnitude of the global products sink is vulnerable to the possible impacts of climate change. It will therefore be necessary to consider projections of near-future forest vegetation dynamics and potential harvesting yields under global change scenarios, as inputs to any assessment of the size of the long-term products sink. For boreal and temperate forests at least, it is probable that there will be a period of increased ecosystem disturbances during which large amounts of biomass (and soil) C will be released, with less raw material consequently being available for harvesting. There are two major directions forest management can take to help counteract this impact: firstly, increased emphasis on protection from fires and insect attacks (see also Chapters 7 and 8); and secondly development of large-scale facilities for handling wood salvaged from disturbed areas.

CONCLUSIONS

Estimates were made of the past and future contributions of forest ecosystem compartments, including forest products, to total C sequestration in a Canadian boreal forest management unit. Given that the management objective of maximum sustained yield is achievable, the products pool may be viewed as a sink for atmospheric C. The finite size and duration of this sink are dependent upon assumptions made concerning the efficacy of forest management in increasing and maintaining forest productivity, and the longevity of harvested products, particularly those discarded in landfills.

In this specific case study, replacing natural disturbances (with a return interval of about 50 years) by sustained yield harvesting (practised on an approximate 80-year rotation), would result in significant increases in total C storage, perhaps exceeding 50% within 150 years, compared to the unmanaged forest, particularly if silvicultural treatments can succeed in maintaining elevated productivity. If the assumptions made here are correct, then for the managed forest ecosystem, the present-day storage in forest products is about 2.5% of total ecosystem C, potentially increasing to 5% over the next 200 years.

The analysis presented here could be applied to other regions of the world. Relationships established among standing forest biomass, soil C storage, sustained yield harvests, and product utilisation, would be used to determine effective upper limits to the regional products sinks, and their approximate saturation times. From such estimates, and information on the forest areas under sustained yield management within each region, it should be possible to make a realistic assessment of the current and potential future roles of the global forest products pool as a sink for atmospheric CO_2.

ACKNOWLEDGEMENTS

We wish to thank David Cheyne and Chris Cieszewski (Canadian Forest Service, Edmonton), Sean Curry (Weldwood of Canada Ltd., Hinton), Jim Goudie (B.C. Ministry of Forests, Victoria), and Ken Hunt (Paprican, Vancouver) for information and assistance.

REFERENCES

Alberta Forestry, Lands and Wildlife (1992) Ecoregions and Ecodistricts of Alberta, Volume 1. Land Information Services Division, Resource Information Branch, Edmonton, AB, 75 pp.

Apps MJ, Kurz WA, Price DT (1993) Estimating carbon budgets of Canadian forest ecosystems using a national scale model. In: Kolchugina T, Vinson TE (eds), Carbon Cycling in Boreal Forests and Subarctic Ecosystems: Biospheric Responses and Feedbacks to Global Climate Change, U.S. Environmental Protection Agency Report EPA/600R-93/084, Office of Research and Development, Washington DC, pp 243–252.

Bergeron Y, Flannigan MD (1995) Predicting the effects of climate change on fire frequency in the southeastern Canadian boreal forest. *Water, Air, Soil Pollut.* 82: 437–444.

Bonnor GM (1982) Canada's Forest Inventory 1981. Canadian Forestry Service, Environment Canada, Forestry Statistics and Systems Branch, Ottawa, 79 pp.

Bonnor GM (1985) Inventory of Forest Biomass in Canada. Canadian Forestry Service, Petawawa National Forest Institute, Chalk River, ON, 63 pp.

Canadian Pulp and Paper Association (CPPA) (1991) A statement by the pulp and paper industry—global climate change. CPPA, Montréal.

Ecoregions Working Group (1989) Ecoclimatic regions of Canada, First Approximation, Ecoregions Working Group of Canada Committee on Ecological Land Classification, Ecological Land Classification Series, No. 23, Sustainable Development Branch, Canadian Wildlife Service, Conservation and Protection, Environment Canada, Ottawa, 118 pp.

Flannigan MD, Van Wagner CE (1991) Climate change and wildfire in Canada. *Can. J. For. Res.* 21: 66-72.

Fleming RA, Volney WJA (1995) Effects of climate change on insect defoliator population processes in Canada's boreal forest. *Water, Air, Soil Pollut.* 82: 445–454.

Forestry Canada (1993) Canada's Model Forest Program—an initiative for sustainable development. Forestry Canada Publications Distribution Centre, Petawawa National Forestry Institute, Chalk River, ON, 12 pp.

Hall DO, Scurlock JMO (1993) Biomass production and data (Appendix). In: Hall DO, Scurlock JMO, Bolhàr-Nordenkampf HR, Leegood RC, Long SP (eds) Photosynthesis and Production in a Changing Environment: A Field and Laboratory Manual, Chapman and Hall, London, pp 425–444.

Harmon ME, Ferrell WK, Franklin JF (1990) Effects on carbon storage of conversion of old-growth forest to young forests. *Science* 239: 42–47.

Keane MG, Weetman GF (1987) Leaf area–sapwood cross-sectional area relationships in repressed stands of lodgepole pine. *Can. J. For. Res.* 17: 205–209.

Kurz WA, Apps MJ (1993) Contribution of northern forests to the global C cycle: Canada as a case study. *Water, Air, Soil Pollut.* 70: 163–176.

Kurz WA, Apps MJ (1994) The carbon budget of Canadian forests: a sensitivity analysis of changes in disturbance regimes, growth rates, and decomposition rates. *Eviron. Pollut.* 83: 55–61.

Kurz WA, Apps MJ, Webb TM, McNamee PJ (1992) The Carbon Budget of the Canadian Forest Sector: Phase 1, Forestry Canada Inf. Rep. NOR-X-326, Northwest Region, Edmonton, AB, 93 pp.

Kurz WA, Apps MJ, Beukma SJ, Lekstrum T (1994) Twentieth century carbon budget of Canadian forests. *Tellus* 47B (in press).

Pearson JA, Fahey TJ, Knight DH. 1984. Biomass and lead area in contrasting lodgepole pine forests. *Can. J. For. Res.* 14: 259–265.

Price DT, Apps MJ, Kurz WA, Wesbrook M, Curry RS (1994) A "model forest model": steps toward detailed carbon budget assessments of boreal forest ecosystems. *World Resour. Review* 6: 461–476.

Salisbury FB, Ross CW (1978) Plant physiology, 2nd edn. Wadsworth, Belmont, CA, pp 218–220.

Siltanen RM, Apps MJ, Zoltai SC, Mair RM, Strong WL (1995) A soil profile database with organic carbon data for non-agricultural mineral soils in Canada, Can. For. Serv., Inf. Rep. NOR-X-xxx, Northwest Region, Edmonton, AB. (Unpublished manuscript).

Singh T (1984) Conversion of tree volume to biomass in the prairie provinces, Canadian Forestry Service, For. Manage. Note No. 28, Northern Forest Research Centre, Edmonton, AB, 7 pp.

Sirois L (1992) The transition between boreal forest and tundra. In: Shugart HH, Leemans R, Bonan GB (eds) A Systems Analysis of the Global Boreal Forest. Cambridge University Press, Cambridge, UK, pp 196–215.

Tait DE, Cieszewski CJ, Bella IE (1988) The stand dynamics of lodgepole pine. *Can. J. For. Res.* 18: 1255–1260.

Van Wagner C (1978) Age-class distribution and the forest fire cycle. *Can. J. For. Res.* 8: 220–227.

Vitousek PM (1991) Can planted forests counteract increasing atmospheric carbon dioxide? *J. Environ. Qual.* 20: 348–354.

Weldwood of Canada Limited (1992) Hinton Division, 1991 Detailed Forest Management Plan, Forest Planning Group, Forest Resources Department, Hinton, AB, 53 pp.

Wellisch M (1992) MB Carbon budget for the Alberni Region, MacMillan Bloedel Research, MacMillan Bloedel Ltd., Vancouver, BC, 111 pp.

24 WG3 Summary: Evaluating the role of forest management and forest products in the carbon cycle

Robert W Matthews[1,2], Gert-Jan Nabuurs[1,2], Vladislav Alexeyev[2], Richard A Birdsey[2], Andreas Fischlin[2], J Piers Maclaren[2], Gregg Marland[2] and David T Price[2]

[1] Co-Chairs of Working Group 3
[2] Addresses are provided in the participant list

INTRODUCTION

Decisions on land-use and forest management can have a significant effect on the global carbon (C) budget. In order to evaluate the effect of land use and forest management decisions on the global C balance, or to assess the potential for positive change, we need not only to evaluate the current and near-term status of forests, but also to understand the processes and sectors, both ecological and industrial, that are affected by forest management decisions.

Forestry policy and management can change the area of forested land and the C content of forest ecosystems. In combination with wider economic and energy policy they can influence the mean life-span of wood products and energy efficiency. At the regional and operational level, a system for evaluating the C sequestration potential of specific forest management options needs to be able to account for all such impacts on the C budget, depending on what has been postulated. The system thus needs to evaluate impacts on forest ecosystems, forest products, wood processing technology and energy supply and consumption.

Many writers have suggested that forest ecosystems and forest products can be managed in order to slow down the accumulation of carbon dioxide in the atmosphere (see for example Marland 1988; Sedjo 1989; Thompson and Matthews 1989). It is difficult, however, to synthesise from the literature a consensus on the role which forests and the forestry sector can play in sequestering C (Chapter 19). The apparent lack of agreement in the literature can be attributed in part to progressive changes of opinion following the gradual improvement in our understanding of forests in the C cycle, and of what is technically, economically and socially possible in terms of forest management and wood processing. It is also the case that some of the divergences of opinion are more apparent than real, in that some authors and reviewers do not distinguish between an inventory of the current status of forests, and an assessment of the larger impacts that can occur when a decision is made to influence that inventory. On the other hand, conflicting viewpoints have also arisen directly from differences in the scope of individual case studies, and in the methodologies adopted by different research groups. Nabuurs (Chapter 20) has argued that efforts should be made to devise an agreed, standard procedure for the estimation of the C sequestration potential of forest management options.

This overview paper sets out some suggestions for guidelines for procedures for evaluating C sequestration potential, and for the presentation of results for specific forestry schemes. Special consideration is given to the role of wood products, and to the likely impacts of alternative methods of forest management and wood use on C sequestration potential. The

NATO ASI Series, Vol. I 40
Forest Ecosystems, Forest Management and the Global Carbon Cycle
Edited by Michael J. Apps and David T. Price

paper draws heavily on material presented in the plenary sessions of the Workshop, and on the discussions involving members of Working Group 3.

PROCEDURES FOR EVALUATING CARBON SEQUESTRATION POTENTIAL

The apparent disagreement in the literature over the potential impacts of forest management on C sequestration may be the result of the adoption of inconsistent methodologies or definitions of the forestry sector for the estimation of C budgets by different research groups. Confusion may also arise from inconsistent styles of presentation, e.g., published statistics on C sequestration potential often consist of a mixture of results for both stocks and fluxes, which can be difficult to interpret. Carbon budgeting is also a relatively new and untried technique, and we are only beginning to appreciate the possibilities and pitfalls. At this early stage of development, it is especially important that analyses make absolutely clear what is and what is not being included.

Much confusion may be avoided if guidelines for the modelling and presentation of C budgets can be formulated and broadly accepted by the scientific community. The specification of such a standard is beyond the scope of this paper, and perhaps beyond current understanding of the problem, but we propose the following five criteria as central to any standard:

1. The procedure should account for impacts on C stocks and fluxes associated with all relevant C pools:

    - Above- and below- ground biomass;
    - 'Necromass' (or 'mortmass') including coarse woody debris, litter, soil organic matter and peat;
    - Wood products;
    - Fossil fuels (including potential savings through direct and indirect substitution).

    It may not always be possible to compute a budget which accounts for impacts on the stocks and fluxes associated with all relevant C pools, and we acknowledge that there is insufficient information for the construction of complete models in all cases. In such cases the omissions should be stated clearly, and the implications discussed.

2. The procedure should account for initial C stocks and expected fluxes in all C pools prior to implementation of any proposed forestry scheme or change in forest management, as well as final conditions.
3. The procedure should assess both the absolute and the marginal C benefits and costs of any proposed forestry scheme or change in forest management.

    The absolute benefits and costs for a given forestry scheme, presented in isolation from alternative options, may appear to overstate the impact of the proposed scheme. For example, a proposed afforestation programme on a certain area of land may be shown to have significant C sequestration benefits in absolute terms, but the marginal benefits of the programme may be small when compared to the alternative of leaving the land area unmanaged and permitting natural forest succession to take place. It is highly desirable, therefore, to compute marginal benefits and costs relative to both non-intervention, and to alternative strategies for intervention. The estimation of initial C stocks (criterion 2 above) is a prerequisite for calculating marginal benefits and costs.
4. The procedure should compute the C budget and present results for a period of at least 100 years from the proposed time of implementation of any forest management scheme.

5. A full and readily understandable description of the procedure should be published. The basic data on which calculations rely should always be included or referenced and be verifiable.

Procedures conforming to the above criteria still encompass a wide range of possible modelling approaches, all of which are equally valid. For example, the procedure could be implemented by carrying out direct measurement of C stocks and fluxes for all C pools, or by constructing an empirical 'book-keeping' model of C pools, stocks and fluxes, or by developing a physically- and physiologically-based model of the forest sector system. Even highly sophisticated procedures may not meet the criteria specified above, thus complexity does not guarantee correct nor consistent results.

The Intergovernmental Panel on Climate Change and Organisation for Economic Co-operation and Development have made an attempt to define a procedure for the estimation and reporting of national statistics on emissions of greenhouse gases, including emissions and/or sequestration attributable to agriculture, forestry and changes in land use (IPCC/OECD 1994).

The IPCC/OECD procedure for the 'forestry sector' concentrates on the estimation of changes in C stocks in the forest biomass and soil pools, with incomplete consideration of the wood products C pool, and no account being taken of links to consumption of fossil fuels. A separate, unrelated procedure for the 'energy sector' is used to monitor emissions resulting from burning of fossil fuels. The purpose of splitting up the overall procedure across a number of discrete sectors of human activity is to reduce the risks of double-counting of C stocks and fluxes in the calculation of national statistics. The importance of a simple methodology is also emphasised, permitting the same procedure to be applied consistently by different nations with widely varying resources and data at their disposal. However, while the IPCC/OECD methodology may provide a useful beginning for national inventories, it does not adequately quantify the impact of certain forestry schemes on the C balance.

Although it might be possible to use the methodology in the role of a 'global carbon tracking system' (Chapter 17), it fails to meet any of the five criteria specified above. Most importantly, the inability of the procedure to account for linkages between forest management and fossil fuel consumption renders it inappropriate for assessing the potential impacts of alternative forest management schemes and policies.

DATA FOR ASSESSMENT OF CARBON SEQUESTRATION POTENTIAL

At the global scale, considerable research effort has been directed towards quantifying stocks and fluxes of C in forest ecosystems. The most recent global estimates are those of Dixon et al. (1994) who estimate 1146 Pg C in the biomass, necromass and soil of forests, and an estimated flux for the year 1990 of 0.9 ±0.2 Pg C to the atmosphere, primarily due to deforestation. Compared to forest ecosystems, little work has been done on the assessment of global C stocks and fluxes associated with the wood products pool, hence estimates are less certain. Estimates of C stocks and fluxes attributable to forest ecosystems and forest management are derived using procedures to quantify the contributions of vegetation and soils, and to evaluate the effects of land-use pressures, management, manufacture of forest products and bioenergy utilization, on the forest C budgets within specific geographic regions. Such regional assessments must be founded on reliable data, which include:

1. Data on areas and C densities of forest ecosystems (biomass, necromass and soil), and changes in areas and C densities associated with forest growth, succession and management.

2. Data on the fate of harvested wood and the impact of harvesting methods and wood utilisation on the size of the wood products C pool.
3. Data on the impacts of different methods of forest management, wood processing and wood use in terms of consumption of fossil fuels.

Data on carbon in forest ecosystems

Forest inventory data provide extensive and detailed information on forest composition, structure and productivity, vital for assessing current and future levels of biomass. A variety of inventory methodologies are in use, based on either direct survey or remote sensing systems. Methods which include the use of yield tables and models have the important advantage that future growth and timber production can be predicted readily. Forest inventory techniques based on remote sensing methods, including the interpretation of satellite images, are currently under development but must still be translated with the help of traditional forest inventory data or models. Forest inventory data must often be adapted or extrapolated to estimate C stocks in forest biomass. Typically, estimates of timber volume have to be expressed in terms of biomass and then inflated to allow for other forest ecosystem components using 'total-to-merchantable ratios'.

Data on soils, ecosystem types, forest health and past disturbances have been collected for some regions of the world, but for many areas data are unreliable, unrepresentative or simply not available. Global databases on soils and ecosystem distribution have been compiled (e.g., Olson et al. 1983; Zinke et al. 1986), although these are necessarily of coarse spatial resolution. Data on soil organic C content, classified according to soil, ecosystem and land use type, are critical to the assessment of the impact of land-use change and forest management on soil C content. Great care needs to be taken to evaluate changes in soil C content where there is reason to suspect that this is not stable. When assessing the impacts of changes in management on soil C content, data are required at the process level on rates of soil organic matter accumulation, decomposition, turnover rates and humification rates, for use in conjunction with models of soil C dynamics.

Forest health or 'condition' data provide information on a range of phenomena that can affect the long term productivity of forests, for example attacks by pathogens or insects, site degradation and incidence of fire or pollution. These data are essential for predicting long-term future C sequestration by forests under different systems of management. It may not always be straightforward to infer the impacts of natural disturbances on C pools from data which may merely indicate the area of land affected. Data on the severity of damage to an ecosystem by fire and subsequent recovery, for example, are difficult to find and usually only apply to specific case studies.

Data on carbon in wood products

Several global estimates of C stocks in this pool were proposed at the Workshop. Birdsey (USDA Forest Service, Portland, 1994, pers. comm.) extrapolated calculations for the USA (from Heath et al., Chapter 22) to the global level, to arrive at an estimate of 4 Pg C in long-lived wood products (greater than 1 year life-span). Houghton (Woods Hole Research Center, Woods Hole MA, 1994, pers. comm.) proposed a global estimate of 15 Pg C in both short-lived and long-lived products. Kurz (ESSA Technologies Ltd, Vancouver, 1994, pers. comm.) noted that estimates for Canada suggested that the C stored in long-lived products constituted approximately 25% of the total stored in all wood products, a result which appears consistent with those of Birdsey and Houghton. Our own calculations, based on Food and

Agriculture Organisation statistics on global wood harvesting, and decay curves for wood products used by Harmon et al. (1990), suggest that global C stocks in long-lived products lie in the range 2–8 Pg C, which encompasses Birdsey's estimate. Comparison of Houghton's estimate with that of Dixon et al. (1994) for forest ecosystem C, suggests that C in wood products amounts to 1.3% of that in forest ecosystems. Similar calculations based on estimates of C in forest ecosystems and in home-grown wood products in Britain (Matthews, Forestry Commission Research Division, Farnham, UK, 1994, pers. comm.) suggests a ratio of about 1.8% for that country. The ratio for Britain might be expected to be higher than the global value, given that 90% of the forest area in Britain is commercially productive. The above estimates are the best currently available, and suggest that C stocks in the wood products pool are small compared to stocks in forest ecosystems.

In many countries it is quite possible to obtain adequate data for the determination of the quantity of C flowing into the wood products pool, broken down by product categories. In contrast, it is very difficult technically to calculate the quantity of C flowing out of the wood products pool. Different research groups use similar methods to estimate the rate of C loss to the atmosphere due to decay or destruction of wood products (see for example Thompson and Matthews 1989; Dewar 1990; Harmon et al. 1990; Marland and Marland 1992; Nabuurs and Mohren 1993; Fischlin and Bugmann 1994; Karjalainen et al. 1994). Curves are constructed, often by guesswork, for estimating the fraction of C remaining in different categories of wood products since time of manufacture. An assumption may then be made that the shapes of these 'decay curves' for different products have not changed over time; given this assumption, releases of C from the wood products pool will reflect the current and previous inputs, allowing changes in the quantity of C in the wood products pool to be estimated. The robustness of such an assumption is unclear, however; for example the 'half-life' of houses built in 1900 is unlikely to be the same as that for houses built in 1994. Houses are often not demolished so much as renovated, and it is difficult to determine the subsequent fate of wood products as they fall out of use and are re-used, burned or otherwise disposed of. For those products that are dumped in landfills, the rate of decay is dependent upon the product type, local climate and other site factors.

Data on wood products and fossil fuel consumption

The relative magnitudes of the global estimates of C stocks for forest ecosystems and wood products might suggest that wood products make an insignificant contribution to the C budgets of forest ecosystems, and offer limited potential for C sequestration. Nevertheless, at the regional and local scale, and for certain types of forests and systems of management and wood use, C sequestered in wood products could over time exceed C sequestered in forest biomass (see for example Thompson and Matthews 1989). Furthermore, in many cases, carbon dioxide emissions saved through direct and indirect substitution by wood products could constitute the single most important element of the C budget of a forestry scheme (see for example Marland and Marland 1992; Chapters 18 and 19). This important component is, however, the least well understood, few supporting data are available and it is not clear how to proceed with calculations.

There are in particular, methodological difficulties in interpreting data on energy and materials consumption for calculating the quantity of C attributable to indirect substitution. Different materials are used in different quantities for the manufacture of the same good, and have different life cycles and life times. For example, per unit mass manufactured wood is less energy-consuming than steel, but a wooden bridge will need a much larger volume of wood than of steel to achieve the same strength. Furthermore, the wooden bridge may not last as long as the steel one (although if well designed it could last longer). Finally, the materials used to make either the wooden or steel bridge may be recycled or re-used in different ways when the bridge is replaced.

Although direct and indirect substitution could potentially constitute the most significant elements of the C budgets for certain forestry schemes, the actual quantity of fossil fuel displaced by wood fuel and energy-efficient wood products often would not be under control in practice but would be determined largely by market forces. In some instances, for example, it could be argued that making additional energy available through efficient use of biomass fuels would simply increase the total supply of primary fuels without actually substituting for current fossil fuel consumption; on the other hand, this fuel would serve to improve lifestyles and must in the long term meet a demand for energy services that would eventually otherwise be met by fossil fuels. Clearly there is a need for more data on the ways in which wood products are used, and on the materials for which they substitute.

POTENTIAL POSITIVE IMPACTS OF FOREST MANAGEMENT ON CARBON SEQUESTRATION

Following on from the discussion of methodologies above, we have reviewed a range of options for changing forest management and forest policy, and have attempted to identify those options with the most potential for increased C sequestration. We must stress, however, that the following discussion is concerned with the impact of various forestry options in terms of C sequestration only, thus excluding wider discussion of social, environmental and economic impacts. It is also important to recognise that the applicability and benefits of the various options may be very site-specific, and that the most appropriate management decision for any given land area will depend on site-specific characteristics such as productivity, the potential efficiency of harvest, and current land use.

Preservation and protection of existing forests

This was considered one of the most important options, not least because of the significant contribution to atmospheric C estimated to be made by deforestation. Clear-cutting of primary forest continues across the boreal, temperate and tropical zones of the world. Where clear-cutting of long-established virgin forest is followed by the establishment of commercial plantation forests or agroforestry systems, it is doubtful that the C released to the atmosphere will ever be fully recovered within the ecosystem. This may not necessarily be the case, however, for forests that are subjected to frequent disturbances, such as interventions by fire (for a specific example see Chapter 23). Marland and Marland (1992) have shown that for sites with large, mature trees and low productivity, it can take many years after harvest to return to the original state in terms of net C sequestered from the atmosphere. In many areas where deforestation is followed by major land degradation, protection of existing forests may be the most significant action that can be taken to maintain and enhance C sequestration potential (P Fearnside, National Institute for Research in the Amazon, Manaus, 1994, pers. comm.).

Afforestation

On many sites, notably certain lands previously denuded of trees or degraded by exploitation for agriculture, afforestation is a significant option for achieving C sequestration. Afforestation can take the form of tree planting or natural succession or some combination of the two. The impact of afforestation is highly site-specific, however. For example, large-scale afforestation for C sequestration is likely to be inappropriate on sites of low potential productivity with an existing high soil C content, notably peatlands, where site preparation and tree planting are likely to result in a significant loss of C from such soils.

Other options

In addition to forest protection and afforestation, there are a number of forest management tools which could also be applied to existing forests and are worthy of consideration as part of a strategy for C sequestration:

- Improved utilisation of wood harvesting and processing residues (e.g., for biomass fuel). This option is only viable if site productivity is not compromised by increased removals of biomass. Purpose-grown wood fuel crops could also be established on ex-agricultural sites. This option is only viable if excessive energy consumption is avoided in fuel processing and fertiliser applications to maintain productivity.
- Longer rotation ages for managed stands of trees ('elevation of standing crops'; Fischlin and Bugmann 1994). The prime objective of this option is to increase the total quantity of C on the forested site. This could involve simply retaining even-aged stands for longer between clear-cuts (Chapter 21), or could involve the introduction of 'continuous-cover' management methods or conversion of coppice to high forest.
- Site preparation for forest establishment, provided that care is taken not to consume excessive fossil-fuel energy in the process nor to disturb existing stocks of C at the site. Some site preparation techniques may produce benefits extending for several rotations.
- Improving site productivity and C density by means such as fertilisation, provided that the fossil fuel energy consumed to make the fertiliser does not negate increased productivity.
- Matching of tree species for planting to site types to achieve optimum productivity and C density.
- Tree breeding to improve productivity and C density.
- Silviculture to increase the proportion of harvested wood in high-value product categories (e.g., thinning, pruning). High value wood products are more likely to be long-lived, thus it may be possible to enhance the accumulation of C in the wood products pool. Perhaps more importantly, high value sawnwood is among the least energy-intensive of materials and can substitute for steel, concrete, aluminium and other materials with resultant energy savings and reductions in consumption of fossil fuel. The manufacture of wood products is often made more energy-efficient by using scrap wood as a fuel source for the manufacturing process.
- Urban forestry. Besides the stored C, trees in urban areas usually improve the city climate and may contribute to reduced consumption of fossil fuels for air conditioning or heating of buildings.

The evaluation of the relative importance of the above options, in terms of their potential for sustaining or increasing C sequestration, requires information not only on the relative effectiveness of each option per unit of land area, but also on areas of land over which each option can be appropriately applied. Not all of this information is currently available but it is apparent from the diversity of ecosystems and types of land use around the globe that there are no blanket prescriptions for forest management and forest policy to achieve an enhancement of C sequestration by forests.

WIDER IMPACTS OF FOREST MANAGEMENT TO ENHANCE CARBON SEQUESTRATION

The wider impacts discussed below are of peripheral relevance to a consideration of methodologies for the estimation of the C sequestration potential of forests and forestry schemes and a more thorough discussion may be found in Duinker et al. (Chapter 32). Nevertheless, it is in this wider context that our results, conclusions and recommendations will be judged and acted on by policy-makers, forest managers and the general public. When evaluating afforestation schemes or other proposals for forest management for their C sequestration potential, there is a risk that the consequences for other benefits accruing from the forest may be neglected. Concentrating on C sequestration alone can be useful for purely analytical purposes, but we have to keep in mind that both negative and positive side effects can occur. In certain circumstances, for example, trying to maximise on-site or off-site C stocks through intensive management of a forest ecosystem could reduce ecological diversity. On the other hand, ecological diversity could be increased if such management was implemented on lands previously managed in marginal agriculture or in other degraded states. If not too much emphasis is put on C sequestration, then it should be possible to reap the traditional benefits of the forest at the same time. A local community in the Sahel region may benefit from shelterbelts as being sources of fuelwood or urban climate may improve because street-side trees have been planted.

In fact, there are a number of socio-economic issues that bear importantly on the use of forests, wood products and biofuels for mitigating the increase of CO_2 in the atmosphere. Perhaps most important is the fact that forest lands serve a variety of public and private purposes including habitation, recreation, aesthetics, protection of watersheds, protecting biodiversity, and so on. There is widespread public sentiment attached to the preservation of forests. It is frequently envisioned that using forests to mitigate atmospheric CO_2 is synonymous with increasing the extent and C density of forests and people do not easily perceive that this goal is often best served by careful and efficient harvest and utilisation of forest resources.

Conflict may arise between short-term and long-term objectives in that, for example, early net losses of C to the atmosphere may be followed by increased sequestration later. To select the forest management strategies that will most effectively serve long-term goals may require planning horizons longer than those generally used in the political process and achieved by setting short-term goals. Long-term planning of forest strategies may also bear on world timber markets and world trade in timber and exploitation of forest resources. The point is that the way forests are managed in one area will have an impact upon forest management in other areas as well. For example, protecting forests in one area may simply hasten their destruction elsewhere and increasing wood production in one area may depress markets elsewhere.

Ultimately, we can expect that any C policy-driven efforts aimed at achieving an increase in forest area or forest products to mitigate increases in atmospheric CO_2 will be strongly influenced by the perceived value of forest land and the demand for forest products. Forest expansion can be stimulated if incentives are offered to encourage displacement of energy-intensive materials by wood products or if a market for wood is generated in the energy sector or if tax or other public policy creates incentives for storing C in forest lands.

Normally, forest policy and forest management are determined primarily by economic considerations, for example an afforestation project may be justified in terms of a cost-benefit analysis. Having demonstrated the apparent viability of a particular forestry project in purely financial terms, the analysis may be supplemented by a consideration of other costs and benefits, including C sequestration. Of course, analysis of the potential damages from climate change may lead us to conclude that there is great economic benefit in using forest land to mitigate net emissions of C to the atmosphere and to adopt public policies that give greater value to this objective.

ACKNOWLEDGEMENTS

The authors acknowledge the advice of two anonymous referees and the comments of the editors.

REFERENCES

Dewar RC (1990) A model of carbon storage in forests and forest products. *Tree Physiol.* 6: 417–428.

Dixon RK, Brown S, Houghton RA, Solomon AM, Trexler MC, Wisniewski J (1994) Carbon pools and flux of global forest ecosystems. *Science* 263: 185–190.

Duinker PN, Sedjo RA, Fearnside PM, Grainger A, Sathaye J, Solberg B, Stewart RB, Strakhov VV, Woodwell GM (1995) Human dimensions of the forest-carbon issue. (Chapter 32, this volume).

Fischlin A, Bugmann HKM (1994) Think globally, act locally! A small country case study in reducing net CO_2 emissions by carbon fixation policies. In: Kanninen M (ed) Carbon balance of the world's forested ecosystems: towards a global assessment. Proc. workshop Intergovernmental Panel on Climate Change, AFOS, Joensuu, Finland, 11–15 May 1992, Painatuskeskus, Helsinki, pp 256–266.

Harmon ME, Ferrell WK, Franklin JF (1990) Effects on carbon storage of conversion of old-growth forests to young forests. *Science* 247: 699–701.

Heath LS, Birdsey RA, Row C, Plantinga AJ (1995) Carbon pools and fluxes in U.S. forest products. (Chapter 22, this volume).

Intergovernmental Panel on Climate Change and Organisation for Economic Cooperation and Development Joint Programme (1994) Greenhouse Gas Inventory. IPCC draft guidelines for national greenhouse gas inventories. 3 vol. IPCC/US-EPA, Bracknell, UK, and Washington, DC.

Karjalainen T, Kellomäki S, Pussinen A (1994) Role of wood based products in absorbing atmospheric carbon. *Silva Fennica* 28: 67–80.

Marland G (1988) The prospects of solving the CO_2 problem through global afforestations. Report for the US Department of energy. Oak Ridge National Laboratory Report DOE/NBB-0082. Oak Ridge National Laboratory, Oak Ridge, TN, 66 pp.

Marland G, Marland S (1992) Should we store carbon in trees? *Water, Air and Soil Pollut.* 64: 181–195.

Nabuurs G-J (1995) Significance of wood products in forest sector carbon balances. (Chapter 20, this volume).

Nabuurs G-J, Mohren GMJ (1993) Carbon fixation through forestation activities. Institute for Forestry and Nature Research Report 93/4. FACE and Institute for Forestry and Nature Research, Wageningen, 205 pp.

Olson JS, Watts JA, Allison LJ (1983) Carbon in live vegetation of major world ecosystems. Oak Ridge National Laboratory Report ORNL-5862. Oak Ridge National Laboratory, Oak Ridge, TN, 152 pp.

Sedjo RA (1989) Forests to offset the greenhouse effect. *J. Forestry* 87: 12–15.

Thompson DA, Matthews RW (1989) The storage of carbon in trees and timber. Forestry Commission Research Information Note 160. Forestry Commission, Edinburgh, 5 pp.

Zinke PJ, Strangenberger AG, Post WM, Emanuel WR, Olson JS (1986) Worldwide organic soil carbon and nitrogen data. Oak Ridge National Laboratory Report ORNL/CDIC-18. Carbon Dioxide Information Centre and Oak Ridge National Laboratory, Oak Ridge, TN.

Part IV

The human dimension: the global carbon cycle and socio-economics

IV The human dimension: the global carbon cycle and socio-economics

Introduction

Michael J Apps and David T Price

Members of Working Group 4 attempted to account for human needs and impacts in the understanding of the role of forests in the global carbon (C) cycle. A major factor in the post-industrial era is the rapidly expanding human population, and the attendant acceleration of land-use change which directly influences C storage in global forests, particularly in the tropics. The important roles of forest ecosystems and their management are increasingly being recognised in global strategies proposed to mitigate the effects of anticipated climate change. Many of these strategies however, focus entirely on the biophysical aspects of the problem while ignoring human needs, and in the opinions of some, ignoring the socio-economic realities facing developing and developed countries alike. If human needs are not fully considered in international policy and management decisions relating to the world's forests, it is likely there will be increased suffering and this will cause greater problems, both within and between countries, hence delaying or even destroying the opportunity for progress.

The papers in Part IV address a diversity of topics, including: methods and results of economic analysis for comparing strategies to mitigate C emissions; socio-economic causes and impacts of deforestation and forest degradation; possible strategies to prevent tropical deforestation; problems associated with adding C sequestration to forest management objectives; international polarisation caused by differences in viewpoints of developed and developing nations; and the need for global action to promote sustainable forest management at local, national and international levels.

Economic analysis to evaluate and compare strategies for mitigating carbon emissions

Methods of economic analysis to compare the costs and benefits of different management options for mitigating C emissions are discussed in depth by Sedjo (Chapter 25), Sathaye (Chapter 26), Solberg and Hoen (Chapter 29) and Duinker (Chapter 30). The available options may be summarised as follows (see also Part III):

1. increase standing stock of C (through forest protection, forestation, agroforestry and community forestry).
2. increase stock of long-lived products (increased efficiencies of harvesting and utilisation, including recycling).
3. use forest products in place of materials that release more C in their manufacture (indirect energy substitution).
4. substitute forest biomass for fossil fuel (direct energy substitution).

NATO ASI Series, Vol. I 40
Forest Ecosystems, Forest Management and the Global Carbon Cycle
Edited by Michael J. Apps and David T. Price

Sedjo points to the relatively few comprehensive cost-benefit analyses of land-use activities for C sequestration, suggesting this is really a problem of *system analysis*. He explains that it is often difficult to determine the most useful boundaries for a given analysis; e.g., when forests are harvested, one must consider the fate of the C removed, as well as biomass regrowth, to get a proper accounting of total C flows (an issue also addressed by Solberg and Hoen). Life-cycle analysis of forests and wood products must be included in comprehensive analyses; e.g., when wood products or residues are burned, to what extent do they offset the use of fossil fuels? (See also Schlamadinger and Marland; Nabuurs; Maclaren, and Matthews et al., Chapters 18, 20, 21 and 24, for more detailed discussion of these topics.)

Current economic thinking centres on the societal value placed on the *damage avoided* by not emitting an additional unit of C. Both Sathaye and Sedjo suggest that it is very difficult to quantify such benefits. The problem has been tackled by considering the notional values proposed for C emission taxes (notably in Norway where such a tax has actually been levied—see Chapter 29 by Solberg and Hoen). Initial estimates for a U.S. carbon tax of 200–250 US\$ Mg^{-1} C (as suggested by some according to Sedjo) would suggest that society as a whole believes it is worth paying at least this much to avoid damage caused by emissions. More recent attempts to quantify the actual damage due to marginal C emissions suggest that a more realistic cost could be as low as 5–20 US\$ Mg^{-1} C.

Solberg and Hoen review two separate techniques. The first of these is *cost efficiency analysis* which allows intercomparison of mitigation strategies both within the forestry sector and with other sectors. This may yield positive indexes where there is sequestration per unit of investment, zero values where the strategy is profitable but with no impact on C sequestration, or even negative values where the strategy releases more C than is emitted. A central problem when comparing between different sectors is the choice of discount rate, which when high tends to favour mitigation strategies with short time-horizons. Another limitation of this technique is that it does not account for *interactions* between different forestry measures, so that only sub-optimal estimates can be obtained.

The second technique is *integrated forest management analysis*—a linear programming method for comparing the performance of forestry with other sectors for C sequestration. It uses costs and benefits related to timber production and CO_2 uptake/loss (including decay from products), to estimate potentials and cost efficiencies for increasing CO_2 uptake through forest management. As suggested by Sedjo, Solberg and Hoen use values for present and future fossil fuel tax derived from a Norwegian government study. Based on 1990 stumpage prices, and the hypothetical C tax, they calculate the future value of 1 m^3 of timber in terms of the monetary value of CO_2 equivalents it contains, finding it to be considerably larger than the present day timber stumpage value. If all CO_2 emissions are taxed and all documented CO_2 accumulations receive a government subsidy, then a typical even-aged stand would have a period of C income (between regeneration and harvesting), followed by a period of C costs from clearcutting until all harvested biomass is decayed. The longer the biomass takes to decay, the longer can the C payments be deferred. Land value can similarly be defined in terms of its capacity to sequester C. The introduction of a direct price on emissions and sequestration allows the model to determine the optimal forest management strategy, i.e., that which maximises the combined net present values (NPV) from both timber and CO_2 sequestration.

Evaluations of different strategies for mitigating carbon emissions

Sathaye reviews the range of cost estimates in the economic evaluation of mitigation options, focusing on developing economies. He uses *cost-effectiveness* as an indicator to rank different mitigation options, but makes the point that since the value of C sequestered—or the costs of future investments—are not discounted to present values, this technique can make no distinction between *plant and store* and *sustainable* approaches to C management. (This differentiation is very similar to one made by Grainger (Chapter 27) for C *stores* and C *sinks*.) Drawing upon published literature, Sathaye finds that *forest protection* costs are typically low (1–15 US$ Mg^{-1} C) and that costs for managed forests are somewhat higher, but still not prohibitive (1–95 US$ Mg^{-1} C, with many options costing less than 15 US$ Mg^{-1} C). These costs do not however include those associated with land, maintenance, overhead, monitoring and evaluation. Such opportunity costs will vary considerably, being very low for developing countries (he cites China as an example where opportunity costs may be effectively zero, compared to close to 150 US$ ha^{-1} for the USA). Conversely, the profit (e.g., from timber production) should exceed opportunity costs significantly in some cases. Hence, Sathaye concludes that the direct costs for reducing net C emissions are low—even before some of the offsetting benefits are included—in comparison to the levels of proposed fossil-fuel taxes being discussed. Sedjo arrives at a similar conclusion in an analysis of C sequestration opportunities in forest plantations in Argentina.

Solberg and Hoen compare the cost efficiencies of 11 possible mitigation strategies applied to forest biomass management in Norway, assuming no future changes in: climate, growth rates, end uses and decay factors for wood products, and the relative costs of goods and services. They find that after 10 years, only four strategies provide significant sequestration benefits (of order 1 Tg CO_2 yr^{-1}), namely fertilisation, afforestation of peatlands, forest drainage and bioenergy in forestry, while total anthropogenic emissions (from Norway) are 34 Tg CO_2 yr^{-1}. In the longer term (40 years or more), other options, such as afforestation programs in remote areas of Norway, should also provide significant benefits. Even though these preliminary estimates were not fully optimised, the cost efficiencies of many forestry measures to store C were found surprisingly competitive with other sectors.

Solberg and Hoen's second study using integrated forest management analysis allowed optimisation, given the objective of maximised C sequestration, under various constraints such as timber harvesting rates, and conservation of some areas for biodiversity or recreation. They find that it is possible to show a profit on many sequestration options even when they are subjected to reasonable management constraints, but list several uncertainties and risks which require further investigation. These are (1) the effects of natural mortality when keeping old-growth forests beyond normal felling age; (2) end use and decay of products (e.g., what proportion of sequestered C is emitted as methane rather than CO_2?); (3) choice of an appropriate discount rate to use in C flow studies; and (4) assessing the impacts on total resource allocation in forestry when C assimilation and emission are priced with different magnitudes of C tax.

Conversely, Sathaye, perhaps reflecting his focus on the developing countries, discusses the needs for better data such as: full cost-benefit analyses, including better estimates of opportunity costs; alternative indicators of cost-effectiveness; and national and sub-national data/analysis on mitigation options. He also suggests that there should be a focused research effort on developing a good C discount rate to account explicitly for the decay of C from the atmosphere. Such a discounting would put increased weighting in favour of projects with near-term benefits.

Sathaye also points to a need for improved estimates of land availability. Grainger addresses this to some extent, by providing area estimates of degraded tropical lands with a view to estimating the availability of these lands for rehabilitation. He distinguishes two main approaches to assessing potential for C mitigation, which he terms the *carbon stores* approach (forests are viewed only as *static* C stores) and the *carbon sinks* approach (forests are viewed as *dynamic*, renewable resources). In the first view, logged forests are commonly thought of as degraded, while in the second they may be treated as consequences, possibly transient, of socio-economic use. (In his paper, he attempts to assess the tropical landbase from both viewpoints.) Grainger also differentiates *technical suitability*, an index based on a comparison of potential biomass density to actual biomass density, from *actual availability* which he attempts to assess by examining the causes of differences in technical suitability. To do this he first proposes a *taxonomy of degraded* forests (for the sub-humid tropics) and uses this to stratify degraded areas. He then examines the type of rehabilitation required, assesses its cost (economic and social) and uses these to gauge its actual availability.

Deforestation and land degradation

Stewart and Maini (Chapter 31) review the causes and extent of global deforestation. They suggest that in general, deforestation and forest degradation can be attributed to mistaken government policies which, though adopted for worthy objectives, in fact lead to environmental deterioration. Deforestation is now considered to be the result of several interacting factors, including government policy, fiscal incentives and the interests of individual land-holders, with examples cited for Amazonia, sub-Saharan Africa and Southeast Asia. Although the major mechanisms are clearance for agriculture and logging, other factors are fuelwood collection, poverty, iniquitous land distribution, and exogenous factors such as political corruption, armed conflict and foreign debt.

In recent years, there have been slight increases in temperate forest cover, and reductions in boreal forest area have stabilised, but tropical forest coverage continues to decline worldwide, fastest in Asia, followed by Latin America and Africa (see also Brown, Chapter 11; Grainger, and Fearnside, Chapter 28). Studies show that deforestation rates may be declining in some areas, notably Brazil and Southeast Asia, but the proposed reasons for this are not reassuring (India being a conspicuous exception), since they relate mainly to near-exhaustion of exploitable forest areas and recent economic slow-downs.

Agricultural expansion, currently proceeding at about 10 Mha yr^{-1}, is a major obstacle to reducing deforestation. Land degradation due to inappropriate agricultural use often follows 5–10 years after clearing the forest, hence only about half the area of cleared land is added to that used for permanent agriculture, while the balance becomes unproductive and abandoned (see also Fearnside). Stewart and Maini therefore argue (as do Brown et al. in Chapter 17) that using existing agricultural land in a more productive and environmentally sustainable manner would be an effective means of reducing tropical deforestation.

Fearnside reviews forest management in the tropics, remarking at the outset:

> *"Unfortunately logging normally functions as a prelude to deforestation, regardless of whether it is stamped as sustainable management in official documentation."*

He argues that the objectives of sustainable forest management, even where technically feasible, are often undermined by socio-economic pressures which reduce the C sequestration benefits of forests. He demonstrates this by examining the theory and examples of various management schemes, drawing strongly on his knowledge of Brazilian Amazonia. Some schemes are not technically viable, he asserts, while for those that are, the claims of sustainability are often grossly exaggerated (see Brown, Chapter 11). Many governments of tropical countries lack means of enforcing adherence to sustainability objectives, and in some cases are subject to corruption.

It is often argued that sustainable forest management will prevent forest destruction because it guarantees long term profitability, but Fearnside claims the reality is that the high discount rates often used in economic analyses generally indicate no profit in sustainability—hence short-term profiteering dominates. Forests are being destroyed faster in Southeast Asia, than in Amazonia, he suggests, precisely because they are more valuable. Large areas of cleared land already exist in Amazonia, but it is cheaper to decree settlement on public (forested) land than to expropriate from private ownership, thereby extending the destruction of native forests.

It is evident both from previous work, and from papers presented in this book, that deforestation will continue to be a significant source of C emissions for the foreseeable future. Stewart and Maini summarise:

"The critical issue ...is how to slow, arrest and eventually reverse this process."

Scientific and socio-economic impacts of deforestation

Stewart and Maini point to the principal effects of deforestation: loss of energy, food, shelter and other resources for local inhabitants. Because of the extent of deforestation and forest degradation, many areas of the developing world already experience acute shortages of fuelwood, and in some tropical regions over-exploitation of forest lands for fuelwood is almost as extensive as deforestation. It must however be recognised that deforestation is often needed to provide land for agriculture and urban expansion with the material benefits these can bring.

Grainger proposes an alternative view. Tropical lands may be considered 'degraded' i.e., *physically and biologically sub-optimal*, and therefore attractive as sites for establishing C plantations. But, Grainger asks, what does *vegetation degradation* mean? Are such lands 'degraded' in the eyes of the people inhabiting them? He suggests there needs to be a way of including socio-economic analysis in the concept, introducing the idea of *departure from cultural normality* as a measure of degradation—where cultural normality integrates both socio-economic and ecological criteria. He argues that there is usually some reasonable balance between an acceptable level of site degradation (i.e., degraded from the ideal or potential state of that site) and the socio-economic benefits obtained by the exploitation of that land (assumed to cause this degradation). He also points out that the intensity of human intervention and the inherent resilience of the ecosystem must be considered together. When human pressure is too intense, the ecosystem loses its capacity to generate the socio-economic benefits and becomes increasingly degraded.

Strategies to prevent and reverse deforestation

In discussing the socio-economic causes and impacts of tropical deforestation, Fearnside reviews the range of possible land management options available to mitigate these, together with assessments of their strengths and weaknesses. Other authors, notably Sedjo and Grainger, contribute further information on particular aspects of managing forests for C sequestration.

Plantations

Sedjo reviews the costs of sequestering C in forest plantations (in biomass and products), using the results of a case study from Argentina. He concludes that where growth rates are high, and C releases associated with conversion from present use are low, plantation forestry can be a cost-effective approach to C sequestration, at least over the first rotation.

Forest plantations however, store relatively little C compared to native forests, but more than agricultural land use. In the tropics, Fearnside thinks that the greatest potential is in using them for bioenergy (see also: Schlamadinger and Marland, Chapter 18; and Nabuurs, Chapter 20). He also draws attention to the *leakage* of C benefits from tropical forest plantations: although they are a technically valid means of increasing C sequestration in cleared areas, their success is often exaggerated by cost-benefit analyses which overlook local socio-economic impacts, and undervalue alternative approaches such as maintaining the standing forest for C storage.

In order to assess the dynamic *carbon sink* approach, Grainger points out the need to account for socio-economic lags and delays. These have both temporal and spatial dimensions. It takes time to create awareness, and provide the necessary infrastructure (e.g., to establish nurseries and prepare planting sites). The diffusion of adoption of afforestation processes also has a spatial component—a given program starts in one or two places and spreads from there. Thus targets for afforestation rates will not be attained quickly because of these delays and because of their influence over long-term patterns of national land-use change.

Agroforestry

Agroforestry is often useful for providing greater economic stability, because it reduces the vulnerability of farmers to risks such as market fluctuations and pest and disease outbreaks. Agroforestry typically maintains larger C stocks than large-scale agriculture (but less than the tropical forest it often replaces), although C benefits are also often claimed because it is assumed to slow deforestation. Fearnside argues such claims are exaggerated because deforestation is often not related to subsistence agriculture; and because opportunities to expand agroforestry will often be limited by available inputs and markets for produce.

Sedjo provides data showing that agroforestry is typically a more cost-effective method of sequestering C in tropical regions than plantations.

Managed secondary succession

Natural forest regrowth on degraded lands (such as abandoned cattle pasture) can theoretically be managed to enhance C storage, as an alternative to replacement for crops or pasture. Fearnside shows that such management of *secondary* forests, in Amazonia, is often an inexpensive ploy used by land owners to avoid their abandoned land being classified as *unproductive*, and thereby subject to higher taxes or expropriation for agrarian reform. Many of these land owners previously benefited from government subsidies to clear the same land and establish pasture! He concludes that managed secondary succession should be left to small farmers.

Non-timber forest products

Fearnside explains that exploitation of forests for Non-Timber Forest Products (NTFPs), such as rubber, is useful in protecting them against commercial timber exploitation. Management for NTFPs is typically labour-intensive, hence providing employment and other collateral benefits, including retention of forest C. With local involvement, the managers of NTFP areas are dedicated to maintaining the forest against encroachment, although there are invariably some local impacts due to the activities of the workforce (e.g., growing food). In Brazil, a relatively new land tenure system which favours long-term sustainable management has reserved a small but important 0.6% of forest area for NTFP extraction.

Protection of standing forest

Fearnside firmly believes that the best opportunities for storing C in the tropics is through maintenance of standing forest. He points out that there are strong environmental benefits, such as maintaining biodiversity and supplying the hydrological cycle (as in Brazil), in addition to C storage. He states: "*Deforestation is strongly influenced by government policy decisions that have little direct connection with financial costs.*" Measures aimed at preventing deforestation (see also Houghton, and Brown et al., Chapters 10 and 17), will have socio-economic costs and benefits, including leakage effects; consequently it is necessary to *scale up* to levels of programs, regions or the globe to assess properly the net impacts of such measures. In concluding, Fearnside suggests a number of ways in which protection of standing forest can generate benefits, at least for some members of the population, including guarding the forest, botanical collection, forest inventory work, and laboratory analysis. Maintaining standing forests promotes significant C storage, and other environmental and socio-economic benefits, for relatively low investment. Reducing deforestation rates would have more immediate benefits than establishing protected areas, but requires better understanding of the causes of deforestation and predicting the impacts of change.

Implications of adding carbon to forest management and policy agendas

Using examples from Canada, Duinker (Chapter 30) develops a unique perspective of the problems likely to occur when foresters are confronted with demands to consider C emission mitigation as an additional management objective. In essence, he believes that the adoption of C management objectives in the near-future is unlikely, given (1) the current state of Canadian national forestry policy, where business relationships between land-owners (the Provincial governments) and tenants (forest products companies) have become complex and difficult; (2) increasing public involvement in forest management and policy which add to the complexity; (3) a myriad of conflicting industrial and environmental management objectives, in which C management is likely to achieve only low priority; (4) lack of public agreement on even basic policy issues, including which areas should be removed from commercial exploitation, future harvest levels (are projections too low or too high?), and whether more or less intensive management is desirable; and (5) increasing conflicts over forest land-use throughout the country. He concludes "*...the present time is simply not right for major changes in forest policy in favour of carbon*". Although domestic and international pressures may force the issue into the public limelight, he believes that if this happens in the near future it will only worsen tensions within the Canadian forestry community.

Duinker also points to the lack of certainty about what should actually be done. He draws attention to the general expert agreement that only a short-term, temporary slow-down of emissions is all that can reasonably be achieved with existing forest ecosystems (e.g., see Schlamadinger and Marland, Chapter 18; and Maclaren, Chapter 21), which runs counter to the long-term management objectives that most foresters embrace. He suggests that short-term transitional policies for C sequestration would only be acceptable if they were accompanied by significant financial incentives designed to compensate the associated reductions in long-term gains from existing management.

Though there are undoubtedly some management actions that would contribute to reductions in fossil C emissions, he points to the new issues that would beset forest managers already dealing with a full array of conflicting objectives. Thus, is it better to protect old growth for its storage potential? (e.g., see Fischlin, Chapter 13; and Fearnside) or to liquidate it into products, capitalising on higher short-term rates of uptake? (e.g., Schlamadinger and Marland, Chapter 18, Nabuurs, Chapter 20; Price et al., Chapter 23). Even where there is no doubt about the potential C benefits of a management practice, Duinker argues that it is virtually impossible to come up with general prescriptions, given the wide diversity of site conditions within and between different forest regions.

Finally, other contemporary forest management objectives may conflict with C. Duinker lists several examples: (1) Maximising fibre production may be consistent with reducing C emissions—if extra fossil energy releases do not exceed net reductions achieved in the products, which depends in turn on how the extra fibre is used (e.g., Schlamadinger and Marland). (2) Obtaining a natural age-class structure (i.e., in contrast to a normal forest), perhaps to ensure areas of old timber are always retained, or to maintain biodiversity, is likely to produce a C-neutral system in the long-term, but the management strategy required to achieve this may release large amounts of C over the first several decades. (3) Allocation of areas as wilderness, left entirely to nature, e.g., to encourage tourism and/or promote biodiversity, would almost certainly increase releases of C in the short-term (in Canada's boreal), because the existing age-class structure is vulnerable to disturbances and/or stand break-up and decay) while removing further opportunities for C management.

Duinker points briefly to possible influences on management of Canadian forests that will favour C sequestration, including: (1) increased management for biomass energy production, since it represents no net additions of C to the atmosphere; (2) increased marketing of wood products, particularly "long-lived products from short-lived trees", which favours farming forests where biomass C is ephemeral, but preservation of forests that will store large pools of C for long periods; (3) increasing forest area through afforestation and restocking of previously harvested land, which seems to be happening both naturally and through human actions in any case; and (4) international pressures which result from the lead role that Canada has adopted in the quest for sustainable management of global forest resources (see also Stewart and Maini, and the Foreword). Such a high international profile, together with considerable attention from international environmental pressure groups, Duinker suggests, will demand that Canada demonstrate real progress toward sustainability at home; in so far as sustainability is commensurate with increased C sequestration, Canadian forest managers will be compelled to consider C as a management objective.

In summarising, Duinker points to the need for convincing economic analyses in support of C sequestration options if changes are to be implemented. He suggests that some C beneficial practices are likely to expand, but concludes that although there are several reasons why C may be considered in the long-term, it is unlikely to be of key importance in the next 10–20 years. He also mentions "the wildcard of climate change" as a major unquantifiable hazard to the ability of Canadian forests to hold on to the C they already contain. He concludes that it will be more pragmatic for Canadian forest managers to concentrate on making better use of harvested biomass, and perhaps increase tree biomass in urban and agricultural areas, than to add to the complexity of existing forest management strategies for the sake of C storage.

Although Canada is certainly not the only country where forests might be used for such purposes, it is surely one where both the extent of forest resources and political and socio-economic conditions should allow C management objectives a reasonable chance of success. Given Duinker's assessment of the situation in Canada, what are the prospects at the global scale? Stewart and Maini confront such questions in Chapter 31, including:

1. how will increasing demands for wood products be met from an ever decreasing global forest land base?
2. what are the likely impacts of *green consumerism*?—i.e., the resistance of consumers to the purchase of wood products obtained from non-sustainably managed forests, a factor already influencing sales of both tropical hardwoods and high-latitude softwoods.
3. what can be done to protect socio-cultural values of forests, such as the rights of indigenous people dependent on forests for survival? (It is being increasingly recognised that such people represent a knowledge base for sustainable management and utilisation of forest resources.)
4. how will geopolitical forestry-related disputes be resolved?—e.g., the trans-boundary impacts of watershed management on soil and water conservation in other countries downstream, or the effects of deforestation in one region on climate in another?

Stewart and Maini remark that the agencies developing international and national polices will increasingly be confronted with the need to balance domestic problems against international responsibilities. Apart from the need to manage global forests for C, other aspects such as biodiversity, cultural and socio-economic forest values must also be considered at all levels, from local to global, while recognising the rights of sovereign nations to meet their own policy objectives.

Differences between developed and developing nations in using forests for mitigating carbon emissions

Sathaye draws attention to the large variation in net forest sector C emissions with latitude. In general, he notes, low-latitude (developing) countries tend to be sources while high latitude countries (developed) tend to be sinks. He also observes that the developed countries have the analytical skills and access to resources but that inadequate attention has been paid to the developing countries (mainly tropical) and compares the resulting bias in scientific effort to the situation of a person looking for a lost key under a lamp, since that is where the light strikes! Solberg and Hoen compare partial cost-efficiencies of C sequestration measures applied to forest ecosystems in Indonesia, and find them considerably more profitable than in Norway. Grainger makes the point that scientific understanding of the physical and biological aspects of global change is currently much better developed than is understanding of socio-economics. We *must* develop models which better integrate all dimensions.

Stewart and Maini also devote considerable attention to the issue of *international polarisation*. The industrialised countries are advocating strong measures to conserve and protect global forests, particularly in the tropics, but the developing countries perceive an equally urgent need for the industrialised nations to reduce fossil fuel emissions. Stewart and Maini agree that both viewpoints are valid—certainly deforestation is not the cause of, and reforestation not the cure for, climate change—but argue that preventing or reversing deforestation would bring several environmental and socio-economic benefits, in addition to combating fossil emissions. Fearnside recognises this argument, cautioning however, that funding from the developed world to support C offset schemes in the tropics should not provoke unacceptable socio-economic impacts in recipient countries—some of the social burden of dealing with global change should be accepted by the industrialised countries, as much as the financial burden. Stewart and Maini also report that many developing countries feel foreign attempts to protect and preserve their forests are an infringement on their sovereign rights to exploit natural resources, and an attempt to suppress economic development. Consequently, developing countries are making their commitments to conserve and expand tropical forests dependent upon receiving funding and technology from the developed countries.

The need for a global policy on sustainable management of the world's forests

Stewart and Maini assess the needs and motivations for a global policy on forests, a problem they state as:

> *"...responding to the demands of an expanding population for more forest-based outputs, while maintaining socio-economic, industrial, and environmental values from forest lands that are globally shrinking and degrading."*

They address the major international problems in implementing sustainable forest management, since these underlie the problems for C management in global forests. An agenda is then presented outlining the actions required to promote sustainability of the world's forests. The objectives are to improve living conditions for millions of poor living in heavily deforested countries, to promote biodiversity in many forested regions and to mitigate some of the effects of anthropogenic greenhouse gas emissions. The agenda lists several actions including:

1. reforestation, including some 800 Mha of degraded land in the tropics.
2. increasing areas of protected forest cover.
3. protection of natural forests, particularly primary or old growth forests.
4. institution of government reporting on national forest status and management progress, including research, and promotion of exported products from sustainably managed forests.
5. making agriculture more sustainable.

To achieve these objectives requires actions at local, national and international levels. At the local level, communities must aim to meet local needs, and have *ownership* of the initiatives if they are to be successful. (Sathaye makes a similar point: *equitable distribution of the benefits* is essential if projects aimed at mitigating C emissions are to survive and the C storage benefits be sustained.) At the national level, land-use policies to promote conservation and sustainable development are needed, both for national development and for international co-operative initiatives. Internationally, governments must develop an effective framework for co-ordinating co-operation—UNCED is but the first major step. Sustainable management policies must be formulated on scientifically-based, internationally-agreed criteria and indicators as a means of promoting multiple benefits for present and future generations.

Tropical and subtropical regions in particular, need funds, technology transfers and training, to assist in many aspects of establishing and maintaining sustainable forests. They conclude:

> *"Goals and guidelines to protect the environment and improve global carbon management emanating from the UN FCCC, although important, are not likely to be successful if they ignore current social and economic values. Only a sustainable development approach that allows for multiple environmental and economic benefits can succeed."*

A discussion of these objectives and some thoughts on action are also presented by Woodwell and Ramakrishna in the Epilogue (Chapter 34). Pointing to the lack of international agreement on a global policy for forests resulting from UNCED, they outline a proposal to achieve global consensus through an independent World Commission on Forests and Sustainable Development. There a two major thrusts: (1) scientific assessment of the role of forests in the biosphere, particularly as human habitat; and (2) development of policies to defend this role against commercial exploitation. Referring to the linkages between global forests and Earth's climate, they point to the urgent need to stabilise atmospheric greenhouse gas concentrations, but recognise that this cannot come about by immediate reductions in fossil energy use. If nothing is done however, we can expect high latitude ecosystems to release vast amounts of additional CO_2 as the climate warms, while impoverishment of the biota and soils will accelerate, particularly at lower latitudes. They observe:

> *"...there can be no solution, interim or long term, without managing forests globally in such a way as to balance the global carbon budget."*

Summarising many of the issues addressed in this book, Woodwell and Ramakrishna outline the challenges to the global community. Scientists do not have all the answers, though they do understand the technical problems. Politicians must move to protect long-term public needs against short-term commercial exploitation by private interests, even when political winds are blowing in the opposite direction.

25 The economics of increased carbon storage through plantations and forest management

Roger A Sedjo

Resources for the Future
1616 P Street NW,
Washington DC 20036

INTRODUCTION

Forest plantations offer what appear to be one of the more attractive approaches to sequester atmospheric carbon (C). They are attractive because: (1) studies have indicated that forests have the potential to sequester large amounts of C; (2) the technology for establishing new forests exists; (3) forests have a number of environmental benefits aside from C sequestration; and (4) most studies indicate that the costs of C sequestration using forests are quite modest.

Practices that increase the forest biomass, increase the stock of C held in trees, other biomass and in the soils. More forest biomass implies a greater sequestration of C. In addition, C can be sequestered beyond the life of a particular tree or forest in long-lived forest products that are either in use or in waste dumps (Row and Phelps 1990; Part III, this volume). Mitigation policies using forests have been estimated to reduce by as much as 80% the overall costs of stabilising U.S. C emissions (Richards et al. 1993).

The purpose of this paper is twofold. First, it will briefly summarise some existing information related to the costs of various activities for sequestering C by increasing forest biomass or forest product stocks. Second, it will examine the benefits of plantation-generated C sequestration using a case study of plantations in Argentina. Although the paper's focus will be global, most of the individual studies discussed will involve project specific or country specific results.

While there is a vast literature examining the C implications of various types of land uses including forests and tree planting, most assessments lack a discussion of economic considerations. Only a relatively few have examined *cost* considerations. Furthermore, even when the financial aspects are examined, often the analysis is partial or incomplete. Additionally, few if any studies have attempted to estimate the *benefits*. This is not wholly the fault of the researchers, however. In the examination of C sequestration, it is often difficult to determine what are the most useful boundaries of an analysis. For example, since forest stands do not last forever, the use of the forest-land after a harvest is very important to the issue of C sequestration. Additionally, the nature of the treatment of the life cycles of wood products is critical to the estimate of forest *system* C sequestration. Finally, if wood or wood residues are burned, are they additional burning or do they substitute for fossil fuel use? These and other aspects of the forest product sector are addressed, in part, in Parts II and III of this volume. All of these considerations contribute to the inherent complexity of estimating the C sequestered by the forest and various forest practices and in drawing out the economic implications of various activities. On the benefits side, there are few estimates of the marginal benefits to C sequestration that are not related to some political willingness to pay.

NATO ASI Series, Vol. I 40
Forest Ecosystems, Forest Management and the Global Carbon Cycle
Edited by Michael J. Apps and David T. Price

Mitigation policies involving the forest sector fall into one of four categories (Binkley and van Kooten 1993). These are those that: (1) increase the standing inventory of forest biomass and therefore sequester C; (2) increase the storage of C in long-lived forest products; (3) substitute wood products for other material that emit more C; and (4) utilise biomass energy crops for fuel. The focus of this study will be on establishing plantations that increase the standing biomass and thereby increase the amount of C sequestered.

The value of sequestering carbon

When undertaking cost-benefit type studies, the benefits are often more difficult to determine than the costs, a point also made in Chapter 24. Nevertheless, at some point the costs are compared with the benefits, either explicitly or implicitly. The value that societies place on the social damages incurred from an additional ton (Mg) of atmospheric C can be assessed in proposals for C taxes. If a country is willing to place a tax of (say) US\$ 100 Mg^{-1} on the emissions of sulphur dioxide, for example, then presumably it must view the damages generated as being at least US\$ 100. Thus, when C taxes are proposed in the range of US\$ 200–250 Mg^{-1} C (Darmstadter 1990; Manne and Richels 1990), as they were initially, the implication is that the society believes the benefits to be at least of that level. In practice, however, the actual taxes imposed have been somewhat different. Although Norway has enacted a tax of about US\$ 148 Mg^{-1} C, most countries have not yet imposed an actual C tax. Additionally, attempts to estimate the actual damages (as opposed to relying on a societal surmise) have tended to get much lower estimates. For example, Nordhaus (1993) estimated the economic damages avoided by the sequestration of 1 Mg C at from US\$ 5 to US\$ 20, with the costs rising from US\$ 5 to US\$ 20 over a period of about a century.

Conversion of marginal crop and pasture lands into trees and forests

Much of the early literature examined the question of the relationship between forest area and/or biomass and amount of C sequestered in the forest, providing valuable information on methods to increase the standing inventory of forest biomass. Marland (1987) pulled together much of the early literature on C sequestration by forest, soils and estimated the relationship between various types of forest conditions and the amount of C sequestered. Sedjo and Solomon (1989) made what appears to be the first estimate of the costs of sequestering major amounts of C by forest establishment, estimating that it would take about 465 Mha of new forest that could be established at a total cost of between US\$ 186–372 thousand million to sequester enough C on an annual basis to offset—for a couple of decades—the net atmospheric increases in C caused by emissions levels at that time. Their estimate of cost was based upon estimates of average land prices in the temperate and tropical regions as well as estimates of the average costs of establishing forest plantations on these lands. Using average costs, they obtained crude point estimates that did not take into consideration increasing costs that are likely to occur as forest plantation establishment moves from land with few alternative uses to more productive land. Darmstadter and Plantinga (1991) derived an estimated cost of US\$ 2–4 Mg^{-1} C from the Sedjo and Solomon gross estimate. This cost level is low compared to cost estimates using other approaches.

A widely cited but extreme early estimate is that of Nordhaus (1990) in which his point estimate of the cost of the removal of C in the U.S. by forest planting was about US\$ 100 Mg^{-1} C. This cost estimate is almost surely too high, at least as an estimate of the current marginal costs of sequestering C by tree planting. The high cost estimate has been challenged, among others, by Cline (1992) who notes that Nordhaus' result is strongly affected by Nordhaus' low estimate of C sequestered per unit area of forest, about 1.0 Mg^{-1} ha^{-1}. A growth rate this low may be common on mature unmanaged stands but are less than one tenth of the estimate used by Sedjo and Solomon (1989) and Moulton and Richards (1990).

A number of recent studies have also examined the question of the costs of sequestering C globally. For example, Winjum et al. (1992) estimated that the costs of sequestering C on a massive world-wide scale would range from less than US$ 1 to about US$ 8 per Mg C sequestered. Dixon et al. (1993) estimated the cost of C sequestration to vary from US$ 1 to US$ 30 Mg^{-1} C for projects proposed by U.S. and European utilities as C offsets outside of their regions.

Other recent studies have estimated the costs of sequestering C via specific forestry activities, usually in developing countries. These include not only forest plantations, but agroforestry activities and forest protection also. Some of these are listed in Table 25.1.

As indicated above, a host of studies have suggested relatively low C sequestering costs through tree planting, in many cases under US$ 10 Mg^{-1} C and rarely over US$ 30 Mg^{-1} C. Other studies with similar results stressing various aspects of the problem include Trexler et al. (1989); Swisher (1991); Winjum and Lewis (1993) and Faeth et al. (1993). Another early U.S. study which included alternative options is that of Dudek and LeBlanc (1990) who examined the use of tree planting under the Conservation Reserve Program (CRP) to sequester C. They estimate this option as costing from US$ 26 to US$ 43 per Mg C removed from the atmosphere. Van Kooten et al. (1992) found similar results for Canada.

Table 25.1. Costs of sequestering carbon (C) through forest projects: some selected cases. (Source: Adapted from Dixon et al. 1993).

	Cost of C sequestration (US$ Mg^{-1} C)				
	Tropical		Temperate	Boreal	
Source	Agroforest	Plantation	Plantation	Plantation	Protection
Andrasko (1993)	3–5	3–6	0–2		
Dixon et al. (1993)	4–16	6–60	2–50	3–27	1–4
Krankina & Dixon (1993)	na[a]	na	1–7	1–8	1–3
Houghton et al. (1991)	3–12	4–37	na	na	na

a na = not applicable.

The Sedjo and Solomon (1989) work and much of the other work cited above provided single point estimates, and thus do not represent the very real problem of rising costs that are associated with involving large land areas. For large projects, the marginal costs would be expected to rise reflecting both increasing land opportunity costs and varying land productivity for tree growing, since low growth rates are associated with low rates of C sequestration. In most of the previous studies, a point estimate was justified since the focus was only a single project. However, point estimates are not appropriate when examining policy options for projects of the scale needed to offset large amounts of C.

In addition, many early cost estimates ignored the use of a discount rate in their calculations or used crude approximations involving averages. However, since the costs of most of the forestry investments are incurred before the C sequestration benefits, some type of discounting procedure is in order and crude approximations may not provide accurate estimates.

Recently, a number of somewhat more sophisticated studies have begun to appear. These include Mouton and Richards (1990) and Adams et al. (1992). These studies have refined the approach of estimating the costs of establishing C sequestering tree plantations in three ways.

First, they estimated a cost function, not a point, that would estimate the rise in costs associated with large scale tree planting. In doing this they recognised that large scale investments in forests to sequester C would face rising costs as forests moved from lands with low productivity and/or low opportunity costs to lands with higher productivity and/or opportunity cost lands. (Opportunity costs refer to the value of the resource in another use. For land the opportunity cost would largely be reflected in land market rents.) Second, they refined the tree plantation establishment cost estimates recognising differences associated with location and site considerations. Third, they built discounting procedures into the methodology.

Mouton and Richards (1990) developed the first cost function for the U.S. by systematically building an upward sloping marginal cost curve that ranges from US$ 5–10 Mg^{-1} C for the first 60 Tg, increasing to US$ 40 Mg^{-1} C as annual sequestration rises to 800 Tg, and total land area involved approaches 340 million acres (about 130 Mha). Thus, the estimates captured the reality that under current conditions the marginal costs of sequestering 1 Mg C in the U.S. is modest, less than US$ 10 Mg^{-1} C and rises gradually to about US$ 40 Mg^{-1} C when large amounts of land, which require higher opportunity costs land to be utilised, are involved.

The estimates of Mouton and Richards (1990) are roughly equivalent to the point estimate of Sedjo and Solomon (1989) and others for low levels of forest plantations but change dramatically for higher levels of tree planting since they incorporate rising marginal costs as lands with higher opportunity costs are brought into forest plantation production. Both approaches used roughly the same assumption regarding the amount of C that would be sequestered per land area planted—about 6 Mg C ha^{-1} yr^{-1}. The work of Adams et al. (1992) is similar to that of Moulton and Richards (1990) in that it estimates a rising cost function. The study estimates of sequestration costs are somewhat higher than Moulton and Richards (1990) being in the range of US$ 16.30 Mg C for low levels of C sequestration to US$ 62 Mg^{-1} C for a U.S. program that sequesters about one-half of U.S. C emissions. In a recent paper, Adams et al. (1993) make modest upward revisions of their earlier estimates but they remain in essentially the same range. All of these studies used Birdsey's (1992) estimates of C sequestered in forest biomass. These estimates, however, have recently been revised downward and this will lead to upward revisions in C sequestration costs in subsequent estimates that are in the process of being made (Richards et al. 1993).

The marginal cost schedule developed in the study is substantially higher than that of either Moulton and Richards (1990) or Adams et al. (1993). The two sets of cost estimates appear in Table 25.2.

It is clear from the dispersion of cost estimates that both the methodology and the empirical estimates are still in a state amenable to further revision. It is probably an accurate generalisation to state that the newer research is tending to find a somewhat steeper increase in costs than did the early studies. In addition, the analysis uses revised C estimates and builds in

Table 25.2. Estimates of cost of carbon (C) sequestered by tree planting: some comparative results for the U.S. (Source: Moulton and Richards 1990, Appendix table 1A; Adams et al. 1993).

	Total C sequestered (Tg C yr^{-1})			
Study	@140 US$ Mg^{-1}	@280 US$ Mg^{-1}	@420 US$ Mg^{-1}	@700 US$ Mg^{-1}
Moulton/Richards (1990)	16.57	20.69	23.24	34.73
Adams et al. (1993)	18.50	25.11	37.21	95.06

greater detail on the opportunity costs of land in developing a long-term marginal cost curve for C uptake. In general, the new results show a much steeper rise than the earlier Moulton and Richards (1990) work, with the costs per Mg C roughly doubling, from about US$ 30 to US$ 60 Mg^{-1} C, for large annual uptakes.

Additionally, the extent to which large scale tree planting can be a permanent solution to the C problem is problematic, since there are presumably limits to the extent to which the terrestrial system can expand forest stocks. Sedjo (1989) noted that it may be most useful to view tree planting as a measure to "buy time" during which more permanent solutions to the basic fossil fuel nature of the C problem are developed. Richards (1993) has been especially sensitive to the inter-temporal nature of the potential sequestration of C, pointing out that the expected life of C sinks may be limited and asking the question of how transitory sinks are to be valued, if at all.

Finally, it should be noted here that the inter-temporal issue rests heavily on the disposition of the trees over time—does this particular stand die and decompose or is the wood primarily used in long-life products such as construction materials and so forth.

Nevertheless, although there are still issues that have not yet been completely resolved, the methodology and empirical estimates have come a long way in the past five years and my judgement is that the next generation of cost estimates of C sequestered by tree growing is likely to be quite accurate.

Carbon sequestration in Argentina: a case study

This section is drawn from a report prepared for the Secretaria de Agricultura, Ganaderia y Pesca, Gobierno de Argentina (Crosson et al. 1994). A proposed Argentine forestry program contemplates the planting of 400 000 ha over a 5 year period. The purpose of the project is multi-dimensional and includes both a desire to produce industrial wood as well as a desire to generate positive environmental benefits. This study focused upon the environmental implications of the establishment of new forest plantations. The plantations were to be established on lands that included grazing lands, unused wetlands, and former forest-lands on which brush has been established, but not new forests. No native forests were to be cleared for the establishment of forest plantations.

The land was to be divided among four priority regions. The discussion focuses on 300 000 ha to be planted in the first and second priority regions. These were the Delta region and a part of Mesopotamia: Entre Rios and south-east of Santa Fe, in which 160 000 ha would be planted. The second priority is given to the rest of Mesopotamia—north-east of Santa Fe, and all of Misiones and Corrientes—and to Buenos Aires Centro and Sur, which would receive 140 000 ha.

Environmental effects examined in the broad study included erosion and flood control, water quality, land fertility, biodiversity and C sequestration. The most specific and detailed assessment of the likely effects were those related to C sequestration. These were also the only effects to which specific values were estimated. This section describes the results of the forest and C sequestration assessment.

Changes in carbon sequestration

In this study C sequestration is the capture of atmospheric C by trees growing in forestry program plantations and by the soils in which the trees are planted. The focus is on net changes in C sequestered by trees and soils when land is shifted from its present use to forest plantations.

Several approaches can be used to estimate the per hectare amount of C that might be sequestered by forest plantation programs. Two were considered. One focuses on the C accumulated over the first growth cycle after planting. The other approach focuses on the steady-state accumulation over an indefinite number of growth cycles. Due to limitations in time and resources, the estimates here are of the amounts sequestered over the first growth cycle after planting. A short-cut estimate of the amounts sequestered in the steady-state approach would be to assume that the forest were a normal regulated forest, with equal representation for each age class. In this situation the average age would be about one-half that of the full rotation and we could attribute about one-half of the total C the first full growth cycle on the assumption that the forest is approximately a regulated forest, with an equal distribution of each age. In addition, the C in undecomposed forest products would need to be added.

To estimate the amount of C sequestered by trees the commonly accepted figure of 0.26 Mg C m^{-3} of total tree volume was used (Marland 1988), recognising that this estimate in fact will vary somewhat with the density of wood. For C sequestered by forest soils in Mesopotamia, information was provided by Birdsey (RA Birdsey, USDA Forest Service, 1994, pers. comm.) with respect to southern Pine (*Pinus elliotii* and *Pinus taeda*) forests in the American south.

In principle, the value of 1 Mg C sequestered by Argentina's plantation forestry program is equal to the value of the resulting reduction in the marginal costs of global warming. In practice, few credible estimates of the reduction in marginal costs are available. A survey of the literature (Sedjo et al. 1994) dealing with the economics found that the cost of sequestering 1 Mg C in plantations commonly ranged from about US$ 5 to about US$ 50. This is consistent with the US$ 5–20 range estimated by Nordhaus (1993). Accordingly, we take US$ 5 and US$ 50 as the range of values of 1 Mg C sequestered by Argentina's plantation forestry program.

Since the interest was in the difference the forestry program would make in reducing atmospheric C, an estimate was made of how much C is being sequestered by current uses of the land that would be included in the program. It was assumed that where the land currently is in crops or pasture there is no net accumulation of C in biomass, so conversion of that land would release no C. Conversion of such land, here assumed to occur only in Mesopotamia, likely would result in a small net increase in C sequestered in the soil. Where the land converted is abandoned crop or pasture land with a regrowth cover of scrubs and small trees, then conversion would release C, reducing the net amount sequestered by the program and a rough adjustment is made for this in the estimates. The wetland soils of the Delta also hold substantial amounts of C that would be released by conversion of the land to forests. Since the amounts of releases were unknown, the discussion of C sequestration notes that the estimate of the value of C sequestered by forest biomass in the Delta is too high by the value of the C released from wetland soils.

The present value is calculated for the amount of C sequestered over one growth cycle on the 300 000 ha of land now planned to be in the program in the Delta and Mesopotamia regions. The cycles assumed vary from 9 years for willow/poplar to 30 years for *Araucaria*. The calculations of present value use 8%, 10%, and 12% discount rates. Other rates, of course, are possible, and may be more appropriate. However, the rates used probably encompass any reasonable range.

It was estimated that the 300 000 ha of plantations would be distributed among tree species and provinces as shown in Table 25.3. Assuming such a distribution of land, that the value of 1 Mg C sequestered ranges from US$ 5 Mg^{-1} C to US$ 50 Mg^{-1} C, and that discount rates of 8%, 10%, and 12% are appropriate, the estimated present values of C sequestered in wood over one growth cycle is shown in Table 25.4.

Table 25.3. Species, area and locations referred to in the case study.

SPECIES	AREA (ha)	PROVINCE
Araucaria	10 000	Misiones
Willow/Poplar	80 000	Delta
Pine	118 000	Misiones/Corrientes/Entre Rios
Eucalyptus	92 000	Misiones/Corrientes/Entre Rios

Table 25.4. Present value of carbon sequestered by the woody biomass of 0.3 Mha of plantation forests in Argentina's forestry program in Mesopotamia and the Delta, by previous land use and species, for discount rates of 8%, 10% and 12%.

	PRESENT VALUE FOR THREE DISCOUNT RATES (US$ MILLION)											
	Araucaria			*Eucalyptus*			PINE			WILLOW/POPLAR[a]		
PREVIOUS LAND USE	8%	10%	12%	8%	10%	12%	8%	10%	12%	8%	10%	12%
Agriculture and pasture[b]												
US$ 5 Mg^{-1}	3.7	3.1	2.7	46.0	41.4	37.7	59.0	51.3	45.4	na[c]	na	na
US$ 50 Mg^{-1}	37.	31.	27.	460.	414.	377.	590.	513.	454.	na	na	na
Abandoned												
US$ 5 Mg^{-1}	1.2	0.6	0.2	23.0	18.4	14.7	29.5	21.8	15.9	na	na	na
US$ 50 Mg^{-1}	12.	6.	2.	230.	184.	147.	295.	218.	159.	na	na	na
Delta wetlands												
US$ 5 Mg^{-1}	na	na	na	na	na	na	na	na	na	36.4	33.6	31.2
US$ 50 Mg^{-1}	na	na	na	na	na	na	na	na	na	364.	336.	312.

a All willow and poplar trees are assumed to be planted in the Delta.
b Estimates for land previously in agriculture or pasture are higher than those for abandoned land because abandoned land has C stored in brush and small trees that would be emitted to the atmosphere when the land is converted to plantations. Net C sequestration by the program on this land, therefore, is less than on land previously in agriculture or pasture. The estimates do not include capture of C in the soils of Mesopotamia, nor the release of soil C from wetland soils in the Delta when land is converted.
c na = not applicable.

The estimates given in Table 25.4 indicate that if all 220 000 ha in the program in Mesopotamia were converted from agriculture and pasture, and the appropriate rate of discount is 10%, then the present value of the C sequestered by wood in that area and the Delta would be US$ 129 million if the "true" value of a Mg of sequestered C is US$ 5 and US$ 1290 million if the true value is US$ 50 Mg^{-1} C. The comparable figures if the 220 000 ha were converted from abandoned land would be US$ 74 million and US$ 740 million. If the appropriate rate of discount were 8% then these totals would be somewhat higher, and if the rate were 12% they would be somewhat lower.

The actual present value of the C sequestered, whether the value per Mg C is US$ 5 or US$ 50, likely would be closer to the higher ends of the ranges than to the lower ends. This is because most of the land in Misiones and Corrientes that might go into the program presently is in pasture rather than being abandoned land. If we assume that 80% of the land going into the program in those provinces, and also in Entre Rios, would be converted from agriculture and pasture, with 20% coming from abandoned land, then the range of estimates of the present value of C sequestered by wood would be from about US$ 118 million to about US$ 1200 million.

It should be understood that these estimates are partial, for two reasons. One is that the estimates do not include the 100 000 ha of forest that would be planted in areas other than in the Delta and Mesopotamia. On this score, therefore, our estimates of the value of C sequestered by the forestry program are understated. The other reason why the estimates are partial is that they do not take account of C either captured or released by the soil when land is converted to forests. Drawing on information provided to us by Birdsey (RA Birdsey, USDA Forest Service, 1994, pers. comm.), we estimate, very roughly, that net C sequestration in the soil resulting from the shift of 220 000 ha of land from agriculture and pasture to forests in Mesopotamia would add about 10% to our estimates above of present values. If the shift were from previously abandoned land, the increase in present value because of soil sequestration of C would be less, probably much less.

The effect of the program on soil C in the Delta looks very different. The wetland soils of this region are carbon-rich. Much of this C would be released to the atmosphere when these soils are drained and converted to plantations. This caveat does not apply to all of the 80 000 ha assumed to be in the forestry program in the Delta because some of that land already is in plantations that after several harvests would be ready for replanting and eligible for inclusion in the program. There would be no significant C release from that land.

To compensate for the release of C from wetland soils in the Delta, the 100 000 ha outside the Delta and Mesopotamia and of C sequestered in the soils of Mesopotamia were excluded from consideration. Consequently, Table 25.4 estimates of the value of C sequestered are more likely to be too low than too high.

These estimates are based on a more detailed set of calculations that we believe will be of interest in understanding how we arrived at the estimates. Field work in the Mesopotamia region suggests that the species of trees used for plantations vary somewhat across the region, with *Araucaria* being used in commercial plantations only in Misiones, and with the use of *Eucalyptus* and *Pinus caribaea* increasing toward the south. Overall, modestly higher per hectare biomass growth can be expected in plantations that are established in the south than in those established in the north of Mesopotamia.

The growth rates and typical rotation period for *Araucaria*, three species of pine, and *Eucalyptus* grown in Mesopotamia are shown in Table 25.5. The table also shows comparable information for poplar and willow, which, for the forestry program, will be grown only in the Delta. The growth rates and rotation periods shown in the table are conservative averages of figures obtained from various sources. They are consistent with growth rates experienced elsewhere in South America with these species (Sedjo 1981). Overall, growth rates in Mesopotamia and the Delta compare favourably with those found almost anywhere in the world. The growth rates rise from north to south in Mesopotamia largely because the lower elevation permits planting of the faster growing species (e.g., *Eucalyptus*).

The growth rates in Table 25.5 are based on the commercial timber within the stand. They do not include growth of branches and roots. A factor commonly used to estimate the total biomass of the forest is 1.6 times the commercial volume. However, for plantations managed to reduce the size of branches, a more conservative value of 1.4 is used. This factor is used to adjust the growth rates of Table 25.5 to estimate the growth of total biomass of the species shown in the table. To estimate the amount of carbon sequestered in the tree biomass, we use a factor of 0.26 Mg m^{-3}, which represents the weight of the carbon in a 'typical' cubic metre of plantation timber. The results of these calculations are shown in Table 25.6.

Table 25.5. Growth rates and usual rotation periods by species.

SPECIES	GROWTH RATE (m^3 ha^{-1} yr^{-1})	USUAL ROTATION (yr)
Araucaria	18	30
Pinus elliotii	25	20
Pinus taeda	28	20
Pinus caribaea	33	18
Eucalyptus	35	13
Poplar/willow	35–40	9

Table 25.6. Carbon (C) sequestered in the woody biomass of plantation forests (per hectare).

SPECIES	GROWTH RATE (m^3 ha^{-1} yr^{-1})	ROOTS AND BRANCHES[a]	PROPORTION C	TOTAL (Mg C ha^{-1} yr^{-1})
Araucaria	18	1.4	0.26	6.6
Pinus elliotii	25	1.4	0.26	9.1
Pinus taeda	28	1.4	0.26	10.2
Pinus caribaea	33	1.4	0.26	12.0
Eucalyptus	35	1.4	0.26	12.7
Poplar/willow	35–40	1.4	0.26	12.7–14.6

a Adjustment factor for mass in roots and branches.

The numbers in the last column of Table 25.6 indicate the amount of C that would be sequestered in one year by one ha of plantation forest planted in the various species listed. However, the numbers would measure net sequestration only if the conversion of the land from its previous use would not release stored C. In Mesopotamia the land likely to be converted now is mostly in pasture, with smaller amounts in annual crops or in no present use, i.e., abandoned land. It can be assumed that land in pasture and annual crops does not accumulate C over time, that is, on land in these uses net C sequestration over time is zero. Abandoned land, however, commonly has some regrowth of scrubs and small trees which do accumulate C that will be released to the atmosphere when the land is cleared for conversion to plantations. The regrowth on abandoned land typically is modest. The land does not seem to revert to tropical forest, perhaps because of a long intervening use in agriculture or pasture and the absence of appropriate tropical forest seeds in the area. In any event, it is assumed that when converted to plantations, abandoned land will release 50 Mg C ha^{-1} in the first year and this amount is subtracted from the estimates of C sequestered by the plantation forests.

Table 25.7 presents estimates of the net C sequestered by the woody biomass of plantations over one growth cycle. The high and low estimates of sequestration per hectare per year are the annual growth rates for the several species shown in Table 25.6, i.e., the lower growth rate of 6.6 Mg C ha^{-1} yr^{-1} is for *Araucaria* and the high growth rate of 12.7 Mg C ha^{-1} yr^{-1} is for *Eucalyptus* and poplar/willow. The high and low estimates of net C sequestered over one growth cycle (the last column in Table 25.7) were found by multiplying annual growth rates by the number of years in one growth cycle and subtracting the estimates of C released from woody biomass on the converted land. The Delta wetlands would capture the least C because the species grown there, poplar and willow, have the shortest growth cycle (9 years).

Table 25.7. Net carbon (C) sequestered by woody biomass of forest plantations over the first growth cycle.

PRIOR LAND USE	CONVERSION C RELEASED FROM WOODY BIOMASS (Mg C)	C CAPTURED IN FOREST[a] ($Mg\ C\ ha^{-1}\ yr^{-1}$) HIGH	LOW	NET CAPTURE OF C ($Mg\ C\ ha^{-1}$) HIGH	LOW
Crops	0	12.7	6.6	204	165
Pasture	0	12.7	6.6	204	165
Abandoned	50	12.7	6.6	154	115
Delta wetlands	0	12.7		114	94

a Does not include C captured by soils in Mesopotamia or released in the Delta when land is converted from present use to plantations.

Table 25.8. Present value per Mg of woody biomass sequestered, by species and previous land use, US$ 5 and US$ 50, and discount rates of 8%, 10%, and 12%.

	PRESENT NET VALUE OF SEQUESTERED BIOMASS (US$ Mg^{-1})											
	Araucaria			*Eucalyptus*			PINE			WILLOW/POPLAR		
PREVIOUS LAND USE	8%	10%	12%	8%	10%	12%	8%	10%	12%	8%	10%	12%
Agriculture and pasture												
US$ 5	370	310	265	500	450	410	500	435	385	na[a]	na	na
US$ 50	3700	3100	2650	5000	4500	4100	5000	4350	3850	na	na	na
Abandoned[b]												
US$ 5	120	60	15	250	200	160	250	185	135	na	na	na
US$ 50	1200	600	150	2500	2000	1600	2500	1850	1350	na	na	na
Delta wetlands												
US$ 5	na	na	na	na	na	na	na	na	na	455	420	390
US$ 50	na	na	na	na	na	na	na	na	na	4550	4200	3900

a na = not applicable
b Conversion of abandoned land to plantations is assumed to release 50 $Mg\ C\ ha^{-1}$ with a present value of US$ 250 to US$ 2500, corresponding to values of US$ 5 Mg^{-1} C and US$ 50 Mg^{-1} C respectively. These amounts are subtracted from the values for land previously in agriculture or pasture to get the present values of carbon sequestered on previously abandoned land.

The final step is to calculate the per hectare present value of the C sequestered by each species of tree over one growth cycle. To do this it is assumed for convenience of calculation that the forest yield function is linear, that is, that the addition of forest biomass, and therefore C, is the same each year over the cycle. For example, *Araucaria* is assumed to sequester 6.6 $Mg\ C\ ha^{-1}\ yr^{-1}$ in its 30 year cycle, *Pinus elliotii* to sequester in its 20 year cycle, and so on (see Tables 25.5 and 25.6). Next, each year's accumulation of C is valued. As noted above, US$ 5 and US$ 50 Mg^{-1} C were selected as the lower and upper bounds of the value per Mg. Using these numbers, two streams of annual values of C sequestered can be calculated for each tree species over its life cycle. Compounded discount rates of 8%, 10%, and 12% were used to calculate the present value of each stream of values. The results are shown in Table 25.8.

Table 25.8 values were multiplied by the number of ha in each species to obtain the estimates of the total present value of C sequestered by the woody biomass in 300 000 ha of plantation forests shown in Table 25.4.

SUMMARY AND CONCLUSIONS

This paper has presented an overview of economic considerations related to C sequestration through plantation forestry. The first section briefly discusses the literature on the costs and benefits of C sequestration using tree planting and plantation forestry. The case study of Argentina estimated the benefits derived from plantations in terms of estimates of the value of the volumes of C sequestered. The results of both the literature review and the case study suggest that plantation forestry, especially where growth rates are high, is a relatively low-cost approach to C sequestration. Although the cost curves of forestry generated C sequestration are rising, the results suggest rather ample opportunities for using forest plantations as a low cost means of C sequestration. Additionally, the discounted present values of carbon sequestered in plantations are large, for feasible amounts of C sequestered in those plantations at maturity.

REFERENCES

Adams RM, Adams DM, Chang CC, McCarl BA, Callaway JM (1993) Sequestering carbon on agricultural land: A preliminary analysis of social cost and impacts on timber markets, *Contemporary Policy Issues* XI.

Adams RM, Chang CC, McCarl BA, Callaway JM (1992) The role of agriculture in climate change: a preliminary evaluation of emission-control strategies. In: Reilly JM, Anderson M (eds) Economic Issues in Global Climate Change: Agriculture, Forestry and Natural Resources. Westview Press, Boulder, CO.

Adams DM, Haynes RW (1991) Softwood Timber Supply and the Future of the Southern Forest Economy, *S. J. Appl. Forestry* 15: 35.

Andrasko KJ (1991) Climate change and global forests: current knowledge and potential effects, adaptation and mitigation options, UN FAO Rome, 78 pp.

Andrasko KJ, Heaton K, Winnett S (1991) Evaluating the costs and efficiency of options to manage global forests: a cost curve approach, Prepared for Technical Workshop to Explore Options for Global Forest Management, Office of the National Environment Board, Bangkok, Thailand, (Unpublished manuscript).

Binkley CS, van Kooten GC (1993) Integrating climate change and forests: economic and ecological assessments. *Climatic Change* (In press).

Birdsey RA (1992) Carbon storage and accumulation in United States forest ecosystems. USDA For. Serv. Gen. Tech. Rep. W059, Washington, DC, 51 pp.

Birdsey RA, Plantinga AJ, Heath LS (1993) Past and prospective carbon storage in United States forests, *Forest Ecol. Manage.* 58: 33–40.

Cline WR (1992) The economics of global warming. Institute for International Economics, Washington, DC.

Congressional Research Service (1990) Near term options for reducing United States carbon dioxide emissions. Environment and Natural Resources Policy Division and Science Policy Research Division, 28 December, 1990.

Crosson P, Frederick K, Sedjo R, Adamilo H (1994) Potential environmental and other external consequences of the program to increase the area in plantation forests in Argentina, report for the Secretaria de Agricultura, Ganaderia y Pesca, Gobierno de Argentina, 47 pp. (Unpublished manuscript).

Darmstadter J, Plantinga A (1991) The economic cost of CO_2 mitigation: A review of estimates for selected world regions, Resources For the Future Discussion Paper ENG91-06.

Dixon RK, Krankina ON (1993) Forest fires in Russia: Carbon dioxide emissions to the atmosphere, *Can. J. For. Res.* 23(4): 700–705.

Dixon RK, Andrasko KJ, Sussman FA, Lavinson MA, Trexler MC, Vinson TS (1993) Forest sector carbon offset projects: Near-term opportunities to mitigate greenhouse gas emissions, *Water, Air, Soil Pollut.* 70: 561–577.

Dixon RK, Winjum JK, Andrasko KJ, Lee JJ, Schroeder PE (1994) Integrated systems: assessment of promising agroforest and alternative land-use practices to enhance carbon conservation and sequestration, *Climatic Change* 27: 71–92.

Dudek DJ, LeBlanc A (1990) Offsetting new CO_2 emissions: A rational first greenhouse policy step, *Contemporary Policy Issues* 8: 29–42.

Faeth P, Cort C, Livernash R (1993) Evaluating the carbon sequestration benefits of sustainable forest projects in developing countries, World Resource Institute Report, Washington, DC.

Houghton RA, Unruh J, LeFebure PA (1991) Current land use in the tropics and its potential for sequestering carbon. In: Howlett D, Sargent C (eds) Proceedings for Technical Workshop to Explore Options for Global Forest Management, IIED, London, pp 279–310.

Krankina ON, Dixon RK (1994) Forest management options to conserve and sequester terrestrial carbon in the Russian Federation. *World Resource Review* 6: 88.

Manne AS, Richels RG (1990) CO_2 emission limits: An economic cost analysis for the USA, *Energy Journal* 11(2): 51–74.

Marland G (1988) The prospect of solving the CO_2 problem through global reforestation. U.S. Department of Energy, Office of Energy Research, Carbon Dioxide Research Division. DOE/NBB-0082, 66 pp.

Moulton R, Richards K (1990) Costs of sequestering carbon through tree planting and forest management in the United States, USDA For. Serv. Gen. Tech. Rep. WO-58. 17 pp+App.

Nordhaus WD (1990) An inter-temporal general-equilibrium model of economic growth and climate change. In: Wood DO, Kaya Y (eds) Proceedings of the Workshop on Economic/Energy/Environmental Modeling for Climate Analysis, Oct. 22–23, 1990, MIT Center for Energy Policy Research, Cambridge, MA, pp 415–33.

Nordhaus WD (1993) *J. Econ. Lit.* 7(4) (Fall 1993).

Richards KM (1993) Valuation of temporary and future greenhouse gas reductions. Presented at Western Economic Association Annual Meeting, Lake Tahoe, NV, June 1993.

Richards KR, Moulton R, Birdsey RA (1993) Costs of creating carbon sinks in the U.S. *Energy Conservation Manage.* (In press).

Row C, Phelps RB (1990) Tracing the flow of carbon through the U.S. forest products sector. Proceedings of the 19th World Congress of the International Union of Forest Research Organizations, August 5, Montreal, QC.

Sedjo RA (1989) Forests: A tool to moderate global warming? *Environment* 31(1): 14–20.

Sedjo RA, Solomon AM (1989) Climate and forests, In: Rosenberg NJ, Easterling WE, Crosson PR, Darmstadter J (eds) Greenhouse Warming: Abatement and Adaptation, RFF Proceedings, Resources for the Future, Washington, DC, pp 105–120.

Sedjo RA, Wisniewski J, Sample A, Kinsman JD (1994) The economics of managing carbon via forestry: assessment of existing studies. (Unpublished manuscript).

Swisher JN (1991) Cost and performance of CO_2 storage in forestry projects, *Biomass and Energy* 1(6): 317–328.

Trexler MC, Faeth PE, Kramer JM (1989) Forestry as a response to global warming: an analysis of the Guatemala agroforestry and carbon sequestration project. World Resource Institute, Washington, DC, 65 pp.

Turner DP, Lee JJ, Koperper GJ, Barker JR (eds) (1993) The forest sector carbon budget of the United States: Carbon pools and flux under alternative policy options, U.S. EPA, Environmental Research Laboratory, Corvallis, OR, 202 pp.

van Kooten GC, Arthur LM, Wilson WR (1992) Potential to sequester carbon in Canadian forests: Economic considerations, *Canadian Public Policy* (June): 105–117.

Winjum JK, Dixon RK, Schroeder PE (1992) An assessment of forest management practices for sequestration of carbon: I. Estimating global potential. U.S. EPA Report, Environmental Research Laboratory, Corvallis, OR.

Winjum JK, Lewis DK (1993) Forest management and the economics of carbon storage: The nonfinancial component, U.S. EPA Research Paper, Environmental Research Laboratory, 11 February, Corvallis, OR.

Wright LL, Hughes EE (1993) U.S. carbon offset potential using biomass energy systems, *Water, Air, Soil Pollut.* 70: 483–497.

26 Costs of forest-sector mitigation options

Jayant Sathaye

International Energy Studies Group, Lawrence Berkeley Laboratory
1 Cyclotron Road, Berkeley, CA 94720 USA

INTRODUCTION

Carbon emissions from the forest sector are unevenly distributed across latitudes. Forests from the high- and mid-latitudes are estimated to be sequestering net C on the order of 0.70 ±0.2 Pg yr^{-1} (Dixon et al. 1994). The low-latitude countries, which are mostly developing nations, are the primary sources of net C emissions, estimated at 1.6 ±0.4 Pg yr^{-1} —a significant amount, which exceeds the net 1990 C emissions from the United States. The forest sector C emissions vary considerably amongst individual countries so that some are a net C sink (e.g., India) and others are a large net C source (e.g., Brazil) (Makundi et al. 1992).

What mitigation options might be pursued to reduce emissions and/or increase C sequestration? How much land is available for C sequestration? What are the costs and benefits of these options? These are important questions to address in devising global and national strategies, policies and measures to reduce net emissions from land-use change in order to stabilise future greenhouse gas (GHG) emissions. In this paper, we (1) review the range of cost estimates (cost per Mg C) that have been reported earlier, and (2) suggest areas for further research in the economic evaluation of mitigation options. The review indicates that considerable C can be sequestered under an initial cost of US\$ 10 Mg^{-1} C.

Three categories of forest sector mitigation options

The forest sector mitigation options refer to those measures and policies which can lead to reduced emissions of GHGs and/or to increased C sequestration in forests, long-term wood products and other tree vegetation. Mitigation options may be classified into three basic categories. The first category of options is to *maintain* the existing stock of C in vegetation and soil and the proportion of forest products currently in use, which keeps GHG emissions from entering into the atmosphere for the duration of the pool maintenance. Maintenance of existing stocks may be achieved through (1) forest protection and conservation, and (2) increased efficiency in forest management, harvesting and product utilisation. These are discussed in various chapters in Parts I and II of this volume.

A second category of options is to *expand* the C sinks in vegetation, soil and wood products pools. Expansion withdraws C from the atmosphere and maintains it in the bio/geosphere. Expansion of stocks may be achieved through (1) afforestation, (2) reforestation, (3) regeneration, (4) agroforestry and (5) urban and community forestry options. Although expansion of sinks and maintenance of C pools in vegetation, soils, and products may be very effective mitigation options, they may be difficult to implement since the alternative use of the land upon which the C is stored is often more valuable to local inhabitants.

A third category of options is to use wood obtained from renewable sources like forest plantations as a *substitute* for non-renewable emission sources, particularly fossil fuels (Hall

NATO ASI Series, Vol. I 40
Forest Ecosystems, Forest Management and the Global Carbon Cycle
Edited by Michael J. Apps and David T. Price

et al. 1991; Chapter 18, this volume). The substitution will lock C in the fossil fuel until the latter resource approaches depletion. Similarly, wood derived from renewable sources, if used as a substitute for wood fuel derived from depletable natural forests, will also delay or stop C release from this non-renewable source (see Chapters 18 and 19, this volume).

ECONOMIC EVALUATION OF MITIGATION OPTIONS

The economic evaluation of a mitigation option requires an accounting of both C and monetary flows within a common system boundary. Ideally, both costs and benefits of a mitigation option should be estimated. Given our current knowledge, however, it is not possible to meaningfully assess the current and future economic damage that might be caused by elevated atmospheric concentrations of C. This precludes us from evaluating the potential monetary benefit of avoided damage. Thus, the economic evaluation of a C mitigation option has focused on the estimation of costs to sequester a unit of atmospheric C or to avoid a unit of C emissions; in other words, on the estimation of the cost-effectiveness of an option. Cost-effectiveness may be expressed using different components of costs and benefits, and of C storage; each expression giving rise to a unique indicator.

Cost-effectiveness indicators

A cost-effectiveness indicator may be used to rank each option. The most commonly reported indicator to evaluate cost-effectiveness is the initial cost per unit of C, which excludes future investments discounted to the present. The indicator provides information on the amount of resources required to initiate a project. Our review focuses on the estimated value of this indicator. Carbon is measured assuming either a *plant and store* or *sustainable rotations* approach, taking into consideration the C stored in live biomass and in soil. Some studies also account for product C and that in detritus–some of these are discussed in Part III of this volume. A useful way to present the initial cost per unit of C or per hectare is to plot it against the total C conserved or saved (Mg C) or the land area (ha), respectively, which results in a cost of conserved C (CCC) curve (Moulton and Richards 1990).

SIGNIFICANTLY IMPROVED COST ESTIMATES SINCE THE 1990 IPCC REPORT

The 1990 IPCC report on Response Strategies to Climate Change (IPCC 1992) noted that slowing deforestation was a low cost option to reducing a unit of atmospheric C. The IPCC report quoted regional average annual costs of about US$ 8 Mg^{-1} C for tropical forestation and reduction of deforestation–this increased to about US$ 28 Mg^{-1} C for forestation in non-US OECD countries. Costs for establishing a forest plantation, excluding the opportunity cost of land, were estimated to range between US$ 230 to US$ 1000 ha^{-1} (Sedjo and Solomon 1989) with an average cost of US$ 400 ha^{-1}.

The unit cost estimates have been improved several ways since the IPCC (1992) report: (1) unit costs have been estimated for individual countries rather than by regions or for the globe as a whole; (2) the estimates have been developed for several types of mitigation options; (3) other cost components–maintenance, land rental (opportunity costs), and monitoring and evaluation–which were not included in the earlier IPCC report are now being addressed; (4) in some studies incremental costs have been estimated; and (5) analytical methods have been improved to provide better insights and techniques for the evaluation of costs and benefits.

Despite these improvements, it should be noted that most unit cost estimates do not discount C flows. Thus net C storage occurring at any time is accorded the same economic value.

PROTECTING FORESTS COSTS ABOUT US$ 3 Mg^{-1} C

The unit cost of reducing C emissions by protecting forests tends to be low (Table 26.1), as their C density tends to be relatively high. Cost estimates, include the small direct cost of protection (often obtained from government budgets), and the comparatively larger opportunity cost of land. For example, we estimate from data of Wangwacharakul and Bowonwiwat (1994), that the direct present value of the cost of protection is US$ 2.1 ha^{-1} in Thailand, but the present value of the opportunity cost of land is US$ 44 to US$ 89 ha^{-1}.

In Brazil, Côte d'Ivoire, and Thailand, agriculture or pasture is the primary alternative land use and the cost per Mg C is lower than in Indonesia where the alternative is logging (Table 26.1). For India, the reported estimate is based on the cost of protection of a tiger habitat, and the higher one for Mexico, on the use of land for a proposed national park. Swisher's (1991) estimates are based on the Tropical Forestry Action Plan (TFAP) proposals for creating forest reserves in Central American countries.

Table 26.1. Cost of protection or reducing deforestation.

COUNTRY	COSTS (US$ PER Mg C)
Brazil	2.3[a], 1 to 4[b]
Côte d'Ivoire	8.0[a]
Indonesia	15.0[a]
Thailand	0.4 to 0.8[c]
Mexico	1 to 6[d]
India	0.1[e]
Central America	1 to 3[f]
Russia	1 to -3[g]

a Darmstadter and Plantinga (1991).
b Cline (1992), based on Reis and Margulies (1991) estimate of the value of agricultural production in the Amazon.
c Author's estimate based on Wangwacharakul and Bowonwiwat (1994), which includes government budget for protection and the opportunity cost of land for agriculture production.
d Author's estimate based on data from Masera, Bellon and Segura (1994). Lower bound is based on government budget for protection and the higher bound on cost of protection of tropical evergreen forests in Tabasco.
e Based on US$ 5 ha^{-1} cost for a tiger sanctuary and 50 Mg C ha^{-1} (a very conservative estimate of C density).
f Swisher (1991) estimate based on cost of protected areas as reported in TFAP proposals for Costa Rica, Honduras and Panama.
g Krankina and Dixon (1994) as reported in Dixon et al. (1993).

The costs reported in Table 26.1 range from a few cents Mg^{-1} C to about US$ 15 Mg^{-1} C. Most of the cost estimates are under US$ 3 Mg^{-1} C. The estimates in Table 26.1 exclude the value of local benefits, such as watershed protection, maintenance of biodiversity, education, tourism, and recreation which may be derived from protected forests. Inclusion of local and ancillary benefits would partially offset the cost reported in Table 26.1 and in some instances result in a net benefit (Dixon et al. 1993). Often the monetary benefits are small and may only be enough to offset the direct cost of protection. For the aforementioned Thai example, the monetary benefits were only US$ 2 ha^{-1}.

50–90% OF TECHNICAL POTENTIAL FOR EXPANSION OF CARBON STOCKS MAY BE TAPPED AT A COST LESS THAN US$ 10 Mg^{-1} C

In Table 26.2, we report a range of country-specific estimates of costs to initiate a forest sector project which generally include establishment, initial tending and supervision costs. These estimates range from as low as US$ 1 Mg^{-1} C for several countries to US$ 31 Mg^{-1} C for Argentina. For the U.S. the estimates range from US$ 12 to US$ 95 Mg^{-1} C.

At a global level, Dixon et al. (1991) reported that the marginal initial cost to sequester global C increased gradually to US$ 10 Mg^{-1} C for a storage level of 70 Pg C, which was about 90% of the identified storage potential. Beyond 70 Pg C, the marginal cost increased rapidly. The corresponding total cost to store 70 Pg C amounted to US$ 230 000 million for an average cost of US$ 3.2 Mg^{-1} C. Country-specific marginal cost estimates for Brazil, Russia and the United States (Dixon et al. 1991) and for India (Ravindranath and Somashekhar 1995), China (Xu 1995) and Thailand (Wangwacharakul and Bowonwiwat 1995) confirm that between 50 and 90% of the C storage potential can be tapped at a cost of less than US$ 10 Mg^{-1} C.

The above estimates do not include the C stored in wood products. For increasing the C stored in wood products, in general, the longer the life-cycle of the wood products and the greater the percentage of the timber that goes into long-lived products, the more effective timber harvests and subsequent replanting of forests are as a technique for sequestration of C. Dewar and Cannell (1992) report that C stored in forest products in the UK can amount to 30–40% of that stored on land. Further, Harmon et al. (1990) found that 57% of the old-growth forest tree boles harvested went into short-term and 43% into longer-term (over 5 years) forest product C pools. By not including the product C, and not considering the possibility of substituting wood from sustainable plantations for fossil fuels, the afore-mentioned studies overestimate the unit costs.

OPPORTUNITY COSTS MAY SIGNIFICANTLY INCREASE ESTIMATED COSTS

The cost components not included in the estimates in Table 26.2 are the opportunity cost of land, annual maintenance costs, overhead, monitoring and evaluation costs. Of these, the latter two are generally a small fraction of the initial cost.

The opportunity cost of land, however, may significantly increase some of the unit cost estimates in Table 26.2. Land rental costs were estimated at US$ 148 ha^{-1} yr^{-1} by Moulton and Richards (1990) for the U.S. and a land purchase price was estimated between us$ 400 and US$ 1000 ha^{-1} by Sedjo and Solomon (1989). Land prices are likely to be lower in the developing countries. For Thailand, Wangwacharakul and Bowonwiwat (1994) report the present value of the opportunity cost of land to be between US$ 44–89 ha^{-1}. For degraded lands that are suitable for reforestation, which are widespread across many countries, the land price may be close to zero. For China, forest lands are already allocated for forest development, while the dry croplands are only suitable for agroforestry development (Xu 1994). Thus the opportunity cost of Chinese non-forestry land or land classified for forestry is very low.

Winjum and Lewis (1993) demonstrated the significance of including the forest stock opportunity costs, which they define as the value of wildlife habitats or visual amenities, recreation, water yield and soil conservation. Using the value of growing stock based on plantation data from Sedjo (1983), they show that without including growing stock costs, the net benefit of storing C is US$ 48 Mg^{-1} C for temperate and US$ 32 Mg^{-1} C for tropical plantations. With the growing stock costs included, the net benefit is lower, but not very different,

Table 26.2. Initial costs of expanding C sinks.

Region/Country		Cost[a] (US$ Mg^{-1} C)	Source
Boreal:	Natural regeneration	5 (4–11)	Dixon and Andrasko (1992) [b,c]
	Reforestation	8 (3–27)	
Temperate:	Natural regeneration	1 (1)	Dixon and Andrasko (1992) [b,c]
	Afforestation	2 (1–5)	
	Reforestation	6 (3–29)	
Tropical:	Natural regeneration	1 (1–2)	Dixon and Andrasko (1992) [b,c]
	Agroforestry	5 (2–11)	
	Reforestation	7 (3–26)	
Central America:	Regeneration	4	Swisher (1991)[e]
	Agroforestry	4	
	Plantations	13	
Argentina:	Reforestation	31	Winjum et al. (1993)[c]
	Afforestation	18	
Australia:	Reforestation	5	Winjum et al. (1993)[c]
Brazil:	Reforestation	10	Winjum et al. (1993)[c]
	FLORAM	3–8[1]	Andrasko et al. (1991)
Canada:	Reforestation	11	Andrasko et al. (1991)
	Natural Regen.	6	
China:	Reforestation	10	Winjum et al. (1993)[c]
	Forest management	3–4	Xu (1995)[d]
	Eucalypt plantations	8	Xu (1995)[d]
	Agroforestry	6–21	Xu (1995)[d]
Germany:	Reforestation	29	Winjum et al. (1993)[c]
India:	Reforestation	15	Winjum et al. (1993)[c]
	Natural regeneration	2	Ravindranath and Somashekhar (1995)[e]
	Teak plantation	3	Ravindranath and Somashekhar (1995)[e]
	Agroforestry	9	Ravindranath and Somashekhar (1995)[e]
Malaysia:	Reforestation	5	Winjum et al. (1993)[b]
Mexico:	Reforestation	4	Winjum et al. (1993)[c]
	Plantations	5–11	Masera et al. (1995)[d]
	Forest management	0.3–3	Masera et al. (1995)[d]
South Africa:	Reforestation	9	Winjum et al. (1993)[c]
Thailand:	Teak plantation	13–26	Wangwacharakul and Bowonwiwat (1995)[d]
	Eucalypt plantation	5–8	
	Agroforestry	8–12	
USA:	Reforestation	5	Winjum et al. (1993)[c]
	Afforestation	2	
USSR:	Reforestation	6	Winjum et al. (1993)[c]
	Natural regeneration	5	Winjum et al. (1993)[c]
Russia:	Plantation	1–8	Krankina and Dixon (1993), as reported in Dixon et al. (1993)
USA: Marginal costs for various options (includes planting and land rental costs)		12–41	Moulton and Richards (1990)[f]
USA: Marginal costs for various options		19–95	Adams et al. (1993)[f]

a Figures vary depending on the costs per ha from US$ 400 to US$ 1000.
b The figures in parenthesis reported by Dixon and Andrasko are interquartile ranges.
c The data reported by Dixon and Andrasko (1992) and Winjum et al. (1993) include only vegetation C and are based on the mean stock of C under a perpetual rotation system.
d The data reported by Xu (1995), Ravindranath and Somashekhar (1995), Wangwacharakul and Bowonwiwat (1995), and Masera et al. (1995) include vegetation and soil C.
e The data reported by Swisher (1991) account for vegetation, soil and litter C.
f Moulton and Richards (1990) (as reported in Cline 1992) and Adams et al. (1993) account for vegetation, soil and litter C and use the "plant and store" approach to estimate total accumulated C, i.e., not the "sustainable rotations" approach.

at US$ 22 and US$ 24 Mg^{-1} C respectively. Inclusion of the opportunity cost of land and of the forest stock thus may increase the cost in developed countries two to three fold but the increase may be smaller in developing countries where degraded lands are available.

MONETARY BENEFITS MAY OFFSET COSTS

The implementation of mitigation options will result in direct benefits derived from timber and non-timber forest products. The commercial value of these products will vary with the site and location of a project, particularly in tropical countries, where non-timber forest products provide sustenance to the local population. Dixon et al. (1993) report on several forest sector C offset projects which have been established world-wide and are maintained at costs well below the estimates reported in Table 26.2 or at a net benefit.

Winjum and Lewis (1993) in their analysis of the data collated by Sedjo (1983), reveal that accounting for the monetary timber benefits alone can often more than offset the project costs. The extent to which the costs are offset varies depending on whether indirect benefits, such as the forest stock value, are included. Other estimates (Ravindranath and Somashekar 1995; Ismail 1995; Wangwacharakul and Bowonwiwat 1995; Xu 1995) of the benefit of C sequestration for developing countries, confirm that product revenue from many plantation and agroforestry projects could completely offset their costs. Schroeder (1993) points out that agroforestry projects could be implemented for C sequestration across many ecozones and practices at a positive present value of net benefit ranging from US$ 54 ha^{-1} to US$ 6000 ha^{-1}. These estimates are based on assumptions of constant real prices in the future. The estimates suggest that it would be possible to achieve C sequestration at a net benefit or negative cost.

EQUITABLE DISTRIBUTION IS NECESSARY FOR PROJECT SURVIVAL

The equitable distribution of benefits will play an important role in assuring the success of either large forestation programs or individual projects to sequester C. Adams et al. (1993) illustrate that a U.S. forestation program to sequester C on agricultural land will bestow higher benefits to agricultural producers and landowners from higher commodity prices but the decline in timber prices will reduce timber sale profits for private forest owners. If the distribution of benefits is to be equitable, Adams et al. (1993) state that the government must compensate private commercial tree planting to prevent farmers from displacing present tree plantations. Saxena (1989) has shown that in the short to medium term, the losers in a rural forestry development program in India will be the local forest officers, while both the rural rich and the poor will gain to varying degree from the non-timber benefits and timber sales. It is thus important that participants in a program aimed at sequestering C gain appropriate and adequate benefits in order to ensure sustained C storage.

CONCLUSIONS

In conclusion, our review of costs shows that the direct costs for reducing net C emissions are lower (about US$ 3 Mg^{-1} C) for forest protection than for expanding tree cover (up to US$ 10 Mg^{-1} C). The latter cost estimates may significantly increase when the opportunity cost of land is included. On the other hand, inclusion of C stored in wood products and the indirect benefits of forests and their stock value will reduce, and may offset, this increase. Even if present cost estimates for C mitigation by forestry were doubled or tripled, they would still be considerably lower than the U.S. fossil-fuel tax of US$ 100–350 Mg^{-1} C suggested by Rubin et al. (1992).

TOPICS FOR FURTHER RESEARCH

Several topics related to the economic analysis of forest sector mitigation options deserve further research. These topics include: (1) better estimates of land availability, (2) consideration of full costs and benefits, (3) evaluation of alternative indicators of cost-effectiveness and (4) better and more detailed national and sub-national data on mitigation options.

The estimation of national or global land availability is a difficult task because of the inconsistency in: (1) land classifications; (2) overlapping definitions of availability such as technical and biophysical as opposed to social and political; and (3) assessment methods such as using government targets, or national or sub-national estimates, or satellite mapping of regions. Earlier and current estimates encompass all these definitions of land availability and provide a mosaic of data points which are indicative but not wholly satisfying in their clarity. One of the recommendations of this paper is that land availability be studied more carefully than has been done thus far. Without reasonably close estimates of land availability it is not possible to estimate the full C sequestration potential of mitigation options. It is thus important that consistent definitions of land classifications and methods for their estimation be established.

It is imperative that more data be collected on the opportunity cost of land and that these be explicitly included in future evaluation of costs. Opportunity costs should include the value of land and its forest stock where appropriate. The consideration of benefits is important since their equitable distribution among the affected participants is the best way to ensure the survival of a plantation. In addition, non-monetary benefits, e.g., biodiversity, should be explicitly stated, if not valued, in addressing an option's economic merits.

Until C damage functions can be sorted out, economic analysts will have to rely on alternative indicators of cost-effectiveness of an option. Various indicators should be prepared which shed light on the policy implications of an option. Indicators should explicitly account for the decay of atmospheric C levels over time and C discounting should be used more frequently so that policy makers become familiar with their use. Discounting C will shift the preference to projects with near-term C benefits. Price (1993) correctly points out that the uncertainty of climate change is a reason for increasing weight on the future and appropriate discounting can help achieve that.

Finally, much of the data and analysis is done for and in industrialised countries which have the analytical skills and access to resources while inadequate work has been done on the tropical countries. This is similar to the situation of a person looking for a lost key under a lamp, since that is where the light shines. More efforts are needed to conduct national and sub-national studies in developing and transition countries in order to validate data and information on land availability, C flows and costs and benefits of alternative options.

ACKNOWLEDGEMENTS

This work was supported by the US Environmental Protection Agency, Climate Change Division through the U.S. Dept. of Energy under Contract No DE-AC03-76SF00098. The author would like to acknowledge the support provided by Ken Andrasko, U.S. EPA and Willy Makundi, LBL.

REFERENCES

Adams R, Adams D, Callaway J, Chang C, McCarl B (1993) Sequestering carbon on agricultural land: social cost and impacts on timber markets. *Contemporary Policy Issues* 11: 76–87.

Andrasko K, Heaton K, Winnett S (1991) Evaluating the costs and efficiency of options to manage global forests: A cost curve approach. In: Proc. of a Technical Workshop to Explore Options for Global Forest Management, April 1991, Bangkok, pp 216–233.

Cline WR (1991) Scientific basis for the greenhouse effect. *The Economic Journal* 101: 904–919.
Darmstadter J, Plantinga A (1991) The economic cost of CO_2 mitigation: a review of estimates for selected world regions. Discussion Paper ENR91-06, Resources for the Future, Washington, DC, pp 54–57.
Dewar R, Cannell M (1992) Carbon sequestration in the trees, products and soils of forest plantations: an analysis using UK examples. *Tree Physiology* 11: 49–71.
Dixon RK, Brown SA, Houghton RA, Solomon AM, Trexler MC, Wisniewski J (1994) Carbon pools and flux of global forest ecosystems. *Science* 263: 185–190.
Dixon R, Andrasko K (1992) Integrated land-use systems: assessment of promising agroforest and alternative land-use practices to enhance carbon conservation and sequestration In: Proceedings of a workshop on assessing technologies and management systems for agriculture and forestry in relation to global climate change. Canberra, January 1992, pp 48–54.
Dixon RK, Andrasko KJ, Sussman FG, Lavinson MA, Trexler MC, Vinson TS (1993) Forest sector carbon offset projects: near-term opportunities to mitigate greenhouse gas emissions. *Water, Air, and Soil Pollution* 70: 561–577.
Hall D, Mynick H, Williams R (1991) Cooling the greenhouse with bioenergy. *Nature* 353: 11–12.
Harmon M, Ferrell W, Franklin J (1990) Effects on carbon storage of conversion of old-growth forests to young forests. *Science* 247: 699–702.
IPCC (1992) Climate Change: The IPCC Response Strategies. Intergovernmental Panel on Climate Change, Island Press, Washington, DC, p 101.
Ismail R (1994) An economic evaluation of carbon emission and sequestration for the forestry sector in Malaysia. *Biomass and Bioenergy*, Special Issue, March 1994.
Krankina O, Dixon R (1994) Forest management options to conserve and sequester terrestrial carbon in the Russian federation. *World Resource Review* 6(1): 88–101.
Makundi W, Sathaye J, Masera O (1992) Carbon emissions and sequestration in forests: case studies from seven developing countries. Volume 1: Summary. Makundi W, Sathaye J (eds) Lawrence Berkeley Laboratory, Report No. LBL-32119, Berkeley, CA.
Masera O, Bellon M, Segura G (1995) Forestry options for sequestering carbon in Mexico's forests. *Biomass and Bioenergy*, Special Issue, March 1995.
Moulton RJ, Richards KR (1990) Costs of sequestering carbon through tree planting and forest management in the United States. USDA For. Serv., Gen. Tech. Rep. WO-58, 48 pp.
Price C (1993) Time, discounting and the valuation of forestry's carbon fluxes. *Commonwealth Forestry Review* 72(4): 265–271.
Ravindranath NH, Somashekhar BS (1995) Potential and economics of forestry options for carbon sequestration in India. *Biomass and Bioenergy*, Special Issue, March 1995.
Reis E, Margulies S (1991) Options for slowing Amazon jungle-clearing In: Dornbusch R, Poterba J, (eds) Global Warming: Economic Policy Responses, MIT Press, Cambridge, Massachusetts, pp 335–374.
Rubin ES, Cooper RN, Frosch RA, Lee TH, Marland G, Rosenfeld AH, Stine DD (1992) Realistic mitigation options for global warming. *Science* 257: 148–266.
Saxena NC (1989) Forestry and rural development. *South Asia Journal* 3: 1,2, Sage Publ., New Delhi.
Sedjo R (1983) The Comparative Economics of Plantation Forestry: A Global Assessment. Resources for the Future/Johns Hopkins University Press, Baltimore, MD.
Sedjo R, Solomon A (1989) Climate and forests In: Rosenberg NJ, Easterling WE, Crosson PR, Darmstadter J (eds) Greenhouse warming: abatement and adaptation. Resources for the Future, Washington, DC, pp 105–120.
Schroeder P (1993) Agroforestry systems: Integrated land use to store and conserve carbon. *Climate Research* 3: 53–60.
Swisher JN (1991) Cost and performance of CO_2 storage in forestry projects. *Biomass and Bioenergy* 1(6): 317–328.
Wangwacharakul V, Bowonwiwat R (1994) Economic evaluation of CO_2 response options in forestry sector: Thailand. *Biomass and Bioenergy*, Special Issue, March 1994.
Winjum JK, Lewi DK (1993) Forest management and the economics of carbon storage: the nonfinancial component. *Climate Research* 3: 111–119.
Winjum JK, Dixon RK, Schroeder PE (1993) Forest management and carbon storage: an analysis of 12 key forest nations. *Water, Air, and Soil Pollution* 70: 239-257.
Xu D (1995) The potentiality of reducing carbon accumulation in the atmosphere by large-scale afforestation in China and related cost/benefit analysis. *Biomass and Bioenergy*, Special Issue, March 1995.

27 Integrating the socio-economic and physical dimensions of degraded tropical lands in global climate change mitigation assessments

Alan Grainger

School of Geography, University of Leeds
Leeds LS2 9JT, UK

INTRODUCTION

Models of global environmental change need to integrate its socio-economic and physical environmental dimensions much better than they do at present. The traditional separation between the social and environmental sciences has hindered development of interdisciplinary methodologies and led to predominantly physical assessments of key problem areas, such as the potential to mitigate global climate change by sequestering more terrestrial carbon (C) in woody biomass.

Degraded tropical lands are often identified as being ideal for additional C sequestration, either by afforesting predominantly deforested lands, or by promoting natural forest regeneration in degraded forests. These lands seem suitable for intervention because they are physically (biologically) sub-optimal, and so rehabilitating them would benefit local as well as global environments. It is often assumed that they are lightly populated and so the lands would be readily available for treatment. However, not only are estimates of the areas of degraded lands very inaccurate, but lands that appear physically degraded may not be equally degraded in the eyes of local people, who in practice will often determine whether or not the land is available for rehabilitation.

This raises some fundamental questions about the nature of vegetation degradation. Most areas of the world are not covered by the climax ecosystems which ideally should be found there. Instead, the vegetation cover has been culturally modified by various forms of human impact. From a physical perspective such vegetation is degraded relative to climax ecosystems, e.g., in having an inferior biomass density. However, from a cultural perspective, since a large proportion of this vegetation is also managed in some way, its current status represents a balance between the properties of human systems and natural systems. What physical degradation does exist may well be an inevitable consequence of continued human intervention. The quality of vegetation may therefore be culturally optimal, or close to it, and attempts to change it will be constrained by the same socio-economic and political factors that determine its present character.

If this is the case, better definitions of degraded land are needed, in which degradation is assessed in relation to cultural norms rather than physical norms alone. Cultural norms would be derived by integrating socio-economic and physical criteria. At some point, the intensity of human intervention may become so high that it reduces the resilience and productivity of the natural ecosystems, the socio-economic benefits that previously outweighed the disadvantages of physical degradation no longer do so, and local people will also regard the land as degraded. The challenge is to find how to assess degradation in this way.

NATO ASI Series, Vol. I 40
Forest Ecosystems, Forest Management and the Global Carbon Cycle
Edited by Michael J. Apps and David T. Price

The first section of this Chapter looks at how to integrate socio-economic and physical dimensions from a static perspective, showing how simple physical assessments of the extent of degraded forests and their suitability for rehabilitation could be modified to estimate their actual availability. The second section examines dynamic aspects, and how societal and market factors could delay programmes to afforest heavily degraded lands.

THE EXTENT OF DEGRADED TROPICAL LANDS

Estimates of the extent of degraded lands in the tropics are very inaccurate. In the 1980s the total area was estimated to be 2077 Mha (Grainger 1988). This consisted of 339 Mha of degraded forests (represented by logged forests and shifting cultivation forest fallows), 87 Mha of deforested watersheds, and 1651 Mha of desertified drylands (Table 27.1).

Table 27.1. Area of degraded tropical lands in 1980 with potential for forest replenishment (Source: Grainger 1988).

	Degraded lands (Mha)	
	All	With potential for replenishment
Forest fallows	203	203
Logged forests	137	137
Deforested watersheds	87	87
Degraded drylands	1651	331
Total	2077	758

The major source of error in this estimate was the area of degraded drylands, which was known to be very inaccurate (see Grainger 1990a for a review of data sources). It was concluded, based on information supplied in the original estimates, that in dry areas only 331 Mha of degraded rainfed and irrigated croplands received enough rainfall to make afforestation successful. Most desertified rangelands would be too dry for this purpose and there was no information on the area of moist rangelands. The figure for deforested watersheds was also very arbitrary, as the FAO/UNEP data on which estimates of degraded forests were based (Lanly 1981) did not refer specifically to montane forest areas.

In response to proposals to assess the feasibility of establishing a large area of new forest plantations to mitigate global climate change, the above data were used to estimate that a maximum of 621 Mha had potential for afforestation (Grainger 1990b), on the assumption that logged forests were best left to regenerate naturally. But since most forest fallows are under agricultural use, this area was later corrected to no more than 420–450 Mha (encompassing all degraded drylands and deforested watersheds and the most degraded forest fallows) (Grainger 1991).

Studies of the potential to sequester more C in forests then became more comprehensive, and now also include the areas of degraded forests with potential for replenishment. Using the above data, 420–450 Mha would be suitable for afforestation, with the rest of the 621 Mha suitable for the protected regeneration of existing forests. But the reliability of this estimate (and others) is severely constrained by the poor quality of the original physical data. The latest estimate of the area of degraded drylands (UNEP 1992) is still very inaccurate (Grainger 1994), as is the latest FAO estimate of tropical forest area (FAO 1993) which is even less disaggregated by forest quality category than the estimates for 1980 (Grainger

1995a). The area of physically degraded tropical forest is probably much greater than currently acknowledged, though how much is culturally degraded is a different matter.

ESTIMATING THE AVAILABILITY OF DEGRADED TROPICAL FORESTS FOR REHABILITATION

Two main methods have been used to assess the potential for forests to sequester more C to mitigate global climate change. The first, which may be termed the *carbon store approach*, views forest only as a static C stock. A net transfer of C from atmosphere to land can therefore be achieved by rehabilitating degraded natural forests, by either promoting natural regeneration or planting trees, so their biomass density moves closer to the potential level for each area. The second, the *carbon sink approach*, takes a more dynamic view of forests as renewable resources, and is in line with the fact that CO_2 and other greenhouse gases continue to be emitted as a result of human activities. With regular harvesting and subsequent industrial wood utilisation, forests can continue to take in C and compensate for annual C emissions from all sources. This approach has been used to evaluate the potential to expand the area of forest plantations on degraded tropical lands (Sedjo and Solomon 1989; Grainger 1990b).

Both approaches are subject to significant socio-economic and political constraints. Degraded forests, such as forest fallow and logged forest, account for just under half of all degraded tropical lands with potential for forest replenishment (Table 27.1). But while from a physical perspective they may be sub-optimal C stores, this may simply be an inevitable consequence of how they are managed. This section shows, using a case study of continental South and Southeast Asia (Iverson et al. 1993) how a more realistic assessment of land availability can be made by discriminating between different types of degraded forest and their socio-economic attributes.

Estimating physical suitability for rehabilitation

The first stage of this study assessed the suitability of degraded forest lands in continental South and Southeast Asia for C enhancement in a purely physical way. The *technical suitability* of land was assumed to be proportional to the difference between its potential and actual biomass density. Potential biomass density was estimated as a function of climate and other environmental variables and represented by a potential carbon sequestration index (PCSI) on a scale of 1 to 100. Actual biomass density was represented by an actual C sequestration index (ACSI) (also on a scale of 1 to 100) estimated in two ways: (a) by a model based on climate and settlement data; (b) by a global vegetation index derived from low-resolution satellite data (the two estimates showed good correlation).

Digital maps of PCSI and ACSI for remaining forests in continental South and Southeast Asia were produced using geographic information system (GIS) software, and digital maps of the distribution of environmental variables and forest (the latter was an FAO map based mainly on the visual interpretation of LANDSAT satellite imagery).

The degree of degradation was then estimated by subtracting ACSI from PCSI using the GIS, leading to a map in which forests were divided into six technical suitability classes (part of the map is shown in Figure 27.1). The more degraded the forest the higher its technical suitability for C sequestration. The biomass density of forest in the lowest class was 50 Mg C ha^{-1} lower than its potential value, while that in the highest class was over 250 Mg C ha^{-1} lower. About 27% of all forest— 45–50 Mha, depending on the scenario chosen—was in the three highest degradation classes. The average degradation was 88 Mg C ha^{-1}, compared with a mean regional biomass density of 185 Mg C ha^{-1}. Two-thirds of all forest could sequester a further 50–150 Mg C ha^{-1}, so the region appears to have significant potential to contribute to climate change mitigation (Iverson et al. 1993).

These results should be treated with caution, as they were based on inaccurate maps of present forest cover and land use, and substantial errors were made in estimating ACSI and PCSI and using regression equations to predict their regional distribution. But the study was the first of its kind in using direct measurements and appropriate models to estimate the large-area distribution of degraded forest.

A new taxonomy of degraded tropical forests

Another limitation of the above study was that technical suitability was estimated solely on the basis of physical criteria. In practice, of all degraded forest of medium or higher technical suitability only some would be available for treatment. If availability depends upon the degree to which the forest is degraded relative to cultural norms, as argued above, physical assessments of degradation must be modified by including socio-economic attributes of vegetation as well.

A first step in doing this is to disaggregate the area of degraded forest into its different sub-types, which by referring to different human impacts or uses may be taken to include some socio-economic attributes. Previously there was little interest in mapping degraded forests, and until recently, vegetation maps focused mainly on the ideal distribution of vegetation in terms of the major climax ecosystems or biome-types. This is in line with the general approach of *natural biogeography*, which is not particularly concerned with the actual distribution of vegetation as modified by people, in contrast with *cultural biogeography*, which is (Simmons 1979). This section outlines an initial taxonomy for distinguishing between the different types of degraded tropical moist forests in Southeast Asia (Grainger 1995b).

Physical degradation indicators

Vegetation degradation has been defined in physical terms as a "*...temporary or permanent reduction in the density, structure, species composition or productivity of vegetation cover*" (Grainger 1992). This leads to three measurable physical indicators of degradation—vertical structure, density, and species composition—which can be used as the basis for a taxonomy.

Vertical structure. Degraded forest often has a different vertical structure from primary forest. Thus, logging primary tropical rain forest removes most of its emergent trees, and forest fallow associated with shifting cultivation has a low flat canopy compared with the high irregular canopy of primary forest.

Density. Degradation also involves changes in vegetation density. There are five main density indicators, though not all are relevant to different forest types. Some degraded forests have a lower percentage *canopy closure* than undisturbed forest, particularly in drier or seasonal areas, although forest fallow has a continuous canopy. Another indicator is *canopy quality*, since drought, fire, insect pests, pollution etc. may cause damage to foliage. Degradation may also lower a forest's *tree density*. Sometimes, overall tree density is reduced; on other occasions, this remains high but the density of large trees (which contain a high proportion of total biomass (Brown et al. 1994)) falls sharply. A reduction in *biomass density* is another indicator, occurring when the overall tree density falls, large trees are removed, or mature forest is replaced by a low bushy forest fallow. Degradation is shown on a wider scale by the *fragmentation* of former homogeneous and contiguous undisturbed forest into complex mosaics of undisturbed forest, degraded forest, gaps and clearances.

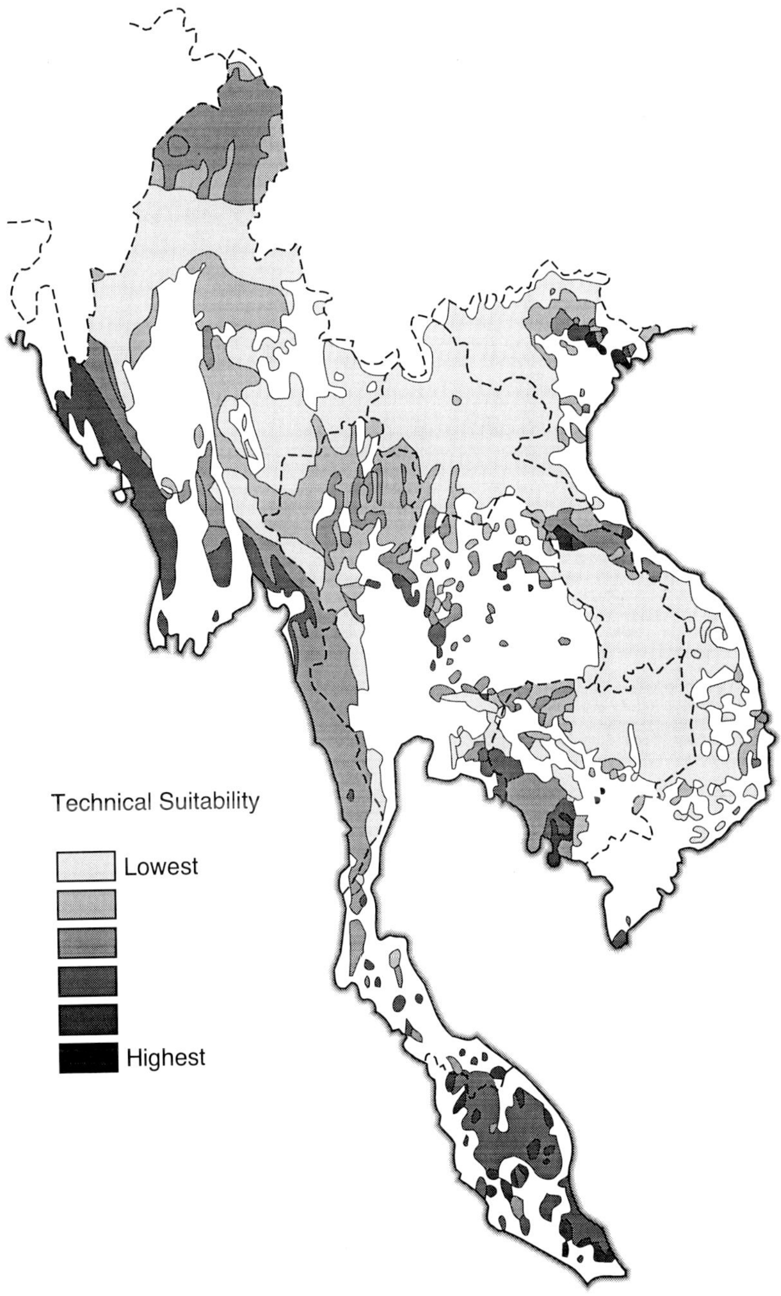

Figure 27.1. Technical suitability of degraded forests in continental South and Southeast Asia for carbon sequestration.

Species composition. Degradation can lead to a shift in species composition, which may be merely temporary or more prolonged. This may result from the extraction of primary species, the regrowth of secondary species, or the planting of monocultures or species mixtures. Animal species composition can also be changed by commercial poaching or the removal of food sources by logging.

An initial taxonomy

Degraded forests can be ordered in an integrated socio-physical manner by grouping them by the degree of disturbance and intensity of human intervention. Six main groups are identified in Table 27.2, with disturbance and intervention intensity increasing down the list.

Table 27.2. A taxonomy of degraded forests in the humid tropics. See text for details (Source: Grainger 1995b).

	CLASSIFICATION GROUP	
1.	Extractive Forest	
	a.	Logged Forest
	b.	Other Extractive Forest
2.	Damaged Forest	
3.	Regenerated Forest	
	a.	Long-rotation shifting cultivation regrowth
	b.	Short-rotation shifting cultivation forest fallow
	c.	Some managed forest
4.	Planted Vegetation	
	a.	Enriched forest
	b.	Agroforestry combinations
	c.	Tree crop plantations
5.	Interrupted Successional Vegetation	
	a.	Encroaching cultivation land
	b.	Weedy grasslands
6.	Dispersed Forest	

1. *Extractive Forest.* Extractive forest has had trees or herbaceous plants removed from it, but has not been cleared. Examples include selectively logged forest, which is temporarily degraded and then subsequently regenerates, and other types of extractive forest exploited for various minor forest products in ways that have a minimal impact on the ecosystem.
2. *Damaged Forest.* Damaged forest has been subjected to various impacts, including drought, wild fires, insect pests and air and water pollution. The effects range from foliar damage to tree mortality, and can be widespread or very localised in scale. For many types of damaged forest, human intervention is indirect and of low intensity.
3. *Regenerated Forest.* Regenerated forest is in the process of regrowth following clearance. The intensity of human intervention varies. Examples include short-rotation shifting cultivation forest fallow, and forest under silvicultural management.
4. *Planted Vegetation.* Another group of degraded forests has been intentionally planted after clearance, so the intensity of human intervention is high. Examples include enriched forest (in which commercial trees species are planted in artificial gaps); agroforestry combinations of various tree crops; and monoculture tree crop plantations.

5. *Interrupted Successional Vegetation.* This group is structurally distinct from the others, being dominated by grasses and shrubs, with only a minor tree component. Forest regeneration is continually interrupted by burning, or constrained when over-cultivation depletes fertility or promotes the spread of weeds.
6. *Dispersed Forest.* Degradation is also apparent on the landscape scale as mixtures of forest trees, tree crops and weedy grasslands in dispersed and heterogeneous mosaics of different types of degraded and non-degraded forest. Overall tree cover is higher than in the previous group; degradation is extensive but not extreme; and land use practices are generally more sustainable. While this group includes members of other groups, it sometimes provides the only indication of degradation in areas where data availability is poor and satellite imagery is the sole data source.

From technical suitability to actual availability

The area of degraded land actually available for C enhancement is generally smaller than the technically suitable area because a good proportion of the latter is not degraded in relation to cultural norms. This section describes the basic principles of the approach being used to refine the assessment of the Iverson et al. (1993) study outlined above. The first stage decides on the kind of enhancing treatment most appropriate to each area and the second incorporates social, economic and political constraints more explicitly.

There are two main kinds of enhancing treatments. Most forest in the low to medium degradation classes is simply degraded secondary or logged forest that is regenerating naturally and already sequestering C. The only management treatment it really needs is *protection.* Forest in higher degradation classes, on the other hand, needs *afforestation* or other forms of intensive management.

A rough estimate of the area of available land can be made by dividing degraded forest into its different sub-types. Land is deemed to be *technically available* either if it is highly degraded, when cultural factors are unlikely to impede its availability, or if it is covered by managed forest, when current management practices should be capable of modification. Land will be technically unavailable if it is only slightly degraded, or if it is being farmed and so does not appear to be culturally degraded.

Land use classes likely to be technically available for protection include Secondary Forest, Logged Forest and Managed Forest. Degraded Forest, Degraded Land, Rangeland and Weedy Grassland are regarded as technically available for afforestation (Table 27.3). Land-use classes not technically available for protection will include Tree Crop Plantations and Forest Fallow. On the other hand, Cropland and Permanent Pasture that have been included by mistake in a map of degraded forests will not be technically available for afforestation. The last four classes of land are being actively farmed.

Only a proportion of the area of technically available land will be *actually available*, so further refinement is necessary to *directly* estimate the influence of social, economic and political constraints. It can be assumed in the first instance that all forest technically available for *protection* will be actually available (although in practice a lot of remaining pristine forest is under threat from illegal clearance). However, much forest land technically available for *afforestation* is likely to be unavailable. Some is too degraded for afforestation to be either physically or economically feasible: for example, on drier lands, or lands which have experienced severe soil degradation, returns will be too low to recoup the high planting and management costs. Other areas of degraded forest will be unavailable for social reasons: a lot of heavily degraded forest fallow (not excluded in earlier stages of refinement) is still used by shifting cultivators, and weedy grasslands and rangelands are often used for grazing. Political constraints will further reduce the actually available area. Procedures are now being developed to include relevant social, economic and political attributes of vegetation in such large-area mapping exercises.

Table 27.3. Technical availability of land by land-use class.

	AVAILABILITY FOR PROTECTION	AVAILABILITY FOR AFFORESTATION
Available	Secondary Forest Logged Forest Managed Forest	Degraded Forest Degraded Land Rangeland Weedy Grassland
Unavailable	Tree Crop Plantations Forest Fallow	Cropland Permanent Pasture

DYNAMIC CONSTRAINTS ON THE AFFORESTATION OF DEGRADED LANDS

The previous section adopted a static approach to assessing the availability of degraded forests for protection or afforestation. This section looks at the possibilities for large-scale afforestation for C sequestration and at how availability might change over time.

Two main sources of delay will make ideal planting-rate targets difficult to achieve in practice. First, any large-scale afforestation programme will be dedicated to meeting demand for a service—C sequestration—which is in addition to current demand for wood products, though it will still produce wood. Without government intervention, world forest products markets will be unable to accommodate a great increase in wood production. Second, societal constraints of various kinds will lead to changes in land availability and afforestation capability over time, and consequently delay the spread of afforestation.

Possible afforestation scenarios for climate change mitigation

A range of afforestation scenarios constructed by this author showed the planting rates required to sequester—using plantations rather than natural forests—different proportions of the current annual net increment in atmospheric CO_2 content from all sources, estimated as 2.9 Pg C yr^{-1} in the 1980s (Detwiler and Hall 1988) (Table 27.4). Sedjo and Solomon (1989) estimated that 465 Mha of new plantations would be needed to sequester the entire increment, but they assumed a uniform plantation growth rate. By taking account of national differences in growth rates it was estimated that 600 Mha of plantations would be needed instead (Grainger 1990b).

Table 27.4. Afforestation scenarios to achieve a range of reductions in present annual atmospheric CO_2 increment over a fifty year period. (Source: Grainger 1990b).

CO_2 INCREASE REDUCTION (%)	TOTAL PLANTED AREA (Mha)	TARGET PLANTING RATES (Mha yr^{-1})
100	600	12
50	300	6
42	250	5
25	150	3
10	60	1

Initial estimates suggested that at least 600 Mha of degraded land were indeed suitable for afforestation (Table 27.1) (Grainger 1988). However, even to afforest this area over 50 years would require 12 Mha of new plantations each year. This was twelve times the annual tropical afforestation rate between 1976 and 1980 and greater than the total area of tropical forest plantations in 1980 (11.5 Mha) (Lanly 1981), and so did not seem a feasible proposition.

It seemed more feasible instead to merely triple the present planting rate to 3 Mha yr^{-1}. This would cut the present net annual CO_2 increment by 25% by 2020, and if combined with a programme to reduce deforestation rates, according to either of two deforestation scenarios modelled by this author (Grainger 1990c), by 2020 C emissions from tropical forests would almost be fully balanced by the C sequestered by the new plantations (Table 27.5).

Impacts on the world forest products trade

These projections assumed that (a) all degraded land was actually available for afforestation, with no social, economic or political constraints; (b) the plantations would be regularly harvested so that new planting on the same sites would allow C sequestration to continue indefinitely; and (c) harvested wood could be stored indefinitely in permanent structures, so the reduction in the rate of increase in atmospheric CO_2 content would not just be temporary.

Table 27.5. Net carbon (C) emissions after combining C emissions by all tropical deforestation (High and Low Scenarios) and C uptake by new tropical forest plantations. This table combines the planting scenarios described in Grainger (1990b) with C emissions scenarios described in Grainger (1990c).

Planting Rate	Net C Emissions (Pg C yr^{-1})									
	High Scenario					Low Scenario				
(Mha yr^{-1})	1980	1990	2000	2010	2020	1980	1990	2000	2010	2020
12	0.757	0.699	0.030	-0.628	-1.269	0.544	0.486	-0.152	-0.832	-1.499
6	0.757	0.728	0.349	-0.018	-0.370	0.544	0.515	0.167	-0.223	-0.599
5	0.757	0.733	0.402	0.083	-0.220	0.544	0.520	0.220	-0.121	-0.449
3	0.757	0.742	0.508	0.286	0.080	0.544	0.530	0.327	0.082	-0.150
2	0.757	0.747	0.562	0.388	0.230	0.544	0.535	0.380	0.184	0.000
1	0.757	0.752	0.615	0.489	0.380	0.544	0.539	0.433	0.285	0.150

The last two assumptions ignored the realities of the world forest products trade. Calculations by Grainger (1990c) also examined the possible impact of 600 Mha of new plantations on the world wood market. The total wood output of about 7000×10^6 m^3 yr^{-1}, when in full production, would be more than four times current world demand for industrial wood, almost three times demand in 2030 according to IIASA projections (Dykstra and Kallio 1987), and more than three times demand in that year as projected by Sedjo and Lyon (1990).

As an alternative, some of the harvested wood could be burnt as fuel. Although, by definition, the half-life of C storage in harvested fuelwood is very short, it would substitute for fossil fuels and reduce their rate of use. (Chapters 18 and 19 in this volume deal with this subject more explicitly.) Table 27.6 shows the outputs for different planting rates in two groups of scenarios: Industrial Scenarios (all wood used as industrial wood) and Mixed Scenarios (half of the wood used for industrial wood, and half for fuelwood).

Table 27.6. Industrial wood production and mixed wood production scenarios to achieve a range of reductions in present annual atmospheric CO_2 increment over a fifty year period. (Source: Grainger 1990b).

CO_2 INCREASE/ REDUCTION (%)	SUSTAINED YIELD (10^6 m^3 yr^{-1})	ANNUAL HARVEST (10^9 m^3) FOR PLANTING PERIOD OF			
		10 yr[a]	20 yr	30 yr	50 yr
A. Industrial Scenarios					
100	6975	10.5	10.5	7.0	7.0
50	3488	5.2	5.2	3.5	3.5
25	1744	2.6	2.6	1.7	1.7
10	698	1.0	1.0	0.7	0.7
B. Mixed Scenarios					
100	3488	5.2	5.2	3.5	3.5
50	1744	2.6	2.6	1.7	1.7
25	872	1.3	1.3	0.9	0.9
10	349	0.5	0.5	0.3	0.3

a Assumes harvest is over 20 yr.

In Industrial Scenarios, even implementing the 25% Reduction Scenario (Table 27.4) over 20 years would increase the annual harvest by 2600×10^6 m^3—equivalent to IIASA's projected world demand for wood in 2030. The 10% Reduction Scenario implemented over 30 years would produce 700×10^6 m^3 more wood every year—half current world demand for industrial wood and a quarter of projected demand in 2030.

In the Mixed Scenario, the 25% Reduction Scenario implemented over 20 years would increase the world industrial wood harvest by 1300×10^6 m^3—equivalent to half projected world demand for industrial wood in 2030. It would also yield an extra 1100×10^6 m^3 yr^{-1} of fuelwood, compared with present total fuelwood consumption of 1700×10^6 m^3 yr^{-1}, though this might be accommodated, since dry and montane tropical areas are expected to have a fuelwood deficit of 607×10^6 m by the year 2000 (FAO 1981; Grainger 1988).

The projections assumed, for the sake of simplicity, that all industrial wood would be permanently stored after harvest, and so did not take account of the differential rates of decay of C from different industrial wood products back into the atmosphere, with short half-lives for paper products and longer ones for solid wood products. Correcting for this would reduce the overall long-term mitigating effect of the C sequestration. (See also Chapters 18, 20 and 23 in this volume.) In the Mixed Scenarios, the short-term impact of fuelwood production on atmospheric CO_2 content was realistically portrayed, though no account was taken of substitution for other fuels.

Thus, even quite modest forest-based mitigation programmes would have significant effects on the world forest products market. The fall in wood prices resulting from a major increase in wood supplies could severely damage the profitability of existing forestry operations and lead to the substantial conversion of forests to other uses, which would be highly counterproductive. Besides providing subsidies for tree planting, inter-governmental intervention would therefore be necessary to support world wood markets, and the possible cost of such intervention should be carefully estimated before proceeding with a large-scale afforestation programme.

Cultural constraints on the expansion of afforestation

Organisational constraints

Whatever target planting rate is set for an international forest-based mitigation programme, considerable organisational delays should be expected. The first form of delay will be institutional. Even if all afforestation were to be carried out by forestry department or other government personnel, delays would occur as policy is transmitted from central government to foresters on the ground. Various intermediate layers of bureaucracy, within the forestry department and in other government organisations, including regional or district offices of the forestry department, provincial governors etc., can influence the speed at which the original policies and plans are implemented. This depends upon the extent to which the different layers are subject to pressures from local interest groups and how they respond to them.

Another set of delays will be caused by the impact of society generally. If the afforestation programme is as big as anticipated it will have to rely greatly on social forestry programmes which involve the voluntary participation of local people, and this will lead to further delays. It takes time to create awareness by educational and publicity campaigns, time to convince people to agree to participate, and time to establish nurseries, financial support mechanisms and other necessary supportive infrastructure (Grainger 1991).

There will also be a lag due to the socio-economic status of participants. Every community has a range of income groups, and experience in social forestry projects in India and elsewhere has shown that it is usually the larger farmers who join such projects first, because they can afford to risk new ventures, whereas smaller farmers, who have a more precarious existence and cannot afford to take large risks, will only join later when the project is visibly successful (Grainger 1990a).

The diffusion of adoption of afforestation practices will have a spatial component too. Any programme will usually begin in one or two locations and spread from there. This is necessary for simple logistical reasons, but also reflects the fact that most successful large-scale afforestation schemes rely on the power of example. When one village is successful, it is visited by people from villages near and far, some of whom will elect to join the scheme, and afforestation spreads.

So even if the areas of actually suitable and technically available degraded lands have been correctly appraised, the actually available area will vary over time, depending upon people's willingness to participate and convert their land to a new use. This, together with institutional delays, will lead to extremely complex spatio-temporal patterns of forest replenishment that need to be allowed for in plans and projections.

National development constraints

Large-scale afforestation programmes or other forms of forest replenishment cannot be treated in isolation from wider aspects of national economic development. A number of generic land-use processes have been identified as part of the long-term evolution of national land-use morphology. They include the *national land-use transition*, which is the period in which a significant proportion of the forest cover in a forested country is removed (Figure 27.2) (Grainger 1993, 1995c). After the transition ends, forest cover in some countries may rise again, as a result of natural forest regeneration and artificial plantation establishment. This *forest replenishment period* is another land-use process (Grainger 1995d) which begins with the *forest transition* (Figure 27.3) (Mather 1992).

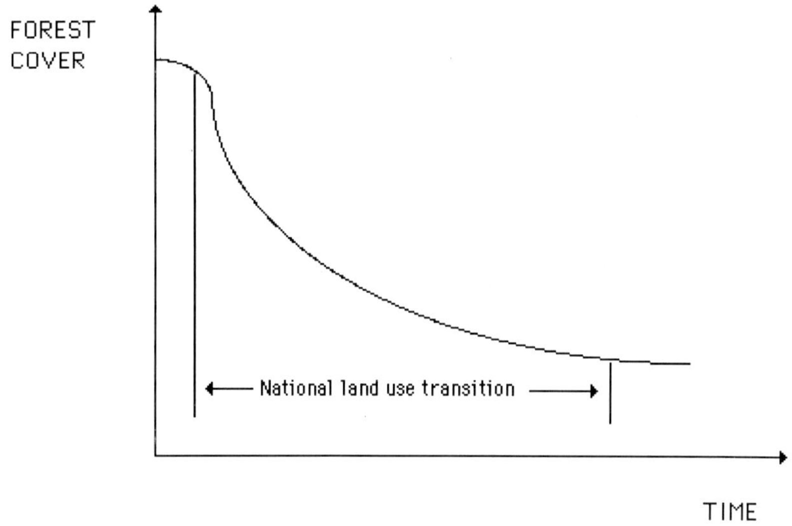

Figure 27.2. The national land-use transition.

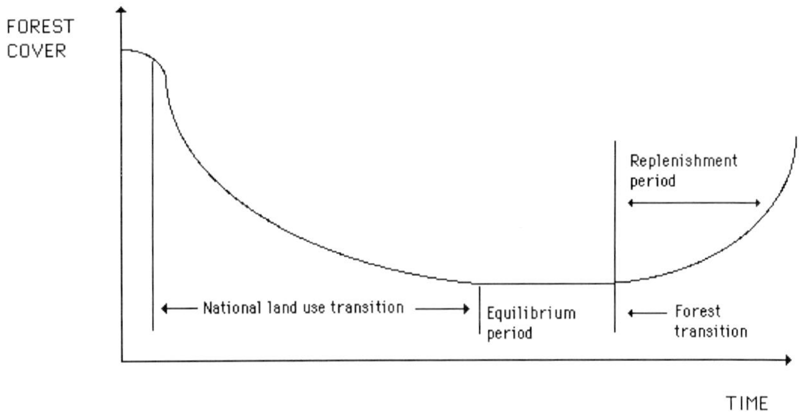

Figure 27.3. Unified model of national land-use transition, forest transition and replenishment period.

If national land-use change were to follow the normative curve in Figure 3, it would be not be possible to gain a net increase in national forest cover until after the national land-use transition had ended. This reflects conditions in the forest and agriculture sectors, which usually make it more economically efficient to exploit natural forests before establishing plantations. It does not preclude prior large-scale plantation establishment, of the kind that has occurred in Brazil and Indonesia, though planting there took place in seasonal areas, in preference to permanently humid areas where extensive forest areas remained and were being cleared. High planting rates and high deforestation rates are therefore not incompatible, but afforestation schemes on the scale needed to mitigate global climate change will be influenced by evolutionary trends in national land use in ways that are not properly understood at present.

CONCLUSIONS

Assessments of the potential for forest-based strategies for global climate change that rely on purely physical appraisals of degraded lands and give simple afforestation rate targets suffer from major limitations. The area actually available for C enhancement is likely to be much smaller than envisaged by estimates of technically suitable land, because availability will be determined by cultural criteria, rather than physical criteria alone. This paper has shown that integrating the physical, social, economic and political attributes of degraded lands might lead to more realistic assessments. Afforestation rate targets will not be attained quickly, because of delays caused by organisational constraints and the influence of long-term patterns of national land-use change, which will cause availability to change over time. These, together with the constraints imposed by the world forest products market, greatly complicate the task of assessment.

REFERENCES

Brown S, Iverson LR, Lugo AE (1994) Land use and biomass changes in peninsular Malaysia during 1972–1982: a GIS approach. In: Dale V (ed) Effects of land use change on atmospheric CO_2 concentrations: southeast Asia as a case study. Springer-Verlag, Berlin, pp 117–143.

Detwiler RP, Hall CS (1988) Tropical forest and the global carbon cycle. *Science* 210: 321–323.

Dykstra DP, Kallio M (1987) Base scenario. scenario variations. In: Kallio M, Dykstra DP, Binkley CS (eds) The global forest sector. An analytical perspective. Wiley, New York, pp 626–672.

FAO (1981) Map of the fuelwood situation in the developing countries. UN Food and Agriculture Organization, Rome.

FAO (1993) Forest resources assessment 1990. Tropical Countries. UN Food and Agriculture Organization, Rome, 59 pp.

Grainger A (1988) Estimating areas of degraded tropical lands requiring replenishment of forest cover. *International Tree Crops Journal* 5: 31–61.

Grainger A (1990a) The spreading desert: controlling desertification. Earthscan Publications, London, 369 pp.

Grainger (1990b) Modelling the impact of alternative afforestation strategies to reduce carbon emissions. In: Proceedings of the IPCC Conference on Tropical Forestry Response Options to Global Climate Change, Sao Paulo, Brazil, 9–12 January 1990. Report No. 20P-2003, Office of Policy Analysis, US Environmental Protection Agency, Washington DC, pp 93–104.

Grainger A (1990c) Modelling future carbon emissions from deforestation in the humid tropics. In: Proceedings of the IPCC Conference on Tropical Forestry Response Options to Global Climate Change, Sao Paulo, Brazil, 9–12 January 1990. Report No. 20P-2003, Office of Policy Analysis, US Environmental Protection Agency, Washington DC, pp 105–119.

Grainger A (1991) Overcoming constraints on assessing feasible afforestation and revegetation rates to combat global climate change. In: Howlett D, Sargent C (eds) Proceedings of the Technical Workshop to Explore Options for Global Forestry Management, Bangkok 1991. International Institute for Environment and Development, London, pp 196–208.

Grainger A (1992) Characterization and assessment of desertification processes. In: Chapman GP (ed) Proceedings of Conference on Grasses of Arid and Semi-Arid Regions, Linnean Society, London, 27 February–1 March 1991. Wiley, Chichester, UK, pp 17–33.

Grainger A (1993) Population as concept and parameter in the modelling of tropical land use change. In: Ness G, Drake WD, Brechin SR (eds) Population-environment dynamics. University of Michigan Press, Ann Arbor MI, pp 71–101.

Grainger A (1994) Review of United Nations environment programme (1992). World atlas of desertification. *Progress in Physical Geography* 18: 621.

Grainger A (1995a) An evaluation of the FAO assessment of tropical forest resources in 1990. Working Paper 95/14, School of Geography, University of Leeds, Leeds, UK.

Grainger A (1995b) The degradation of tropical rain forest in Southeast Asia: taxonomy and appraisal. In: Eden ME, Parry JT (eds) Land degradation in the tropics. Mansell Publishers, London, (In press).

Grainger A (1995c) National land use morphology: patterns and possibilities. *Geography* 20 (In press).
Grainger A (1995d) The forest transition: an alternative approach. *Area* 27 (In press).
Iverson R, Brown S, Grainger A, Prasad A, Liu D (1993) Carbon sequestration in south/southeast Asia; an assessment of technically suitable forest lands using geographic information systems analysis. *Climate Research* 3: 23–38.
Lanly JP (ed) (1981) Tropical forest resources assessment project (GEMS): Tropical Africa, tropical Asia, tropical America. 4 Vols. UN FAO/UNEP, Rome.
Mather AS (1992) The forest transition. *Area* 24: 367–379.
Sedjo RA, Lyon KS (1990) The long-term adequacy of world timber supply. Resources for the Future, Washington DC. (Unpublished manuscript).
Sedjo RA, Solomon RA (1989) Climate and forests. In: Rosenberg NJ, Easterling WE III, Crosson PR, Darmstadter J (eds) Greenhouse warming: abatement and adaptation. Resources for the Future, Washington DC, pp 105–119.
Simmons IG (1979) Biogeography: natural and cultural. Blackwell, Oxford, 400 pp.
UNEP (1992) World atlas of desertification. Edward Arnold, London, 69 pp.

28 Socio-economic factors in the management of tropical forests for carbon

Philip M Fearnside

National Institute for Research in the Amazon (INPA)
C. P. 478, 69011–970 Manaus, Amazonas, Brazil

INTRODUCTION

Tropical forest management response options to global warming include sustained harvest of timber, extraction of non-timber forest products, silvicultural plantations, agroforestry, managed secondary succession and forest maintenance (including both reserve protection and policy changes affecting deforestation). Socio-economic factors affect C management projects, and *vice versa*, and can negate C benefits and cause hardship for local populations. Forest maintenance has significant C benefits, as well as other environmental and social advantages. Prerequisites include understanding causes of deforestation.

TYPES OF FOREST MANAGEMENT FOR CARBON

Timber management

Carbon benefits of timber management

Forest management most commonly refers to timber management. In the tropics, this includes selectively logging a forest at a specified intensity on a cycle of sufficient duration to maintain a sustainable flow of harvestable wood. Theoretically, standing biomass of natural forests under these regimes can keep C out of the atmosphere that would otherwise be released through deforestation. Some modifications in management practices can contribute to maximising the standing stock of biomass, for example, by refraining from thinning stands of non-commercial species through poison girdling, and by allowing trees to grow larger before harvesting them. Programs to reduce logging impacts on unharvested trees also help reduce emissions that occur through logging damage (Putz and Pinard 1993). In addition to C held in forest biomass and soil, wood products derived from logging represent pools of C kept out of the atmosphere for periods of years or decades, depending on the end use of the wood. Hardwoods for furniture and construction have the longest lifetimes, and therefore the greatest C benefits. Unfortunately, logging normally functions as a prelude to deforestation regardless of whether it is stamped as *sustainable management* in official documentation.

Socio-economic effects on management

Socio-economic factors can undermine sustainability of management schemes and thereby reduce their true C benefits. Estimates of the probability of socio-economic or political factors interrupting a C sequestration management program would of necessity be approximate, but this kind of estimate is nevertheless routinely produced for commercial

NATO ASI Series, Vol. I 40
Forest Ecosystems, Forest Management and the Global Carbon Cycle
Edited by Michael J. Apps and David T. Price

decisions. Adjusting C calculations would require weighting each year's expected sequestration by the expected probability of its taking place in practice, in a manner similar to that used to adjust for risk and uncertainty in Bayesian calculations of expected monetary value (e.g., Raiffa 1968). Experience indicates that timber management plans have low chances of actually producing the C benefits expected. Accepting calculated C benefits for such proposals at face value is equivalent to expecting to win a million-dollar lottery by buying a one-dollar ticket, neglecting to allow for the minuscule probability of winning the prize.

Socio-economic factors are often critical in preventing management of tropical forests for timber production from being sustainable in practice, even if silvicultural parameters indicate technical viability. Logging roads provide access routes for pioneer farmers who enter to clear land for agriculture, often outside of government control. Logging firms themselves can sacrifice future sustained yield by acceleration and/or abandonment of management cycles. Because a number of countries (including Brazil) require plans for sustainable management as a precondition for granting logging permits, a strong motivation is set in place for logging firms to promise the government anything it wants to hear, even if they have no real intention of following the planned management system over the long term.

Many systems of sustainable timber management proposed and/or implemented in tropical countries are technically unpromising. An example is a proposal for sustainable management announced in 1993 by the governor of Brazil's State of Amazonas that would cut trees leaving 60 cm-high stumps—theoretically to resprout. In Malaysia, a country that claims all forest exploitation is *already* sustainable, a high probability of forest management systems proving unsustainable is suggested by a comparison of timber cut with the approximate amounts that could be produced sustainably in the *permanent production forests*. Wood removal exceeds sustainable yield by 39–85% in Peninsular Malaysia, 96–161% in Sabah and 77–236% in Sarawak (calculated from Burgess 1989). Although over-harvesting is likely as at least part of the explanation for these discrepancies, some wood removal also comes from conversion of forest to agriculture.

The argument is frequently made that sustainable forest management (usually taken to mean logging) prevents forest from being wilfully destroyed. It is argued that tropical countries must obtain a financial return from their forests, otherwise they will replace them with agriculture. Logging must therefore be encouraged, together with research to find uses for woods from more tree species and to identify sustainable harvest rates and cutting cycle lengths. The potential long-term profits from logging are expected to result in the countries following these sustainable techniques.

The hope placed in the beneficial effect of increasing profits to loggers is based on two expected chains of events. First, increased profit to loggers results in increased tax revenues, employment, and other benefits to governments; the governments are therefore expected to ensure the long-term continuation of these benefits by instituting restrictions on logging intensity. The second chain of events expected is that greater profit will motivate the loggers themselves to take an interest in guaranteeing continuation of the income stream, leading to investment in long-term production by restriction of logging intensity.

Unfortunately, these two chains of events represent an incomplete view of the real-world system. Increasing profit to loggers has other effects that act in the opposite direction, leading to increase of logging intensity and destruction of the resource. Increased profit to loggers also leads to an increase in area logged. Only long-term profits act to lower logging intensity. Wood harvested by increasing logging intensity swells short-term profits to loggers, which motivates loggers to invest in further increasing logging intensity. This reduces to two opposing positive feedback relationships. One, acting through long-term profits, leads logging intensity to be maintained at a reduced level indicated by the maximum sustainable harvest derived from the growth rate of the trees; the other, acting through short-term profits, leads to greater logging intensity. Such a situation is invariably unstable, leading to one extreme or the other. Which way the balance goes depends on the strength of the forces on each side. However, it is not a mystery as to which side is the stronger, as indicated by the obvious lack of commercial-scale forest management systems in the world today. Poore et al. (1989) surveyed management throughout the tropics and concluded that

> *"...the extent of tropical moist forest which is being deliberately managed at an operational scale for the sustainable production of timber is, on a world scale, negligible."*

The root of the problem lies in the rapid discounting of future returns applied in financial calculations, leading to decisions to harvest natural populations at unsustainable rates. This occurs when the discount rate is more than twice the maximum reproductive potential of the population (see Clark 1973, 1976 for mathematical proof). Growth rates of tropical trees are controlled by biological factors having nothing to do with rates of financial return obtainable on investments in other parts of the economy. These biological limitations place sustainable management for timber at an inherent disadvantage (Fearnside 1989a).

The contrast between Southeast Asia and Amazonia indicates a lack of factual basis for the theory that increasing profits to loggers leads to sustainable management. For various reasons, forests of Southeast Asia are commercially more valuable and easier to manage than those in Amazonia (Fearnside 1989a). If raising the commercial value of forest leads to sustainable management, one would expect Southeast Asia to be a paradise of sustainability. On the contrary, Southeast Asian forests are being destroyed more rapidly than those in Amazonia precisely because Asian forests are more valuable. Higher value increases motivation to destroy the forest more than it increases motivation to sustain production.

In addition to the problem of discount rates, most forest in Brazilian Amazonia is effectively an open-access resource, repeating the tragedy of the commons at each site brought under exploitation. Sawmills in Amazonia can be moved when forest is exhausted in any particular place. Many sawmill operations migrated from Espírito Santo to northern Mato Grosso after the Atlantic forest dwindled in their former home (only about 4% of the Atlantic forest remains). Sawmills are now moving from northern Mato Grosso to other parts of the Amazon region.

The danger of forestry management plans being used to legitimise activities that in reality will lead to destruction of forest (and to greenhouse gas emissions) is increased by the presence of corruption. Papua New Guinea is the best documented example (Marshall 1990). The political value of offering forests for destructive use also contravenes any management scenario that might be devised on the basis of data on silviculture and markets. Making global warming response proposals on the assumption that corruption and local politics are irrelevant is exceedingly naive.

Corruption, although the subject of minimal quantitative study and little open discussion, is a critical socio-economic factor in determining the effectiveness of global warming response options in the forest sector. Why, for example, is Costa Rica the focus of so much more international interest for C offset projects than Zaire, even though Zaire is a much more important country in terms of tropical forests? The notoriety of Zaire for corruption (e.g., Witte 1993) is surely an important part of the answer.

Socio-economic impacts of management

Timber management precludes use of the land for agriculture. This means that agricultural populations must have alternative locations to cultivate, or must turn to other professions to support themselves. It is worth noting that large areas of already cleared land exist in

Brazilian Amazonia, and that the tendency to establish agricultural settlement areas in forests on public lands is explained by political expediency rather than physical limits. The path of least resistance is to decree settlement areas on public land, most of which is forested, rather than to expropriate private lands. Even in other parts of the tropics, where agricultural populations are proportionally much larger, the argument that maintaining native forest represents a threat to the poor is fallacious (Fearnside 1993a).

Recommended role of timber management

Timber management offers some opportunities for C offsets, as in reducing logging impact on the remainder of the forest. However, the more central question of promoting expansion of timber management into presently undisturbed areas requires considerable caution. In practice, plans for sustainable management of tropical forests for timber frequently result instead in destruction of forest. Better results may be obtainable by trying to substitute plantation-grown wood as much as possible for natural forest logging. Rather than by promoting timber management, the key to maintaining C stocks in natural forests is likely to lie in designing systems to provide compensation for the environmental services they provide, including C storage. Maintenance of standing forest as a form of management will be discussed later.

Non-timber forest products (NTFPs)

Carbon benefits of management for NTFPs

Non-timber forest products, or NTFPs, are an important source of revenue and of unique products. Most have the great advantage over timber of not destroying or significantly damaging forest when extracted. Proposals for managing forests for these products are of two types: extractive reserves, where only NTFPs may be harvested, and mixed management systems where both timber and NTFPs are exploited. Brazil has a system of extractive reserves in which populations of rubber tappers and other extractivists are granted use rights to forest on the condition that only NTFPs be removed, with an allowance for a limited amount of subsistence agriculture (Allegretti 1990; Fearnside 1989b). The proposal for extractive reserves originated with the extractivists themselves, rather than being handed down from above as is the norm in Amazonian development planning. The financial value of the NTFPs sold from the reserves, while very important to sustaining the extractivist population, is not the rationale for the government's creation of these reserves. The reserves are justified as a means of maintaining the forest for its environmental functions, which is why the reserves are created by the Brazilian Institute for the Environment and Renewable Natural Resources (IBAMA) rather than by the National Institute for Colonisation and Agrarian Reform (INCRA). Brazil's extractive reserves, important as they are, occupy a minuscule fraction of the forest area—about 0.6% as of 1993 (Brazil IBGE 1993). Carbon stored in the reserves is one argument for their expansion (Brown et al. 1992).

Mixed management systems for timber and NTFPs have been initiated in several locations, including an experimental system under study by the International Tropical Timber Organisation in Brazil's state of Acre. NTFPs can be critical factor in making combined timber/NTFP management financially attractive (Perez et al. 1993).

Socio-economic effects on management

When NTFPs are harvested by a resident population, as in the case of Brazil's extractive reserves, these people are an integral part of the management system. This has the advantage of providing a dedicated interest group to defend the forest against encroachment. It also implies a certain level of impact on the forest through expansion of the area each household uses for subsistence agriculture and through increase in the number of households through reproduction or immigration of population. The population is also subject to the ever present temptation to produce cash crops from agriculture or to sell timber. How these forces are handled by local associations of extractivists and by government agencies that oversee reserves will have important impacts both on the management of the existing reserves and on the extent to which this land use eventually expands. The land tenure arrangement of Brazil's extractive reserves removes the possibility of land becoming a commodity rather than an input to production; were this precaution not taken, financial returns of extractive use would compare unfavourably with deforestation (Hecht 1992). Since the first extractive reserve was created in 1988, the results of this land use have been much better in terms of maintaining forests, with their C stocks, than have common alternatives such as logging and/or cattle ranching.

Socio-economic impacts of management

Management of forests for NTFPs can have positive effects on local populations, as collection occupies a substantial amount of labour and more financial returns accrue to the local population than is usually the case with timber. In the case of extractive reserves in Brazil, the social organisation required to request, establish and manage the reserves has numerous collateral benefits for the population involved by allowing them to improve other social services, such as education and health care.

Recommended role of NTFPs

Collection and management of NTFPs are beneficial additions to timber management schemes. However, intense controversy arises when the reverse suggestion is made: to add timber harvesting to projects designed for sustainable extraction of NTFPs. This is because, in practice, adding timber harvesting to extractive reserve management plans can lead to destruction or degradation of forest for the same reasons that forests are destroyed or degraded through pure timber management. In Brazil, the National Council of Rubber Tappers (CNS) has therefore opposed moves to allow timber management in extractive reserves. Allowing timber harvesting undermines the principal argument upon which the creation of extractive reserves is based, which is environmental benefits of the reserves rather than commodity production.

Silvicultural plantations

Carbon benefits of plantations

Silvicultural plantations are classified as *managed forests* by the Intergovernmental Panel on Climate Change (IPCC) (Kupfer and Karimanzira 1991). Plantations have been the focus of most response options undertaken in the forestry sector in tropical countries, such as those funded by the Global Environment Facility (GEF). Plantations maintain some C in standing biomass of planted trees and also direct C to wood product pools.

In the humid tropics the amount of C held in standing stock is invariably much less than native forest, but more than in uses such as agriculture or pasture. The greatest potential C benefits of plantations, however, are by means of fossil fuel substitution when biomass is used as fuelwood, charcoal or, in the future, liquid biofuels such as methanol (Fearnside 1995).

Socio-economic effects on plantations

Socio-economic factors can act in various ways to cause *leakage* from plantation projects, or the negation of C benefits by events that the project sets in motion beyond its defined borders.

One example is provided by the controversy surrounding plantations that the World Bank is considering funding to supply charcoal to pig iron smelters in Brazil's Grande Carajás area. Private pulp mills that are being set up in the area are likely to offer a higher price for plantation-produced wood than can be expected from charcoal makers. Subsidised plantation owners would be likely to sell their wood to pulp mills. The pig iron smelters, whose licensing and access to subsidies are legitimised by the plantation plans, would then obtain charcoal made from native forest wood, provoking C emissions and other impacts. The inherent attraction of free wood from native forest makes charcoal manufacture for pig iron a continual threat to remaining forests in eastern Amazonia (Anderson 1990; Fearnside 1989c).

Population displacement can lead to *leakage* of C benefits. If former residents of plantation areas move to clear new plots in tropical forest, they will provoke substantial C emissions. This was one of the concerns affecting a proposed C offset plantation in Ecuador, from which the Global Environment Facility withdrew its commitment of support in 1993.

Markets for wood products exert a strong influence on plantation operations and their C benefits, including *leakage* of the benefits of C offsets in wood product pools. This is because expansion of plantations at one location will subsequently increase global supplies of wood products, thereby making prices lower than they otherwise would be, and at other unsubsidised locations would cause marginal plantations to be abandoned and/or would discourage initiation of new plantations. To the extent that the economic *invisible hand* functions as expected, no net global gain in C stocks would result from subsidising plantations other than the small gain from time lags as the subsidised plantations grow prior to their first harvest, and as additions to wood product pools and areas of unprofitable unsubsidised plantations remain in place for a time while a new equilibrium is established.

Socio-economic impacts of plantations

Conversion of land to plantations can deprive local populations of their means of support (Barnett 1992). Plantations can involve displacement of local populations occupying the sites prior to initiating the schemes. Depending on the social system surrounding use of plantation output, socio-economic conditions that develop can be highly undesirable. In the case of plantations for charcoal in Brazil, the industry's competitiveness depends on maintaining most of the labour pool under sub-human conditions through a system of debt slavery (Pachauski 1994; Ribeiro 1994; Sutton 1994).

Plantations in some parts of the world take over commons that are traditionally used by local populations. In India, for example, *social forestry* programs have often benefited wealthy land-holders and paper mills at the expense of rural poor (Centre for Science and Environment 1985; Shiva et al. 1985). These plantations often occupy public roadsides or unplanted portions of private land-holdings that traditionally provide firewood and animal fodder to poor villagers. In India, the choice of *Eucalyptus* deprives the poor of useful supplementary products such as foliage for fodder (Saxena 1989). Poor people have sometimes reacted by ripping *Eucalyptus* seedlings out of the nurseries (Joyce 1988). India's social forestry program was launched with the avowed objective of helping the poor (Eckholm 1979).

In Thailand plantations on common lands could result in eviction of hundreds of thousands of people if industrial plans are fully realised (Koohacharoen 1992). Employment in plantations can only support about one local family per 20 ha of trees, according to an estimate by Royal Dutch Shell in reference to its planned 200 000 ha of plantations in Thailand (Lohmann 1990). Silviculture expansion in Thailand illustrates well the gulf between analyses of such programs from the standpoint of C benefits as opposed to their impact on the local population: the Thai program has been lauded as a model for C sequestration throughout the tropics because of its low establishment costs (Dixon et al. 1994).

Recommended role of plantations

Plantations have a legitimate place in efforts to combat global warming. The place of plantations is constrained, however, by the comparative costs and benefits of other options, such as maintenance of native tropical forests, and by the socio-economic impacts of plantations. For a variety of reasons, the tendency has been to overestimate the C benefits of plantations and underestimate their social impacts, while the reverse tendency applies to forest maintenance calculations (Fearnside 1995).

Agroforestry

Carbon benefits of agroforestry

Agroforestry refers to the combination of trees (either planted or unplanted) with other trees, arable farming, or grazing. This land use maintains a larger stock of C than pasture or arable farming. However, if native tropical forest is sacrificed to implant agroforestry, then the effect on C stores would be negative. A large C credit is often claimed for agroforestry on the basis of its assumed role in slowing deforestation. These benefits are often exaggerated, because much deforestation is not related to subsistence farming, especially in Brazil (Fearnside 1992a).

Socio-economic effects on agroforestry

The extent to which agroforestry can be expanded is severely limited by markets for the products. Were any significant portion of Brazilian Amazonia converted to agroforestry, for example, markets would be quickly saturated. Input requirements also limit expansion (Fearnside 1992a).

Socio-economic impacts of agroforestry

Agroforestry projects generally have greater benefits for local population than do silvicultural plantations of trees such as *Eucalyptus*. Agroforestry systems have the great advantage of being appropriate for small farmers and of producing a variety of products throughout the year. This maximises direct use of the products and use of family labour. The diversity of crop species planted minimises effects of swings in commodity market prices and risks of biological problems (such as insect outbreaks or disease). Offering a stable economic base for small farmers is an important objective for social reasons, independent of environmental benefits.

Recommended role of agroforestry

Agroforestry has real C benefits. The priority attached to it for C sequestration, however, depends strongly on the forces underlying deforestation in each location. In Brazilian Amazonia, the priority of promoting agroforestry would be low as an anti-deforestation measure (Fearnside 1992a).

Managed secondary succession

Carbon benefits of managed succession

Managed secondary forests have been suggested as an option for use of degraded lands, such as the growing areas of abandoned cattle pastures in Brazilian Amazonia. The C store in managed secondary succession is greater than what would be present if the secondary forests were cut for arable crops or pasture. However, immediate cutting of secondary vegetation is often not the most likely alternative in the absence of management programs.

Socio-economic effects on management

Management of secondary succession implies a reclassification of this vegetation from an *unproductive* to a *productive* status. Such a reclassification, both in the legal sense and in the public relations sense, can be very useful to firms (and nations) eager to improve their image as destroyers of tropical forests. In Brazil, ranch owners are anxious not to have their abandoned pastures considered as *unproductive* because land so classified is subject to higher taxes, and because it increases chances that their land will be taken by the government for agrarian reform. If secondary forest can be reclassified as *managed* with a minimal investment, it would be much cheaper as a means of maintaining claim to these large land-holdings than would the most common alternative at present: investing in either reclearing and burning or bulldozing and fertilising to replant cattle pasture. These considerations would be likely to result in a willingness to embrace managed secondary succession projects greatly in excess of what might be justified by expected financial returns from selling the commodities produced, or even from expected environmental return from C storage.

Socio-economic impacts of management

Secondary forests can produce biomass and other products useful to humans (Brown and Lugo 1990). This option has the advantage of requiring little investment and physical inputs. In Brazilian Amazonia, however, it should be remembered that most secondary forests are growing in degraded cattle pastures and produce less in all respects than do similar forests in shifting cultivation fallows.

An important issue in deciding policy on encouraging different land-use systems is the question of who is to be benefited. Most secondary forests in Brazilian Amazonia are in the hands of large ranchers, many of whom have long enjoyed generous government subsidies for clearing forest and planting pasture. Should these same ranchers receive additional subsidies? The prospect of granting additional subsidies to these land-holders to manage the secondary succession that now occupies their degraded pasture sites implies an official commitment to perpetuating the existing highly skewed distribution of land tenure in the region. At the time of the last agricultural census, 62.3% of private land in the region was in properties over 1000 ha in area, while only 11.1% was in properties under 100 ha in area (Brazil IBGE 1989). The socio-economic impacts of this distribution of land tenure are many (Fearnside 1985).

Recommended role of managed succession

Any promotion of managed succession in Brazilian Amazonia should be restricted to small farmers. Better options exist for C storage through maintenance of standing native forest.

Maintenance of standing forest

Carbon benefits of forest maintenance

Capturing the value of environmental services of standing forest, including the value of storing C, must be viewed as a form of forest management. In addition to their role in averting global warming, forests have great value (for which no one is paying) in maintaining biodiversity and, in the case of Brazil, in supplying the hydrological cycle that provides rainfall to much of the country. Measures designed to slow deforestation can easily be justified on the basis of C benefits (Fearnside 1992b). These measures also have socio-economic consequences.

Socio-economic effects on forest maintenance

Socio-economic factors can lead to *leakage,* negating the benefits of forest protection achieved through reserve establishment and through some types of changes in deforestation reduction policy. If establishing a reserve simply means that potential deforesters move elsewhere to continue clearing, then little or no net C gain is achieved. The benefits of forest maintenance proposals can therefore only be assessed at the level of programs, nations, or the world—not with project-level analyses. Little progress has been made on estimating the most likely costs of forest maintenance, let alone assessing the uncertainty attached to such an estimate. Deforestation is strongly influenced by government policy decisions that have little direct connection with financial costs (Fearnside 1987).

Socio-economic factors increase uncertainty of both costs and expected benefits of forest maintenance projects. This is true of both reserve establishment and programs to reduce deforestation, but is especially important for the latter. Unlike plantations, for which accumulated experience makes the costs and benefits relatively well known, forest maintenance is fraught with unknowns. Many depend on the outcome of struggles between opposing political and economic interest groups. For example, if heavy taxes were applied to speculative profits from land sales in order to remove one of the primary forces behind deforestation, those expecting to realise such profits would surely object. The fact that the main impediments to forest maintenance are in the realm of political will rather than financial expense makes this option attractive from the monetary cost-effectiveness standpoint for C offsets. The same fact also explains why more has not been done to slow deforestation. Political barriers, while they must not be underestimated, should not simply be accepted as immutable. The loss that forest destruction represents needs to be translated into a force of appropriate strength directed at changing the key policies that lie within the government's control (see Fearnside 1989d).

Socio-economic impacts of forest maintenance

One of the most persistent myths about deforestation in Brazilian Amazonia is that it is primarily done by the poor. In 1990 and 1991, only 30.5% of the clearing was done by small farmers (defined in Brazilian Amazonia as having less than 100 ha of land), while almost 70% was done by medium and large ranchers (Fearnside 1993b). Distribution of property sizes alone explains 74% of the variance in deforestation rates among the nine Amazonian states. This means that deforestation in Brazil could be slowed tremendously at minimal social cost—up to 70% without even touching any small farmers. It should be recognised that Brazil is different from most other tropical areas, and that such great gains could not be so easily achieved elsewhere without either inflicting hardship on the poor or providing alternative means for their support.

Employment is often the first question raised in discussions of forest preservation in protected areas. Would it not be better to hand out the land as agricultural lots to support part of the unemployed population? The answer to employment depends very much on what is to be done with money that is brought in by the environmental services of the forest. If the sums involved are large, as the true importance of the services implies they should be, then there is substantial scope for creating employment. One form of employment is guarding the reserves themselves. It is important to realise that this form of employment can only sustain a limited number of people, and that these are not the same people who would receive lots if the land were to be handed out for agricultural settlement instead of being made into a reserve. However, for the true local inhabitants (rubber tappers, etc.) already in the interior, this is an important option. Often these people would not have other opportunities for employment. Additional rural employment could be generated in scientific research, for example, if programs were established to botanically collect, map, and measure trees in large areas in the reserves, followed by monitoring of tree mortality, regeneration, phenology and other factors. In addition, population distribution in Brazilian Amazonia is now predominantly urban. Employment in urban centres is, in some ways, easier to create. Activities linked to forest maintenance would be preferable. For example, laboratories could be set up in Amazonian cities to analyse plant secondary compounds obtained from forest reserves.

Recommended role of forest maintenance

The large potential C benefits, parallel benefits of other environmental services, and relatively low financial requirements of forest maintenance all indicate this as the top priority for forest management for C. If done with a view to maximising benefits for local populations, socio-economic effects could be positive, both in the short and long term. Policy changes to slow deforestation would have more immediate payoffs in reducing C emissions than establishing protected areas. A prerequisite for evaluating benefits of such changes is a sound understanding of causes of deforestation and incorporation of this understanding into models capable of generating reliable forecasts under different policy scenarios.

THE PLACE OF SOCIO-ECONOMIC FACTORS IN CARBON OFFSETS

Socio-economic factors constrain the potential scale to which forest management options for C can expand. These factors also affect the C benefits if the projects are successful, and the probability of success. Much depends on social costs of the management schemes. Insufficient attention to socio-economic factors has resulted in a tendency for C proposals to underestimate social costs in some cases (such as plantations) and overestimate them in others (such as controlling deforestation in Brazilian Amazonia). The benefits of plantations are often overestimated because *leakage* caused by socio-economic factors is ignored. Tendencies in evaluating forestry response options are summarised in Table 28.1.

Table 28.1. Tendencies in evaluating forestry response options. Table symbols indicate whether the evaluation tends to be overestimated (+), underestimated (–), or OK (0).

Response options	C benefits if successful	Potential scale	Probability of success	Social costs
Timber management	+	0	++	0
Non-wood products	–	+	0	0
Plantations	+	+	0	–
Agroforestry	+	+	0	0
Managed succession	+	0	0	0
Forest maintenance:				
a) Protected areas	+	+	+	0
b) Deforestation reduction policies	–	–	+	+

Most of the foregoing discussion has dealt with negative aspects of relations between socio-economic factors and forest management: the impediments socio-economic concerns pose to management and the detrimental effects of management on local populations. One might be led to believe that management of native forest and establishment of plantations should automatically be ruled out as global warming response options. I hasten to add that these impediments and impacts are only part of the suite of considerations that must be weighed in judging proposed expansion of these activities. Also important are the supply of each nation's domestic demand for wood products, the avoidance of global warming impacts, and the impacts of other alternatives for supplying wood and combating the greenhouse effect.

While response options in the forestry sector can have harmful socio-economic impacts, it should never be forgotten that inaction also has impacts. Impacts of global warming are not restricted to damaging the economies of a few rich countries, even if this constitutes a major motivation behind the willingness of industrialised nations to invest in response options around the world, including forestry options in the tropics. Effects of global warming will also be felt each time a tropical storm hits the mudflats of Bangladesh or a drought hits famine-prone areas of Africa. Global warming could result in millions of deaths in these places over the next century (Daily and Ehrlich 1990).

Global warming must be addressed on a scale sufficient to solve the problem: it is not reasonable to conclude that all possible countermeasures have undesirable effects, therefore we will do nothing. Forestry responses in the tropics cannot substitute for the large reductions that must occur in emissions from fossil fuels burned in the industrialised countries. The interest of industrialised countries in funding management in tropical forests as a response to the greenhouse effect offers an opportunity for tropical countries to achieve a variety of environmental and social goals in addition to those related to avoiding global warming impacts. However, extreme care is needed, both on the part of international funders and recipient nations, to ensure that forest activities implemented under the banner of global warming abatement do not provoke unacceptable socio-economic impacts. Because global warming responses have social as well as financial costs, it is essential that there be international equity in sharing social costs. It is not enough for the wealthy nations to be willing to pay a large part of the financial costs of combating global warming: they must also be willing to accept some of the social costs.

CONCLUSIONS

Forestry responses to global warming in the tropics can have substantial socio-economic consequences, some are beneficial to local populations but many are detrimental. Socio-economic factors also strongly influence the attractiveness of different response options and the likelihood of their achieving expected levels of C benefit. In general, slowing deforestation has been underrated while other options have been overrated as a means of countering global warming. Plantations have greater social impacts and fewer economic benefits than many proponents believe. Slowing deforestation has substantial ancillary benefits in maintaining other environmental services of the forest. In the case of Brazil, most deforestation is done by cattle ranchers, and great reduction of clearing could be achieved with no effect on feeding local populations. In many other parts of the tropics, where the role of small farmers is greater, alternative means of supporting local populations must accompany programs designed to slow forest loss. Decision-making on management of tropical forests for C must give proper weight to socio-economic factors if forestry projects are to benefit local people and to function effectively to avert global warming.

ACKNOWLEDGEMENTS

Funds from the Pew Scholars Program in Conservation and the Environment were used for travel in Asia. I thank SV Wilson and four anonymous reviewers for comments on the manuscript.

REFERENCES

Allegretti MH (1990) Extractive reserves: An alternative for reconciling development and environmental conservation in Amazonia. In: Anderson AB (ed) Alternatives to deforestation: steps toward sustainable use of Amazonian rain forest. Columbia University Press, New York, pp 252–264.

Anderson AB (1990) Smokestacks in the rainforest: Industrial development and deforestation in the Amazon Basin. *World Devel.* 18: 1556–1570.

Barnett A (1992) Desert of trees: The environmental and social impacts of large-scale tropical reforestation in response to global climate change. Friends of the Earth, London, 62 pp.

Brazil IBGE (Instituto Brasileiro de Geografia e Estatística) (1989) Anuário Estatístico do Brasil 1989. IBGE, Rio de Janeiro, 716 pp.

Brazil IBGE (Instituto Brasileiro de Geografia e Estatística) (1993) Recursos Naturais e Meio Ambiente. IBGE, Rio de Janeiro, 154 pp.

Brown IF, Nepstad DC, Pires IO, Luz LM, Alechandre AS (1992) Carbon storage and land use in extractive reserves, Acre, Brazil. *Envir. Cons.* 19: 307–315.

Brown S, Lugo AE (1990) Tropical secondary forests. *J. Trop. Ecol.* 6: 1–32.

Burgess PF (1989) Asia. In: Poore D, Burgess PF, Palmer J, Ritbergen S, Synnott T (eds) No timber without trees: sustainability in the tropical forest. Earthscan, London, pp 117–153.

Centre for Science and Environment (1985) Social forestry. In: The State of India's Environment 1984–85: The Second Citizens' Report. Centre for Science and Environment, New Delhi, pp 51–62.

Clark CB, (1973) The economics of overexploitation. *Science* 181: 630–634.

Clark CB (1976) Mathematical bioeconomics: The optimal management of renewable resources. Wiley-Interscience, New York, 352 pp.

Daily GC, Ehrlich PR (1990) An exploratory model of the impact of rapid climate change on the world food situation. *Proc. Royal Soc. London (Biol.)* 241: 232–244.

Dixon RK, Winjum JK, Andrasko KJ, Lee JJ, Schroeder PE (1994) Integrated land-use systems: assessment of promising agroforest and alternative land-use practices to enhance carbon conservation and sequestration. *Clim. Change* 27: 71–92.

Eckholm E (1979) Planting for the future: forestry for human needs. Worldwatch Paper No. 26. Worldwatch Institute, Washington DC, 64 pp.

Fearnside PM (1985) Agriculture in Amazonia. In: Prance GT and Lovejoy TE (eds) Key environments: Amazonia. Pergamon, Oxford, pp 393–418.

Fearnside PM (1987) Causes of deforestation in the Brazilian Amazon. In: Dickinson RF (ed) The geophysiology of Amazonia: vegetation and climate interactions. Wiley, New York, pp 37–61.

Fearnside PM (1989a) Forest management in Amazonia: The need for new criteria in evaluating development options. *For. Ecol. Manage.* 27: 61–79.

Fearnside PM (1989b) Extractive reserves in Brazilian Amazonia: An opportunity to maintain tropical rain forest under sustainable use. *BioScience* 39: 387–393.

Fearnside PM (1989c) The charcoal of Carajás: pig-iron smelting threatens the forests of Brazil's eastern Amazon region. *Ambio* 18: 141–143.

Fearnside PM (1989d) A prescription for slowing deforestation in Brazilian Amazonia. *Environment* 31: 16–20, 39–40.

Fearnside PM (1992a) Agroforestry in Brazil's Amazonian development policy: The role and limits of a potential use for degraded lands. In: Aragón LE (ed) Desenvolvimento Sustentável nos Trópicos. Úmidos. Associação de Universidades Amazônicas (UNAMAZ)/Universidade Federal do Pará (UFPA), Belém, Brazil, pp 417–433.

Fearnside PM (1992b) Greenhouse gas emissions from deforestation in the Brazilian Amazon. Carbon emissions and sequestration in forests: case studies from developing countries. Volume 2. Climate Change Division, Environmental Protection Agency, Washington DC, and Energy and Environment Division, Lawrence Berkeley Laboratory, University of California, Berkeley, CA, 73 pp.

Fearnside PM (1993a) Forests or fields: A response to the theory that tropical forest conservation poses a threat to the poor. *Land Use Policy* 10: 108–121.

Fearnside PM (1993b) Deforestation in Brazilian Amazonia: The effect of population and land tenure. *Ambio* 22: 537–545.

Fearnside PM (1995) Global warming response options in Brazil's forest sector: Comparison of project-level costs and benefits. *Biomass and Bioenergy* (In press).

Hecht SB (1992) Valuing land uses in Amazonia: Colonist agriculture, cattle, and petty extraction in comparative perspective. In: Redford KH, and Padoch C (eds) Conservation of neotropical forests: working from traditional resource use. Columbia University Press, New York, pp 379–399.

Joyce C (1988) The tree that caused a riot. *New Sci.* 18: 95–100.
Koohacharoen O (1992) Commercial reforestation policy. In: Leungaramsri P and Rajesh N (eds) The Future of People and Forests in Thailand after the Logging Ban. Project for Ecological Recovery, Bangkok, pp 56–78.
Kupfer D, Karimanzira R (1991) Agriculture, forestry, and other human activities. In: Climate Change: The IPCC Response Strategies. IPCC, Island Press, Covelo, CA, pp 73–127.
Lohmann L (1990) Commercial tree plantations in Thailand: Deforestation by any other name. *The Ecologist* 20: 9–17.
Lugo AE, Brown S (1986) Brazil's Amazon forest and the global carbon problem. *Interciencia* 11: 57–58.
Marshall G (1990) The political economy of logging: The Barnett inquiry into corruption in the Papua New Guinea timber industry. *The Ecologist* 20: 174–181.
Pachauski F (1994) *Trabalha, escravo.* Isto É (Brasília) 4 May 1994, 32–35.
Perez MR, Sayer JA, Jehoram SC (1993) El Extractivismo en América Latina. World Conservation Union (IUCN), Gland, Switzerland, p. 99.
Poore D, Burgess PF, Palmer J, Ritbergen S, Synnott T (1989) No timber without trees: sustainability in the tropical forest. Earthscan, London, 252 pp.
Putz FE, Pinard MA (1993) Reduced-impact logging as a carbon-offset method. *Cons. Biol.* 7: 755–757.
Raiffa H (1968) Decision Analysis: introductory lectures on choices under uncertainty. Addison-Wesley, Reading, MA, 312 pp.
Ribeiro A Jr. (1994) Carvoeiros são 'escravos' em MG. A Folha de São Paulo, 31 July 1994, pp 1–1, 1–12.
Saxena NC (1989) Forestry and rural development. *South Asia Journal* 3: 59–89.
Shiva V, Bandyopadhyay J, Jayal ND (1985) Afforestation in India: problems and strategies. *Ambio* 14: 329–333.
Sutton A (1994) Slavery in Brazil—A link in the chain of modernization. Anti-Slavery International, London.
Witte J (1993) Deforestation in Zaire: logging and landlessness. In: Colchester M and Lohmann L (eds) The struggle for land and the fate of the forests. Zed Books, London, pp 179–197.

29 Economic aspects of carbon sequestration—some findings from Norway

Birger Solberg and Hans Fredrik Hoen[1]

European Forest Institute, Torikatu 34,
Fin-80100 Joensuu, Finland

INTRODUCTION

Increased forest biomass to sequester carbon (C) has been one of several measures proposed to reduce the concentration of greenhouse gases in the atmosphere (cf. Dyson 1977; Sedjo 1990; EPA 1990; Cline 1992).

One major issue in this respect is to analyse how desirable are measures in forestry, and to compare them with measures in other sectors of the economy. Such types of analyses started in Norway in 1990, and since then different types have been performed—from rather simple cost-efficiency analyses to analyses integrating biological and economic factors in relatively large optimisation models.

The purposes of this paper are: (1) to give an overview of analyses done in Norway regarding economics and sequestration of C in forest biomass; and (2) to discuss our experiences with the different types of analyses that were carried out.

OVERVIEW OF CARBON SEQUESTRATION ANALYSES DONE IN NORWAY

Economic analyses of C sequestration in forest biomass done in Norway have so far been based on two different approaches: (1) comparing cost-efficiency of increased C sequestration based on measurement of C flows (benefits) in physical units; and (2) maximisation of net present value (NPV) when C flows (benefits) are measured in monetary units by applying a politically determined shadow price on CO_2 emissions to the atmosphere.

Cost efficiency analyses

Independent partial analyses

The first type of analysis done in Norway was what we could call independent partial analyses of different measures in forestry (and bioenergy) to sequester C.

By *cost efficiency* we mean benefit per net NOK invested (1 NOK = 1 Norwegian krone; US$ 1 ≈ NOK 6.50 using the exchange rates of April 1992). The benefit was measured as the net amount (kg) of CO_2 sequestered. This is equivalent to the criterion used by The State Pollution Agency (SFT) in their analysis of strategies against greenhouse gases (GHGs) (SFT 1990). With this criterion it is possible to compare strategies to reduce the concentration of

1 The Norwegian Forest Research Institute,
P.O. Box 5044, N-1432 Ås, Norway

NATO ASI Series, Vol. I 40
Forest Ecosystems, Forest Management and the Global Carbon Cycle
Edited by Michael J. Apps and David T. Price

greenhouse gases not only in forestry, but also to compare strategies in forestry with strategies in other sectors. Specifically the cost efficiency E was calculated as:

$$E = \frac{M}{K} \tag{1}$$

$$M = \sum_{t=0}^{T} R_t (1+r)^{-t} (1+p)^t \tag{2}$$

$$K = \sum_{t=0}^{T} (C_t - U_t)(1+r)^{-t} \tag{3}$$

Here, R_t is the net fixation (positive or negative) of atmospheric CO_2 (kg ha^{-1} yr^{-1}) at time t, r is the discount rate, p is the change per year in the relative importance over time of the net fixation of 1 marginal kg of CO_2, C_t is gross costs (negative numbers) of the project at time t, U_t is the market price of the gross benefits (e.g., timber harvest) at time t, and T is the time until 95% of the C sequestered has returned to the atmosphere.

The different strategies can be classified into three categories according to their cost efficiency. First of all we have the category where the cost efficiency is positive, e.g., 5. this means that the sequestration is 5 kg CO_2-equivalents per net invested NOK. The higher the number, the more cost efficient is the strategy. The second category is where no net cost is connected to the strategy. These strategies will be profitable with no concern to CO_2 sequestration. These strategies are denoted “no cost”. The third category consists of strategies where there is no net CO_2-sequestration. Strategies in this category emit more CO_2 than they sequester. In our context this category of strategies is of no interest.

Both benefits and costs are discounted. One basic problem is then the choice of discount rate. The problem lies in that costs and benefits occur at different times. Most of the strategies in forestry have a long time horizon (60–100 years), while strategies in other sectors have a rather shorter time horizon (5–10 years). When applying a high discount rate e.g. 7% p.a., strategies in forestry will be relatively less favourable than strategies with a shorter time horizon. SFT has used 7% p.a. in their analysis of the cost efficiencies for different strategies against GHGs. The same discount rate is chosen here but when ranking the strategies in forestry, a lower discount rate (e.g., 3–5%) would probably be more correct as this is the range used for investments in this sector in Norway today. In this analysis, p is assumed to be zero.

While cost efficiencies can be used to rank the different strategies, they give neither information about the size of sequestration nor information about conflicts with other environmental concerns.

Assumptions that should be kept in mind when judging the results include (Lunnan et al. 1991):

- Future biomass growth equals the present growth potential. This means that we assume approximately the same climate in the future as today.
- Future prices and costs are the same as today (in real terms). This means that a constant relative price development is assumed for all products evaluated in the analysis.
- The rate of fixation of C in forest biomass is as described in Blingsmo (1990).
- The end use and decay time of forest products are as present, and described in Aalstad (1990a, b) and Hoen and Solberg (1994).
- For the rest of the factors used in the analysis, and which are associated with uncertainty, the most probable estimates are used, as described *op. cit.*

Table 29.1 shows:

- The 11 different strategies that are evaluated in the analysis.
- The estimated cost efficiencies.
- The area available for the different strategies. The available area is divided into two categories—maximum available area, and area available with little degree of conflict regarding environmental interests other than CO_2 sequestration (e.g., recreation and conservation of ecosystems).
- Amount of CO_2-equivalents (Tg) that the different strategies can sequester per year, 10, 20, 30 and 40 years after the strategies are implemented.

As can be seen from Table 29.1, only strategies 3, 6, 7 and 10 can sequester more than 0.9 Tg yr^{-1} CO_2-equivalents after 10 years. The other strategies are more long ranged; they often have relatively low effect in the first 30–40 years. Only one strategy is free (no net cost) concerning CO_2 sequestration. Some of the others (1a, 3, 8 and 10) can be free. Strategies 2a, 3, 6 and 7 have high cost efficiency ratings. It is worth noticing that increasing the rotation age 10 years longer than the economic optimum (strategy 4) is seen to have a high cost efficiency. Long rotations are one main reason why we have had a relatively large increase in the standing volume during the last decades. A further increase in the use of this strategy will probably lead to negative consequences for prices, income and employment in the forest sector (cf. Note f, Table 29.1); furthermore, the estimates assume that imported timber is not used as a substitute for domestic timber.

All strategies in Table 29.1 are, in principle, independent. The sum of the four right rows give the total amount of CO_2 that can be sequestered. As can be seen this is of the order of 9, 17, 20 and 28 Tg CO_2 equivalents per year in years 10, 20, 30 and 40 respectively after the strategies are implemented. The use of the strategies to this maximum extent is unrealistic. If we look at what it is possible to sequester with little degree of conflict, Table 29.1 shows that this is approximately 4, 6, 8 and 10 Tg CO_2 equivalents per year in years 10, 20, 30 and 40 after implementation. For comparison, the total anthropogenic emission of CO_2 to the atmosphere in Norway in 1990 was 34 Tg CO_2 (SFT 1990).

Integrated forest management analyses

This study started in 1990 and results have been reported by Hoen (1990) and Hoen and Solberg (1992, 1994). Hoen and Solberg (1994) provide a discussion of methodological issues which have to be considered when analysing how preferable measures in forestry are compared to measures in other economic sectors to decrease the atmospheric concentration of GHGs. The most important issues are related to how the objective criteria and the intertemporal evaluation of costs and benefits, related to timber production and CO_2 fixation and emission, are to be considered. These two issues are, of course, quite strongly interrelated. The second objective was to estimate the potentials and cost efficiency of increasing future net CO_2 fixation by changing forest management, within a given period of time, on an initially forested area. It was of particular interest to look at a large forested area and estimate the potentials given the initial state of the forest.

The basic idea was to investigate the production possibilities frontier of the forest area in question, and provide numbers regarding CO_2 emission and assimilation as well as the conventional forestry parameters such as timber inventories, periodic fellings, net payment flows and net present value (NPV). To do this we first had to develop a model for estimation of C flows in and out of forest biomass related to a stand treatment schedule.

Table 29.1. Cost efficiency and inputs of various strategies to sequester atmospheric CO_2 in forest biomass in Norway.

	Cost efficiency[a]		Area available[b], 1000 ha		Net sequestration[c], Tg CO_2 equivalents yr^{-1}			
Strategy	Mean	Interval	Maximum	Small conflicts	Year 10	Year 20	Year 30	Year 40
1a) Afforestation on western coast[d]			172	26	0.01 (0)	0.8 (0)	1.8 (0.3)	2.5 (0.4)
- forest roads have to be built	4	(1–7)						
- forest roads already exist	17	(2–no cost)						
b) Afforestation in northern Norway			442	81	-0.6 (-0.1)	1.2 (0.2)	1.5 (0.3)	4.6 (0.8)
- forest roads have to be built	2							
- forest roads already exist	5							
2a) Afforestation on marginal agricultural land	8		284	143	0.04 (0)	1.5 (0.8)	3.4 (1.7)	4.3 (2.2)
b) Afforestation under electrical lines	4		50	50	0.01 (0.01)	0.1 (0.1)	0.2 (0.2)	0.4 (0.4)
3 Fertilisation of forests (age-class IV)								
a) spruce (site class G17 and lower)	>22		[c]	[c]	1.7 (0.3)	2.1 (0.3)	0.7 (0.1)	0.4 (0.1)
b) pine (site class F14 and lower)	>17		[e]	[e]	1.2 (0.2)	1.4 (0.2)	0.4 (0.1)	0.2 (0.1)
4 Postponing harvest (10–20 years)	>100		[f]	[f]				
5 Higher density reforestation								
- 1500–2000 plants ha^{-1}	8							
- 2000–2500 plants ha^{-1}	5							
- 2500–3000 plants ha^{-1}	1		30	30[g]	0 (0)	0.1 (0.1)	0.1 (0.1)	0.3 (0.3)
6 Ditching, fertilisation and afforestation on peatland[h]								
a) soligenous	14	(10–18)	296	22	1.6 (0.1)	2.6 (0.2)	3.0 (0.2)	3.3 (0.2)
b) ombrogenous	5	(4–9)	103	7	0.1 (0)	0.4 (0)	0.6 (0)	1.2 (0.1)
c) soligenous naturally seeded	>14		158	12	0.8 (0.1)	1.4 (0.1)	1.6 (0.1)	1.8 (0.1)
d) ombrogenous naturally seeded	>5		215	16	0.2 (0)	0.6 (0)	1.3 (0.1)	2.6 (0.2)
Total peatland			772	57	2.7 (0.2)	5.0 (0.3)	6.5 (0.4)	8.9 (0.6)
7a) Ditching, poorly drained forest[i]	28		144	15	0.9 (0.1)	0.9 (0.1)	0.9 (0.1)	0.9 (0.1)
b) Reditching	28		150	150	0.9 (0.9)	0.9 (0.9)	0.9 (0.9)	0.9 (0.9)
8 Bioenergy—increased use of border zones	no cost		70	30	0.2 (0.1)	0.2 (0.1)	0.2 (0.1)	0.2 (0.1)
9 Bioenergy—more effective stoves	5	(2–no cost)			0.5 (0.5)	0.9 (0.9)	1.1 (1.1)	1.1 (1.1)
10 Bioenergy—wood residues and thinnings	0.5	(0.1–0.9)			0.5 (0.5)	0.8 (0.8)	1.2 (1.2)	1.6 (1.6)
11 Bioenergy—mini-rotation forestry	2	(0.1–no cost)	50	50	1.4 (1.4)	1.4 (1.4)	1.4 (1.4)	1.4 (1.4)
Total Tg CO_2-equivalents					9 (4)	17 (6)	20 (8)	28 (10)

Notes to Table 29.1.

a Measured as NOK per net CO_2 sequestered, cost and efficiency is dependent on factors such as site class cost and energy prices. Interval covers the most likely area. The estimation of the mean is based on the most probable estimates of site class, space, prices, etc.

b As estimated in Appendix 2 in Lunnan et al. (1991)

c As estimated in Chapter 4 *op. cit.* The first figure reflects the maximum available area while the other reflects the area available where conflicts with other interests (recreation, preservation, etc.) are valued as small. We assume that each strategy will be started in year 0 on all area.

d At time 0 this strategy will give emission of 36 tons CO_2 equivalents per ha, because of burning of firewood. This is an upper estimate based on today's wood stoves. If a better technology could be used the emission would decrease, as shown *op. cit.* in chapter 4.8.

e Between year 15 and 20 the annual sequestration will be less. It is assumed that equal areas will be fertilised every 20 years.

f Increased rotation period (that is increased harvesting age) is the main reason why we have got much mature forest in Norway. Increased use of this strategy will yield considerable consequences for the timber price, income and employment in the forest sector of Norway. If the forest industry increases imports of timber to equal the reduction in the national harvest, there will be no net effect on global CO_2-sequestration. It is obvious that if we increase the rotation period too much, this will lead to cost (caused by rot, wind and snow), and the cost efficiency will be low. For these reasons this strategy is not included in the table when it comes to the potential sequestration in forests.

g The area used is the annual reforestation area, estimated as the mean reforestation area in the 1980s.

h Same terminology as used by Braekke (1990).

i We assume that the harvest will take place 40 years after the strategy is started.

In addition to the harvesting volume, the decay of the harvested volume had to be included in the model, i.e., the end use of the forest products as well as their anthropogenic life-time and decay time, as described in Aalstad (1990a, b) and Hoen and Solberg (1994).

The. CO_2 fixation and emission model was linked to a long range forest management planning (LFMP) model, called GAYA-LP. The model is reported by Hoen (1990) and Hoen and Eid (1990), and is of a model I type based on stand-wise simulations and solution of the inter-temporal forest management planning problem by linear programming (LP). The growth and yield functions were not adjusted in order to reflect possible changes in future growth rates, due to potential changes in the climate.

GAYA-LP projects the development of a forest stand in 5 or 10 year periods. The stand simulator module is used as a matrix generator to calculate the input/output coefficients related to each stand treatment schedule. The calculations are based on the existing forest growth data for Norway as described in Hoen (1990). Growth responses to commercial nitrogen fertilisation were estimated using the functions of Rosvall (1979). The present value of the flow of net timber payments NPV_{NOK}, is calculated according to a modified Faustmann formula. The development of the stands was projected beyond the planning horizon until an exogenously defined and economically efficient rotation age, depending on tree species, interest rate and site class, was reached. In addition, the value of a steady-state rotation, repeated indefinitely, was calculated for each stand. Thus the value of the forest area for perpetual rotations, evaluated at the end of the last 5 or 10 year period, was calculated and included in NPV_{NOK} and NPV_{CO2}. A similar approach has been proposed by Hardie et al. (1984) for net present value optimisation of timber production.

The general LP formulation for a forest management problem consisting of k stands (calculation units) can be written as follows:

$$\max Z_p = c^T X \qquad [4]$$

subject to:

$$A_1 X \leq b_1 \qquad [5]$$

$$A_2 X = b_2 \qquad [6]$$

$$X \geq 0 \qquad [7]$$

where:

Z_p = The objective function; in this example either NPV_{NOK} or NPV_{CO2} is related to each stand treatment schedule.

X = N × 1 (column) vector of decision variables, i.e., the number of hectares to be treated by a specific management schedule (activity).

c^T = 1 × N (row) vector of known marginal net income related to each activity represented in X. The inter-temporal dimension is taken care of as the marginal net income is represented as the net present value (NPV) of the sequence of net payments related to the activity.

A_1 = M × N (M rows and N columns) matrix of input–output production coefficients, e.g., the total harvested quantity, volume of saw timber/pulpwood harvested, net payment or other state or flow variables related to the development on the forest area.

b_1 = M × 1 (column) vector of resource constraints, one constraint related to each factor (row).

A_2 = K × N (K rows and N columns) matrix of area-related coefficients, which is 1 for element (i,j) if activity j belongs to timber class i, and otherwise is 0.

b_2 = K × 1 (column) vector consisting of the initial area within each timber class.

Hoen and Solberg (1994) made calculations for a planning period of 30 years, divided into six 5-year periods. Thus, the simulated stand treatment schedules consist of different combinations of the available treatment options over the six 5-year periods. All treatments were assumed to take place in the middle of each period. Calculations were done for the productive forest area of the county of Buskerud located in the south-eastern part of Norway. According to the inventory done by the Norwegian Institute of Land Inventory (NIJOS) in 1983–1984, the productive forest area of Buskerud covers 575 000 ha. The volume of the growing stock is estimated to be 58 975 × 10^6 m^3 WB (with bark). Nearly 50% of the growing stock is old growth timber, covering nearly 30% of the area. The inventory consists of 2237 observations, each representing a temporary relascope sample-plot, systematically distributed along the sides of an inventory-quadrate (cluster sampling). The sample plots were aggregated into 40 *type-stands* by cluster analysis as described in Hoen (l.c.). All economic data refer to the price level of the harvesting season 1987/88 in Norway.

Simulation of stand treatment schedules were carried out allowing for 1 out of 7 available treatment options to be applied in each period. The management options were: (1) no treatment i.e., continued growth; (2) release thinning in young growth; (3) thinning; (4) fertilisation; (5) clear felling; (6) clear felling with retention of seed trees; (7) planting or natural regeneration depending on the cutting regime. *A priori* feasibility requirements were defined for each of the management options except (1), in order to prevent the calculation of totally unrealistic treatment schedules. A total enumeration of all combinations of allowed treatments, over six 5 year periods, resulted in 253 stand treatment schedules.

Initially two different management problems were defined, named the BASE problem and the CO_2 problem. The BASE problem was defined in order to estimate the "business as usual" development for the productive forest area in Buskerud, assuming no changes in current forest policy. The harvest levels in each 5-year period were restricted to equal 2 650 000 m^3 WOB (without bark) of saw timber and 2 320 000 m^3 WOB of pulpwood. This represents the mean of the actual harvests in the 8 logging seasons from 1980/81 to 1987/88 based on statistics from the Central Bureau of Statistics (CBS). This harvest level represents 60% of the current total increment and 70% of the harvest level in a non-declining flow forest management strategy. The management objective was to maximise NPV_{NOK}, restricted by the harvest level constraints. We thus assumed a fixed demand for saw timber and pulpwood throughout the 30-year period.

The CO_2 problem differed from the BASE problem only regarding the formulation of the objective function. Maximisation of NPV_{CO2} was the objective, again restricted by the harvest level constraints. The harvest level constraints were applied in order to maintain a reasonable activity among the forest owners and to meet an expected demand for timber from the sawmilling and pulp and paper industry, thus avoiding the question of substitution of the timber supply from other geographical regions. The difference in NPV_{CO2} between the CO_2 problem and the BASE problem would thus give an estimate of the maximum increase in discounted net CO_2 fixation obtainable, given the management options, the initial state of the forest and the fixed timber demand level. The estimates of the maximum increase in net CO_2 fixation was used to map the production possibilities frontier of NPV_{NOK} and NPV_{CO2}. This was done by adding NPV_{CO2} as an additional constraint (≥) to the BASE problem. Six new LP-problems were specified: right hand side (RHS) values of 100%, 80%, 60%, 40%, 20% and 0% of the difference in NPV_{CO2} between the CO_2 and BASE problem, added to the NPV_{CO2} level obtained in the initial BASE problem. The solution of these problems provided us with the shadow prices, i.e., the marginal costs in terms of NPV_{NOK}, of producing the different levels of NPV_{CO2}.

Analysis when carbon fixation has an economic value

The Norwegian Ministry of Environment (Miljøverndepartementet 1991) reported the results from an interdepartmental study of the effects of, and measures against, the greenhouse effect. In this report, macro-economic analyses of the effect of a fossil fuel tax are presented. To keep the man-made CO_2 emissions in Norway at 1990 levels in year 2000, an objective stated by the Norwegian Parliament, these macro-economic analyses show that a CO_2 price (tax) has to be introduced in 1995 and increased to a level of approximately NOK 650 Mg^{-1} CO_2 (1990 price level) in year 2000 and up to NOK 1450 Mg^{-1} CO_2 in year 2025 (1990 price levels). These CO_2 price levels correspond to international studies (Nicolaisen et al. 1991). In Norway there is already a separate tax on mineral oil, e.g., gasoline, intended to reflect the social costs of emission of CO_2 to the atmosphere. In 1992, this CO_2 tax was NOK 0.80 L^{-1} (or US$0.50 per gallon) of petrol, which according to NOU (1992), corresponds to a price of NOK 250 Mg^{-1} CO_2 emitted to the atmosphere.

Given the current and expected price levels on CO_2 emissions, and using this as an estimate of the shadow price on CO_2 emission, it is possible to calculate the marginal value of timber related to CO_2 assimilation/emission, as outlined in Solberg (1991).

Let us first look at the value of a marginal m^3 of biomass sold as timber in Norway. A reasonable average price level (in 1990 prices) for Norway spruce (*Picea abies* L. Karst.) delivered to roadside, is NOK 350 m^{-3}, with corresponding harvesting costs of NOK 140 m^{-3}. Harvesting costs include administration, felling, debranching and terrain transportation to forest road. This gives a net marginal value (stumpage value) of NOK 210 m^{-3}.

One m^3 of timber represents approximately 1.95 m^3 (Marklund 1988; Blingsmo 1991) of biomass when top, branches, bark, root and root system are included. The average density of Norway spruce is assumed to be 0.42, while the biomass consists of 50% C (Nagoda 1982). The relative weight of a CO_2 molecule in relation to a C atom is 44/12. Thus, 1 m^3 of timber represents 1.5 Mg CO_2. The value of a marginal m^3 of timber in terms of CO_2 assimilation should then be NOK 975 m^{-3} in year 2000 and NOK 2175 m^{-3} in year 2025 in 1990 prices, given a NOK 650 Mg^{-1} CO_2 (year 2000) to NOK 1450 Mg^{-1} CO_2 (year 2025) price scenario. The CO_2 benefit is obtained without any costs related to felling or transportation and is thus a net benefit. The CO_2 price level is from 4.6 to 10.4 times the timber (stumpage) value of 1990.

Obviously, one should expect that introduction of such a shadow price level will have large impacts on forest management if put into practice.

Let us assume that all activities causing emissions of CO_2 to the atmosphere are taxed according to a general CO_2 price reflecting the social marginal cost of increasing the accumulation of CO_2 in the atmosphere. We further assume that a corresponding net payment, a subsidy, is paid by the government for a documented net accumulation of CO_2 during a specified time period (estimated on an areal basis). An even-aged timber rotation will typically represent a biomass growth period, with net CO_2 assimilation, and a biomass decay period, with net CO_2 emission. The annual cash-flow related to the CO_2 shadow prices will consist of a number of years with positive net payments (income), from regeneration to clearcutting, followed by a number of years with negative net payments (costs), from clearcutting until the biomass is decayed. This differs somewhat from a traditional timber production cash-flow where net negative payments (costs) are incurred at different points in time during the rotation period, depending on the investment intensity, and the income is obtained when the timber stand is harvested.

Given a positive price for CO_2 assimilation, the value of land in terms of forest biomass production, *V(T)*, can be stated:

$$V(T)=\left[\int_0^T f'(n)\times p_1(n)\times e^{-rn}dn+\int_0^{T+d} g'(m)\times\left[-p_1(m)\right]\times e^{-rm}dm\right]\times\left(1-e^{-rT}\right)^{-1} \qquad [8]$$

where $f(n)$ defines the total quantity of CO_2 fixed in the stand at time n, $f'(n)$ gives the annual assimilation of CO_2 in a stand n years after regeneration, $p_1(\nu)$ is the marginal value of atmospheric CO_2 reduction at the corresponding point of time in prices at time 0 (ν is n or m), r is the real rate of discount (RRD), $g'(m)$ is the emission of CO_2 from the biomass m years after regeneration and d is the decay lag, i.e., the time it takes from harvesting until all the biomass is decayed and the CO_2 is emitted back to the atmosphere (the decay lag can alternatively represent the time it takes until a fraction of the biomass is left, e.g., when 90% of the biomass has decayed). The decay lag will be a function of the timber quality when harvested. Thus when timber is used for purposes like bioenergy or pulp and paper production, this implies a short decay lag (e.g., from 0 to 5 years), while end uses based on saw timber indicate a substantially longer decay lag (e.g., from 20 to 200 years).

The approach described under "Integrated forest management analyses" made it possible to produce results regarding: (a) differences in the physical flow of C, expressed in CO_2-equivalents, as a result of different forest management objectives; (b) analysis of cost-efficiency of different forest management practices in order to increase net CO_2-assimilation in forest biomass; and (c) the shadow price of increased CO_2-assimilation caused by altered management of the forest area in question, for different RHS values of the NPV_{CO2} constraint.

The analyses were extended in Hoen and Solberg (1992) by introducing a shadow price level related to emission of CO_2 into the atmosphere and calculating CO_2 incomes and costs related to biomass production and decay in forest production. Thus we could solve for the "optimal" forest management strategy, based on the maximisation of the NPV of the flow of net timber payments *plus* the flow of net CO_2 "payments".

Based on information from Norwegian government authorities (Miljøverndepartementet 1991), three different CO_2 price scenarios were defined:

1. The current CO_2 tax level, i.e., NOK 250 Mg^{-1} CO_2
2. A constant price of NOK 650 Mg^{-1} CO_2
3. A price of NOK 650 Mg^{-1} CO_2 in the first 5-year period increasing linearly to NOK 1450 Mg^{-1} CO_2 in the sixth 5-year period.

The solution of the BASE problem, when the NPV_{NOK} is the sum of timber and CO_2 net payment flows, is plotted in Figure 29.1 as the points A, B, and C, referring to each price scenario respectively. The optimal strategy under price scenario C lies down and to the left of the NPV_{NOK}/NPV_{CO2} frontier. This result is due to the increasing CO_2 price over the first 30 years in scenario C, while NPV_{CO2} is the discounted flow of physical CO_2-units (i.e., with a price level equal to 1 per Mg CO_2).

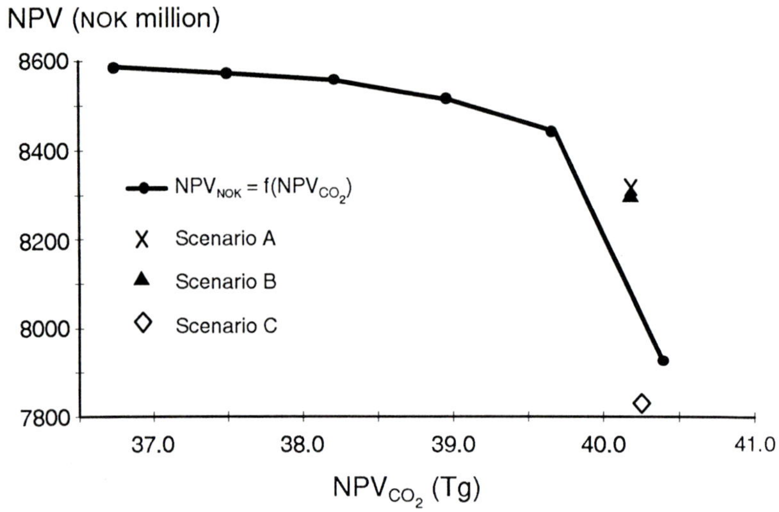

Figure 29.1. The frontier between NPV_{CO2} and NPV_{NOK}. The real rate of discount is 3% yr^{-1}.

MODELLING EXPERIENCES

The first cost-efficiency study with the model described earlier under "independent partial analyses" provided a rough overview of the potential C sequestration measures to be undertaken in forestry and their relative importance. It was found (with some surprise) that the cost efficiency of many of the forestry measures were highly competitive with measures in other sectors of the economy.

The methodology used was easy to implement, and to understand by policy makers.

The drawback with this model was mainly that no mutual interactions between the forestry measures were possible to consider. One knew that sub-optimality (or second-best) estimates were arrived at. (This indicated on the other side that the calculated cost-efficiencies were low estimates, all other factors assumed equal.)

The study based on the integrated forest management and C flow model addressed this problem, as it estimated the total optimal forestry management measures to be undertaken given the objective of maximised C sequestration (cf. above). In addition this model made it possible to do that under various types of constraints regarding, for example, timber harvesting volume of different types and species, conservation of certain types of areas for biodiversity or recreation, etc. The effects of different types of objective functions could also be analysed, and the shadow prices (dual values) related to CO_2 constraints could directly be compared with politically determined C taxes.

In the last model described, the introduction of a direct price on the assimilation and emission of C made possible a direct optimisation of the two outputs: timber and C sequestration. This provided for clearer analysis and easier interpretation of the results with respect to forest management. In addition the above-mentioned constraints regarding conservation of certain areas, etc., could be obtained.

Nearly all factors involved in the above type of analyses are burdened with uncertainty/risks. At present we feel that with respect to these types of analyses in Norway, the uncertainty is highest, and research most needed, for the following factors:

- Natural mortality when keeping old-growth forests beyond normal clearcutting age.
- End use and decay over time of various forest products, and how product life time can be influenced by a changed forest product mix. (One aspect here is, e.g., how much of the sequestered C is emitted as methane instead of CO_2, as methane is a much more potent greenhouse gas than CO_2).
- Which real rate of discount to use in C flow studies (cf., Cline 1992).
- Impacts on total resource allocation in forestry and the forest industry, when C assimilation and emission are priced with various magnitudes on the C tax.

FINAL REMARKS

In addition to the above mentioned studies, Solberg and Svendsrud (1991) and Hoen (1992) have performed analyses on how to include C sequestration in forests in the sector's contribution to the net national product of Norway, and the order of magnitude of this contribution. The partial cost-efficiency method was applied for various forest ecosystems in Indonesia as reported by Solberg and Puranama (1994). The main result was that measures to sequester C in forests wereconsiderably more profitable in Indonesia than in Norway.

As long as increased concentration of greenhouse gases is considered a problem, the sequestration of C in forest biomass will be of high public interest, and will, most likely, influence forest policy in many countries. Research in forest biology, wood technology and forest economics on these issues seems to be of increasing interest.

REFERENCES

Aalstad T (1990a) Decay of organic material in forests, with particular reference to decay time and the consequences of increased temperatures on growth and decay. [Nedbryting av organisk materiale i skog, med vekt på nedbrytingstid og følgene av temperaturøking på vekst og nedbryting.] Notat pr. 15.07.90. Department of Forestry, Agricultural University of Norway, Ås, Norway 19 pp [In Norwegian]

Aalstad T (1990b) Consumption of timber in Norway, what are the end-uses and how long lifetime does the products have? [Forbruk av trevirke i Norge, hva går trevirket til og hvor lang levetid har produktene?] Department of Forestry, Agricultural University of Norway, Ås, Norway, 15 pp [In Norwegian].

Blingsmo KR (1990) Transformation from production figures to biomass figures. [Omregning fra produksjonstabeller til biomassetabeller.] Delrapport. Norwegian Forest Research Institute. 11 pp+app. [In Norwegian].

Cline WR (1992) The economics of global warming. Institute for International Economics. Washington DC, 416 pp.

Dyson FJ (1977) Can we control the carbon dioxide in the atmosphere? *Energy* 2: 287–291.

EPA (1990) Policy options for stabilizing global climate change. Vol I and II. Environmental Protection Agency, Office of Policy, Planning and Evaluation. Washington DC.

Hardie IW, Daberkow JN, McConnell KE (1984) A timber harvesting model with variable rotation lengths. *For. Sci.* 30(2): 511–523.

Hoen HF (1990) Theoretical and empirical studies of long range forest management planning. Agricultural University of Norway, Department of Forestry. Doctor Scientiarum Theses 1990 23 Ås-NLH, 144 pp.

Hoen HF (1992) On the optimal economic rotation of a forest stand when CO_2 fixation has a positive price. Department of Forestry, Agricultural University of Norway, Ås, Norway. (Unpublished manuscript).

Hoen HF, Eid T (1990) A model for analysis of treatment strategies for a forest applying stand simulations and linear programming. Research Paper 9/90 from the Norwegian Forest Research Institute. 24 pp+app. [In Norwegian, with English abstract].

Hoen HF, Solberg B (1994) Potential and economic efficiency of carbon sequestration in forest biomass through silviculture management. *For. Sci.* (40)3: 429–451.

Hoen HF, Solberg B (1992) Impacts of a positive price on CO_2-fixation in forest production. Paper presented at Forest Sector, Trade and Environmental Impact Models: Theory and applications. International symposium, Seattle, Washington, April 30–May 1, 1992. 8 pp.

Lunnan A, Navrud S, Rørstad PK, Simensen K, Solberg B (1991) Forestry and forest production in Norway as a measure against CO_2-accumulation in the atmosphere. [Skog og skogproduksjon i Norge som virkemiddel mot CO_2-opphopning i atmosfæren.] Aktuelt fra Skogforsk nr. 6-1991. Institutt for Skogfag, NLH. 67 pp+appendices. [In Norwegian].

Marklund LG (1988) Biomass functions for pine, spruce, and birch in Sweden. Department of Forest Survey Report 45. Swedish University of Agricultural Sciences, Umeå, Sweden, 15 pp+appendix. [In Swedish with English summary].

Miljøverndepartementet (1991) The greenhouse-effect, effects and policies. [Drivhuseffekten, virkninger og tiltak.] Report from the Inter-Departemental Working Group on climate issues. Oslo, 213 pp [In Norwegian].

Nicolaisen J, Dean A, Hoeller P (1991). Economies and the environment: A survey of issues and policy options. OECD Economic Studies No. 16, Paris.

Rosvall O (1979) Prognosis-functions for calculation of fertilization response. [Prognosfunktioner för beräckning av gödslingseffekter.] in Särtryck ur Föreningen Skogträdsförädling och Institutet för skogsförbättring, årsbok 1979, pp 71–109 [In Swedish].

Sedjo R (1990) The global carbon cycle. *J. For.* 88(10), 33–34.

SFT (1990) An overview of potential actions in Norway for reduction of "green-house"-gases. [Tiltakskatalog for reduksjon av klimagasser i Norge]. Contribution to the inter-ministerial climate survey. State pollution control authority. Oslo [In Norwegian].

Solberg B (1991) On the economic value of sequestering atmospheric CO_2 in forest biomass. Preliminary report. Department of Forestry, Section of Resource Economics and Planning, Agricultural University of Norway, Ås, Norway, 8 pp [In Norwegian].

Solberg B, Hoen HF (1992) The use of taxes to reduce net CO_2 emissions to the atmosphere. [Bruk av avgifter for å redusere nettoutslippene av CO_2 til atmosfaeren.] Department of Forestry. Agricultural University of Norway, Ås, Norway. (Unpublished manuscript). [In Norwegian].

Solberg B, Purnama B (1994) CO_2-sequestration and cost efficiency. In: Climate change and forestry Indonesia—Ecostrategies for terrestrial CO_2-fixation. Report TE 580, Directorate for Nature Management, Trondheim, Norway, pp 68–91.

Solberg B, Svendsrud A (1992) Environmental factors and national accounts of forestry—some findings from Norway. In: Solberg B (ed) Theoretical and methodological aspects of policy evaluations. European Forest Institute Proceedings No. 4, 1995, Joensuu, Finland. (In press).

30 People and forests in Canada: fitting carbon into a perplexing future

Peter N Duinker

Faculty of Forestry
Lakehead University
Thunder Bay
Ontario P7B 5E1 Canada

INTRODUCTION

> *"Almost certainly, Canadian foresters will be required to adopt carbon storage as an objective for forest management."* (Pollard 1991; p. 40)

This is a strong statement, and no doubt there are many who believe it to be true. Indeed, there are likely many who want to see carbon (C) take a meaningful place within the substance of management objectives for Canadian forests. However, Pollard himself expresses some doubt in his conclusion:

> *"It is unlikely that the forest sector will take major steps in the near future in anticipation of a change in climate. Whether public or political pressure to manage forests for carbon storage will become sufficient to influence forest policies in the near future remains to be seen, although there are signs that such pressure is developing."* (Pollard 1991; p. 41)

Upon examining the issue as reported in the literature, and reflecting on the current and possible future status of forest management and policy in Canada, I am not optimistic for C. My thesis is that there is little room on the current and near-future forest-policy agenda in Canada, and thus among forest-management objectives, for C. Lines of argument in support of this thesis, and which are developed below, include the following:

1. Not only is the forest-policy agenda packed full of other issues that will dominate for the next decade or two, but forest-policy development is in a state of confusion and flux. New issues such as C will have to wait.
2. What a forest manager should do in managing forest C (presumably to raise net uptake rates from the atmosphere) seems far from clear for many forests, especially boreal forests which dominate Canada's landscape.
3. Experts seem to agree that forests can in most respects serve only as a temporary retardant to atmospheric build-up of C. While this may be important, foresters are likely to favour management objectives on longer-term values.
4. Some of the current trends in forest management may be at odds with managing for C. Such situations will lead to even more conflict in Canadian forest management than at present.

Despite my gloomy outlook for C as a management objective for Canadian forests, there are some factors which may make Canadian forests evolve in a direction favourable for atmospheric C. These include, among others: (a) potential expansion of the use of biomass

NATO ASI Series, Vol. I 40
Forest Ecosystems, Forest Management and the Global Carbon Cycle
Edited by Michael J. Apps and David T. Price

energy; (b) expanded global markets for wood-based products; (c) increasing forest area; and (d) international pressure to manage forests for C storage and uptake. On the other hand, though, there are some serious potential threats to the ability of forests even to retain the C they now have. The main such threat for Canadian forests is global climatic change. I will examine both these kinds of factors in the essay, too.

WHY CARBON WILL NOT BECOME A KEY MANAGEMENT OBJECTIVE FOR CANADIAN FORESTS

Forest-policy agendas in Canada

In my view, forest-policy agendas in Canada are overflowing with issues (e.g., Anonymous 1992; Ontario Forest Policy Panel 1993) and in a state of confusion. The scene is one of unprecedented change and shifting ground. Here are some trends I see in forest-policy development in Canada:

1. Relationships between the traditional players are changing. Provincial governments, as owners of most of Canada's forest land, and large private enterprises, interested in economical wood supplies, have traditionally dominated forest management, the former mainly as regulator and the latter mainly as timber harvester. As time has progressed, the relationships have become those of landlord and tenant, and the responsibilities and financial obligations have become strongly intertwined and complex. Problems have become so severe that the Government of Ontario has begun a set of negotiations to determine a new business relationship between the province and the forest-products industry. I expect continuing flux in the government–industry relationship for many more years.
2. New players, many with considerable influence, have entered the scene. Of great importance is the growing role of interested publics in forest management and policy decision-making. In some parts of Canada, that role is just beginning to be exercised, mainly as requests by foresters for public reaction to draft plans. However, in other areas, public participation takes the forms of advisory committees, co-management boards, problem-solving teams, round-tables, and negotiations groups (Johnson and Duinker 1993; Higgelke and Duinker 1993). Strong, formal mechanisms for engaging broader groups of stakeholders and members of the public in forest decision-making include Canada's Model Forests (of which there are ten formed under Canada's Green Plan), and community forests (Duinker et al. 1994; Harvey and Hillier 1994). Partnerships in forest decision-making are seen these days as the right philosophy (e.g., OMNR 1991), and they will get more numerous and complex with time.
3. The agenda is overflowing with issues to be addressed. Issues move from society's systemic agenda to government's institutional agenda in a myriad of complex means (Ellefson 1992). Those means are certainly being taxed in Canada at the moment. A look at forest policy some 20 years ago would reveal relative simplicity and an overwhelming emphasis on things like timber production, wildlife production, protection of trees from fire and insects, and community stability. Today, the forest-policy agenda has a whole new vocabulary for an ever-expanding array of issues. Examples include biodiversity, old growth, forest sustainability, ecosystem health and integrity, landscape ecology, community empowerment and self-determination, habitat supply, endangered species, endangered spaces, natural heritage, wilderness, roadless

areas, watershed protection, regeneration backlog, Aboriginal treaty rights and land claims, remote tourism, climate change, C storage, air pollution, ozone depletion, urban forests, agroforestry, tenure, rights and responsibilities—the list seems endless.

Obviously, societies and their governments have to set priorities if they are to make any progress in resolving forest-policy issues. While priority setting can take major and rapid shifts (e.g., with changes in government, or with shifts in societal preferences), my assessment of the relative importance of C among the other issues listed above is that C is currently low and will remain low until other issues get sorted out.

4. The directions in which forest policy should go are less than clear. I sense several tensions in current forest-policy discussions. Debates in policy literature and discussion forums on unresolved issues are healthy, but they point to a lack of clear direction. Here are three examples:
   a. Many Canadians seem to support increases in the forest area removed from industrial activity, but which lands and how much are hotly debated.
   b. The long-term sustainable timber harvest levels from Canada's forests remain subject to considerable disagreement. Some (e.g., Reed 1991) believe the levels could be doubled over the next several decades, as long as adequate silviculture is practised to support the increases. Projections of the world market demands for traditional wood-based products are such that markets for higher timber production in Canada should materialise, as long as we are cost-competitive. However, others warn of wood-supply crises (e.g., Swift 1985) and local "fall-downs" (e.g., Dellert 1991), suggesting that Canadian timber harvests may have been too high.
   c. In some quarters, there are arguments for widespread intensification of silviculture across Canada (e.g., the 1991 Silviculture Conference in Vancouver), while others present a case for less-intrusive forest management (e.g., Booth et al. 1993). There is increasing appeal for a management system including preserved areas (no industrial activity), extensively managed areas (gentle interventions for multiple use), and intensively managed areas (timber farming), but such an approach to forest land allocation has yet to be agreed upon and implemented in Canada.
5. Conflict over forest allocations among competing uses and interests is abundant. If there are disagreements over forest policies, as depicted above, there are many more conflicts at the local level (Johnson and Duinker 1993). World-wide, people are aware of forest and land-use conflicts in British Columbia, but similar conflicts are brewing across the country. Ontario, too, has its fair share; conflict issues include timber management in remote tourism landscapes, expansion of protected areas, Aboriginal land claims and treaty rights, conservation of old growth, allocation of scarce, high-quality sawlogs, and others.

Johnson and Duinker (1993) concluded that local conflicts pertaining to forest use and management in Canada will increase in the future, and it will take special efforts by all stakeholders to bring some stability and civility to relationships among each other and with the forests.

Thus, the forest management and policy scene in Canada has been depicted as full, confused, conflict-ridden, and unstable. I believe it will resist advocacy that tries to push C onto centre stage. Trexler's (1991) conclusion for the United States applies to Canada as well: "*Reliably increasing the size of . . . biotic carbon sinks requires major changes in public policy*". Unless C sequestration measures are fully consistent with current trends and pressures regarding forest management in Canada (i.e., no-regret options), the present time is simply not right for major changes in forest policy in favour of C. Perhaps C could be forced onto centre stage if it is seen internationally to be a key element of achieving forest sustainability, especially if buttressed by forces such as international conventions and forest-product certification schemes. If this would come soon, though, it is likely to increase the tensions within Canada's forest community.

Uncertainty about how to manage forests for carbon

There seem to be some forest-related actions that unquestionably would help the carbon cause. Afforestation is one set of such actions—establishing forest ecosystems where they do not now exist (presumably in ecosystems with no vegetation or dominated by non-woody vegetation) would surely increase the C pools of such ecosystems (Sedjo and Solomon 1989). As well, directing forest biomass either into long-lived forest products (e.g., books, building materials) or into energy facilities (as long as fossil fuels were thus displaced) are steps in the right direction (Kurz and Apps 1992).

However, what can one say to the forester who now tries to manage a million hectares of Canadian boreal forest for timber and an increasing array of other uses and values including recreation, tourism and wildlife habitat? For example, is retention of so-called old growth the right thing to do? Really old growth may be prevalent in Canada's west-coast coniferous rainforest or south-central Great Lakes/St. Lawrence hardwood forest, but old stands in the boreal forest are by comparison young and transient (Blachford and Duinker 1992). Under natural conditions, much of the boreal forest burns roughly every 50–100 years, in addition to succumbing to insect infestations and blow-downs that can cause widespread tree mortality. As Kurz and Apps (1993) pointed out, the C dynamics of a boreal forest depend strongly on the age-class structure of the forest. Since the age-class structure can change substantially over the planning horizons of forest management (e.g., 100 yr), C is likely to be oscillating between coming and going, even in a net whole-forest context, to and from a boreal forest over the decades. It may be influenced somewhat by the particular management regime chosen, but is unlikely to be controlled by that management over the longer term.

Management strategies such as regeneration using advanced understorey growth may seem unequivocally beneficial to uptake of atmospheric C in managed forest ecosystems (Kurz and Apps 1992), and indeed such techniques have other benefits as well. However, in my view it is hardly possible today to advise those who manage timber-producing boreal forests at this time how to manage their forests to take up and store more C. Indeed, it would be silly to prepare general advice, because what may be sensible in one forest, due largely to its own age and type structures and site productivities, at a particular time is likely to be quite different than what is sensible in another forest at the same time. Given the rapid development of forest-planning tools such as GIS-based forest inventory projection models (e.g., Moore and Lockwood 1990), the increased use of such models for planning for non-timber values (e.g., Kayll 1994; Duinker et al. 1993), and the growing understanding of C cycling in forests, it should be possible soon to address the question posed by Kurz and Apps (1992) at that meaningful level of the working forest: what are the C consequences of alternative forest-management strategies? However, as indicated above, while we will be in a position soon, from a technical point of view, to do this, will there be room for C on the already overburdened forester's agenda?

Transience of the importance of forest carbon management

Increased forest area and use of forest biomass to displace fossil fuels may have long-term and lasting benefits for the global C cycle. However, activities dedicated to helping the C cause within existing forest ecosystems can only temporarily slow the build-up of C within the atmosphere (e.g., Marland 1988). By pointing this out, I do not wish to diminish the importance of transitional strategies while humanity works out long-term mitigations for global climate change. But I am hard pressed to imagine that foresters will jump at the opportunity to add a complex value to their forest management objectives for some decades, after which they can drop it from their portfolio. They might be tempted to co-operate if it means that significant new monies for implementing forest management come with the obligation to manage for C storage and uptake for the next half to full century. Society will have to change in some pretty serious ways before I see that happening!

Conflicting objectives

Suppose that, notwithstanding all the arguments above, Canadian forest-policy makers decided that forests would be managed for C uptake and storage, and foresters agreed to give it an honest effort. How might C fit with the host of other forest-management objectives? Would it be in harmony or discord? To set a context for exploration, let us assume that a forester in charge of a boreal forest is asked to do two things: (a) have the forest ecosystems store as much C as possible, and (b) have them take up as much atmospheric C as possible. First of all, are these two sub-objectives compatible? Perhaps not for aboveground living biomass. Carbon uptake peaks while trees are in pole stages, and C storage peaks while trees are mature (Kurz and Apps 1992). It is difficult in the boreal forest to store C alive in trees for more than 100–150 years. Stands that escape fire, insect infestation, windthrow, and harvest for timber usually break up due to senescence, and young stands establish in the understorey of the dying overstorey.

What about other forest-management objectives? I will enumerate several, and reflect on how they mesh with managing for C according to the objectives above.

1. *Maximise fibre production.* On a tree farm, emphasis is placed on C uptake, not long-term C storage (Marland 1988). The biomass pool, and perhaps even the non-living pools, of ecosystem C may be comparatively small under a timber-production strategy, but rates of C fixation from the atmosphere would be relatively high. This will have a net positive effect on atmospheric C only if, on one hand, the fossil-fuel C expended in raising timber-production rates beyond natural unassisted rates were less than that in the new timber produced, and, on the other hand, the new timber produced were used for long-lived products or for biomass energy to displace fossil fuels. However, these considerations are beyond the purview of the forester, and a tree-farm scenario is one where the forester is turning lots of atmospheric C into wood. High fibre production from forest ecosystems is consistent with the objective of C uptake, and may be consistent, depending on the type of forest and the management strategies adopted, with the objective of C storage.
2. *Manage for natural levels and forms of biodiversity.* There is still much uncertainty as to what it means to manage a forest this way (Duinker 1994), but for pioneers like the planning team for the White River Forest (a boreal forest near Wawa, Ontario), it means managing for natural landscape patterns (see Wedeles et al. 1994). This in turn means managing the age-class structure so it has the more natural shape of a reverse-J curve rather than either the imbalance toward older age classes characteristic of so many boreal forests in Ontario, or the flat age-class structure the foresters consider ideal from a timber-producing point of view. Such an age-class structure has large areas of stands

in regeneration and juvenile stages, and increasingly small portions in mature and declining stages.

Without quantitative forest-specific analysis, it is impossible to say whether such a management scheme would set up the forest to be a net C source or sink. Even more interesting a question, given that most timber-producing boreal forests in Canada have age-class structures quite unlike a natural one, is what the C implications would be of managing a century-long transition from a current age-class structure favouring regeneration and mature/overmature stages to a more natural one. Surely the mix of disturbances used/allowed to shape the forest age-class structure to a natural pattern is a strong determinant of the C stocks and flows associated with the forest. Nevertheless, it seems reasonable to posit that many boreal forests would, during the first decades of such transitions, lose C in biomass storage and gain C through an overall increased uptake rate. However, if managing for natural levels and forms of biodiversity means matching natural landscape patterns, over time the scheme would likely be carbon-neutral with respect to the atmosphere.

3. *Manage for old growth and* the *species that depend on old growth.* In most of the boreal forest zone, it may be possible to keep stands old for some decades (e.g., by successfully combating fire and insect infestations and preventing timber harvest), but nature dictates that stand break-up will ultimately occur and old stands will develop into young stands. Perhaps the only way to manage boreal old growth is to ensure that some meaningful proportion of the forest area, no matter where, is always in the old-growth stage (Blachford and Duinker 1992). This indeed is one of the age-class objectives for the White River Forest (Wedeles et al. 1994), but it is part of the overall objective of keeping a natural age-class structure for the whole forest. Thus, managing for old growth in the boreal forest is much like managing for natural levels and forms of biodiversity, as described above.
4. *Set forest area aside as wilderness, so the forces of nature can function freely.* Increases in forest wilderness, whether to satisfy existence (Duinker and Plinte 1994), tourism (Duinker 1991), biodiversity (Noss 1991) or other values, are hotly advocated and pursued in Canada (e.g., Hummel 1985). This too would ultimately lead to more natural forest age-class structures, but it would occur without timber harvests, and may have to occur with progressively less intense fire management (walking away from most timber-producing boreal forests today may lead to quite erratic natural disturbance regimes). Thus, any C that had been accumulating in wood and was planned to feed a mill or displace fossil fuels in energy production would now be destined for the atmosphere either rapidly through fire or slowly via decomposition on the forest floor. I conclude that, given boreal-forest dynamics as they are, during the transition period of the next half-century when forest management may be of any significance in alleviating atmospheric build-up of C, setting aside boreal forests as wilderness will remove significant mitigation options.

Clearly, it will only be possible to tell with any confidence if forest C objectives clash with or complement other forest-management objectives by undertaking forest-specific analysis using appropriate forest-level simulation models. While the potential for significant complementarity seems strong, I hypothesise that the potential for significant clashes is equally large.

Summary

There are significant obstacles to wide-scale adoption of C as a key forest-management objective in the near future. They include, among possible others: (a) overloaded, confused, conflict-ridden, and unstable forest-policy agendas; (b) uncertainty about how to manage forests for C; (c) knowledge that forest C management may be no more than a transient help to the dilemma of increased atmospheric C; and (d) possibilities that C objectives will clash with other important forest-management objectives.

FACTORS FAVOURING CARBON UPTAKE AND STORAGE IN CANADA'S FORESTS

So far I have related factors which will mitigate against C becoming a serious management objective for Canadian forests. There are, however, some significant forces in carbon's favour relative to forests. I discuss four of these below.

Potential increased use of biomass energy

Already, roughly half of the world's timber harvest is used as fuelwood (FAO 1991). This use is strongly regionally distributed across the globe, with the greatest amounts occurring in the developing countries of South America, Africa and Asia (Mather 1990). Developed countries, especially those of North America and Europe, use small proportions of their total wood harvests for energy purposes. Canada uses only about 4% of its harvested wood volume as fuelwood, and this represents about 1% of the country's total energy budget (FAO 1991).

In Canada, interest in wood for energy, as well as other forms of renewable energy, rose dramatically in the 1970s as oil prices increased and the potential of fossil-fuel scarcity loomed. The interest waned when oil prices stabilised at relatively low levels in the 1980s. I believe that the interest will again be piqued. Reasons will be many, and among them may be the fact that biomass energy, to the extent that it replaces fossil fuels, represents no long-term net addition of C to the atmosphere (as long as biomass regrows where it was harvested for energy).

Canada's forests have considerable potential to supply energy (Carlisle 1976; Love and Overend 1979; Henry et al. 1982; McDaniels 1982). Should Canadian society move in such a direction that it wishes to use the forests as a major long-term energy source, then the role of Canadian forests in ameliorating C build-up in the atmosphere may be significant.

Expanded global markets for non-fuel wood-based products

Recent forecasts for the long-term global consumption of wood-based products show modest increases for sawnwood and rapid rises for wood-based panels and paper products (e.g., FAO 1991). Canadian forest-products manufacturers, to say the least, would like to increase their share of the world market and satisfy the growing demands (e.g., Forest Industry Action Group Steering Committee 1993).

While I find it helpful in many respects to separate the realm of forest management from the manufacture and use of wood-based products, in the case of forests and C a link should be made. A compelling argument can be made that, from the C point of view, it would be advantageous to manufacture long-lived wood products from short-lived trees (rather than short-lived products from long-lived trees) (Kurz and Apps 1992). This suggests preserving forests that can hold large C pools for long periods of time, and farming forests where biomass C pools are ephemeral. However it works, if Canadian manufacturers can increase the movement of ephemeral forest C into long-lived forest products, then Canadian forests will make an increased contribution to alleviating the atmospheric C problem.

Increasing forest area

Regarding increased forest area as an option for mitigating increasing atmospheric C, people think first of afforesting open land (e.g., Sedjo and Solomon 1989), and second of regenerating the so-called "backlog" of "not sufficiently restocked" (NSR) land (e.g., Van Kooten et al. 1992). Across much of Canada's boreal forest, NSR does not mean non-forested, but rather means insufficiently stocked to commercially valuable tree species. The Ontario Independent Forest Audit Committee (1992) found that virtually all of the clearcuts made in Ontario's boreal forests in the 1970s had regenerated well to boreal tree species, but not necessarily to the species harvested nor to commercially valuable species. This suggests that, in broad terms, trees were again accumulating C, but the C was in species not considered valuable from a timber perspective. The issue of NSR in Canada's boreal forest and C accumulation is moot, in my view.

In gross terms, I reckon that Canada's forest area is not declining. In many areas over the past half century or so, it has surely increased, especially with the abandonment of marginal agricultural land. In some expanding urban areas, it may have declined. So forest area, as traditionally measured, is also not an issue with respect to C accumulation in ecosystems. Unfortunately, standard forest area measurements do not account for trees in settlement areas (e.g., along residential streets) and farmlands (e.g., along fences, as field windbreaks, along streams). Some people (e.g., Dwyer et al. 1992) believe that additional trees growing in these environments can make real contributions to C uptake from the atmosphere. It is hard to estimate whether the biomass in Canada's urban and agricultural trees is increasing or decreasing, but my sense is that there are strong signs of increases. Urban forest programs in Canadian cities and towns are expanding, as are programs to establish more trees on farms (e.g., National Community Tree Foundation 1994).

The Ontario Forest Policy Panel (1993) urged people to include urban and farm trees when considering the total forest heritage. If we take a view of forests as including all trees, then I believe that future increases in forest area (i.e., ground above which trees are growing) will be strongest in Canada in urban and agricultural areas, in single-tree and row configurations.

International pressures

Canada is participating strongly in international processes focused on sustainable development. This presence and leadership seems to be as strong regarding sustainable forest development as it is for any other ecosystem type or development sector. Witness the statements in the national forest strategy (Anonymous 1992) committing Canada to action, and subsequent activities, on the international stage with respect to United Nations conventions on climate change, biodiversity, and sustainable forest development. Canada is also sponsoring Model Forests abroad in Mexico, Russia and Malaysia, and is a leader in developing criteria and indicators for sustainable forest management (Criteria and Indicators Secretariat 1994).

Such a high profile in international forest deliberations, coupled with a focus on Canadian forest management practices by international environmental groups such as Greenpeace, make it imperative that Canada demonstrate real progress towards forest sustainability at home. If such progress will be gauged partly in terms of the contribution of Canada's forests to C sequestration, then it may become unavoidable for forest managers to consider C as a management objective. For the moment, our national forest report cards (e.g., Forestry

Canada 1992, 1993; Canadian Forest Service 1994) keep reiterating that Canada's forest sector (including both forest ecosystems and forest products) is (or at least was in 1986) a net C sink. Kurz and Apps (1993), however, suggested that that situation may reverse in coming decades, such that the Canadian forest sector becomes a net source. Should this happen, and Canada has committed on the international scene to trying to keep its forests as a net sink, then forest managers may be pressed into changing their strategies to try to make forests retain and draw in more C.

Summary

So far, I have mentioned little about rational policies based on economic analysis of alternative C sequestration alternatives available to a society like Canada. If governments decide that they must take *some* actions to lower their net C emissions to the atmosphere, then it seems sensible to weigh the potential cost-effectiveness of forest measures against other mitigation measures, and also the cost-effectiveness of various forest measures. Methods for such analysis are being developed and applied (see various papers in this volume). If such analyses show that forests ought to be managed differently to sequester C because such a mitigation approach is cost-effective, then the analyses will have to be very convincing—people can not directly see C benefits from forest management like they can timber or wildlife benefits. In these times of short supplies of money, unwillingness to pay may be the norm.

The forces above, all of which will tend to favour the role of Canadian forests, broadly defined, in taking up and storing atmospheric C, have significant potential to be realised. However, some are likely to occur for reasons quite unrelated to C (e.g., biomass for energy, increased urban and farm tree biomass). Also, some may be yet some distance in the future, and are unlikely to play a significant role in the next 10–20 years (e.g., biomass energy, increased demand for non-fuel wood products). I think that, for the next decade or two, the forces keeping C off the forest-management agenda will overwhelm those that favour it being there.

THE WILDCARD OF CLIMATE CHANGE

Much scientific agreement is converging on an expectation of significant global climate change during the next century because of increased atmospheric concentrations of radiatively active gases (e.g., Houghton et al. 1990, 1992). Some authors (e.g., Michaels 1989, 1993) reject these predictions as premature, but I believe they are sufficiently firm and alarming for foresters to take serious note, and consider insurance-oriented strategies (Duinker 1990, 1992; Pollard 1991).

How might climatic change, if it comes to Canadian forests the way many general circulation models of the atmosphere suggest it might, influence the forest C budget? I lean for guidance here upon projections of transient ecosystem responses to shifting climate (e.g., Solomon 1986; Solomon et al. 1984; Solomon and West 1985, 1986) rather than with projections of forest states once the climate has stabilised and the forests have come to equilibrium with the new climate (e.g., Kauppi and Posch 1988). If climate change comes strongly over a few 21st-century decades in Canada, climate resources for forests will shift hundreds of kilometres north of where they now exist (e.g., Wheaton et al. 1987). Modelling of transient forest responses to such shifts generally indicates declines of current forests, and ecosystem confusion until the forests (and soils, for that matter) can adjust to the new conditions. Apps et al. (1993) argued that the most likely scenario for forest C in Canada under such circumstances is the net release of C to the atmosphere. In fact, Apps et al. (1993) pointed out that Canadian boreal forests, even without climate change, could switch from being a net sink of atmospheric C to being a net source, mostly due to successional processes.

If the above scenario happens, what hope would there be for foresters to intervene and reverse the increased release of forest C to the atmosphere? Again, Apps et al. (1993) argued that little could be done, mainly because forest managers treat so little of the total forest area. The Canadian Forest Service (1994) reported that 57% of Canada's 416 Mha of forests are capable of producing timber commercially, but that only 28% are currently managed for timber. I would add to this that much of the Canadian boreal forest is simply unattractive, for reasons of remoteness and low productivity, from a commercial timber point of view. In sum, if climate change were to mean that Canadian forests would send huge amounts of C to the atmosphere, there is little to be done about it.

CONCLUSIONS

The future of Canada's forests is perplexing. The forests are like a carriage hitched to dozens of teams of horses all pulling in different directions. While this is happening, the carriage is hardly moving but it sure is changing, in unknown and unknowable ways, by being variously pushed and pulled in conflicting manners.

How can there sensibly be room for C among the management objectives for Canadian forests? Across the whole country, we practice silviculture in so little of the total forest area. And in the areas where we do, foresters have yet to learn how to manage for two or three major objectives properly, let alone all the other values society would have them take into explicit account. I suggest that more would be gained by focusing our forest-carbon lenses on the uses we make of wood biomass (e.g., biomass energy and long-lived wood products), and increased tree biomass in urban and agricultural forests.

REFERENCES

Anonymous (1992) Sustainable Forests: A Canadian Commitment. National Forestry Strategy. Canadian Council of Forest Ministers, Hull, QC, 51 pp.

Apps MJ, Kurz WA, Luxmoore RJ, Nilsson LO, Sedjo RA, Schmidt R, Simpson LG, Vinson TS (1993) Boreal forests and tundra. *Water, Air, and Soil Pollution* 70: 39–53.

Blachford C, Duinker PN (1992) On Planning for Boreal Old Growth. A draft discussion paper of the Chair in Forest Management and Policy, School of Forestry, Lakehead University, Thunder Bay, ON, 42 pp.

Booth DL, Boulter DWK, Neave DJ, Rotherham AA, Welsh DA (1993) Natural forest landscape management: A strategy for Canada. *Forestry Chronicle* 69: 141–145.

Canadian Forest Service (1994) The State of Canada's Forests 1993. Fourth Report to Parliament Minister of Supply and Services Canada, Ottawa, 112 pp.

Carlisle A (1976) The Utilization of Forest Biomass and Forest Industry Wastes for the Production and Conservation of Energy. Petawawa Forest Experiment Station, Canadian Forestry Service, Chalk River, ON, 54 pp.

Criteria and Indicators Secretariat (1994) *CI Newsletter*, Volume 1, Nos 1 and 2 Canadian Forest Service, Hull, QC.

Dellert LH (1991) What is British Columbia's timber supply forecast to the year 2050? In: Brand, DG (ed) Canada's Timber Resources. For. Can. Inf. Rep. PI-X-101, Petawawa National Forestry Institute, Chalk River, ON, pp 157–163.

Duinker PN (1990) Climate change and forest management, policy and land use. *Land Use Policy* 7: 124–137.

Duinker PN (1991) Tourism and Timber in the Granite-Hill/Obakamiga Lakes Area Near Hornepayne: Approaches to Breaking the Impasse. Report prepared for the Nagagami Forest TMP Steering Committee, Wawa, ON, 44 pp.

Duinker PN (1992) Climate Change and Forest Planning and Policy in Eastern Canada. In: Qureshi, A (ed) Forests in a Changing Climate, Climate Institute, Washington DC, pp 346–357.

Duinker PN (1994) Managing biodiversity in Canada's public forests. Paper presented at International Forum on "Biodiversity: Science and Development—Towards a New Partnership", September 1994, International Union of Biological Sciences, Paris.

Duinker PN, Higgelke P, Bookey N (1993) Future habitat for Moose on the Aulneau Peninsula, Northwestern Ontario Paper presented at: GIS '93: Eyes on the Future, 7th Annual Symposium on Geographic Information Systems in Forestry, Environment and Natural Resources Management. February 1993, Forestry Canada, Vancouver, BC.

Duinker PN, Plinte RM (1994) Measuring Up: Indicators of Forest Sustainability. Vols I and II Draft Reports to the Canadian Forest Service, Sault Ste Marie, ON.

Duinker PN, Matakala PW, Chege F, Bouthillier L (1994) Community forests in Canada: an overview. *Forestry Chronicle* 70: 711–720.

Dwyer JE, McPherson G, Schroeder HW, Rowntree RA (1992) Assessing the benefits and costs of the urban forest. *Journal of Arboriculture* 18: 227–234.

Ellefson PV (1992) Forest Resources Policy: Process, Participants, and Programs. McGraw-Hill, New York, 504 pp.

FAO (1991) Forestry Statistics Today for Tomorrow: Wood and Wood Products. Food and Agriculture Organization of the United Nations, Rome, 39 pp.

Forest Industry Action Group Steering Committee (1993) Hard Choices—Bright Prospects: A Report and Recommendations from Labour, Industry and Government to the Ontario Minister of Natural Resources. Ministry of Natural Resources, Toronto, ON, 78 pp.

Forestry Canada (1992) The State of Canada's Forests 1991. Second Report to Parliament Minister of Supply and Services Canada, Ottawa, 85 pp.

Forestry Canada (1993) The State of Canada's Forests 1992. Third Report to Parliament Minister of Supply and Services Canada, Ottawa, 112 pp.

Harvey S, Hillier B (1994) Community forestry in Ontario. *Forestry Chronicle* 70: 725–730.

Henry G, Hanson AJ, Freedman B (1982) The renewable energy potential of Nova Scotia's hardwoods. *Biomass* 2: 139–151.

Higgelke PE, Duinker PN (1993) Open Doors: Public Participation in Forest Management in Canada. Report to the Canadian Pulp and Paper Association and Forestry Canada, School of Forestry, Lakehead University, Thunder Bay, ON, 45 pp.

Houghton JT, Jenkins GJ, Ephraums JJ (eds) (1990) Climate Change: The IPCC Scientific Assessment. Cambridge University Press, Cambridge, UK, 365 pp.

Houghton JT, Callander BA, Varney SK (eds) (1992) Climate Change 1992: The Supplementary Report to the IPCC Scientific Assessment. Cambridge University Press, Cambridge, UK, 200 pp.

Johnson PJ, Duinker PN (1993) Beyond Dispute: Collaborative Approaches to Resolving Natural Resources and Environmental Conflicts. School of Forestry, Lakehead University, Thunder Bay, ON, 51 pp.

Kauppi P, Posch M (1988) A case study of the effects of CO_2-induced climatic warming on forest growth and the forest sector: A Productivity reactions of northern boreal forests. In: Parry ML, Carter TR, Konijn NT (eds) The Impact of Climatic Variations on Agriculture Volume 1 Assessments in Cool Temperate and Cold Regions, Reidel, Dordrecht, The Netherlands, pp 183–195.

Kayll AJ (ed) (1994) Forest Planning: The Leading Edge. Ontario Forest Research Institute, Sault Ste Marie, ON, 169 pp.

Kurz WA, Apps MJ (1992) Atmospheric carbon and Pacific Northwest forests. In: Wall G (ed) Implications of Climate Change for Pacific Northwest Forest Management. Occasional Paper No 15, Department of Geography, University of Waterloo, Waterloo, ON, pp 69–80.

Kurz WA, Apps MJ (1993) Contribution of northern forests to the global C cycle: Canada as a case study. *Water, Air, and Soil Pollution* 70: 163–176.

Love P, Overend R (1979) Tree Power: An Assessment of the Energy Potential of Forest Biomass in Canada. Report ER 78-1, Energy, Mines and Resources Canada, Ottawa, 35 pp.

Marland G (1988) The Prospect of Solving the CO_2 Problem through Global Reforestation. DOE/NBB-0082, United States Department of Energy, Washington DC, 66 pp.

Mather AS (1990) Global Forest Resources. Belhaven Press, London, 341 pp.

McDaniels TL (1982) Forest Biomass Energy in British Columbia: Opportunities, Impacts, and Constraints. Forestry Canada Information Report BC-X-226, Pacific Forest Research Centre, Canadian Forestry Service, Victoria, BC, 154 pp.

Michaels PJ (1989) The Greenhouse Effect: Chicken Little and our response to "global warming". *Journal of Forestry* 87(7): 35–39.

Michaels P (1993) Global Warming: Failed Forecasts and Politicized Science. Policy Study No 117, Center for the Study of American Business, Washington University, St Louis, MO, 23 pp.

Moore TGE, Lockwood CG (1990) The HSG Wood Supply Model: Description and User's Manual. Forestry Canada Information Report PI-X-98, Petawawa National Forestry Institute, Chalk River, ON, 31 pp.

National Community Tree Foundation (1994) Annual Report 1993–94. National Community Tree Foundation, Ottawa.

Noss RF (1991) Sustainability and wilderness. *Conservation Biology* 5: 120–122.

Ontario Forest Policy Panel (1993) Diversity: Forests, Peoples, Communities. Report of the Ontario Forest Policy Panel, Ontario Ministry of Natural Resources, Toronto, ON, 147 pp.

Ontario Independent Forest Audit Committee (1992) A Report on the Status of Forest Regeneration. Queen's Printer for Ontario, Toronto, ON, 117 pp.

Ontario Ministry of Natural Resources (1991) Direction 90's. Queen's Printer for Ontario, Toronto, ON, 14 pp.

Pollard DFW (1991) Climate change as a current issue for the Canadian forest sector. *The Environmental Professional* 13: 37–42.

Reed FLC (1991) Canada's timber situation in 2050. In: Brand DG (ed) Canada's Timber Resources. Forestry Canada Information Report PI-X-101, Petawawa National Forestry Institute, Chalk River, ON, pp 135–143.

Sedjo RA, Solomon AM (1989) Climate and forests. In: Rosenberg NJ, Easterling III WE, Crosson PR, Darmstadter J (eds) Greenhouse Warming: Abatement and Adaptation. Resources for the Future, Washington DC, pp 105–119.

Solomon AM (1986) Transient response of forests to CO_2-induced climate change: simulation modeling experiments in eastern North America. *Oecologia* 68: 567–579.

Solomon AM, West DC (1985) Potential responses of forests to CO_2-induced climate change. In: White MR (ed) Characterization of Information Requirements for Studies of CO_2 Effects: Water Resources, Agriculture, Fisheries, Forests and Human Health. DOE/ER-0236, US Department of Energy, Washington DC, pp 145–169.

Solomon AM, West DC (1986) Atmospheric carbon dioxide change: agent of future forest growth or decline? In: Titus JG (ed) Effects of Changes in Stratospheric Ozone and Global Climate; Volume 3: Climate Change. United Nations Environment Programme, and US Environmental Protection Agency, Washington DC, pp 23–38.

Solomon AM, Tharp ML, West DC, Taylor GE, Webb JW, Trimble JL (1984) Response of Unmanaged Forests to CO_2-induced Climate Change: Available Information, Initial Tests, and Data Requirements. DOE/NBB-0053, US Department of Energy, Washington DC, 93 pp.

Swift J (1985) The Canadian forest establishment: world laggards. In: Fulthorpe R, Buffinga A (eds) Woodshock: The Forestry Conference—The Perception and Use of Forested Land in Canada. Sierra Club of Ontario, Toronto, ON, pp 48–50.

Thompson S, Webb A (eds) (1994) Forest Round Table on Sustainable Development: Final Report. National Round Table on the Environment and the Economy, Ottawa, 44 pp.

Trexler MC (1991) Minding the Carbon Store: Weighing US Forestry Strategies to Slow Global Warming. World Resources Institute, Washington DC, 81 pp.

Van Kooten C, Arthur LM, Wilson WR (1992) Potential to sequester carbon in Canadian forests: some economic considerations. *Canadian Public Policy* 18(2): 127-138.

Wedeles CHR, Watt WR, Daniel C, Duinker PN (1994) Forest Ecosystem Management in the Northeast Region: Draft Workshop Report. Northeast Science and Technology, Ontario Ministry of Natural Resources, Timmins, ON.

Wheaton EE, Singh T, Dempster R, Higginbotham KO, Thorpe JP, Van Kooten GC, Taylor JS (1987) An Exploration and Assessment of the Implications of Climatic Change for the Boreal Forest and Forestry Economies of the Prairie Provinces and Northwest Territories: Phase 1. SRC Technical Report N 211, Publication No E-906-36-B-87, Saskatchewan Research Council, Saskatoon, SK, 282 pp.

31 Forests and global carbon management: a policy perspective

Robert B Stewart and Jagmohan S Maini

Natural Resources Canada, Canadian Forest Service
351 St. Joseph Boulevard, Hull, QC K1A 1G5, Canada

INTRODUCTION

The human population has increased from 1.6×10^9 at the beginning of the 20th century to some 5.6×10^9 today. Associated with this increase, industrialisation and economic growth has increased per capita consumption of natural resources leading to increased pollution of air, soil and water resources. Increasing levels of environmental degradation have raised concerns for environmentally sustainable development. Of particular concern is the loss and rate of loss of global forests and their ability to meet the increasing needs of an expanding population. Because of this concern forests have emerged as one of the key items on national and international political and policy agendas.

A number of conferences and international fora have proposed the adoption of a convention on global forests, forestry principals (UNCED 1992) and the development of sustainable forestry criteria to conserve and sustain forests world-wide (Maini 1991; Maini and Ullsten 1993). At the same time, the Framework Convention on Climate Change, signed in 1992 at the Earth Summit in Brazil, includes forestry as an integral component in terms of enhancing and sustaining forests as sinks and reservoirs for atmospheric carbon (C). The focus of this workshop on C management and how we might conserve and sustainably develop our forests and their role as sinks and reservoirs for atmospheric CO_2 is therefore an important element in the current international deliberations aimed at setting goals and standards and the development of programs to mitigate against potential climate change.

For forests the problem facing the international community lies in responding to the demands of an expanding population for more forest-based outputs, while maintaining socio-economic, industrial, and environmental values from forest lands that are globally shrinking and degrading. Loss and degradation of forest resources world-wide have significant implications for the role of forests as C sinks and reservoirs in terms of the global C cycle. This paper reviews some of the central issues surrounding the conservation and sustainable development of forests world-wide and their linkage to global C management as a means of addressing the climate change issue. Potential policy responses are offered for local, national, and international levels. This paper builds and expands upon views expressed in previous work by Maini (Maini 1991a, 1991b; Maini 1992; Maini and Ullsten 1993) concerning the causes of global deforestation and effort being directed at making global forestry more sustainable.

The term *environmentally sustainable economic development*, more commonly known as *sustainable development*, was first defined by the World Commission on Environment and Development (WCED) as:

NATO ASI Series, Vol. I 40
Forest Ecosystems, Forest Management and the Global Carbon Cycle
Edited by Michael J. Apps and David T. Price

> *"...economic development that meets the needs of the present without compromising the ability of future generations to meet their own needs"* (WCED 1987).

In this paper, sustainable development of forest land, and its multiple economic and environmental values, involves maintaining indefinitely, without unacceptable impairment, the productive and renewal capacities as well as the species and ecological diversity of forest ecosystems (Maini 1992). Expressed in a slightly different way it means *improving the quality of life while living within the carrying capacity of forest ecosystems* (IUCN/UNEP/WWF 1991; Cairns and Meganck 1994).

THE ISSUES

In the context of sustainable development, the central issues related to global C management are linked to the difficulties in implementing sustainable forestry. These can be discussed in terms of broad ecological, trade, social-cultural and geopolitical perspectives.

Ecological issues

Forests are a major renewable resource presently covering some 28% of the global land area and are a principal source of the world's terrestrial biodiversity which is important to the proper functioning and health of the Earth's biosphere. At the global, regional and local level they influence hydrological and atmospheric conditions and are important in maintaining water tables and regulating stream flows as well as preventing soil degradation and erosion. Forests also represent a major pool of C containing on the order of 460 Pg C in standing biomass and 400–755 Pg C in their soils. This compares to an estimated 560 Pg C for all terrestrial vegetation and 1500–1600 Pg C in the soils (Melillo et al. 1990). By exchanging carbon dioxide with the atmosphere, through processes involving photosynthesis, respiration, decomposition and burning, forests play an integral role in the global C cycle.

The key issues affecting the role of forests as C sinks and reservoirs are deforestation and forest degradation. In this paper, the term *deforestation* refers to a change in land use (i.e., the removal of the forest and its replacement by another use, be it shifting or permanent agriculture, ranching, roads, cities, dams etc.), whereas *forest degradation* involves a reduction or loss of species diversity, productive and renewal capacity, and ecological integrity of forest ecosystems.

Deforestation

Deforestation has local, national, trans-boundary and global implications for the environment and the economy. Environmental impacts include soil erosion (Fearnside 1979; Hecht 1985; Serrao and Toledo 1990), shifts in the hydrological regime; siltation of dams, rivers and lakes (Salati and Vose 1984); reduced soil water holding capacity; sharp changes in microclimates including reduced rainfall, increased frequency and severity of floods, and warmer temperatures (Houghton 1993). At the same time, from a social and economic perspective, deforestation has been linked to the loss of energy, food, shelter, and other resources for local inhabitants. These effects are of major concern in developing nations where much of the population depends directly on the forest for their survival. On the other hand, in many parts of the world, conversion of forest land to other uses is necessary to meet the demand for food from increasing populations and to assist in economic development.

The human population is expected to approach 10×10^9 sometime in the next century. The increase in global population and economic development associated with this growth in many parts of the world has put increasing pressure on the forest resources to provide land for other uses and to provide more products and services. Much of this pressure has been to convert forest land to other uses—most of this for agricultural purposes. Other pressures including gathering of fuelwood, food and fodder, conversion to pasture, poverty, lack of property rights and land tenure, and a lengthy list of exogenous factors such as political corruption, foreign debts etc., have added to the loss and degradation of forests and forest lands.

Deforestation has being occurring for a long time. During the past ten thousand years, global forest cover has been reduced by about one third, from an estimated 6.3×10^3 Mha (Maini and Ullsten 1993) to about 3.4×10^3 Mha (FAO 1993); representing a decrease in global land area of 42–44% to 28%. Most of this reduction has taken place in the temperate and boreal regions. Recent estimates, however, indicate that the boreal forest cover has stabilized while temperate forest cover is expanding slightly (Allan and Lanly 1991; FAO 1991). Tropical forest area, on the other hand, is declining world-wide. Development pressures on this forest in many respects are similar to those experienced in temperate and boreal regions in the course of their development.

Of major concern to the international community is the magnitude and rate of loss which has accelerated dramatically over the last 20 to 30 years. Studies of the rates of deforestation by FAO/UNEP (1981, 1991) for the world's tropical forests suggest that the rate averaged 11.3 Mha yr^{-1} during the 1970s accelerating dramatically during the 1980s to as much as 17 Mha yr^{-1} (Table 31.1). Similar but higher findings have been reported by Myers (1989). Three key points emerge from these studies: first, that the deforestation rate over the last 20 years has increased dramatically, nearly doubling over the period 1970–89; secondly, that it has increased world-wide as opposed to regionally; and third, approximately half of the deforestation is taking place in tropical America, much of this occurring in the Amazon basin in South America. Of the three tropical regions, the rate of deforestation over the period 1981–90 was highest in Asia (1.2%), followed by Latin America (0.9%) and Africa (0.8%).

The magnitude of the deforestation problem is further illustrated by the high rates of deforestation occurring in a large number of countries. For example, for the period 1980–89, 24 countries had rates greater than 1.1% of their total forest covers (FAO 1991). Haiti with a rate of 4.0% had the greatest rate of deforestation; four countries including Paraguay, El Salvador, Brunei and Gambia had rates of between 3–3.9%; and 10 countries between 2.0 and 2.9%. Results of investigations by Myers (1993) suggest that the majority of rapid deforestation is taking place in a relatively small expanse of the forest. He suggests that 43% of the deforestation is happening in an area covering 25% of the tropical forests.

Table 31.1. Estimated rates of tropical deforestation (10^3 ha yr^{-1}).

CLOSED AND OPEN FORESTS	1976–80 [a]	1981–90 [b]
Tropical America	5611	8300
Tropical Africa	3675	5000
Tropical Asia	2016	3600
Total	11 300	16 900

a Source: FAO/UNEP 1981as cited in Houghton (1993).
b Source: FAO 1991 as cited in WRI/UNEP/UNDP (1992, Table 8.2).

More recent studies suggest that deforestation rates may be declining in some parts of Southeast Asia (Allan and Lanly 1991; Houghton 1993), Africa (Allan and Lanly 1991) and the Amazon (Fearnside 1993; Hecht 1993; Skole et al. 1994). In India conservation and large scale afforestation efforts appear to have stabilised the area under forest at about 64 Mha since 1980 (Ravindranath and Hall 1994). In the rest of southeast Asia and west Africa (Côte d'Ivoire and Nigeria) the decline in deforestation rate has been attributed to the significant loss in total forest cover and the fact that what little forest remains, is largely inaccessible (Allan and Lanly 1991; Houghton 1993). In the Amazon basin, however, the decline in the rate of deforestation has been linked to the global and regional economic recession (Fearnside 1993; Hecht 1993; Skole et al. 1994). Hecht (1993), in her analysis of the relationship between the livestock industry and deforestation indicates that the recent decline in the deforestation rate in Amazonia is only temporary and that as the economy improves the deforestation rates will begin to escalate again.

The level of deforestation that has taken place in the tropical forests is also of major concern. Analyses by Fearnside (1993) and Skole et al. (1994) of the Brazilian Amazon and by Kummer et al. (1994) for the Philippines in southeast Asia further illustrate the seriousness of the deforestation problem. Fearnside, for example, estimates that, as of 1991, 426 000 km^2 (10.5%) of the forested portion of Brazil's Legal Amazon Region (the total area is 5×10^6 km^2) have been cleared. A similar value has been derived by Skole et al. (1994). Kummer and Turner (1994) in their analysis of deforestation in the Philippines indicate that since 1948, 84 900 km^2 of forest land has been converted to other land uses representing a loss in forest area of approximately 56%.

World-wide forests represent a C reservoir estimated to be in the range of 450 Pg C (Melillo et al. 1990; Bolin et al. 1986; Ajtay et al. 1979; Olson 1983). As the global forest area decreases, C from this reservoir is being released to the atmosphere and adding to the potential for global warming. Carbon emissions to the atmosphere through loss of forest cover are estimated to be on the order of 1–3 Pg C yr^{-1} (Houghton 1993); in comparison human use of fossil fuels is estimated to be contributing approximately 6.0 Pg C yr^{-1} to the atmosphere (IPCC 1992). Houghton's analysis further suggests that over the last 20 years emissions from biomass burning have been increasing at a faster rate than emissions from fossil fuel burning. In terms of global warming forcing over the last century more than half has been accounted for by increasing concentrations of atmospheric CO_2 associated with fossil fuel burning and deforestation—the remainder is associated with increases in other greenhouse gases including methane, ozone, CFCs and nitrous oxides. Fossil fuel burning and deforestation are expected to account for approximately 55% of the additional forcing over the next century (Houghton 1990, 1993). From these estimates it is evident that deforestation is going to continue being a significant factor associated with the increase in atmospheric CO_2 concentrations for the foreseeable future. The critical issue in terms of global C management in relation to the climate change issue is how to slow, arrest and eventually reverse this process.

Agriculture

A major cause of deforestation is related to the need to increase agricultural cropland to provide the food and fibre needs of an ever increasing global population. Analysis by Allan and Lanly (1991) indicates that expansion of the world's agricultural croplands by about 10 Mha yr^{-1} is required to meet the food needs of current increases in global population. Current estimates of deforestation suggest that the rate of conversion of forest land to other uses is about twice this level. Analysis by the FAO (1990a) and Houghton (1993) indicate that nearly half the lands converted annually can be accounted for by the increase in abandoned lands—lands composed of degraded croplands and pastures that no longer support crop or livestock production. Degraded lands are not usually the immediate result of the conversion of forest land to agricultural use, but typically are the result of an inappropriate agricultural use leading to a degraded impoverished condition over time. Because of this, only half of the area of tropical forest converted annually to agricultural uses is added to the area used for permanent agriculture. The other half is used only temporarily, often for periods of 5 to 10 years, after which it becomes waste fit neither for agriculture nor forest (Houghton 1993; Hecht 1994; Skole et al. 1994). Since much of the land is abandoned in this way, it would seem evident that making existing agriculture lands more productive and environmentally sustainable would be an effective approach to reducing tropical deforestation, and maintaining them as major sinks and reservoirs for atmospheric C.

Fuelwood and fodder

Wood in many parts of the world represents the primary source of energy. For example, figures reported by the WRI/UNEP/UNDP indicate that traditional fuels supplied 37% of the total energy required in Africa, 40% in Central America (excluding Mexico), 20% in South. America, and approximately 32% in tropical Asia. In terms of wood production much of the current production is being utilised to support daily energy needs and requirements of the local population (Tho 1991; WRI 1992). For the period 1987–89 figures suggest the proportion was 89% for Africa, 80% for Central America, 70% for South America, and 84% for tropical Asia. In many areas of the developing world because of the extent of deforestation and forest degradation, acute shortages of fuelwood already exist which is further accentuating current deforestation trends (WRI/UNEP/UNDP 1992; Ravindranath 1993). Estimates by the WRI/UNEP/UNDP (1992) suggest that in Africa and Central America over-exploitation of forest lands for fuelwood is almost as great as deforestation as a cause of land degradation (Table 13.2). In Asia and South America over-exploitation of forest land for fuel is also important. In contrast, the opposite is the case for developed countries where the bulk of the wood produced is for industrial and value-added uses.

Table 31.2. Human causes of soil degradation, 1990 (Source: WRI/UNEP/UNDP 1992).

	DEGRADED AREA		NOMINATIVE CAUSE				
REGION	Total (Mha)	Vegetated[a] (%)	Deforested (%)	Fuelwood[b] (%)	Overgrazed (%)	Agriculture (%)	Industrialised (%)
World	1964.4	17.0	30	7	35	28	1
Africa	494.2	22.1	14	13	49	24	-
Asia	747.0	19.8	40	6	26	27	-
South America	243.4	14.0	41	5	28	26	-
Central America and Mexico	62.8	24.8	22	18	15	45	-
Oceania	102.9	13.1	12	-	80	8	-
Europe	218.9	23.1	38	-	23	29	9
North America	95.5	5.3	4	-	30	66	-

a Proportion of vegetated land.
b Land that has been over-exploited for fuelwood.

The causes of deforestation and forest degradation are many and wide ranging, depending on the region. For example, in the temperate and boreal regions the major problem has been attributed to clearcut harvesting practices, transboundary air pollution, and fire. In the

Mediterranean zone, the most serious threats combine overgrazing and fire (Allan and Lanly 1991). In Africa, the cause has been associated with weather factors (drought, floods), unfavourable terms of trade, unwise use of loans and grants, rapid population growth, corrupt leadership, civil strife and land tenure (Stahl 1994; World Bank 1989), over-exploitation for fuelwood and fire (Alan and Lanly 1991). In Southeast Asia it has been primarily logging for the international market followed by agricultural expansion (Kummer et al. 1994). In tropical America it has been agriculture expansion, pasture conversion for livestock, and land tenure (Fearnside 1993; Skole et al. 1994; Hecht 1993). Most, if not all of the above causes can be attributed to government policy which influences the uses of forest lands both intentionally and unintentionally (Repetto 1993). Repetto suggests that governments are aggravating the loss of forests under their stewardship through mistaken policies. Policies that, by and large, have been adopted for worthy objectives (i.e., industrial or agricultural growth, regional development, job creation, or poverty alleviation) but, have typically not been realized or have been attained only at excessive cost—in this instance environmental degradation or deforestation.

In the past the causes of deforestation were thought to be relatively straightforward and driven by single factors such as population growth (Allen and Barnes 1985; Reis and Margulis 1990). In other cases, such as in Central America, the prime reason for deforestation has been linked to the international beef market (Nigh and Nations 1980; Myers 1981). The causes of deforestation, however, are much more complex and it is increasingly being reported as the result of multiple interacting factors. For example, Hecht (1985, 1993) in a study of cattle ranching in the Amazon suggests that government policy, fiscal incentives, and the nature of individual farmer decisions in an inflationary economy are more significant determinants of deforestation than are demographic considerations. Similar conclusions have been drawn in recent studies of declining wood stocks in sub-Saharan Africa (Anderson 1986; Anderson and Fishwick 1984) and deforestation in Southeast Asia (Kummer 1992; Kummer and Turner 1994).

Understanding the causes of deforestation and forest degradation in many respects is the key to sustainable forest development and improvement in their use as C sinks and reservoirs. Only by determining and understanding the root causes can solutions be devised to slow, arrest and eventually reverse current deforestation trends which will preserve global forests as sinks and reservoirs for atmospheric C.

International polarisation

The role of forests under the FCCC is receiving considerable attention in international deliberations concerning the use and preservation of forests as sinks and reservoirs for atmospheric C. From a policy perspective discussion is not advancing much because of polarisation of views and positions (WRI/UNEP/UNDP 1992; Maini and Ullsten 1993). Polarisation appears to focus along three themes.

First, industrialised countries are advocating strong measures to conserve and protect the world's forests particularly in tropical areas. Developing countries, on the other hand, are concerned that the industrialised countries' preoccupation with tropical deforestation is inconsistent with the amount of attention being paid to prevent global warming through reduced fossil fuel. They see an equally urgent need to reduce the use of fossil fuels by industrialised countries. In the context of the climate change issue it is recognised that deforestation is not the sole cause, nor reforestation, the main solution. Reducing deforestation, implementation of large scale reforestation programs, including rehabilitation of derelict lands, and intensifying sustainable forest management, however, would contribute to mitigation of some of the potential global warming. More important, these actions at the same time would provide an array of other environmental, economic, social, and cultural benefits in the context of sustainable development (Maini and Ullsten 1993).

Second, many developing countries feel attempts to protect and preserve tropical forests as "locking up" their forests. It is perceived as an infringement on their national sovereignty and right to develop their forest resources to meet their own economic development needs. Others suggest this as an attempt by the developed world (the North) to impede the economic development of developing countries (the south).

Finally, the commitment of developing countries to conserve and expand the world's forests and to enhance them as C sinks and reservoirs and to practice environmentally sustainable forestry remains conditional on obtaining additional funding and technologies at favourable terms from developed countries. Basically what is being demanded of developed countries is that they compensate developing countries for both the economic opportunities foregone and the impact of global warming that will occur as a consequence of developed countries past deforestation and economic development activities considered to be the source of much of the increase in atmospheric CO_2 concentrations.

International trade

World production of wood products was valued at US$ 386 000 million in 1989 (FAO 1991). International trade in forest products represented about 3% of total world trade and amounted to approximately US$ 101 000 million in 1989 (FAO 1991). The rate of increase in demand for forest products usually exceeds the rate of population increase (Maini and Ullsten 1993). Current FAO projections suggest that over the next three decades, the increase in demand for forest products will be on the order of 3% per year. At the same time world forest area is declining. As a result of the loss of forest area a number of developing countries that were formerly net exporters of forest products are now net importers and this trend is increasing Maini and Ullsten (1992). Based on these trends it is apparent that the global forestry community is faced with a major challenge—meeting increasing demand for a wide range of forest products from a shrinking resource base. The implications of this challenge for sustainable forest development, global C management and a number of other global change issues are evident.

Green consumerism

Green consumerism in international trade in forest products is another factor that is beginning to have an impact on the conservation and sustainable management of forests world-wide (Maini and Ullsten 1993). Many consumers in industrialised countries are choosing to buy forest products from sustainably managed forests and manufactured by environmentally sound processes and technologies. Another way green consumerism is being felt is through the threats of boycotts of certain products (WRI/UNEP/UNDP 1992). For instance consumers are being encouraged by environmental groups and activists to boycott tropical hardwood products because of the unsustainable manner in which they are being produced. In order to protect their international trade, countries exporting forest products are responding to these aspects through the development of internationally recognised criteria for sustainable forest development (Maini 1991b, Maini 1993) and the application of these criteria to their domestic forestry practices.

Social and cultural issues

Forests provide food, employment and generation of capital and have played a wide ranging social and cultural role in most regions of the world. Following the publication of the Brundtland Report (World Commission on Environment and Development 1987), conservation and sustainable development for the present and future generations emerged as a global ethical issue particularly in developed countries (Maini and Ullsten 1993). The issue of sustainable forest development, however, attracts limited attention in developing countries, where day-to-day survival is the major and over riding preoccupation.

Closely related to the ethical issue of inter-generational responsibility is the intra-generational need to protect the rights of forest dwellers, indigenous people and communities living in and around forests who depend on them for their existence (Maini 1991a; Maini and Ullsten 1993). It is increasing recognised that over the millennia these peoples have developed sustainable and ecologically beneficial farming and agroforestry practices that have sustained their way of life and cultures for generations (WRI/UNEP/UNDP 1992). Posey (1993) suggests that native peoples understand how to use and conserve the forest and that current deforestation and forest degradation is occurring, in part, because of non-forest dwellers' lack of knowledge about how best to exploit their vast diversity of medicines, foods, natural fertilisers and pesticides. As forests, particularly tropical forests, continue to disappear, we are learning that their inhabitants may offer the best instruction in survival and forest use (Posey 1993). Resolution of issues associated with the transfer of traditional knowledge will be essential if we are to succeed in maintaining much of the world's forest and is a key aspect of ongoing discussions and international deliberations (WRI/UNEP/UNDP 1992; Maini and Ullsten 1993; Posey 1993; Cairns and Meganck 1994).

Geopolitical issues

Forests represent a major problem in terms of resolving global environmental issues because they are physically located within the territories of sovereign nations, yet, their environmental role extends well beyond national borders (Maini 1991b, Maini and Ullsten 1993). Management or mismanagement of watershed forests of international rivers, for example, often has trans-boundary implications in terms of soil and water conservation in neighbouring countries. Similarly, deforestation in one part of the globe may contribute to climate change in another, while reforestation in one area could contribute to mitigation in another. The role of forests in the global C cycle and its importance to the world's climate system illustrates clearly the environmental significance forests can have well beyond the boundaries of the nations where they are located. Consequently, national and international policy making communities are faced with the need to reconcile national interests (i.e., reduce poverty, reduce unemployment, improve balance of trade, or payment of international debt) with international responsibilities (i.e., global C management to mitigate climate change, international watershed management to reduce flooding or prevent water shortages) and to develop appropriate policy responses, instruments, and institutions to accomplish this balance.

Conservation and enhancement of forests world-wide as C sinks and reservoirs as a means of reducing or mitigating potential climate change have emerged as a major issue on the international policy agenda. However, it is only one of several aspects that must be considered within the context of sustainable forest development. Other aspects including biodiversity, cultural, social and economic perspectives must also be considered in relation to the potential values and benefits to be derived from the forests at local, national, and international levels while at the same time acknowledging the sovereign rights of nations to conserve and utilise their forests to meet their domestic policy objectives.

SUSTAINABILITY OF GLOBAL FORESTS: AN AGENDA FOR ACTION

Improving the sustainability of global forests is important for a number of reasons. First, it is a way to enhance the livelihood of millions of poor in countries heavily affected by current deforestation trends. Second, it is a means of protecting and preserving the wide variety of biological life which global economies are dependent upon. Last, it is a means of mitigating some of the potential unwanted changes in climate that may result from human induced changes in the global C balance. In the context of the global C cycle improving forest sustainability would contribute to the enhancement of these forests as sinks and reservoirs for atmospheric C.

In an early paper Maini and Ullsten (1993) proposed a number of actions that would increase the sustainability of global forests. These actions still remain current and are outlined in the following.

The first action is to increase total forest cover, embracing all kinds of forests—natural, replanted, and new plantations. It would involve planting programs and the rehabilitation of degraded lands. Estimates by Grainger (1988) and Houghton (1993) suggest that 765 Mha to 865 Mha of degraded grazing land and land in the fallow cycle of shifting cultivation is available for reforestation in the three major tropical areas of Latin America, Asia, and Africa. Reforestation of these lands will be difficult and require considerable time, however, it is essential that the effort be made to recover these resources for future generations.

The second action is to increase permanent forest cover, including those areas within individual countries which have been designated through national land use policies to be permanently kept under forest cover, rather than left open for conversion to other kinds of land use. This target would include forest land slated for commercial exploitation under management regimes that permit regeneration of the forest, in order to maintain its status as productive forest land.

The third action is to protect natural forests, embracing examples of representative and unique forest types protected from commercial exploitation in order to conserve C, biodiversity and other ecological values. Protection of primary or old growth forests is a good example. Primary forests for the most part contain much more C than secondary or plantation forests on a per hectare basis. Carbon storage in plantations does not approach that in old-growth for at least 200 years (Harmon et al. 1990; Cairns and Meganck 1994). While secondary forest can sequester more C in the short run, primary forests on a per hectare basis conserve much more C preventing its release to the atmosphere. Preserving old growth forests will help conserve the terrestrial C pool, whereas, plantation forests established on degraded or rehabilitated lands can increase the size of the overall terrestrial C pool.

Fourth, a clear institutional focus on the national level should be developed with the responsibility and accountability for forest-related policies and programs. Government agencies should produce timely and reliable reports of the state of forests at national levels. These reports would include economic, environmental, social, and cultural dimensions of forests; they would not merely offer a traditional forest inventory with an industrial-use perspective on forests. Governments must also invest in forest research, particularly in the management of forest ecosystems for multiple values and in the rehabilitation of degraded forest lands. In addition, promoting exports of forest products from sustainably managed forests needs urgent attention.

A fifth action which can be added to the four proposed by Maini and Ullsten would be to make agriculture more sustainable. As noted previously, approximately 10 Mha per year must be added to current agricultural lands globally to meet the basic food requirements of the increases in human population. This represents nearly half of the current deforestation rate. Approximately 50% of the forest lands converted to agricultural uses each year are abandoned within 5 years. Subsequently, if agriculture could be made more sustainable deforestation rates could be nearly cut in half. The implications of this for conserving global forests as sinks and reservoirs for C are immense.

The above actions in many respects are comparable to the recent integrated forestry plan proposed by Cairns and Meganck (1994) that recognises the interdependence and collateral benefits of addressing simultaneously the three global issues of biodiversity, climate change and sustainable development. Their plan has four main thrusts: (1) land-use planning emphasising primary forest preservation and management; (2) intensified use of non-timber biotic resources including both forested and nonforested lands; (3) agroforestry applications; and (4) plantation forestry in appropriate situations. Land use planning, tailored to local conditions and emphasising forest preservation and management within an economic development context, is essential to the success of their plan. Cairns and Meganck's plan emulates closely the views expressed by many others including Allan and Lanly (1991), Maini and Ullsten (1993), Houghton, (1993), Fearnside (1993), Stahl (1993) and Hecht (1993).

To be successful implementation of the above action plans will require a concerted effort globally involving co-operative actions at the local, national and international levels. The following outlines some of the more important actions needed at each level.

Local actions

Communities must plant trees and forests to provide immediate benefits and meet local needs for food, fibre, fodder, shelter, and economic development. As well, local individuals and communities should be involved in developing and implementing action plans. They must have *ownership* of these initiatives if they are to be successful (Maini and Ullsten 1993).

National actions

National land-use policies should be developed and implemented that promote conservation and sustainable development of forests. Of particular importance is the development and implementation of policies and programs for the establishment of permanent forest cover that have clear targets and timetables for their achievement.

At the national level conservation and sustainable development of forests must be on the political agenda and demonstrated by clearly enunciated policies that are supported by effective national programs and institutional arrangements (Maini 1991). This factor is critical to the success of international co-operative efforts.

International actions

At the international level, governments must establish an effective policy and institutional framework to facilitate and co-ordinate international co-operation on the conservation and sustainable development of forests. Development of an international consensus on the guiding principles applicable to the conservation and sustainable development of all types of forests world-wide within the UNCED process has been a major first step. It is not enough, however. The formulation of policies based on scientifically based, internationally agreed, criteria and indicators for sustainable forest management is a crucial next step to ensuring multiple benefits of forests for present and future generations. This is the subject of considerable international deliberation currently. When completed, an international forest convention would be the final step. A Convention would provide the international policy and institutional framework to strengthen multi and international co-operation, and national policies and institutions, dealing with forest-related issues and opportunities (Maini and Ullsten 1993).

Strong international institutions are needed to provide leadership on forest-related issues, to manage international co-operation on forests, to assess the state of world's forests, and to provide timely and credible reports on the progress toward the conservation and sustainable development of forests at regional and global levels. Enhanced regional co-operation, particularly among contiguous states sharing a watershed, a particular forest type, or a unique forest ecosystem, offers an important complement to international co-operation. Appropriate funding and co-operation in sharing technology and other kinds of knowledge that will help accomplish the above noted actions must also be arranged.

The actions proposed above represent an important opportunity for the international community to progress toward a wide range of forest management goals that could contribute to a host of environmental and socio-economic issues. It is evident that the socio-economic and policy context for sustainable forest development in the industrialised boreal and temperate regions is different from that of the developing tropical regions. It is important to recognise these differences in order to ensure the success of co-operation between the donor and the recipient countries.

The interactions among trade, environmental degradation, and economic development are increasingly becoming the subject of considerable international scrutiny. Trade of forest products is perhaps one of the more salient topics within these issues. Green consumerism has and continues to increase in the industrialised countries with consumers preferring to buy forest products obtained from sustainably managed forests and manufactured with environmentally acceptable technologies. Threats of boycotts of tropical hardwood products by consumers, however, are counterproductive for one reason. Forests without economic, social, environmental, or other values are unlikely to be conserved or sustainably managed. Movements toward the conservation and sustainable management of forests are likely to succeed through profits, benefits, and incentives, not by punitive action (Maini and Ullsten 1993). International trade in forest products from sustainably managed forests should be promoted. Yet a necessary prerequisite is the formulation of internationally accepted criteria for sustainable forest development (Maini and Ullsten 1993).

Although conservation and sustainable development of all types or forests world-wide is in the collective interest of the community of nations, the primary responsibility for formulating and implementing appropriate policies and practices lie with national governments. The establishment and implementation of strong national forest policies and a clear institutional focus for forest-related activities would permit the international community to assist individual nations to meet nationally established targets and timetables (Maini and Ullsten 1993).

The tropical and subtropical regions need urgent collective action in terms of funds, technology, and training to assist in social forestry, agroforestry, afforestation, reforestation, sustainable forest development, forest biodiversity conservation, and rehabilitation of degraded lands. Planting individual trees or afforestation programs, as well as conservation and sustainable development of forests, will succeed only if these activities benefit local communities and national interests. Strategies that protect and conserve the forest and enable people to sustainably meet essential needs are likely to be successful in conserving the forest as sinks and reservoirs for C amongst other environmental and economic benefits (Cairns and Meganck 1994).

CONCLUSIONS

The conservation and sustainable development of forests world-wide provides a more cohesive and comprehensive framework within which to address a wide range of issues and opportunities associated with forests. First and foremost, forests world-wide represent a renewable resource of immense importance to the economic development of many countries. Therefore, any non-sustainable use of forests, particularly the conversion of forest to abandoned waste land, translates into an economic loss. In this regard for many developing countries, where the destruction of forests is taking place at an alarming pace, deforestation is both an economic and environmental disaster. From the developing countries' perspective however, the economic rather than the environmental aspects of forest management usually receive the highest priority. In the context of sustainable forest development we cannot understate the significance of ecological considerations. Forests will benefit the environment as well as the economy if they are conserved and sustainably managed to meet multiple needs.

In the context of sustainable development, the world forestry community faces two major challenges: first, given the expanding world population and the anticipated increase in demand for wood and non-wood products, how to meet this future demand without degrading the forest resource base and forest environment; and second, what technical, financial, institutional and political means to tap in order to promote sustainable development of forests world-wide. Both of these challenges impact directly on the conservation and enhancement of forests world-wide as sinks and reservoirs for C as a means of mitigating against potential climate change. Increasing and enhancing global forest cover through rehabilitation of degraded forest lands, conservation and preservation of existing forest resources, implementation of sustainable forestry practices, and improving the sustainability of agriculture world-wide is essential if we are to be successful. To accomplish this will require national and international commitment and co-operation, an appropriate policy and institutional framework, and access to appropriate technical and financial resources.

The concept of sustainable development has instigated considerable discussion of both the public and politicians at local, regional, national, and international levels. Goals and guidelines to protect the environment and improve global C management emanating from the Framework Convention on Climate Change, although important, are not likely to be successful if they ignore current social and economic values. Only a sustainable development approach that allows for multiple environmental and economic benefits can succeed.

REFERENCES

Allan T, Lanly JP (1991) Overview of status and trends of world forests. In: Hewlett D, Sargent C (eds) Proceedings of the Technical Workshop to Explore Options for Global Forestry Management Bangkok, 24–30 April 1991. London International Institute for Environment and Development, London, pp 17–39.

Allen JC, Barnes DF (1985) The causes of deforestation in developing countries. *Ann. Assoc. Amer. Geographers* 75: 163–184.

Anderson D (1986). Declining tree stocks in African countries. *World Development* 14: 853–863.

Anderson D, Fishwick R (1984) Fuelwood Consumption and Deforestation in African Countries. World Bank, Washington, DC.

Atjay GL, Ketner P, Duvigneaud P (1979) Terrestrial primary production and phytomass. In: Bolin B, Degens E, Kempe S, Ketner PI (eds) The Global Carbon Cycle. SCOPE 13, Wiley, Chichester, UK, pp 129–182.

Bolin B (1986) How much CO_2 will remain in the atmosphere? The carbon cycle and projections for the future. In: Bolin B, Döös BR, Jäger J, Warrick RA (eds) The Greenhouse Effect, Climatic Change, and Ecosystems. SCOPE 29, Wiley, Chichester, UK, pp 93–156.

Cairns MA, Meganck RA (1994) Carbon sequestration, biological diversity, and sustainable development: Integrated forest management. *Environ. Manage.* 18(1): 13–22.

Food and Agricultural Organization of the United Nations (FAO) (1993) Forest resources assessment. *FAO Forestry Paper 112*, Rome.

FAO (1990a) Interim Report on Forest Resources Assessment. 1990 Project Committee on Forestry, Tenth Session. Rome: Food and Agriculture Organization. COFO–90/S(a).

FAO (1990b) 1989 Production Yearbook. Rome: Food and Agriculture Organization. 1991. Second Interim Report on the State of Tropical Forests. Presented at the Tenth World Forestry Conference. Paris, September 1991.

FAO/UNEP (1981) Tropical Forest Resources Assessment Project. Rome: Food and Agriculture Organization.

FAO (1993) Forestry statistics today for tomorrow: 1961–1991..2010. FAO, Rome.

Fearnside PM (1979) Cattle yield prediction for the Trans-Amazon Highway of Brazil. *Interciencia* 4: 338–343.

Fearnside PM (1986) Brazil's Amazon forest and the global carbon rroblem: Reply to Lugo and Brown. *Interciencia* 11: 58–64.

Fearnside PM (1993) Deforestation in Brazilian Amazonia: The effect of population and land tenure. *Ambio* 22(8): 537–545.

Grainger A (1988) Estimating areas of degraded tropical lands requiring replenishment of forest cover. *Intern. Tree Crops J.* 5: 31–61.

Harmon ME, Ferrell WK, Franklin J (1990) Effects on carbon storage of conversion of old-growth forests to young forests. *Science* 247: 699–702.

Hecht SB (1985) Environment, development and politics: capital accumulation and the livestock sector in eastern Amazonia. *World Develop.* 13: 663–684.

Hecht SB (1993) The logic of livestock and deforestation in Amazonia. *BioScience* 43(10): 687–695.

Houghton RA (1993) The role of the world's forests in global warming. In: Ramakrishna K, Woodwell GM (eds) World Forests for the Future: Their Use and Conservation. Yale University Press, New Haven and London, pp 21–58.

IPCC (1992) 1992 IPCC Supplement: Scientific Assessment of Climate Change. Intergovernmental Panel on Climate Change—Working Group I Report. Cambridge University Press, New York. 24 pp.

IUCN/UNEP/WWF (The World Conservation Union/United Nations Environment Programme/World Wide Fund for Nature) (1991) Caring for the Earth. A strategy for sustainable living. IUCN/UNEP/WWF, Gland, Switzerland.

Kummer DM (1992) Deforestation in the Post-war Philippines. University of Chicago Press, Chicago, IL.

Kummer DM, Turner II BL (1994) The human causes of deforestation in southeast Asia. *BioScience* 44(5): 323–328.

Maini JS (1990a) Forests: barometers of environment and economy. In: Mungall C, McLearen CD (eds). Planet under Stress. Oxford University Press, pp 168–187.

Maini JS (1990b) Sustainable development in the Canadian forest sector. *Forestry Chronicle* 66(4): 346–349.

Maini JS (1991a) Guiding principles towards a global consensus for the conservation and sustainable development of all types of forests world-wide. Department of Forestry, Ottawa.

Maini JS (1991b) Towards an international instrument on forests. In: Hewlett D, Sargent C (eds) Proceedings of the Technical Workshop to Explore Options for Global Forestry Management. Bangkok, 4–30 April 1991. International Institute for Environment and Development, London, pp 278–285.

Maini JS (1992) Sustainable development of forests. *Unasylva* 43(169): 3–8.

Maini JS (1993) Sustainable development of forests: A systematic approach to defining criteria, guidelines and indicators. In: Proceedings of the Seminar of Experts on Sustainable Development of Boreal and Temperate Forests: Technical Report—Annex I. Conference on Security and Cooperation in Europe, pp 61–70.

Maini JS, Ullsten O (1993). Conservation and sustainable development of forests globally: issues and opportunities. In: Ramakrishna K, Woodwell GM (eds) World Forests for the Future: Their Use and Conservation, Yale University Press, New Haven and London, pp 111–119.

Melillo JM, Callaghan TV, Woodward FI, Salti E, Sinha SK (1990) Effects on ecosystems. In: Houghton JT, Jenkins GJ, Ephraums JJ (eds) Climate Change: The IPCC Scientific Assessment, Cambridge University Press, Cambridge, pp 283–310.

Myers N (1981) How Central America's forests became North America's hamburgers. *Ambio* 10(1): 3–8.

Myers N (1989) Deforestation rates in the tropical forests and their climatic implications. Friends of the Earth, London.

Myers N (1993) Tropical forests—the main deforestation fronts. *Environ. Conserv.* 20(1): 9–16.

Nations JD, Nigh RB (1980) The evolutionary potential of Lacandon Maya sustained-yield tropical forest agriculture. *J. of Anthropological Res.* 36: 1–30.

Olson JS, Watts JA, Allison LJ (1983) Carbon in live vegetation of major world ecosystems. DOE/NBB-0037 National Technical Information Service, Springfield, VA, 152 pp.

Posey DA (1993) Indigenous knowledge in the conservation and use of world forests. In: Ramakrishna K, Woodwell GM (eds) World Forests for the Future: Their Use and Conservation. Yale University Press, New Haven and London, pp 59–77.

Ravindranath NH (1993) Biomass energy conservation options for forest conservations. In: Hewlett D, Sargent C (eds) Proceedings of the Technical Workshop to Explore Options for Global Forestry Management, Bangkok, 24–30 April, 1991. International Institute for Environment and Development, London, p 286.

Ravindranath NH, Hall D (1994) Indian forest conservation and tropical deforestation. *Ambio* 23: 521–523.

Reis EJ, Margulis S (1990) Economic perspectives on deforestation in the Brazilian Amazon. Project Link Conference, Manilla.

Repetto R (1993) Government policy, economics, and the forest sector. In: Ramakrishna K, Woodwell GM (eds) World Forests for the Future: Their Use and Conservation, Yale University Press, New Haven and London, pp 93–110.

Salati E, Vose T (1984) Amazonia: a system inequilibrium. *Science* 225: 129–138.

Serrao EA, Toledo JM (1990) The search for sustainability in Amazonian pastures. In: Anderson A (ed) Alternatives to Deforestation. Columbia University Press, New York, pp 195–215.

Skole DL, Chomentowki WH, Salas WA, Nobre AD (1994) Physical and human dimensions of deforestation in Amazonia. *BioScience* 44(5): 314–322.

Stahl M (1993) Land degradation in East Africa. *Ambio* 25(8): 505–508.

Tho YP (1991) Tropical moist forests—facts and issues. In: Ramakrishna K, Woodwell GM (eds) World Forests for the Future: Their Use and Conservation, Yale University Press, New Haven and London, pp 43–62.

Ullsten O (1991) Keynote speech. In: Hewlett D, Sargent C (eds) Proceedings of the Technical Workshop to Explore Options for Global Forestry Management. Bangkok, 4–30 April 1991. International Institute for Environment and Development, London, pp 10–13.

WCED (World Commission on Environment and Development) (1987) Our Common Future (The Brundtland Report). Oxford University Press, Oxford, UK, 383 pp.

Williams M (1990) Forests. In: Turner II BL, Clark WC, Kates RW, Richards JT, Mathews JT, Meyer WB (eds) The Earth as Transformed by Human Action. Cambridge University Press, New York, pp 179–201.

World Bank (1989) Philippines: Environment and Natural Resource Management Study. World Bank, Washington DC.

WRI/UNEP/UNDP (1992) World Resources 1992–1993: A Guide to the Global Environment: Toward Sustainable Development. Oxford University Press, New York, 385 pp.

The views expressed in this paper are personal views of the authors and do not necessarily reflect official positions of their organisation or of the Canadian government.

32 WG4 Summary: Human dimensions of the forest-carbon issue

Peter N Duinker[1,2], Roger A Sedjo[1,2], Philip M Fearnside[2], Alan Grainger[2], Jayant Sathaye[2], Birger Solberg[2], Robert B Stewart[2], Valentin V Strakhov[2], and George M Woodwell[2]

[1] Co-Chairs of Working Group 4.
[2] Addresses are provided in the list of participants

INTRODUCTION

The preceding discussions have dealt with basic biophysical processes associated with forests and the global carbon (C) cycle, and the technical feasibilities of intervening in those processes so that forests play a stronger role in sequestering C from the atmosphere. Such discussions are a vital component of a full understanding of how humanity might mitigate what we shall call the C problem—what will it take to manage and use forests to lessen climatic change due to increasing atmospheric CO_2? Any meaningful answer requires that we put the technical possibilities into the social, economic, institutional and political contexts within which all human endeavour takes place. Are these contexts—current and future—facilitative of strong implementation of the technically feasible actions, or will they pose constraints? If the latter, what will it take to remove the barriers?

These and related questions were the subject of vigorous discussion and debate by members of the Working Group on the Human Dimensions of the Forest-Carbon Issue. We were charged in our deliberations as follows:

1. define conceptually the relationships between key human-dimension themes and the forest-carbon issue (i.e., how does each theme relate to forest management and the global C cycle?);
2. outline the information and knowledge needed to develop and implement policies related to forest management and the global C cycle (i.e., what key questions urgently need research?); and
3. summarise each contributed paper and relate its key messages to the themes.

The subject matter assigned to the group is conceptually difficult. Frameworks for understanding the relationships between human dimensions and the forest-C issue are ill-defined and complex. In that context, we offer the following encapsulation of our short meeting with the modest objective that the reader might find it a helpful synthesis of ideas on the broad links between forests and the C problem. References to literature beyond the papers contributed at the Workshop are scant; readers will find extensive lists in those papers to back up most of the statements made in this summary.

NATO ASI Series, Vol. I 40
Forest Ecosystems, Forest Management and the Global Carbon Cycle
Edited by Michael J. Apps and David T. Price

ISSUES, RESEARCH NEEDS, AND THE CONTRIBUTED PAPERS

Assessments: modelling, monitoring, and integration

Modelling

The transfer of C from vegetation to the atmosphere is mainly a result of land-use change and forest exploitation. In the last decade there has been growing activity in modelling the underlying socio-economic and policy mechanisms that drive land-use change and logging, in the tropics in particular. In addition, certain long-term generic land-use processes, e.g., land-use and forest transitions, have been identified and provide a basis for modelling long-term change. Grainger (Chapter 27) has described the *national land-use transition* as the process in which much of a country's forest cover is removed from the land. The *forest transition* occurs when, after some equilibrium period, the forest area of a country begins to rise through a combination of natural regeneration processes and plantation establishment.

Programs intended to increase forest cover may be constrained by the particular phase of land-use and forest transitions which a country has reached. Thus, studying them is vital for confident modelling of future C emissions trends and for planning mitigation strategies.

Besides the clearance of forest caused by land-use change, retained forest can be degraded by logging and other types of exploitation. Degraded forest is often characterised by the partial regrowth that follows many clearances. Degraded forest takes on many forms in a complex spatial mosaic (see Chapter 27), and its relationships with underlying mechanisms are still poorly understood.

Several research priorities can be identified to contribute essential knowledge for modelling the roles of forest degradation and land-use change in the global C cycle. The following topics require research and development:

1. spatio-temporal patterns of land-use change and its underlying mechanisms, at national or sub-national levels;
2. factors bringing deforestation under control when it reaches an advanced stage in a particular area;
3. factors initiating and driving forest replenishment;
4. extension of present non-spatial national/global deforestation models by addition of spatial attributes;
5. interfacing of spatial deforestation models with spatial databases to simulate trends in CO_2 emissions resulting from deforestation;
6. interfacing of spatio-temporal deforestation models with general circulation models of the atmosphere; and
7. case studies of the so-called national land-use and forest transitions in tropical countries.

Monitoring

Solving the C problem depends, in part, on conservation and sustainable management of global forests. Sustainable forest management is possible only by ensuring balanced use of forest resources, functions and properties by humankind both now and in the future. This balance, in turn, requires a thorough knowledge of forest-C phenomena, among others. One essential means of creating the required knowledge is systematic monitoring of global vegetation cover. From a forest point of view, such monitoring includes both creation of a global forest inventory system and establishment of credible estimates of national forest inventories. Of vital importance here is the creation of scientifically valid and globally consistent standards of forest resource accounting and monitoring (see Parts II and III of this volume).

An operational and high quality global forest monitoring system would generate data indispensable for addressing a wide range of pressing questions related to the C problem. It would permit increased accuracy in estimates of the role of forests in global environmental change and the C cycle. It would increase insights into how humanity can achieve a balance between acceptable forest conditions for the present generation and the possible needs of future generations. In addition, it would enlighten program design for control of deforestation and afforestation processes and for improving logging practices.

To make global forest monitoring effective, certain basic requirements are evident. First, it is necessary to compile and publish global forest monitoring data at regular periodic intervals. Second, the physical forest data must be complemented with information on the socio-economic attributes of forest land in a spatially consistent format.

In a background paper and in the working group sessions, Strakhov and colleagues (1994) summarised some of the problems facing assessments of forest accounts in Russia. These difficulties include that of interpreting data in Russia's forest information system, including basic forest inventory data, and how these data may be related to the unfolding system of forest tenure and management planning. Several shortcomings of the traditional information system were identified, including unknown yet suspected significant biases and other errors, lack of geo-referencing at the local level, and inability to serve in management and policy decision-making on non-timber values such as biodiversity, C storage, and, more generally, sustainable forest management. It appears that the information system needs to be overhauled to serve contemporary needs. The overhauls include use of modern computer systems, including geographic information systems, automatic interpretation of remotely sensed data, mathematical modelling of forest succession and utilisation, problem-oriented expert systems, and the latest communications technology.

The contributed papers

Chapter 28 by Fearnside, entitled *Socio-economic Factors in the Management of Tropical Forests for Carbon*, combines a broad literature review with the author's personal observations of tropical forest development to caution against overly narrow assessments of the feasibility and the costs and benefits of proposed forest-management solutions to the C problem. Fearnside discusses a range of commonly proposed solutions such as *sustainable* logging in natural forests, plantation development, agroforestry, and protected areas. He observes that most assessments of such schemes fail to account properly for socio-political and income distribution effects, especially as they pertain to the rural poor. Political corruption and local socio-economic constraints are seen as particularly strong frustrations to well-meaning but naive C-management proposals involving forests. Fearnside's greatest hope is that tropical deforestation can be curbed, if not halted. This would bring not only strong C benefits, but many other benefits as well. A key factor in slowing deforestation would be proper valuation of the environmental services of tropical forests.

Grainger's main point in Chapter 27 (*Integrating the Socio-Economic and Physical Dimensions of Global Climate Change Mitigation Assessments*) is that purely physical assessments of the potential mitigation of atmospheric C build-up using forests are overly optimistic because of inevitable constraints and delays due to actual cultural, social, economic and political factors. He illustrates this point with case studies dealing both with physical improvement of existing degraded forests and with afforestation. His analysis of existing forests is buttressed with a new taxonomy of degraded tropical forests. Grainger extends his basic analyses by examining the potential effects of forest C sequestration schemes on global wood products markets. He argues that, without intervention in those markets, the schemes could have significant effects by making more wood available than the expected demand. Grainger concludes that improvements to existing forests and afforestation will occur in complex spatio-temporal patterns, often slower than planners would like. For realism, assessments of feasible programs to mitigate the C problem using forests must be characterised by strong integration of physical, social, economic and political attributes of land.

Economics and markets

Economic evaluation of mitigation options

There is no global consensus at present on the monetary value of reducing a unit of atmospheric C. Most economic studies therefore focus on estimating cost-effectiveness, i.e., the cost of sequestering a unit of C or avoiding a unit of C emissions, which is often expressed as US\$ Mg^{-1} C.

The 1990 IPCC report on Response Strategies to Climate Change (Houghton et al. 1992) noted that slowing deforestation was a low-cost option for reducing a unit of atmospheric C. The IPCC report quoted regional average annual costs of about US\$ 8 Mg^{-1} C (at the time of publication) for tropical forestation and reduction of deforestation, which increased to about US\$ 28 Mg^{-1} C for forestation in non-US OECD countries. Costs for establishing a forest plantation, excluding the opportunity cost of land, were estimated to range between US\$ 230 to US\$ 1000 ha^{-1} (Sedjo and Solomon 1989) with an average cost of US\$ 400 ha^{-1}.

Cost estimates for atmospheric C reduction have been improved in several ways since the IPCC (1992) report: (1) unit costs have been estimated for individual countries rather than by regions or for the globe as a whole; (2) the estimates have been developed for several types of mitigation options; (3) other cost components (e.g., maintenance, land rental (opportunity costs), and monitoring and evaluation, which were not included in the earlier IPCC report) are now being addressed; (4) incremental costs have been estimated by some researchers; and (5) analytical methods have been improved to provide better insights and techniques for the evaluation of costs and benefits.

The improved cost estimates suggest that 70 Pg C, i.e., between 50% and 90% of the technical C sequestration potential, may be tapped at a cost under US\$ 10 Mg^{-1} C. While this cost of C sequestration may appear reasonable, many issues related to the evaluation of costs, and the use of this indicator to reflect cost-effectiveness, are yet to be resolved. In addition, monetary and other non-C environmental benefits also need to be considered. Evaluating cost-effectiveness requires that both physical flows and stocks of C and monetary flows be evaluated in a comparable manner, which raises the following issues that require further research and resolution.

Methodological consistency: The anthropogenic interventions necessary for C sequestration through land-use and forest management need to be identified and described to estimate their C flows. The associated monetary flows also need to be defined and estimated over time. At present, there is inadequate reporting of what is included in the monetary and C evaluation of a forest project. On the C side, it is often unclear whether litter and product C are included. For monetary flows, it is often not clear whether the evaluation excludes harvesting and transport costs. A common and more precisely defined system boundary is thus needed to delineate the items to be included in the evaluation of costs and C flows.

Incremental costs of projects: The Framework Convention on Climate Change (FCCC) calls for the evaluation of *total agreed incremental costs* of a project. To determine the incremental funds that a recipient country may be justified in seeking, the incremental cost estimation requires definition of baseline costs and domestic and global benefits. New methods and data are needed to pursue such estimation.

Time preference: Is C sequestered today worth the same as that in the future? Economic evaluation lowers the value of future damage avoided if a positive discount rate is used to value future benefits. The evaluation thus accords a lower value to future generations, which raises questions about equitable distribution of resources among generations. However, if early C mitigation and sequestration activities are given an advantage when compared with later activities, while a positive discount rate applies to the costs of sequestration, there is no economic incentive for early action. Cost minimisation rationality will provide incentives to delay C mitigation to the last moment.

Similar issues may also be raised across populations living in different parts of the world. Should the estimation of benefits and costs be adjusted through reducing the discount rate or through changing the shadow price of C? Should one account for the decades-long life of C in the atmosphere? Is the selection of projects significantly affected by the explicit consideration of these issues? Answers to these questions will need to be sorted out using alternative indicators discussed below.

Alternative indicators of cost-effectiveness: The US$ Mg^{-1} C cost indicator has limitations. Obviously, costs are just one side of a cost–benefit assessment. However, the benefits associated with C sequestration are usually articulated simply as mass of C sequestered. Unfortunately, such an aggregate approach does not provide information on income distribution implications, which may be important in some countries. Additionally, early studies of the costs of sequestering C through forest management activities often ignored some of the costs, especially future costs, and almost all studies have ignored the non-C benefits of forests and forest practices. Few studies have estimated the net costs of sequestering C with forests, after inclusion of the non-C benefits, as in an application of the highly touted "no regrets" approach. Nevertheless, early studies generally suggested that the costs of sequestering C through forests were relatively low and additional investigation has tended to confirm that finding. Subsequent work has shown considerable progress in achieving a proper rendering of the costs of sequestering a unit of C through various forest practices. Such information is needed for a useful comparison of various sequestration and mitigation activities, as well as for a determination of the most cost-efficient mix of these activities.

A complete cost–benefit treatment awaits more definitive estimates of the damage functions related to increased atmospheric C. A complete cost–benefit analysis would need to relate the time stream of benefits associated with a reduction in the increase of atmospheric C with the time stream of sequestration and its associated discounted costs.

Evaluation of environmental benefits: Most forest projects will have environmental benefits in addition to C sequestration. Their inclusion in the estimation of the economic benefit of a project has the potential to make a project significantly more attractive. Unfortunately, these benefits are difficult to express in monetary terms. Researchers have estimated the existence and use values of forests which illustrate the substantial benefits to be derived from a forest project. Better techniques and data are needed to improve the consistency of monetary estimates of forest benefits.

Explicit consideration of project risks: The above issues become more complex when the probabilities of C sequestration and of achieving other environmental benefits are taken into consideration. Projects may be pursued with expected high net benefits, but with a low likelihood of implementation and survival. The evaluation of cost-effectiveness should, where feasible, consider the probability of the expected outcomes of a project.

Role of global forest products

From a C point of view, forest products can be viewed as a human-created pool of products that function as a C reservoir and can be a source or sink depending upon the fluxes into and out of the reservoir. The size of the pool increases as products are added to the forest products stock and is reduced as products leave the pool and their C released via various types of destruction and decomposition.

The forest products industry affects the global C cycle in a number of ways. The global market for industrial wood produces and consumes approximately 1.6×10^9 m^3 of industrial roundwood annually in recent years, with something over one-half going into solid wood products. While the process of timber harvest itself releases C, some portion of the C in the harvests remains sequestered in stocks of forest products.

The market for forest products is a key incentive to replant trees in harvested areas and to establish industrial forest plantations, often on lands not recently in forest. About 1 Mha of plantations are established annually in the tropics, and about 4 Mha in the temperate world. Much of this occurs on lands not recently in forest cover, thus creating additional C sinks.

Wood products, especially as construction materials, require relatively little energy to manufacture. Thus, to the extent that wood substitutes for other energy-intensive materials, total C releases may be reduced. Furthermore, in some regions plantation forests deflect logging pressures on native forests, thereby contributing to the maintenance of C in the native forest biomass.

International trade mechanisms may frustrate or even eliminate intended beneficial effects of local or national actions taken to mitigate the C problem. For example, a country may limit timber harvests in an attempt to reduce global C emissions. Operation of international trade can easily offset the C benefits of this action by increasing timber harvests elsewhere and reallocating the trade flows of forest products. Thus, reduced local harvest will likely result in increased harvests elsewhere and increased forest products imports (or reduced exports) by the country that has reduced its harvests.

Green consumerism in international trade in forest products is another factor that is beginning to have an impact on the conservation and sustainable management of forests world-wide, and thus on the role of forests in mitigating the C problem. Many consumers in industrialised countries are choosing forest products from sustainably managed forests and manufactured by environmentally sound processes and technologies. In the same vein, consumers occasionally threaten to boycott certain products, such as tropical hardwood products, because of the alleged unsustainable manner in which they are produced. To protect their international markets, countries exporting forest products are responding through development of internationally recognised criteria for sustainable forest development and the application of these criteria to their domestic forestry practices. In most cases, such criteria address C sequestration to some degree.

Further research is needed to determine more accurately: (a) the C implications of manufacture and use of various materials including wood; (b) the life cycles of various wood products and their effects on the forest products C stock; (c) the extent to which the forest products industry can effectively produce wood-based products and accomplish significant C sequestration at the same time; and (d) the potential effects of green consumerism on forest practices aimed at mitigating the C problem.

The contributed papers

In Chapter 26 on *Economic Evaluation of Forest-Sector Mitigation Options*, Sathaye explores concepts and approaches for estimating costs of forest-oriented options for mitigating the C problem. Such options include methods to maintain existing C stocks, expand forest C sinks, and substitute for fossil fuels in energy production. When it comes to forest plantations, the author supports the concept of sustainable logging over that of *planting and storing* C (i.e., with no timber harvest). Analyses of land available for plantations lead Sathaye to the conclusion that there is more land, from a technical point of view, available for plantation establishment around the world than earlier estimates suggested. He points out that benefit estimation regarding sequestered C is fraught with uncertainty, and the most promising economic analysis is that of cost-effectiveness. However, Sathaye suggests that, in many cases, the non-C benefits of plantation establishment may serve to cover most or all of the costs of plantation establishment, making the cost near zero or even negative for C sequestration via plantations.

In Chapter 25 (*The Economics of Increased Carbon Storage through Plantations and Forest Management*), Sedjo's objectives are to summarise the findings of recent studies on the costs of sequestering C through increases in forest biomass and forest product stocks, and to report on the benefits of plantations for sequestering C based on an Argentine case study. Sedjo notes that forest plantations are attractive for this purpose because of four factors: (a) they have potential to sequester large amounts of C; (b) they are readily established from a technical point of view; (c) they have many product and service functions beyond C; and (d) their costs are modest compared to other means of addressing the C problem. Both his literature review and case study confirm these general statements, particularly the final one about modest costs.

Solberg and Hoen report on recent developments in calculating economic performance indicators for C sequestration in Norwegian forests in Chapter 29, entitled *Economic Aspects of Carbon Sequestration: Some Findings from Norway*. Two approaches to such calculations are summarised: one based on cost-efficiencies of increasing C sequestration rates, and one based on maximising net present value when C sequestration rates are monetarily valued. The former approach provides a rough overview of potentials, and demonstrates that forest measures are highly competitive with measures in other sectors of the economy. In a comparison with Indonesia, the approach also shows that forest measures to mitigate C build-up in the atmosphere are much more profitable in Indonesia than in Norway. The latter approach corrects a weakness in the former by accounting for mutual interactions among forest measures. Solberg and Hoen flag several key uncertainties in such studies for Norway: (a) rates of natural mortality when stands are permitted to develop beyond normal clearfelling age; (b) end uses and decay times for various forest products; (c) real discount rates for C flow studies; and (d) effects of pricing C flows through taxes on resource allocations in the forest sector.

Forest policy and institutions

Equity in sharing costs and benefits

Paramount among concerns regarding the social, economic and political contexts of forest sector responses to the C problem are questions of international equity and the distribution of costs and benefits among sub-national regions and social actors. Mitigation responses to global warming can have substantial undesirable social impacts. While such impacts must be reduced, any successful fight against global warming will not be painless. The social impacts of combating global warming, not only the financial costs, must be shared fairly among nations. This is particularly important given the likelihood of impacts in developing countries as a consequence of a global problem that is, in fact, largely due to activities in developed countries.

The blame for greenhouse gas emissions, as well as the credit for emissions reductions, must be identified not only by nation but also by sub-national region and by the various social actors within nations. Only with this kind of information do opportunities become apparent for emissions reductions with minimal social cost. An example is provided by the tremendous discrepancy between the emissions of a small number of large-ranch owners in Brazilian Amazonia and those of the rest of the country's population (see Fearnside, Chapter 28).

The FCCC is based on the principle that individual nations can decide for themselves how emission reduction targets are to be met. Nations may choose, for example, to restrain fossil fuel combustion, reduce deforestation, or plant trees. A country may well decide to meet its emissions reduction commitments by converting lands to *Eucalyptus* plantations, a likely consequence of which would be displacement of rural poor people. Such social impacts must be understood and considered in allocating international financing to projects proposed as response options.

Trade-offs

Although forests are a major global C sink, many useful outputs are produced by forests and their ability to sequester C is not entirely independent of their production of other outputs. These outputs include timber, recreation, wildlife, and biodiversity, among others. However, although the production of these involves some complementarity, it can also involve conflict. As a hypothetical example, biodiversity in Canadian boreal forests may well be reduced if they were managed to maximise C sequestration. Similarly, trade-offs between industrial wood and C sequestration in the forest are also often likely to occur. However, careful management of C, typically going well beyond the individual forest or even the aggregate global forest, can offset, at least in part, some of the adverse C consequences. For example, activities such as sequestering C in long-lived wood products, burning wood instead of fossil fuels, and establishing new industrial plantation forests may contribute to achievement of non-C forest objectives without compromising global C objectives.

In many respects, the new demands placed on forests exacerbate the complexities associated with the multiple-use management paradigm developed two decades ago. Some trade-offs between C and other outputs appear inevitable, and politicians, the public and forest managers must recognise this reality. These trade-offs will manifest themselves at both broad policy and local management levels. At the broad scale, governments will allocate funds to a variety of activities that may conflict with C sequestration, and in such a way at least to shorten the supply of funds to allocate directly to mitigation of the C problem. Similarly, forest managers will be expected to deliver on complex management objectives which often conflict with each other. Research is needed to develop more effective ways to characterise and address these trade-offs and conflicts.

Institutional effectiveness

Strong international institutions are needed to provide leadership on forest-related issues, to manage international co-operation on forests, to assess the state of the world's forests, and to provide timely and credible reports on progress toward the conservation and sustainable development of forests at regional and global levels. However, there is no single global agency responsible for forests. The Food and Agriculture Organization (FAO) of the United Nations has some forest-related responsibility, but concentrates much of its efforts on world agriculture. Most international organisations are relatively narrow in scope; for example, the International Tropical Timber Organization (ITTO) addresses market aspects of tropical timber and the International Union of Forest Research Organizations (IUFRO) directs its efforts at co-ordinating forest research.

At regional and national levels, strong forest-oriented organisations exist. To increase the effectiveness of such institutions in achieving global forest sustainability, efforts are needed on several fronts. For one, efforts are needed to ensure development and implementation of rigorous national and sub-national policies for forest conservation. Although conservation and sustainable development of all types of forests world-wide is in the collective interest of the community of nations, the primary responsibility for formulating and implementing appropriate policies and practices lies with national and sub-national governments. Any international deliberations must recognise the sovereign rights of nation states to conserve and use their forests to meet domestic policy objectives.

However, global forest conservation requires a consistent set of national land-use policies that promote sustainable forest development and that have clear targets and timetables for achievement of adequate levels of permanent forest cover. Such policies must, as recent experiences world-wide have demonstrated, foster involvement of local people and communities in developing and implementing forest plans. In developing countries in particular, local people must gain ownership of any initiatives designed to provide for their needs for food, fibre, fodder, shelter, and economic development. Combined with a clear institutional focus for forest-related activities, the elevation of forest sustainability on the political and public agenda, and establishment and implementation of strong national forest policies, would encourage international co-operative efforts, particularly international assistance to individual nations to meet their targets and timetables.

Another key front where progress is required is in international co-operation and collaboration of national institutions. The difficulty, of course, is how to reconcile national interests with international responsibilities and to develop appropriate policy responses, instruments, and institutions to accomplish this balance. The socio-economic and policy context for sustainable forest development in the industrialised boreal and temperate regions is different than that for the developing tropical regions. It is important to recognise these differences to ensure the success of co-operation between donor and recipient countries. An international convention addressing the conservation of all types of forests world-wide is an important step to improving international co-operation (see below).

A third area needing improvement to secure more-effective institutions is the procedures for assessing socio-economic impacts. Currently, these procedures seldom fulfil their intended function of protecting the interests of the affected populations. An example is small-holding farmers who may be displaced by a large plantation scheme. Ironically, all those involved in assessing the scheme (except the displaced people) may be happy with the outcome. Consultants who have held nothing back in reporting doubts about proposed schemes have clear consciences, bankers who have moved large volumes of funds are happy, and governments of recipient countries and any firms or special interest groups that profit from implementing the schemes are content.

The inadequacy of current safeguards to prevent social disasters is evident in the case of the World Bank, the recent history of the Narmada Dams in India speaking far louder than any announcement of Bank policies (see Fearnside, Chapter 28). Since the World Bank administers the Global Environmental Facility (GEF), which is charged with distributing funds for combating global warming, the procedures for including socio-economic considerations must be improved not only so that these considerations are better assessed, but also so that they are given the weight they merit when decisions are taken. This is needed both for decisions regarding broad strategies for responding to global warming and for approval of individual projects.

International directives and initiatives

A number of conferences and international fora have proposed adoption of a convention on global forests, a set of world-wide forestry principles, and sets of criteria for forest sustainability. At the same time, the FCCC and the Framework Convention on Biological Diversity were signed in 1992 at the Earth Summit in Brazil, and are now in the process of being implemented. Each of these framework conventions include forests as an integral component.

Of particular importance to global C management is activity that may be undertaken under the FCCC. The primary objective of the FCCC is to influence human activities to stabilise concentrations of greenhouse gases in the atmosphere before they upset the natural climate system beyond the ability of the ecosystems to evolve and adapt to the changed climate. Signatory countries are obliged to maintain inventories of greenhouse gas emissions, undertake mitigation programs, and collaborate in developing means of reducing emissions and conserving sinks and reservoirs of greenhouse gases. This latter aspect pertains almost exclusively to forests and is an important element in current international deliberations on setting goals and standards as well as on the development of programs to mitigate against potential climate change.

The Framework Convention on Biological Diversity is aimed at protecting nations' and other interests' supply of genetic materials for commercial exploitation. This most definitely implicates forests in a major way, especially tropical forests. Unfortunately, the Convention misses the point of avoiding systematic, cumulative and irreversible biotic impoverishment of the Earth as a whole (see Woodwell and Ramakrishna, Chapter 34).

A major failure of the Rio Conference was the inability to create a convention specifically on forests. Aside from a number of political reasons, agreement on such a convention was not forthcoming because of an insufficient recognition and understanding of the role of forests in maintaining Earth's life support systems and meeting human needs. The Conference did witness development of an international consensus on guiding principles for conservation and sustainable development of all types of forests world-wide. This is not enough, however. The formulation of policies based on scientifically sound, internationally agreed criteria and indicators for sustainable forest management is a crucial next step to ensure forest sustainability (see Stewart and Maini, Chapter 31).

To move strongly towards a global forest convention, two realms in particular need attention: (a) technical and scientific analyses of the importance of forests in the biosphere; and (b) the political innovations needed to raise forest sustainability to the highest level on all agendas. Establishment of a World Commission on Forests and Sustainable Development is a logical step (see Woodwell and Ramakrishna, Chapter 34). The purpose would be to advance forest issues as part of the mainstream of the current international progress in conservation of the biosphere. The Commission would also champion continued efforts toward development of a global forests convention.

The contributed papers

Duinker, writing about *People and Forests: Fitting Carbon into a Perplexing Future* in Chapter 30, takes a pessimistic view that C as a forest management objective is unlikely in the near term to take a front-and-centre seat among other values for which forests are now being managed. He notes, based on the literature and his experience in forest policy dialogues in Canada, that the forest policy agenda is already packed with other issues, that directions for managing existing forests for C are far from clear, that changes in forest management can only be temporary or short-term aids to the C problem, and that management for other forest goods and services may be at odds with managing for C. On the other hand, Duinker identifies several factors that may hasten the day when C takes a strong place among forest management objectives, even if indirectly. These factors include expanded use of biomass energy, expanded markets for all wood products, increasing forest area, and international pressures to solve the problem of increasing C in the atmosphere.

Stewart and Maini provide a thoughtful review of policy considerations, particularly at the national and global levels, of forests and the C problem in Chapter 31 on *Forests and Global Carbon Management: A Policy Perspective*. The authors show how sustainable forest management fits squarely within the context of the framework conventions on climate change and biodiversity which were developed at the Earth Summit in Rio in 1992. Policy responses they identify for achieving sustainable forest development, and thus for addressing the C problem using forests, include: (a) increase the total global forest cover, particularly that which is permanent; (b) protect natural forests from commercial exploitation; (c) clarify national responsibilities for forests, including preparation of state-of-the-forest reports and strengthening research funding: and (d) increase the sustainability of food production. Stewart and Maini conclude that the world forest community faces two major challenges in determining how to meet the rising world demand for forest products and services without degrading forest ecosystems, and how to martial the technical, financial, institutional and political means to secure sustainable forest development world-wide.

The epilogue (Chapter 34) by Woodwell and Ramakrishna, entitled *Forests and the Human Habitat: The Case for a World Commission on Forests*, is a thought-provoking combination of both inductive and deductive reasoning that advocates strongly for a proper global dialogue about forests. The dialogue would not assume the customary focus of forest products, but rather would emphasise forest services in making the global environment an hospitable, even pleasant, place to live. Woodwell and Ramakrishna both justify and sketch the dimensions of a World Commission on Forests and Sustainable Development, which they suggest be modelled after the World Commission on Environment and Development of the late 1980s. The authors see the work of the Commission on two fronts. One is scientific, where the key task is to reduce uncertainties in the collective knowledge of how forests interact with the other major components of the biosphere and how humans might direct their use and development more favourably for acceptable functioning of the human habitat. The other is political, where decision-making forces at all levels (e.g., political leaders, business leaders, the citizenry of all countries) need to discuss the service functions of forests and ascribe them their due in the overall affairs of society. Woodwell and Ramakrishna urge prompt establishment of the Commission, because the current courses on which humankind finds itself all point to rather unpleasant collisions. In connection with the C question, the authors provocatively state that:

> *"...there can be no solution, interim or long term, without managing forests globally in such a way as to balance the global carbon budget".*

SUMMARY AND CONCLUSIONS

As the foregoing discourse shows, the relationships between key social, economic and political dimensions related to forests and the forest-C issue are complex and at best poorly understood. Moreover, due mainly to the constraints imposed by a short (however intense) period of collaborative discussion at the Workshop, we have given here an incomplete and partial view of these relationships. In so doing, though, we have sketched a wide range of research needs and questions that the scholarly community needs to address to elucidate more clearly how forests world-wide can contribute to resolving the problem of increasing C content in the atmosphere.

If forest and land-use management is to be designed explicitly to address C objectives, changes in policies at all levels—local, sub-national, national, international—will in most cases be necessary. Considering the complex and sometimes bewildering array of values for which today's societies are insisting that forests be managed, it becomes more important than ever that forest management policies be both effective and efficient.

Technical possibilities, as this book clearly shows, hold much promise for substantial resolution of the C problem through appropriate and directed use and management of current forests and forestable land. Achievement of significant proportions of the technically possible depend on marshalling instrumental combinations of human, technical, financial, knowledge, biophysical, and political resources in creative and collaborative new ways. Failure to do so will doubtless render a future for the biosphere many would wish to have avoided.

REFERENCES

Duinker PN (1995) People and forests in Canada: fitting carbon into a perplexing future (Chapter 30, this volume).

Fearnside PM (1995) Socio-economic factors in the management of tropical forests for carbon (Chapter 28, this volume).

Grainger A (1995) Integrating the socio-economic and physical dimensions of degraded tropical lands in global climate change mitigation assessments (Chapter 27, this volume).

Houghton JT, Callander BA, Varney SK (eds) (1992) Climate Change 1992: The Supplementary Report to the IPCC Scientific Assessment. Cambridge University Press, Cambridge, UK. 200 pp.

Sathaye J (1995) Costs of forest-sector mitigation options (Chapter 26, this volume).

Sedjo RA (1995) The economics of increased carbon storage through plantations and forest management (Chapter 25, this volume).

Sedjo RA, Solomon AM (1989) Climate and forests. In: NJ Rosenberg, WE Easterling III, PR Crosson, J Darmstadter (eds) Greenhouse Warming: Abatement and Adaptation. Resources for the Future, Washington, DC, pp 105–119.

Solberg B, Hoen HF (1995) Economic aspects of carbon sequestration—some findings from Norway (Chapter 29, this volume).

Stewart RB, Maini JS (1995) Forests and global carbon management: a policy perspective (Chapter 31, this volume).

Strakhov VV, Filipchuk AN, Shvidenko AZ (1994) Structure of forest account works in Russia: current state and outlook. Presented at NATO Advanced Research Workshop, "The Role of Forest Ecosystems and Forest Management in the Global Carbon Cycle", held 12–16 September 1994, Banff, Canada, (Unpublished manuscript).

Woodwell GM, Ramakrishna K (1995) Forests and the human habitat: the case for building a global consensus (Chapter 34, this volume).

Concluding remarks

33 Summary

David T Price and Michael J Apps

Natural Resources Canada, Canadian Forest Service, Northwest Region,
Northern Forestry Centre, 5320 - 122 Street, Edmonton, AB T6H 3S5, Canada

Global forests are complex dynamic systems that both respond to, and create physical feedbacks to, the global environment. Systematic global change imposed upon the biosphere will therefore undoubtedly cause responses in forest ecosystems, with potential consequences that cannot be predicted in detail. These consequences will almost certainly contribute to further environmental and socio-economic impacts, of which some will be deleterious for the human habitat. Among these will be major changes to global forest carbon (C) dynamics, with regional shifts in C sequestration and release, linked to gains or losses in net ecosystem productivity (NEP). It is doubtful whether global NEP would be truly zero when averaged over the planetary surface and over long time periods—even in the absence of the human species. It is therefore virtually certain that, given the current situation of accelerating anthropogenic changes, global average NEP is also undergoing significant change. Resolving the magnitude and direction of this change however, even for specific geographic regions, will continue to be a complex and challenging problem.

Effects of climatic change on forest ecosystems and carbon dynamics

The global biosphere is clearly very sensitive to climatic perturbations (Fung, Chapter 2; Woodwell and Mackenzie 1994), but our present understanding of the interactions among changes in temperature, precipitation, elevated CO_2 concentration and nutrient dynamics limit our ability to make confident predictions of the precise responses of forest ecosystems to projected changes in climate. Individual responses of C dynamics to nitrogen perturbations, for example, are very variable, depending both upon the ecosystem type and environmental factors (Townsend and Rastetter, Chapter 3). For forested peatlands, changes in subsurface hydrology are potentially more important than warming *per se* as a factor influencing C cycling (Zoltai and Martikainen, Chapter 4). These factors and many others contribute to the challenges of predicting ecosystem responses, but some significant advances have been made in recent years. Increasingly sophisticated models indicate variable ecosystem responses to climate change. Thus, for example potentially significant biomass increases are expected in northern ecosystems currently limited by low temperature, while decreases are more likely at mid-latitudes in the short term (e.g., Kellomäki and Karjalainen, Chapter 5; Sykes and Prentice, Chapter 6). Models of physiological processes suggest that increased CO_2 concentrations may increase biomass productivity (or offset potential decreases). On the other hand, models of ecosystem level processes suggest that adaptation, changes in species competition and other factors complicate the situation so that the net responses may be short-lived (e.g., Kellomäki and Karjalainen).

Some of the greatest uncertainties arise in the assessment of boreal ecosystem C dynamics, especially in soils and wetlands. This is particularly worrisome because GCM projections generally indicate that some of the greatest changes in climate will occur in boreal regions,

NATO ASI Series, Vol. I 40
Forest Ecosystems, Forest Management and the Global Carbon Cycle
Edited by Michael J. Apps and David T. Price

and because these regions are being subjected to increasing industrial development. Large-scale disturbances greatly add to the complexity of C dynamics in boreal ecosystems, but important progress in understanding their effects has been made (e.g., Shvidenko et al., Chapter 12; Kurz and Apps, Chapter 14). Boreal forests are particularly susceptible to large wildfires (Stocks et al, Chapter 8) and to widespread insect pest outbreaks (Volney Chapter 7), both of which are strongly influenced by climatic conditions. Climate change projections for boreal North America suggest greater frequency and larger areas lost to fires and insect activity, raising the possibility of significant biospheric feedbacks. As a consequence of the asymmetry in rates of ecosystem change between mortality and regrowth (Kurz et al. 1994), C would tend to be released faster than it could accumulate—at least during the transitional period. Other direct anthropogenic factors, particularly atmospheric pollution (Galinski and Witowski, Chapter 15), can exacerbate and accelerate these feedback mechanisms. Such changes in boreal ecosystems may prove to be some of the more serious ecological consequences of climate change on global forests (see also Woodwell and Mackenzie 1994).

Forests that are currently sinks may become sources as the age-class structure alters in response to external changes (Kauppi, Chapter 16; Maclaren, Chapter 21). This phenomenon may already be occurring on a large scale for the major boreal regions (Shvidenko et al.; Kurz and Apps). Will similar effects begin to appear in other forest biomes?

Forested peatlands

Methane emissions from forested wetland areas can be expected to increase with warming where there is no associated change in the water table. The net impacts of climate change on peatland NEP (and hence CO_2 exchanges), however, are more difficult to predict with confidence. In general, where lowering of the water table occurs, emissions of GHGs from peat decomposition will be offset to some extent by increases in aboveground biomass production, and by reductions in methane production. In the short to medium term however, some authorities have suggested that the drying down of southern peatlands may occur more slowly than previously thought, because feedback mechanisms in the peat ecosystems tend to stabilise changes in the water table. Where melting of permafrost occurs in northern wetlands, greater increases in C accumulation would also be expected (Zoltai and Martikainen). Assessment of the net GHG exchanges induced by global warming at regional scales must also account for extension of peat-forming areas at higher latitudes which will compensate for losses of appropriate environments in more southerly locations (Zoltai 1994)

Human impacts and the missing carbon sink

Strong evidence for the existence of an unidentified terrestrial C sink has been derived indirectly from analyses of atmospheric geochemistry and its spatial variation (e.g., Keeling et al. 1989; Tans et al. 1990; Sarmiento et al. 1992). These studies indicate a need for terrestrial sinks, particularly in the land-dominated mid-latitudes of the Northern Hemisphere. Various causal mechanisms have been proposed—of which CO_2 fertilisation has been a favoured candidate for many years (e.g., Taylor and Lloyd 1992). The observed latitudinal gradients in CO_2 concentrations have lead many authors to conclude, however, that CO_2 fertilisation cannot be the complete explanation (Keeling et al. 1989, 1995; Tans et al. 1990; Enting and Mansbridge 1991; Friedlingstein et al. 1995).

The C flux required to balance the global budget is similar in magnitude to the net C release estimated to result from land-use changes—but of opposite sign. This leads Houghton (Chapter 10) to state two important conclusions if global terrestrial ecosystems are to account for the global budget discrepancy. Firstly, regardless of the causal mechanism, the total magnitude of the missing sink must be approximately *twice* the size of the currently observed discrepancy. Secondly, the sink is likely to be primarily associated with *natural* ecosystems (i.e., those not the result of human-caused changes in land use) because the dynamics of the directly modified ecosystems have already been taken into account.

It seems probable that the missing C sink will not be found in any single region or be due to any one mechanism. Thus, for example, Houghton notes that increased forest growth, estimates of which are derived mainly from inventory data, cannot be the sole explanation for a terrestrial missing C sink. The sink is more likely to be due to a number of different biological and abiotic phenomena which contribute to changes in ecosystem structure and function across several biomes. A number of alternative mechanisms for terrestrial C uptake are presented in the chapters of this book, which are summarised here briefly.

In the tropics, re-establishment of substantial forest biomass density may be occurring over large areas of land previously considered deforested, according to Brown's assessment of recent inventories (Chapter 11). Thus in these regions, where deforestation has been a major cause of past C releases, recovery of forest structure may now be adding to any enhancement of forest function—such as increased NPP due to CO_2 fertilisation.

Kauppi observes that harvesting removals have decreased, while average increment appears to have increased, in many temperate forests of Europe and the USA. He therefore suggests that current management trends, combined with changes in the global environment, have caused increases in forest NEP in much of the temperate zone. Hence both structural and functional changes appear to have caused these forests to act as a C sink. Increased increment may be due in part to eutrophication through air-borne pollutants, but whether this effect will continue—and for how long—depends on future population pressure and the achieved levels of environmental controls. Some forested areas in central Europe may already have become net sources of C due to severe pollution impacts (Galinski and Witowski).

Afforestation programs in Europe will not contribute significantly to the temperate C sink effect until early in the 21st Century. On the other hand, in New Zealand, for example, the created sink is already important and likely to increase for the next few decades (Kauppi; Maclaren).

In the boreal regions (Chapters 7, 8, 12, 14), the C budgets of natural forest ecosystems are markedly affected on time scales of decades to centuries by fluctuations in disturbance regime. Kurz and Apps have shown that significant variations in the large-scale C flux of Canadian forests took place during the 70-year period prior to 1990. Their results indicate significant changes in the C dynamics of Canadian boreal forests in the most recent 20 years, with the small C sink of the previous decades being replaced by a small net C source. These changes in the C dynamics were caused by a sharp change in the disturbance regime that began in the early 1970s (Kurz et al. 1995). Scenarios of future trends in disturbance regimes, reported elsewhere by these authors (Kurz and Apps 1995) indicate that the C dynamics of Canadian forests will change again during the next few decades. Kurz and Apps also suggest that similar dynamics may be affecting boreal forests in Alaska and Russia. Assessments of the vast areas of the former Soviet Union (FSU) by Kolchugina and Vinson (1993) yielded net C uptake estimates significantly greater than previous official ones (ECE/FAO 1992). The more recent work reported by Shvidenko et al., however, draws different conclusions and supports Kurz and Apps' hypothesis, while highlighting the uncertainties in such analyses.

Biomass productivity and mortality, with their associated influences on litter and soils, are fundamental to determining the C dynamics of forests. Thus Shvidenko et al. point to the urgent need for better knowledge of the processes governing biomass productivity and mortality, particularly in the forests of the FSU. Kurz and Apps point out however, that for boreal forests at least, the frequency and intensity of natural disturbance regimes are key factors influencing mortality at larger scales. As already discussed, these factors are certainly linked to climatic variations (Kurz and Apps), and could therefore contribute significantly to the global atmospheric CO_2 anomalies recently reported by Keeling et al. (1995). This notion also seriously questions the tacit assumption that the world's natural forest ecosystems are in steady state with respect to atmospheric C exchange. Drawing on Keeling et al.'s observations, it seems highly possible that cool and warm episodes, caused by factors such as El Niños or volcanic eruptions affecting planetary albedo, lead to coherent global-scale variations in the rates of uptake and release of atmospheric CO_2 by terrestrial ecosystems—of which large-scale changes in disturbance regimes could be an important component. When these variations continue for periods of several years or more, they result in marked changes in forest age-class structure, as demonstrated in Chapter 14, with significant impacts on global C cycling that persist for decades. We speculate that additional changes in NEP (associated with altered nutrient cycles for example) may amplify such changes, but evaluation of such effects immediately presents the scaling challenges mentioned in Chapter 1.

Carbon budget methodologies

Matthews (Chapter 19) notes an emerging trend toward greater comprehensiveness in C budget analyses to account for the dynamics of all important C pools. He finds that many of the conclusions drawn from individual assessments are sensitive to the C budget methodology adopted. It is therefore important to maintain a clear picture of the entire system (e.g., Figure 1.1) to avoid making errors in assessments. For example, much of the apparent disagreement over the potential for enhancing C sequestration in forest systems can be explained by differences in the analytical approach and C budget methodology employed (Matthews; Nabuurs, Chapter 20; Matthews et al., Chapter 24). Unambiguous assessment of the forest management influences on C sequestration requires clear definition and acceptance of systems conventions and analysis by scientists, foresters and policy-makers alike. Guidelines from IPCC (1993) for conducting national C budgets provide a useful starting point, but require improvement (Kellomäki and Karjalainen; Matthews et al.). The diversity of forest ecosystems, and of their environmental and socio-economic roles, remain as major obstacles to standardisation of C budget methods (Matthews; Nabuurs).

The use and management of forest products can potentially affect present and future anthropogenic contributions to atmospheric C levels. Carbon pools and fluxes in wood products can be significant in the context of both national and global forest ecosystem C budgets. They must be considered, therefore, when tracking C fluxes resulting from forest management and utilisation policies, including the use of wood as bioenergy (Heath et al., Chapter 22). In particular, consideration of direct and indirect *energy substitution* by wood products, can result in significant effects on the long-term C budget. Changes in wood utilisation could well be more important than changes in forest management for reducing anthropogenic C emissions in the long term.

Treating wood products as a C sink or source, however, requires care. If considered part of the terrestrial ecosystem, then wood products can only release C to the atmosphere, hence they should be treated as a C source. Increases in the total C stored in products may represent an increase of terrestrial C but it is only the *net* change (product additions minus releases) that can loosely be termed a 'products sink' (Houghton).

The current trend appears to be toward better integration of physical and ecosystem process information to enable the effects of climate variability, and its interactions with key factors such as atmospheric CO_2 concentration, site nutrient dynamics and exogenous disturbances, to be considered. Better knowledge of these processes is needed to understand the temporal record of past changes in the atmosphere (see also Chapters 2 and 3), and may prove essential for predicting the future. One of the most pressing challenges still remaining, is to be able to distiguish between *direct* and *indirect* impacts of human activities. Assessments of the effects on terrestrial C balances due to environmental changes, as distinct from those due to changes in land use, are only now being attempted. It has proved possible to quantify land-use change impacts reasonably well because the areas affected are generally documented, and the resultant changes in C density are relatively large. In contrast, the impacts of environmental change are often associated with relatively small changes in C density over much wider and more poorly defined areas. Separating direct land-use impacts from the indirect and inadvertent impacts of global change will help to isolate the human contribution to global C dynamics from 'natural' variability. It appears that a quantitative assessment of these distinct terms may be essential both to close the global C budget, and to predict future biospheric responses more reliably (Houghton).

Monitoring

The complexities of forest systems also present significant challenges for the monitoring of regional C reservoirs and fluxes. Monitoring change in natural and managed ecosystems is clearly required for a number of policy reasons at regional, national and international scales. It is also fundamental to obtaining improved regional carbon budgets and to linking these to global assessments. Part of the problem in obtaining accurate C budget assessments in many regions appears to be that statistics on past changes in land use have not been adequately ground-truthed. For example, land formerly classified as deforested may have recovered to the point where it now harbours significant biomass (Brown). Not accounting for such changes will lead to errors in assessing the current situation, including underestimation of regional C uptake and overestimation of land available for rehabilitation (see also Sathaye, Chapter 26; and Grainger, Chapter 27).

Better observations of intrinsic ecosystem fluxes (i.e., among biomass, forest floor litter, and soils), as well as of the transfers from forest ecosystems to the forest product sector, are also part of the required monitoring system. Current regional assessments however, point to additional key requirements, such as systematic monitoring of the extent and intensity of forest disturbances (Shvidenko et al.). Consistent monitoring of forest growing stocks and other indexes of ecosystem carbon storage is required, particularly in parts of the boreal and tropical regions. Detailed forest biomass inventories have typically been maintained in many temperate-region countries for some time, making the need for additional monitoring in these regions less acute (Brown et al., Chapter 17; Sathaye). Even in the more developed countries however, where maintenance of forest inventory data is routine, better documentation of the non-biomass carbon reservoirs in soils and detritus is almost certainly required.

Monitoring programs should also support our ability to predict future C flows and be capable of detecting ecosystem responses to global change. They must therefore be backed up by parallel development of process-based models in order to understand, and ultimately predict, forest ecosystem responses to global change at all scales (Fung). Assessments of the fate of soil organic C due to changes in disturbances, for example, can be improved both by systematic monitoring and better process understanding at local and regional scales (Shvidenko et al.). As a further example, changes in permafrost regions and wetlands—both being particularly vulnerable to climate warming effects (Zoltai and Martikainen)—can be monitored at various scales and the observations used to develop process-based models, carbon assessments and predictions of future change.

Use of forestry and wood products to mitigate greenhouse gas emissions

There appear to be two strongly overlapping ways in which forest resources may be used to mitigate anthropogenic carbon dioxide emissions. The first is to *sequester atmospheric* CO_2—either in forest ecosystems (including protection of existing forests) or in harvested products, or both. The second approach is *reduction of C emissions*, through the use of harvested products as direct or indirect *substitutes* for fossil energy sources. Both have an important role to play. In a given set of circumstances, the most appropriate method may depend on several factors, including the forest ecosystems under consideration and the time frame over which greenhouse gas (GHG) mitigation benefits are sought. Where wood production (for construction or as fuel) is not an important management objective, protection costs are low (e.g., because of a highly developed road network), and where maximum C sequestration in the shortest period is the goal, the best results are likely to be obtained by protecting the forest from disturbances or other losses, and allowing it to achieve maximum forest biomass density (Fischlin, Chapter 13). Conversely, where wood production is a primary objective and long term C benefits are sought, intensive management aimed at maximising output of *long-lived* wood products and/or *fossil energy substitutes* is likely to be the better policy. Evaluation of such strategies, and the many possible combinations, requires an accurate accounting of the C flows resulting from each of the various management options.

Sequestration of atmospheric CO_2 can be achieved by increasing the total C stored both in forest ecosystems and in the forest economic system (see Figure 1.1). Harvest withdrawals from the forest ecosystem reduce its C stock, but contribute to C storage in the forest economic system thereby increase that sector's influence on net C exchange (by increasing C in products and landfills or by reducing fossil fuel emissions through substitution). Hence, in assessing the net C sequestration in the complete forest sector, the balance between the proportion of total C stored in forest products (whose dynamics may be controlled by socio-economic factors) and the ecosystem (under the control of 'natural' processes) is a key consideration. Nabuurs estimates that the C stored in wood products typically represents about one-third of forest biomass C (excluding soil C) in managed plantations. In many extensively managed systems, however, the fraction retained in forest products will be much smaller (Price et al., Chapter 23). When energy substitution is not considered a factor, maximum long-term C storage in wood products, forest biomass and soils is achieved with long rotations, so as to maintain forest biomass as close to the site carrying capacity as possible (Kellomäki and Karjalainen; Fischlin). If energy substitution programs can be implemented, however, inclusion of these benefits in the analyses leads to a different strategy: significantly greater reduction of GHG emissions is obtained with short rotation, high productivity fuelwood plantations. Further increases in the mitigation benefits can be obtained by recycling of used wood products, either as construction material or for bioenergy (Kellomäki and Karjalainen; Schlamadinger and Marland; Nabuurs).

Management of forest for commercial timber exploitation may also bring significant increases in total ecosystem C storage compared to the unmanaged forest in certain instances. Enhanced growing stocks through silvicultural improvements is an obvious example, although in assessing the C benefits of such improvements, acounting of all vegetation biomass and not just the commercially valued components must be made. Price et al. also point out that increased C storage in both the forest ecosystem and in the forest product sector could be obtained through effective protection programs which *extend* the rotation period. This is particularly true if forest protection from natural disturbances can significantly increase the chances of stands achieving harvestable age and thereby more closely approach the site carrying capacity for biomass, litter and soils.

Over the longer term, the best mitigation strategies appear to be associated with the use of biofuels to displace fossil fuels, and the indirect substitution achieved through use of wood products in place of other, more energy-consumptive materials. Using forest resources in this way provides a cumulative and very long term C storage benefit over and above the storage in forest ecosystems and in forest products themselves. Although there can be significant C storage in existing or newly-established forest stands in the short term, sooner or later the rate of carbon uptake by any unharvested stand tends to zero, after which it provides no further net C sequestration. Harvesting, regrowth and the wise use of forest products can add to the ecosystem storage function (Schlamadinger and Marland; Maclaren; Heath et al.). Even where there is a short-term loss of C from newly established plantations (Nabuurs), biofuel strategies combined with other forest product uses can result in cumulative long-term C benefits (Schlamadinger and Marland).

The overall efficiency with which biofuels replace fossil fuels plays a major role in determining how significant will be the net gains. It is important therefore to estimate the additional GHG emissions associated with this aspect of forest sector activities. Full product life-cycle analyses including all energy costs and benefits, however, are difficult to perform. Documentation of the GHG emissions resulting from forest management operations, harvesting, wood processing and recycling of used materials is difficult to obtain (not least because of the proprietary nature of the information) and this leads to some uncertainty in estimating the true net benefits. The general consensus appears to be that forest sector emissions associated with product generation are relatively small compared to the increased sector storage, at least over one or two rotations (Kellomäki and Karjalainen; Schlamadinger and Marland; Nabuurs; Maclaren; Price et al.; Matthews et al.). One area that appears to have been overlooked is the potential energy use of methane from landfills derived from discarded forest products. (As an aside, the decay of products in landfills is itself poorly documented.) Assessing the longer term benefits of indirect substitution is also a difficult aspect of product life-cycle analysis still in its infancy—for example, how do the C fluxes associated with manufacture and recycling of steel and aluminium, compare to those resulting from harvesting, milling and recycling of wood products? Many of these issues are complicated by the need for comparative life-cycle analyses of different construction materials. To account for differences in the timing of C flows (see below), they also touch upon various economic factors (such as discount rates).

Current land use may also be an important consideration affecting net C balance in the long term. Forest management strategies which look attractive in C terms on agricultural land may be far less relevant for existing forest land. Scaling up from small-scale case studies to assess global mitigation opportunities has been complicated, and error prone, due to the fact that most studies have tended to focus on ecosystems of high productivity, where more data tend to be available. Such ecosystems are unrepresentative of available lands and the results cannot be extrapolated easily to larger areas of lower average productivity.

Socio-economic effects on forest carbon storage

The underlying causes of systematic changes in ecosystem C storage are largely socio-economic in all geographic regions. Though some are deliberate, many are inadvertent consequences of economic development. For example, it has already been noted that low levels of environmental pollution may enhance forest productivity, all other factors being equal (Kauppi). There is, however, a real danger that developing countries undergoing rapid economic growth, such as mainland China, may create high levels of industrial pollution with serious localised damage. A further aspect of this problem is the tendency for environmental stresses, both natural and anthropogenic, to cause loss of forest health, hence reducing their resistance to the effects of pest and disease organisms (Galinski and Witowski). The impacts of attacks by forest pathogens can therefore be greatly amplified by inadvertent human perturbations such as air pollution. Global change is likely to lead to increases in many such environmental stresses, with increasingly important socio-economic impacts on all forest ecosystems.

Extensive human-caused degradation of tropical forests has been occurring for more than 100 years (e.g., in tropical Asia), in addition to deforestation. Brown identifies the main causes as selective logging (sanctioned and unsanctioned), and temporary conversions to agricultural uses including slash-and-burn, grazing, and shifting cultivation. The net results are conversion of closed forest to open forest, increased fragmentation, and decreases in average biomass C density. These contribute to changes in both the temporal and spatial patterns of atmospheric C exchange.

Brown shows that in general, there is a steep response of forest biomass degradation to low population densities with a levelling off at higher densities. Part of this response is probably due to initial 'selective' removal of the relatively few large stems which significantly reduces total biomass density. There is however, evidence that some forests are becoming less fragmented and better managed, with resultant increases in biomass density. Higher population densities in urbanised areas may halt and reverse the degradation trend, as has happened in many developed countries that have progressed beyond the exploitive period of what Grainger terms the *national land-use transition*.

The transition from exploitation to conservation is typically associated with higher standards of forest management, where at least some level of sustainability is an objective. This is itself a socio-economic effect on forest C density, and deserves to be recognised as such. For example, afforestation programs can have important impacts on regional forest C dynamics in temperate and tropical regions (Kauppi; Maclaren; Matthews et al.; Sedjo, Chapter 25; see also Fearnside, Chapter 28). But when confronted with the potential impacts of environmental change, many of which will directly increase the future susceptibility of forests to natural catastrophes, the question must be asked: do we have adequate economic and scientific resources to succeed?

As a case in point, it is clear that preservation of the large C stores in boreal ecosystems is desirable. Vast areas of forest in the FSU, Canada and Alaska have very low population densities and limited infrastructure to support such large-scale protection efforts. A warmer climate is likely to increase losses in the remoter regions where naturally caused wildfires already destroy large areas of forest. Economic constraints are likely to reduce protection efforts rather than expand them (Stocks et al.). A similar difficulty in balancing socio-economic needs and ecosystem realities applies to protection from insect damage. Endemic insect populations impose powerful constraints on both structural and functional aspects of NEP and hence C dynamics (Volney), while epidemic insect outbreaks can cause significant disturbance events leading to marked changes in forest age-class structure (Chapters 1 and 14). Volney points out however, that despite significant effort, deliberate attempts at large-scale control of insect populations have not been particularly effective.

Duinker (Chapter 30) suggests that the technical possibilities for using forests in Canada (and by implication, in other developed countries) to mitigate GHG emissions are in any case fraught with problems because of conflicting socio-economic interests and other management objectives. In Canada at least, he argues, C sequestration is not high on the list of societal priorities for the managment of forest resources. More is potentially to be gained by better wood utilisation (i.e., as biofuels and in expanded production of long-lived products), and by managing woodlands in urban and agricultural environments. Although there seems to be general acceptance that C is likely to become an increasingly important future consideration that will affect other policy objectives, it is clear that forest management decisions will not and should never be based on C sequestration or offsetting benefits alone (Nabuurs; Price et al.; Matthews et al.; Duinker).

Wetlands

Conversion of peatlands, particularly to agricultural use, is a serious threat to both a valuable natural resource and an important C sink. Because such changes are largely irreversible,the most effective C management for peat-forming wetlands is preservation (Zoltai and Martikainen). A very high proportion of wetlands have already been exploited in Europe however, and the trend is expected to continue elsewhere unless concerted efforts are made to preserve remaining peat-forming environments. It is generally held that global change pressures could contribute to greater losses of northern wetlands in the future (Gorham 1991), particularly if climate warming drives agricultural production areas to higher latitudes (Zoltai and Martikainen).

Economic analysis of greenhouse gas mitigation options

Political awareness of the possible economic consequences of increasing atmospheric GHG concentrations has steadily increased in the last decade. This has led some to the optimistic view that forests and the forest industry can provide an effective means of offsetting anthopogenic emissions. Are these expectations too high? Are the costs prohibitive? The debate surrounding such questions has generated a need for better methods of assessing the economics of GHG mitigation options. Forestry economics has traditionally been plagued by the disparate time scales of (immediate) investment costs and (future) economic returns. Including GHG mitigation costs and benefits introduces another layer of complexity, not the least due to the difficulty of quantifying these costs and benefits (Sedjo; Sathaye; Solberg and Hoen, Chapter 29). Costs may be ascribed to the environmental damage *caused* by increased atmospheric CO_2 concentrations, while equivalent values can be awarded to any process which *avoids* or *reduces* this damage. Since living forests provide a means of sequestering atmospheric CO_2, any practice which increases the C stored will clearly add to their existing economic value (Sedjo; Solberg and Hoen). Simply put: if the C stored in wood also *adds* to its value as construction timber or other forestry output, then the GHG mitigation benefits can add significantly to the economic attractiveness of investing in forestry. Determining an accurate estimate of the true damage cost of global change, however, is beyond our present capability. Until this can be done, other economic indicators, such as calculated reductions in atmospheric CO_2 and C discounting, may be used (Sathaye). Solberg and Hoen also argue that discounting of C costs and benefits is necessary to permit meaningful comparison of the long-term C gains and losses in forestry options with GHG mitigation approaches in other sectors.

When damage costs and sequestration revenues are accounted for, several options in the forest sector are found to be surprisingly effective for mitigating C emissions. This is potentially true in all geographic regions, but particularly so in the tropics. For example, Sedjo identifies afforestation of marginal and unproductive agricultural lands in Argentina as a relatively low-cost approach to C sequestration. Where site productivity is high, the discounted present values of future C sequestration can be very attractive, although there may be significant associated opportunity costs.

Full and accurate economic comparisons of different GHG emission mitigation approaches require rigorous life-cycle analyses of C flows in products, and honest weighting of the anticipated socio-economic benefits. Sathaye argues that developed countries have better access to analytical tools and data and that insufficient effort is therefore being devoted to developing countries where it may be most needed. He identifies needs for realistic assessments of land availability, evaluation of true social benefits and opportunity costs, and agreement on rates of return for C discounting calculations. Non-monetary values, such as protection of wildlife habitat and biodiversity, must be explicitly stated, if not quantified, when assessing economic benefits. The uncertainty of climate change adds a further level of complexity to the economic assessments, pointing to the need for greater interaction between economists and ecosystem scientists.

Estimation of land availability is currently difficult because of inconsistency in classifications, overlaps in definitions and differences in assessment methods. Grainger criticises appraisals of forest sector mitigation options that rely solely on the biophysical capability of degraded land and technical assessments of afforestation potential. He argues that when cultural constraints are considered, the areas of land classified as available are likely to be greatly reduced. Hence, integration of the biophysical, socio-economic and political attributes of degraded lands should lead to more realistic assessments of afforestation potential. Additionally, long-term trends in land-use change will generally reduce land availability for afforestation projects still further. Grainger also adds the caution that rapid expansion of afforestation programs cannot be achieved in any case because of political and cultural inertia.

In many tropical areas, special difficulties arise when attempting to assess the socio-economic benefits of C sequestration projects, because of numerous conflicting interests surrounding land use. Fearnside explores many of these issues qualitatively. He believes some forest sector mitigation options proposed for tropical regions may have socio-economic benefits, but that many are harmful. The GHG-mitigation benefits of slowing deforestation have been under-estimated, while those of other options are frequently overestimated. Thus, Fearnside argues, plantations have greater social impacts and lower economic benefits than many proponents believe. Conversely, slowing deforestation will have many benefits, in addition to maintaining the environmental values of forests. If further land-clearing for non-sustainable agriculture is to be avoided, however, achieving sustainability of food production must be a high priority in these regions (Houghton; Brown et al.; Fearnside; Stewart and Maini, Chapter 31).

Sustainable development of global forest resources

It is becoming generally recognised that the Earth is undergoing changes due to human activity which are systematically reducing its capacity to support the expanding population. As Woodwell and Ramakrishna remind us in Chapter 34, although the essential role of the world's forests in maintaining this human habitat is often accepted intuitively, it is difficult to demonstrate scientifically. It is therefore even more difficult to convince policy makers of the need for action. Stewart and Maini identify two major policy challenges: (1) how to meet expanding demand for forest products without further degrading existing forest resources; and (2) how to mobilise scientific, fiscal and political resources to promote sustainability. The level of success in meeting these challenges will have direct impacts on the future role of forest ecosystems in the global C cycle. Increasing both the quality and quantity of global forest cover, through rehabilitation of degraded land, conservation of existing forests and implementation of sustainable land-use practices, are all essential if we are serious in our desire to use global forests to increase sequestration of atmospheric CO_2. These and other mitigation measures will require international political, technical and financial commitment and cooperation at all levels of human organisation—a major challenge for the international policy community.

Admittedly, it will be very difficult to persuade a global community of diverse political opinions to embark on vast concerted programmes to halt and reverse human degradation of the global environment, including preservation of the world's forests. Yet responsible action at the highest policy levels is almost certainly necessary if we are to have any chance of preserving global forests, and human habitats generally, from further damage. To succeed, such programmes must have widespread, popular support. Hence the implementation of such policies and actions must involve both democratic debate on the principles of sustainable forest development and continuous evaluation of its practice. A coordinated international programme can succeed, but only if multiple environmental and socio-economic values are considered. The development of such a programme must be pursued as a matter of great urgency.

ACKNOWLEDGEMENTS

We greatly appreciated the comments of RB Stewart on a previous draft of this summary chapter.

REFERENCES

ECE/FAO (1992) The Forest Resources of the Temperate Zones. Main Findings of the UN-ECE/FAO 1990 Forest Resource Assessment. ECE/TIM/60, United Nations, New York, 32 pp.

Enting IG, Mansbridge JV (1991) Latitudinal distribution of sources and sinks of CO_2: results from an inversion study. *Tellus* B 43: 156–170.

Friedlingstein P, Fung I, Holland E, John J, Brasseur G, Erickson D, Schimel D (1995) On the contribution of the biospheric CO_2 fertilisation to the missing sink. *Global Biogeochemical Cycles*, (In press).

Gorham E (1991) Northern peatlands: role in the carbon cycle and probable responses to climatic warming. *Ecol. Applic.* 1: 182–195.

IPCC (1993) Intergovernmental Panel on Climate Change, Draft Guidelines for National Greenhouse Gas Inventories. Vol 1–3.

Keeling CD, Bacastow RB, Carter AF, Piper SC, Whorf TP, Heimann M, Mook WG, Roeloffzen H (1989) A three dimensional model of atmospheric CO_2 transport based on observed winds: 1. Analysis of observational data. In: Peterson DH (ed) Aspects of Climate Variability in the Pacific and the Western Americas. *Geophys. Monogr.* 55, American Geophysical Union, Washington, DC, pp 165–236.

Keeling CD, Whorf TP, Wahlen M, van der Plicht J (1995) Interannual extremes in the rate of rise of atmospheric carbon dioxide since 1980. *Nature* 375: 666–670.

Kurz WA, Apps MJ, Beukema SJ, Lekstrum T (1995) Twentieth century carbon budget of Canadian forests. *Tellus* 47B: 170–177.

Kurz WA, Apps MJ, Stocks BJ, Volney WJA (1995) Global climate change: disturbance regimes and biospheric feedbacks of temperate and boreal forests. In: Woodwell GM, Mackenzie FT (eds) Biotic feedbacks in the global climate change system will the warming speed the warming? Oxford University Press. Oxford, pp 119–133.

Kolchugina TP, Vinson TS (1993) Carbon sources and sinks in forest biomes of the former Soviet Union. *Global Biogeochemical Cycles* 7: 291–304.

Sarmiento JL, Orr JC, Siegenthaler U (1992) A perturbation simulation of CO_2 uptake in an ocean general circulation model. *J. Geophys. Res.* 97: 3621–3645.

Smith TM, Shugart (1993) The transient response of terrestrial carbon storage to a perturbed climate. *Nature* 361: 523–526.

Tans PP, Fung IY, Takahashi T (1990) Observational constraints on the global atmospheric CO_2 budget. *Science* 247: 1431–1438.

Taylor JA, Lloyd J (1992) Sources and sinks of atmospheric CO_2. *Austral. J. Bot.* 40: 407 418.

Woodwell GM, Mackenzie FT (1994) Biotic Feedbacks in the Warming of the Earth: Will the Warming Feed the Warming? Oxford University Press, New York, 416 pp.

Zoltai SC (1994) Impact of climate change on Canadian peatlands. In: Grünig A (ed) Mires and Man, Swiss Fed. Inst. for Forest, Snow and Landscape Res., Birmensdorf, Switzerland, pp 315–319.

34 Epilogue: Forests and the human habitat: the case for building a global consensus

George M Woodwell and Kilaparti Ramakrishna

The Woods Hole Research Center
P.O. Box 296, Woods Hole, MA 02543, USA

FORESTS AND THE GLOBAL ENVIRONMENT

In the midst of the heady political changes that have marked the early years of this decade, including the end of the cold war, the reunion of the two Germany's, and the rekindling of dormant ancient hatreds and ambitions in Yugoslavia, Rwanda–Burundi and elsewhere, the world has discovered a series of emergent global environmental problems for which there is no precedent and no cure short of a degree of global co-operation never previously sought or achieved. Success in addressing them offers the world virtually unlimited prospects for further improvement of the human enterprise; failure offers a bleak future beset by intensified tribal clashes over the bare necessities of life as the human habitat becomes progressively impoverished.

In acknowledgement of this dilemma, articulated so sharply as the need for *sustainable development* by the Brundtland Commission, the United Nations Conference on Environment and Development attempted to prepare three international conventions for adoption at the June 1992 meetings in Brazil. The topics were the atmosphere, biodiversity, and forests. As we outline below, success in addressing the atmosphere and biodiversity depends on success in addressing forests, the topic on which the least progress was made.

The most important effort there, and the greatest success, led to the advancement for signature by all nations of a Framework Convention on Climate Change. The Convention was substantially weakened by opposition from the US under the Bush administration and as a result contains no quantitative goals or timetable for progress. It does, however, state the purpose as

> *"...stabilisation of greenhouse gas concentrations...at a level that would prevent dangerous anthropogenic interference with the climate system...to allow ecosystems to adapt naturally...to ensure that food production is not threatened..."*

Those purposes, if realised, will ultimately require abandonment of fossil fuels as the source of energy driving industrialisation. The sooner that transition occurs, the easier it will be and the greater the margin of security. The transition will also require a system of stabilising the large pool of carbon held in forests and their soils globally. Forests and atmosphere are intimately linked ... and ever will be.

So, too, forests are the major reservoir for life on land, and any effort at protecting biodiversity must protect forests if it is to be effective. The treaty on biological diversity was aimed not at that purpose, but at offering protection to nations or other interests that might supply genetic material for commercial exploitation. It missed the central point, the need for moving quickly internationally to avoid further systematic, cumulative and irreversible biotic impoverishment of the Earth as a whole. That process is far more general and comprehensively threatening than the loss of species, serious as that process is. Those drafting that treaty overlooked

NATO ASI Series, Vol. I 40
Forest Ecosystems, Forest Management and the Global Carbon Cycle
Edited by Michael J. Apps and David T. Price

the lessons of history, the fate of at least a score of garden-empires such as those of the Tigris and Euphrates, Minoa, Greece, Rome, and ancient Mexico that have crashed successively as biotic impoverishment in all its various forms—in particular through the destruction of forests—overtook and destroyed political ambition and human potential (Perlin 1989).

When preparations for the Rio Conference turned to forests, a treaty proved impossible. There was insufficient agreement as to the purpose of a treaty, insufficient understanding of the role of forests in maintaining the human habitat, and undisguised suspicion of the objectives of the North in advocating an international convention that might have profound effects on the South's exploitation of what the South envisions as vast resources in tropical forests. The best that could be advanced by the Rio Conference was a set of *Non-Legally Binding Principles*, a simple exhortation that carried little weight. The principles did, however, demonstrate the need to expand the discussion of forests to put it again on the international agenda. The preparations for the Rio Conference foundered on the inability of official national delegates to reach beyond national interests to address the emergent international and global issues effectively. Our objective is to describe a specific effort to advance a global consensus through, apart from other initiatives, an independent World Commission on Forests and Sustainable Development (WCFSD).

EMERGENT GLOBAL ENVIRONMENTAL ISSUES CONCERNING FORESTS

There are compelling reasons for early progress on all three topics addressed in the Rio Conference despite the limitation on progress there. Time is short in a world whose human population can double in 35 years and whose primary resource, a living habitat, is threatened globally through rapid climate change combined with a host of cumulative disturbances including the accumulating toxic effects of the wastes of a crude but still burgeoning fossil-fuel powered technology. But the issues converge on forests because of their role as a large biotic flywheel that keeps the Earth a place suitable for life and as a habitat for people.

The failure in Rio to address global interests in forests effectively in contrast to the progress made on climate suggested that the extensive scholarly review of climate that had occurred over two decades and more carried great weight among the nations. So, too, had the report of the World Commission on Environment and Development. A recent review by Alan T Durning (1993) of the role of forests in human affairs has shown that while much is known about forests, the information has emphasised private interests in timber and profits; public interests in the human habitat have received much less attention from science and in public affairs. There is a clear need for a far more extensive review of the role of forests in further economic development and in the essential shift toward long-term management of the human habitat.

This need for a review of the role of forests has been clear throughout human history (Perlin 1989) but it became especially conspicuous during the preparatory meetings months before the Rio Conference. To examine the question of whether and how to proceed, a conference with representatives of the FAO, the UNDP, various governments, non-governmental organisations, and scientific and scholarly institutions was held in Woods Hole in October of 1991. The conference report (WHRI 1991) outlined the need to identify the newly emergent global interests in forests and to set them in the context of classical interests in commercial exploitation. Two realms require strengthening now as in the past: the technical and scientific analyses of the importance of forests in the biosphere as a whole with special emphasis on the human habitat; and the political innovations that might be appropriate in defending these public interests as opposed to commercial interests. The special contemporary aspect of these issues is that they are now global and there is no room for failure. The conference recommended the establishment of an independent commission in the model of the World Commission on Environment and Development chaired by Gro Harlem Brundtland of Norway or an intergovernmental panel in the model of the Intergovernmental Panel on Climate Change (IPCC). The purpose, in either case, would be to advance forest issues as part of the mainstream of the current international developments in addressing the public interests that are now emergent globally.

The world community is now concerned about the demise of natural forests globally. Recent estimates show that about 17 Mha of forests are destroyed annually in the tropics. The last of the primary forests of North America are being destroyed currently and extensive inroads are being made on boreal forests across Asia. The loss of forest results in high social, economic, and environmental costs, with serious local, regional, and global implications. The most important global concerns are:

1. the warming of the Earth through the accumulation of heat trapping gases, especially carbon dioxide and methane, in the atmosphere;
2. progressive, irreversible biotic impoverishment;
3. soil erosion and desertification; and
4. loss of cultural diversity.

Large gaps exist in both our knowledge of these topics and in our ability to define long-term solutions. Our understanding of these issues is further complicated by the dynamic and complex interactions that occur between the natural and socio–economic processes.

The warming of the Earth and the role of forests

The review that led to the above conclusion was extensive. It considered, among other issues, the warming of the Earth that is clearly to be expected as the effect of current additions of heat-trapping gases are felt. The warming is a general and increasing burden on civilisation. Its major effects include the destruction of forests and other natural communities, as well as the expansion of the arid zones into the currently arable land of the Northern Hemisphere and the destabilisation of climate globally.

Changes in the area of forests are both cause and effect of the warming. The major cause of the warming is the accumulation of carbon dioxide and methane and other heat-trapping gases in the atmosphere. Carbon dioxide is responsible for about 50% of the total warming potential of the heat trapping gases added to the atmosphere by human activities. About 6 Pg carbon are released annually as carbon dioxide through burning fossil fuels. Another approximately 1.6 Pg are released by destruction of forests globally.

A less generally recognised effect of the warming, recently emphasised but long known (Woodwell and Mackenzie 1994; Woodwell 1993), is the potential for any general trend of temperature change to trigger biotic changes that augment the change. As a warming continues, the organic matter in forests and soils decays at a more rapid rate and releases additional carbon dioxide and methane into the atmosphere. A cooling, whatever the cause, will reduce the rates of decay and the release of these gases to the atmosphere. The process has been observed over the past several years in response both to the extraordinary warming experienced during the 1980s and throughout 1991, a period in which the rate of accumulation of carbon in the atmosphere increased to as much as 5 Pg annually. As the climate cooled in the following years in response to the introduction of particulate matter including sulphate aerosols into the stratosphere from the 1991 explosion of Mt. Pinatubo in the Philippines, the rate of accumulation of carbon in the atmosphere also dropped, as would be expected. The Earth appears now (1995) to be warming again and the rate of accumulation of carbon is expected to increase again as well. The result of all these factors has been the annual accumulation over a decade and more of 3–4 Pg carbon in the atmosphere as carbon dioxide.

If we were to decide to stabilise the carbon dioxide content of the atmosphere immediately as the first step in reducing human influences on climate, we would have to remove 3–4 Pg carbon from the current annual release. We have three tools: a reduction in the use of fossil fuels; a reduction in deforestation; and the possibility of restoring forests to deforested land, thereby accumulating carbon in trees and soils for the period during which the forest is expanding in area and stature. There is no possibility of avoiding, even temporarily, restrictions in the use of fossil fuels. The immediate reduction necessary in use of fossil fuels would approach 50%, no matter what steps were taken in management of forests.

Delay increases the difficulty in the transition because the warming, as it proceeds, speeds the further release of carbon from soil by speeding the decay of organic matter. The problem is especially important in the middle and higher latitudes of the Northern Hemisphere where there is a large pool of carbon in the soils of temperate zone forests, the boreal forests, and in the highly organic peats of bogs and tundra. This effect, now gradually being recognised as underway in response to the warming of the last two decades, has the potential for overwhelming the human capacity for reducing releases enough to stabilise the composition of the atmosphere. The longer the delay in reducing the accumulation in the atmosphere, the greater the difficulty of control. The problem adds a new dimension to the urgency attached to progress in arriving at an international agreement on forests consistent with the needs of forest dwellers and others who depend on forests for food and fibre, with the continuing interests of all nations in commerce, and with efforts in controlling the warming of the Earth.

Biotic impoverishment

Public interests lie in a stable human habitat that retains the capacity for providing for the diverse needs of industrial and agrarian human activities which are increasingly mutually dependent. What is too commonly overlooked is the fact that both are dependent on the continued functioning of a biosphere that provides, substantially free of human effort, the essential needs for life. That circumstance is now in question globally, threatened by global warming, by toxification of the human habitat, by the progressive loss of plants and animals and the services they provide. The process has been seen as a loss of *biodiversity*, but it is more than a mere loss of species. It is a loss of the potential for support of life, better recognised as progressive biotic impoverishment. The change is not sudden—the transit of a threshold, at least as far as we can tell at the moment—but a continuum, a progressive, systematic, cumulative loss of structure and function within the human habitat: the accumulation of barren land, no longer suitable for agriculture because of salinization or toxification or erosion to rock; the loss of whales and fisheries, tree species and whole forests; as loss of the potential for a family to survive easily on the resources of its immediate environment. It is the force that has generated the immediate crises of Somalia and El Salvador and Honduras and Haiti and the continued migration from the farms of Brazil to the *favellas* of Rio.

Unfortunately, in a world that appears to accept continued, unbridled growth in the human population, there is little vision as to the alternative. On the other hand, the causes of biotic impoverishment are known and measurable and curable. While there are many causes, one of the most threatening and certain is the warming of the Earth.

Soil erosion and desertification

In many areas throughout the world, uncontrolled deforestation results initially in a loss of soil fertility and in soil erosion, and eventually in desertification. Although this problem is common in many countries, it is becoming a serious problem in the Sahelian countries in sub-Saharan Africa and in many developing countries in Latin America and Asia. Soil erosion and desertification impede sustainable agriculture and undermine the food security in many developing countries. Without conservation measures and better management of forests, this problem will be exacerbated, especially as population increases significantly over the next several decades.

Loss of cultural diversity

As many as 500 million people throughout the world depend on forests directly for part or all of their habitat and livelihoods. Many of these are forest dwellers. Others live in adjacent areas but depend on forests for food, fodder, and fibre. Many, perhaps most, have special knowledge of these forest ecosystems and how to use them on a sustainable basis. Their special knowledge is an important resource both locally and for all of humanity, knowledge that is in jeopardy in proportion as the forests are in jeopardy.

Some requirements for management: scientific and political

The warming of the Earth can be controlled only by limiting—ultimately abandoning—the use of fossil fuels. There can be no solution, interim or long term, without managing forests globally in such a way as to balance the global carbon budget. Management of forests and all other resources is, of course, all local, but the sum of local management must be a global solution. The challenge is substantial. What will constitute a *solution*? Answers require, unfortunately, a broadly held understanding of complicated details of the functioning of the biosphere and in broadly acceptable political innovations not yet imagined to accommodate those details. The circumstance is one that appears to require a broad consensus derived from research. Such a consensus was available for the Intergovernmental Panel on Climate Change in 1988 when the Panel was established. The consensus was the product of research over a century supplemented by a score of recent international meetings and reviews by the scientific community. The political community, in contrast, was unprepared for the challenge and remains so on this and other topics.

No consensus on forests was possible before or at the Rio Conference. Tensions between North and South were too great. Facts were few and are still disputed. Issues were tortured and all hope of progress denied by what appeared to be outrageous assumptions on the part of some of the less developed countries, abetted by what was clearly outrageous behaviour on the part of the United States. The United States was advocating restraint in the destructive exploitation of forests elsewhere, but unwilling to bridle the exploitive interests in the Pacific Northwest and in Alaska, where the last of the largest trees on the continent outside a few parks remained. There was a clear need for sober reconsideration and analysis outside immediate public scrutiny and the vagaries of politics. All indicated the need for a new forum, for new information, new reviews and much more discussion. The suggestion that an independent World Commission on Forests be established, as suggested at the Woods Hole conference, opened these possibilities, was well received then and has been well received subsequently.

By addressing the issues outlined in this section, the global focus will yield several bases for addressing critical questions of how much forest is necessary in the world, and how these forests should be used and managed. Acquiring these data will require new research, new syntheses of existing data and experience, as well as dissemination of the information.

THE LOCAL ISSUES AND WHAT MUST BE DONE

Whatever the global considerations, forests are managed locally—locally taken here to mean at the community, national, or regional levels. Moreover, the socio–economic benefits of forests and forest resources (or costs from their loss) are felt most profoundly and directly at the local level. For tens of thousands of years, in tropical, temperate, and boreal zones, people have been dependent on forests for basic needs of fuel, food, medicine, and shelter.

The current crisis of forests is the product of domestic and international industrialisation and commercialisation of forest products, and a continuous search for new agricultural land and fuelwood. Forests are now intimately woven into the economies of virtually all nations, and the immediate value in commerce is what commonly dominates decisions in management. National and international policies affect daily decisions, which in turn lead to both local and global consequences.

The issue is not only trees for their commercial and economic value; it is also forests, natural forests of each region because these are the assemblages of plants and animals that have demonstrated their capacity for survival in the long term in that place. The survival and health of diverse species directly affect the well-being of all people, whether forest-dwellers or not.

A challenge facing any global effort in addressing the crisis of forests, be it through an independent commission or through intergovernmental channels, will be finding mechanisms of communication among local, regional, and international levels. Filling in gaps of knowledge alone will not lessen deforestation. Information generated through the research agenda must return to the individuals and governments which daily decide the fate of forests. Participatory planning, education, and extension programs will be necessary to develop concrete implementation strategies. It is important to carry out comparative studies of common elements of successes and failures across localities and regions.

PUBLIC AND PRIVATE PURPOSES: INDEPENDENCE IS ESSENTIAL

Forests perform many functions that are commonly overlooked or ignored, sometimes at great cost. They stabilise the landscape against erosion, control the flows of water in streams and groundwater, control the temperature and chemistry and biology of water, maintain the habitat of the plants and animals of naturally forested regions (biodiversity), and influence the accumulation and retention of the nutrient elements essential for life. Now, suddenly, we recognise that forests, globally, not only influence the energy budgets of the human habitat but also affect the composition of the atmosphere profoundly. All these functions and more are widely recognised as *public service functions*. They are commonly taken for granted as beyond human influence or the necessity for concern or control, yet they are now clearly in jeopardy locally and globally.

Private interests are far more parochial, easily focused and defended. They are often completely defined in the one-time harvest of fibre for short-term profits. Protection of the public interests, especially interests in the public service functions, is the job of government, but the temptations of commerce often prove corrupting, even to the point where the destruction of forests is advanced as the purpose of government and the process is not only defended but subsidised. Such is the case on certain of the publicly-owned lands of the U.S. at present. Between these two extremes there is the simple fact that 500 to 1000 million people depend directly or indirectly on forests for their habitat and livelihood.

The linkage between what is called *biodiversity* and forests is conspicuous. So is the uncertainty as to how to proceed effectively in both the continued use of forests and in their conservation for future use. If forests are important to the normal functioning of the biosphere, a question far from resolved in the eyes of many, how important are they and how rigorously must they be preserved? Where are forests needed? What forests and for what purpose?

These are technical questions whose answers have profound political and economic consequences in a world that is progressively packed with people. The scientific community, preoccupied with its own interests, not entirely public, has neglected such topics and has little to say at the moment. A conservative judgement, based on first principles, might land heavily on the side of preserving public interests at the expense of the private, but the surge of *neo-conservatism* now gripping the world and especially the United States, does not embrace such thinking. (The term *neo*-conservatism is used as opposed to the more classical conservatism implied by the word itself and embodied in the actions of political leaders such as Theodore Roosevelt and many others who developed a profound respect for the need to protect nature from progressive destruction in the name of human progress.)

The challenge is substantial. Scientists do not have a consensus, or even strong advice, on the importance of forests, the fraction of the landscape that would be forested in a landscape designed for sustained use in support of people over centuries. We do know what has failed and why. The keys to success are often lost in the scramble for short-term advantages justified on the economists' assumptions of the *exchangeability* of resources based on prices. The evidence is that the free market does not protect scarce resources but simply raises its price and assures that the resources will be destroyed as those who can afford a luxury indulge themselves down to the very last specimen. The topic will require substantial review in a world where there are no new places to go for fresh habitat when the last one has been exhausted. The facts argue for innovations in economics and government not yet considered.

CONCLUSIONS

The issues surrounding the conservation and better utilisation of forests globally obviously require continuing review, both in science and in government. The World Commission on Forests and Sustainable Development has been proposed to provide sufficient opportunity for review of both realms. Scientists have not yet defined in any quantitative or otherwise compelling way the importance of forests globally in maintaining the human habitat, a technical and scientific question that underlies all current discussions of forests. The widely accepted general assumption that forests are good and necessary has never been adequate to preserve them (Perlin 1989), nor have the quality and continuity of the human habitat been equated so clearly with the continuity of forests globally. Secondly, it is now clear that innovations in international and national governance will be appropriate if we are to be successful in protecting the human habitat from further erosion. While the changes may be global, the causes and effects are local and the solutions will require agreements as to how best to manage locally to avoid an apparently more remote but real and threatening communal problem.

It is the emergence and general recognition of such issues of the public interest globally that mark this moment as different from any previous moment. The Earth is clearly undergoing changes due to human activities that are accruing as systematic reductions in its capacity for sustaining the present civilisation. A new focus on the details of the preservation of the human habitat is appropriate. The technical and scientific challenge is straightforward and simple by comparison with the political, but the scientific community is barely in the earliest stages of addressing it. Both will require careful cultivation outside the normal channels of formal intergovernmental meetings. The political community found how difficult the ground is in the preparations for the Rio meetings when talks on forests simply collapsed. There was no agreement, or even insight, as to what a convention might be expected to offer. The circumstance is not different now. A recent week-long meeting of scholars and political representatives arranged by the new Center for International Research on Forestry (CIFOR) on a ship in the Java Sea in December 1994 to address research on international forest issues was paralysed by narrow political considerations. Official governmental representatives could, and did, exercise veto power over the agenda. One purpose, defended adamantly by one or more representatives in a meeting that was focused innocently enough on plans for research, was to deny that forests have any attributes that can be said to be *international* or *global.*

Such representations prevented progress leading to the Rio Conference and can be expected to prevent free discussion in other official meetings of nations, including meetings under the auspices of UN agencies such as the Food and Agriculture Organisation and the Commission on Sustainable Development. The recent experience under the auspices of an organisation as potentially independent as CIFOR simply confirms the need for an independent agency competent to open and explore any topic of interest, free of national dogma or purpose. An independent World Commission on Forests[1] is but one of several steps needed to stimulate the scholarly and political reviews now needed.

ACKNOWLEDGEMENTS

The work in this article has been supported in part by grants from the W Alton Jones Foundation, the Rockefeller Brothers Fund, the Pew Charitable Trusts, the John Merck Fund, the John D and Catherine T MacArthur Foundation, and the Woods Hole Research Center.

REFERENCES

Perlin J (1989) A Forest Journey. Norton WW, New York, 445 pp.

Durning AT (1993) Saving the Forests: What will it take? Worldwatch Institute Paper #117, Worldwatch Institute, Washington, DC, 51 pp.

Ramakrishna K, Woodwell GM (1991) Report of the Workshop on the Conservation and Utilization of Forests, 17–20 October, 1991, Special Report, The Woods Hole Research Center, Woods Hole, MA, 26 pp.

Woodwell GM (1983) Biotic effects on the concentration of atmospheric carbon dioxide: A review and projection. In: Changing Climate, NAS Press, Washington, DC, pp 216–241.

Woodwell GM, Mackenzie FT (1994) Biotic Feedbacks in the Warming of the Earth: Will the Warming Feed the Warming? Oxford University Press, New York and London, 416 pp.

[1] The World Commission on Forests and Sustainable Development has been established by a decision of the InterAction Council of Former Heads of State and Government, which is chaired by former German Chancellor Helmut Schmidt. The InterAction Council has asked and obtained the consent of the Honourable Ola Ullsten of Sweden and the Honourable Emil Salim of Indonesia to be the co-chairmen of the Commission. The Commission is expected to complete its deliberations and report by the 1997 review of the United Nations Commission on Sustainable Development (UNCSD).

Glossary

Afforestation: establishment of forest on land previously used for some purpose other than growing trees. Differs from *reforestation* (q.v.).

Agroforestry: combination of trees (of natural or planted origin) with other trees, arable farming, horticulture or grazing.

Biodiversity: range of organisms (life-forms, phyla and/or species) inhabiting a particular site, ecosystem or landscape. Though difficult to quantify, biodiversity is often used as an index of ecosystem stability, and as a means of monitoring changes therein.

Biosphere: portion of the Earth and its atmosphere capable of supporting life. The parts of the global carbon cycle that include living organisms and derived organic matter (e.g., litter, detritus, soil).

Bioenergy: energy derived from the combustion of biological material (not including fossil fuel) which is assumed to be renewable through regrowth. In this book, *peat* (q.v.) is not considered a source for bioenergy because it is not renewable in an appropriate time frame.

Business-as-usual scenario: particular *scenario* (q.v.) which defines conditions associated with continuing into the future what has been done in the past, usually applied to human actions. Often used as a base-line against which to compare the possible outcomes of alternative scenarios, e.g., effects of management actions or global change on the present-day biosphere.

Carbon balance (carbon budget): a conceptual mass-balance equation accounting for inputs, outputs and storage of carbon (and its compounds) in the *reservoirs* (q.v.) or compartments of a biological system and its physical environment. The system boundaries may be drawn at many scales, from individual organelles (e.g., chloroplasts), through single organisms (e.g., individual trees), to ecosystems (e.g., forest stands, forest canopies, lakes) and entire geographic regions including the planet as a whole. Some authors in this volume use the term *carbon balance* to indicate the net exchange of a terrestrial system with the atmosphere such that a positive balance implies net carbon uptake by that system.

Carbon cycle: the continuous transfer of carbon among the main *reservoirs* (q.v.) or *pools* (q.v.) inter-connected by exchange pathways. At the global scale, the four main reservoirs are: the atmosphere, terrestrial *biosphere* (q.v.), oceans and sediments (including fossil fuels). Each of these global pools may be subdivided into smaller pools ranging from individual communities to the total of all living organisms.

Carbon flux: the rate of mass transfer of carbon between two *reservoirs* (q.v.).

Carbon pump hypothesis: theory that annual *net primary production* (q.v.) contributes to a small but continuous net accumulation of carbon in other components or reservoirs such as forest soils. Also applied to the use of forests to manufacture wood products which may provide longer-term storage of carbon than would occur in unharvested stands.

Carbon stock: the absolute quantity of carbon stored within a pool at a specified time.

Clear-cutting: common forest logging practice where all trees that remain standing on the site at the time of final harvest, are felled at once, such that substantial areas (i.e., not small patches) are devoid of trees until regeneration occurs.

Climate change: the changes in Earth's climate that are expected to be caused by the *enhanced greenhouse effect* (q.v.). This is also referred to as *global warming* but includes changes in precipitation, circulation patterns, and other climate variables besides temperature. Separate and additional effects are associated with changes in the chemical composition of the atmosphere (e.g., *CO_2 fertilisation*, q.v.).

Climate variability: natural changes and variations in Earth's climate that are not associated with the *enhanced greenhouse effect* (q.v.) or other consequences of human activities.

CO_2 fertilisation: increased photosynthetic uptake of atmospheric CO_2 by vegetation due to the steeper leaf–air CO_2 concentration gradient that arises as a result of increased ambient CO_2 levels. Also used to refer to hypothesised changes at larger scales that may have ecological significance.

Deforestation: removal of forest cover followed by conversion of land to another use such as grazing land or urban expansion. Differs from normal forest harvesting practices such as *clear-cutting* (q.v.), in that the removed forest is not regenerated (by natural or artificial means) and felled timber is usually burned or left to decompose on-site rather than used in the manufacture of wood products.

Direct substitution: use of harvested forest biomass as a direct, renewable source of energy, replacing the combustion of fossil fuels. Such substitution can be regarded as directly reducing net emissions of CO_2 to the atmosphere.

Disturbance: any relatively discrete event in time that disrupts ecosystem, community or population structure and changes resources, substrate availability or the physical environment. For forests, disturbances include events such as wildfire, insect epidemics and wind-throw.

El Niño event: large-scale, coherent variations in ocean current off the west coast of South America during January–March. Occasionally leads to major and prolonged changes in sea surface temperature, and a collapse of the trade winds over the Pacific Ocean. These effects sometimes result in extreme climatic events in other regions, and consequent impacts on global biospheric carbon exchange.

Enhanced greenhouse effect: increased heating of Earth's surface resulting from increases in the atmospheric concentration of *greenhouse gases* (q.v.) due to human activities.

Forest: area of tree-covered land typically consisting of hundreds or thousands of individual *stands* (q.v.).

Forest degradation: temporary or permanent reduction in species diversity, ecological integrity, productivity and renewal capacity, of a forest ecosystem. It may be manifested as changes in three measurable physical indicators: *vertical structure*, *vegetation density*, and *species composition*, which can be used as the basis for a taxonomy of degradation.

Forest floor: surface material covering the mineral soil under forests. It includes coarse woody debris; a surface *litter* layer of recently dead leaves, needles, twigs and understory detritus; a middle *fermentation* layer of somewhat decomposed litter; and a lower *humus* layer consisting of well-decomposed plant material (humus). Such layers occur in various proportions, according to vegetation type, stand condition, soil moisture and soil nutrient conditions.

Forest product: commodity derived from harvesting of forests, generally restricted to *wood products* (q.v.), *fuelwoods* (q.v.), and *peat* (q.v.) but may include other organic materials such as latex, cork and mosses, as well as pharmaceuticals and food items (of animal and plant origin).

Forestation: any land management procedure aimed at establishing new forest on land not currently under tree cover. Includes *afforestation* (q.v.) and *reforestation* (q.v.).

Fossil fuel: any organic fuel derived from carbonaceous sedimentary deposits of organic material accumulated during the geologic past, including coal, crude oil and natural gas. Human extraction and burning of fossil fuels is the major cause of increases in the atmospheric concentration of CO_2 and other *greenhouse gases* (q.v.) during the past two centuries.

Fuelwood: Forest biomass used as *bioenergy* (q.v.), usually harvested for this specific purpose.

General Circulation Model (GCM): physically-based dynamic simulation of the planetary atmosphere (AGCM) or oceans (OGCM) which predicts horizontal and vertical distributions of meteorological quantities (temperature, pressure, atmospheric vapour pressure, etc.) as a function of time.

Global change: integrated impacts of human influences on the global environment, including changes in land use due to demographic, socio-economic and technological pressures, and changes in atmospheric composition which may have direct impacts on the *biosphere* (q.v.) including any resultant alterations in climate.

Greenhouse effect: warming of the planetary surface and troposphere due to downward longwave (4–25 μm waveband) radiation from the atmosphere. The presence of *greenhouse gases* (q.v.) in the atmosphere causes it both to absorb upwelling longwave radiation emitted by the surface, and to re-radiate it, both downwards to the surface and upwards (into space). This process maintains the global average temperature approximately 32 K warmer than it would be in the absence of the atmosphere. The term derives from the conceptual similarity to the glass of a greenhouse, which also absorbs longwave radiation, but is misleading because the warming effect in a greenhouse is due primarily to the reduction of convective losses by the enclosure.

Greenhouse gas (GHG): atmospheric gas constituent which absorbs electromagnetic radiation in the longwave band (approximately 4–25 μm wavelength), and hence contributes to radiative warming of the planetary surface by the atmosphere (the *greenhouse effect*, q.v.). Naturally-occurring GHGs include water vapour, CO_2 , methane and ozone. Since the industrial revolution when *fossil fuel* (q.v.) consumption became significant, anthropogenic emissions of GHGs such as CO_2, methane and chlorofluorocarbons, have contributed to an *enhanced greenhouse effect* (q.v.).

Gross Primary Production (GPP): total plant biomass production that occurs from gross photosynthesis during a defined time period, usually measured on a unit ground area basis.

Indirect substitution: use of manufactured wood products to replace equivalent items made from other materials such as aluminium, steel, concrete and plastics and which generally require more energy for their production and/or disposal. Where indirect substitution is practical, associated reductions in fossil fuel consumption are to be expected.

Life-span (wood products): period of time elapsed between manufacture of wood products and their final disposal in landfills, combustion or through decomposition. Some authorities recognise individual product-use categories, having separate life-spans, and the transfer of biomass from one product type to another (as in recycling). Life-spans of wood products can vary from a few days for products such as newsprint, to several centuries for the wood used in historic buildings.

Litter layer: predominantly undecomposed plant residues on top of mineral soil, and distinct from the fermentation and humus layers of the *forest floor* (q.v.). Some authors include coarse woody debris in the litter layer, others account for it separately. The thickness of the litter layer can vary from less than 1 cm to more than 1 m depending on ecosystem, climate, litterfall rates, and disturbance regime.

Merchantable Volume: volume of wood which is potentially saleable for any purpose. Merchantable volume is invariably less than measured stem volume because of losses due to distorted tree stems, damage during harvesting, and some trees being too small to harvest economically. Merchantable wood volume may remain unharvested if there is no market within economic distance of the forest.

Net Ecosystem Production (NEP): the difference between *Net Primary Production* (q.v.) and heterotrophic respiration (of live and dead organic material both above- and below-ground), i.e., the net change in organic matter mass within the entire ecosystem.

Net Primary Production (NPP): that part of the *Gross Primary Production* (q.v.) that remains stored in green plants after deducting the amount used during growth and maintenance respiration. Allocation of remaining carbohydrates after deducting respiration thus results in build-up of plant tissues.

Normal forest: A theoretical *forest* (q.v.) in which there are equal areas of stands within each annual age-class, and where the age of the oldest age-class equals the rotation age. The oldest age-class is felled every year and immediately replanted.

Old growth: term often used in North America for *primary* forest, i.e., forest which has established following natural disturbance and succession, and consisting of stands that usually exceed the normal managed rotation age, sometimes containing multiple canopy layers, and trees of several species and ages. Such forests are typically assumed to be relatively unaffected by past human interference.

Peat: organic material, consisting of plant remains in various states of decomposition, preserved under anaerobic conditions. The inorganic (ash) content is generally low, but may reach 25% by weight.

Peatland: peat-forming *wetlands* (q.v.) (such as bogs, mires, and fens) where only a portion of the ground biomass is decomposed under aerobic conditions, the rest becoming waterlogged and subject to anaerobic decomposition with the result that a continuous accumulation of peat occurs.

Plantation: forest stand established mainly by physical planting of tree seedlings.

Pool: a *reservoir* (q.v.). Any system component which has the capacity to accumulate or release carbon. Examples of carbon pools are forest biomass, wood products, organic soils, fossil fuel deposits and the atmosphere.

Reforestation: establishment of forest on land previously under some form of forest cover, which may have been harvested for merchantable timber, or merely cleared to allow establishment of more valuable species. Differs from *afforestation* (q.v.).

Reservoir: a *pool* (q.v.). Component or components of the *biosphere* (q.v.) or physical climate system where C—or more generally a greenhouse gas, or precursor of a greenhouse gas—is stored.

Scenario: set of prescribed or hypothetical conditions, generally specified for some time in the past or future, which may be used to investigate possible responses of a system, such as an ecosystem or an economy. Often used as a means of system sensitivity analysis.

Sequestration: the process of increasing carbon content of a C *reservoir* (q.v.).

Sink: any process, activity or mechanism which removes a greenhouse gas, an aerosol or a precursor of a greenhouse gas from the atmosphere. A given reservoir can be a sink for atmospheric carbon if, during a given time interval, (1) more carbon is flowing into it than is flowing out, and (2) the changes in the remainder of the connected system are such that there is a net withdrawal of C from the atmosphere.

Soil organic matter: all organic matter from plant residues, not necessarily completely decomposed, that is intimately mixed with mineral soil components. See also *peat* (q.v.).

Source: opposite of *sink* (q.v.). A carbon *reservoir* (q.v.)can be a source of carbon to the atmosphere if, during a given time interval, (1) less carbon is flowing into it than is flowing out, and (2) the changes in the remainder of the connected system are such that there is a net release of C to the atmosphere.

Stand: area of *forest* (q.v.) comprising trees of similar species composition, age-structure and management regime.

Stemwood: portion of a tree's total wood volume contained in the aboveground stem, excluding branches, from which most wood products are manufactured. Estimates of forest stemwood volume are typically derived from field measurements of stem diameter and height, and form the bases for many inventories of forest biomass and carbon content.

Terrestrial biosphere: the biota (plants and animals) living either on land or in non-marine aquatic environments, together with the associated litter and soils. See *biosphere* (q.v.).

Wetland: land area transitional between terrestrial and aquatic ecosystems, where the water table is usually at or near the surface, or the land is covered by shallow water. Accumulation of *peat* (q.v.) may, but need not, occur.

Wood products: harvested forest fibre, usually *stemwood* (q.v.), that is converted into a product after processing the raw wood material. In carbon balance studies, wood products are usually divided into product groups according to application or life-span, e.g., pulpwood and construction timber, or short- and long- lived products.

Yield Tables: Output from models of forest stand growth and yield, reporting periodic estimates of tree volume and other stand parameters, which can be used as the bases for forestry planning and economic analysis. Yield tables are usually derived from statistical models, which predict typical stand growth for a particular species or mixture of species, under specified combinations of site conditions and management.

REFERENCES

The Editors acknowledge the contributions of all the workshop participants in helping to provide definitions for the glossary, in addition to the following sources:

CDIAC (1990) Glossary: carbon dioxide and climate. Carbon Dioxide Information Analysis Center, Oak Ridge National Laboratory, Oak Ridge, TN, 59 pp.

Cowardin LM, Carter V, Golet FC, LaRoe ET (1979) Classification of wetlands and deepwater habitats of the United States. U.S. Dept. Interior, Fish and Wildlife Serv., FWS/OBS-79/31, Washington DC, 103 pp.

Keeling CD, Whorf TP, Wahlen M, van der Plicht J (1995) Interannual extremes in the rate of rise of atmospheric carbon dioxide since 1980. *Nature* 375: 666–670.

Houghton JT, Jenkins GJ, Ephraums JJ (eds) (1990) Climate Change: the IPCC Scientific Assessment. Intergovernmental Panel on Climate Change, WMO/UNEP, Cambridge, UK, 365 pp.

Pickett STA, White P (1985) The Ecology of Natural Disturbances and Patch Dynamics, Academic Press, Orlando FL.

Steffen WL, Walker BH, Ingram JSI, Koch GW (1992) Global change and terrestrial ecosystems: the operational plan. International Geosphere-Biosphere Programme, Report No 21, Stockholm, 95 pp.

UNCED (United Nations Conference on Environment and Development) (1992) Non-legally binding authoritative statement of principles for a global consensus on the management, conservation and sustainable development of all types of forests, United Nations, New York.

List of acronyms and abbreviations

CIFOR	Center for International Research on Forestry
DBH	Diameter at Breast Height
DOE	Department of Energy (USA)
EFI	European Forestry Institute
ENSO	El Niño–Southern Oscillation
EPA	Environmental Protection Agency (USA)
FAO	Food and Agriculture Organisation
FCCC	Framework Convention on Climate Change
FSU	Former Soviet Union
GEF	Global environment facility
GCM	General circulation model
GDD	Growing degree day
GDR	German Democratic Republic
GFDL	Geophysical Fluid Dynamics Laboratory (USA)
GIS	Geographic information system
GISS	Goddard Institute of Space Sciences (USA)
GPP	Gross primary production
GWP	Global warming potential
IIASA	International Institute for Applied Systems Analysis
INPE	Instituto Nacional de Pesquisas Espaciais (Brazil)
IPCC	Intergovernmental Panel on Climate Change
IUCN	World Conservation Union
MAI	Mean annual increment
MSY	Maximum sustainable yield
NATO	North Atlantic Treaty Organization
NPV	Net present value
OECD	Organisation for Economic Cooperation and Development
ORNL	Oak Ridge National Laboratory (USA)
TFAP	Tropical Forestry Action Plan
UKMO	United Kingdom Meteorological Office

UNCED	United Nations Conference on Environment and Development
UNCSD	United Nations Commission on Sustainable Development
UNDP	United Nations Development Programme
UNEP	United Nations Environment Programme
USDA	United States Department of Agriculture
WCED	World Commission on Environment and Development
WCFSD	World Commission on Forests and Sustainable Development
WRI	World Resources Institute
WWF	World Wildlife Fund

List of reviewers

Andrasko KJ, USA
Auclair A, USA
Beck J, Canada
Beren M, UK
Binkley C, Canada
Birdsey RA, USA
Botkin DB, USA
Brown S, USA
Bunnell F, Canada
Callander B, UK
Chapin R, USA
Dewar RC, Australia
Dixon RK, USA
Duinker PN, Canada
Eriksson H, Sweden
Fearnside PM, Brazil
Fischlin A, Switzerland
Fulton M, USA
Fung I, Canada
Galinski W, Poland
Gilles K, USA
Grainger A, UK
Hall CAS, USA
Hall DO, UK
Harmon ME, USA
Helms J, USA
Houghton RA, USA
Humble L, Canada
Ismail R, Malaysia
Kasischke ES, USA
Kauppi P, Finland
Kellomäki S, Finland
King A, USA
Kurz WA, Canada
Lorrain-Smith R, UK
Lugo AE, USA
Maclaren JP, NZ
Maini JS, Canada
Marland G, USA
McNabb D, Canada
Myers N, UK
Nabuurs G-J, Netherlands
Nalder IA, Canada
Ojima D, USA
Overdieck D, Germany
Penner M, Canada
Price C, UK
Rastetter EB, USA
Rojkov VA, Russia
Roland J, Canada
Sampson RN, USA
Sathaye JA, USA
Sedjo RA, USA
Shvidenko AZ, Russia
Solberg B, Finland
Solomon AM, USA
Sykes MT, Sweden
Townsend AR, USA
Trexler MC, USA
Volney WJA, Canada
Wein R, Canada
Woodwell GM, USA
Zoltai SC, Canada

Author index

Subject index

D

E

H

I

L

M

N

T

U

V

W

Y

Z

The ASI Series Books Published as a Result of Activities of the Special Programme on Global Environmental Change

This book contains the proceedings of a NATO Advanced Research Workshop held within the activities of the NATO Special Programme on Global Environmental Change, which started in 1991 under the auspices of the NATO Science Committee.

The volumes published as a result of the activities of the Special Programme are:

Vol. 1: Global Environmental Change.
Edited by R. W. Corell and P. A. Anderson. 1991.
Vol. 2: The Last Deglaciation: Absolute and Radiocarbon Chronologies.
Edited by E. Bard and W. S. Broecker. 1992.
Vol. 3: Start of a Glacial. Edited by G. J. Kukla and E. Went. 1992.
Vol. 4: Interactions of C, N, P and S Biogeochemical Cycles and Global Change.
Edited by R. Wollast, F. T. Mackenzie and L. Chou. 1993.
Vol. 5: Energy and Water Cycles in the Climate System.
Edited by E. Raschke and D. Jacob. 1993.
Vol. 6: Prediction of Interannual Climate Variations.
Edited by J. Shukla. 1993.
Vol. 7: The Tropospheric Chemistry of Ozone in the Polar Regions.
Edited by H. Niki and K. H. Becker. 1993.
Vol. 8: The Role of the Stratosphere in Global Change.
Edited by M.-L. Chanin. 1993.
Vol. 9: High Spectral Resolution Infrared Remote Sensing for Earth's Weather and Climate Studies.
Edited by A. Chedin, M.T. Chahine and N.A. Scott. 1993.
Vol. 10: Towards a Model of Ocean Biogeochemical Processes.
Edited by G. T. Evans and M. J. R. Fasham. 1993.
Vol. 11: Modelling Oceanic Climate Interactions.
Edited by J. Willebrand and D. L.T. Anderson. 1993.
Vol. 12: Ice in the Climate System. Edited by W. Richard Peltier. 1993.
Vol. 13: Atmospheric Methane: Sources, Sinks, and Role in Global Change.
Edited by M. A. K. Khalil. 1993.
Vol. 14: The Role of Regional Organizations in the Context of Climate Change.
Edited by M. H. Glantz. 1993.
Vol. 15: The Global Carbon Cycle.
Edited by M. Heimann. 1993.
Vol. 16: Interacting Stresses on Plants in a Changing Climate.
Edited by M. B. Jackson and C. R. Black. 1993.
Vol. 17: Carbon Cycling in the Glacial Ocean: Constraints on the Ocean's Role in Global Change.
Edited by R. Zahn, T. F. Pedersen, M. A. Kaminski and L. Labeyrie. 1994.
Vol. 18: Stratospheric Ozone Depletion/UV-B Radiation in the Biosphere.
Edited by R. H. Biggs and M. E. B. Joyner. 1994.
Vol. 19: Data Assimilation: Tools for Modelling the Ocean in a Global Change Perspective.
Edited by P. O. Brasseur and J. Nihoul. 1994.

Vol. 20: Biodiversity, Temperate Ecosystems, and Global Change.
Edited by T. J. B. Boyle and C. E. B. Boyle. 1994.

Vol. 21: Low-Temperature Chemistry of the Atmosphere.
Edited by G. K. Moortgat, A. J. Barnes, G. Le Bras and J. R. Sodeau. 1994.

Vol. 22: Long-Term Climatic Variations – Data and Modelling.
Edited by J.-C. Duplessy and M.-T. Spyridakis. 1994.

Vol. 23: Soil Responses to Climate Change.
Edited by M. D. A. Rounsevell and P. J. Loveland. 1994.

Vol. 24: Remote Sensing and Global Climate Change.
Edited by R. A. Vaughan and A. P. Cracknell. 1994.

Vol. 25: The Solar Engine and Its Influence on Terrestrial Atmosphere and Climate.
Edited by E. Nesme-Ribes. 1994.

Vol. 26: Global Precipitations and Climate Change.
Edited by M. Desbois and F. Désalmand. 1994.

Vol. 27: Cenozoic Plants and Climates of the Arctic.
Edited by M. C. Boulter and H. C. Fisher. 1994.

Vol. 28: Evaluating and Monitoring the Health of Large-Scale Ecosystems.
Edited by D. J. Rapport, C. L. Gaudet and P. Calow. 1995.

Vol. 29: Global Environmental Change Science: Education and Training.
Edited by D. J. Waddington. 1995.

Vol. 30: Ice Core Studies of Global Biogeochemical Cycles.
Edited by R. J. Delmas. 1995.

Vol. 31: The Role of Water and the Hydrological Cycle in Global Change.
Edited by H. R. Oliver and S. A. Oliver. 1995.

Vol. 32: Atmospheric Ozone as a Climate Gas.
Edited by W.-C. Wang and I. S. A. Isaksen. 1995.

Vol. 33: Carbon Sequestration in the Biosphere.
Edited by M. A. Beran. 1995.

Vol. 34: Climate Sensitivity to Radiative Perturbations: Physical Mechanisms and Their Validation.
Edited by H. Le Treut. 1996

Vol. 35: Clouds, Chemistry and Climate.
Edited by P. J. Crutzen and V. Ramanathan. 1996.

Vol. 36: Diachronic Climatic Impacts on Water Resources.
Edited by A. N. Angelakis and A. S. Issar. 1996.

Vol. 37: Climate Change and World Food Security.
Edited by T. E. Downing. 1996.

Vol. 38: Evaluation of Soil Organic Matter Models.
Edited by D. S. Powlson, P. Smith and J. U. Smith. 1996.

Vol. 39: Microbiology of Atmospheric Trace Gases.
Edited by J. C. Murrell and D. P. Kelly. 1996.

Vol. 40: Forest Ecosystems, Forest Management and the Global Carbon Cycle.
Edited by M. J. Apps and D. T. Price. 1996.

Vol. 41: Climate Variations and Forcing Mechanisms of the Last 2000 Years.
Edited by P. D. Jones, R. S. Bradley and J. Jouzel. 1996

Springer-Verlag and the Environment

We at Springer-Verlag firmly believe that an international science publisher has a special obligation to the environment, and our corporate policies consistently reflect this conviction.

We also expect our business partners – paper mills, printers, packaging manufacturers, etc. – to commit themselves to using environmentally friendly materials and production processes.

The paper in this book is made from low- or no-chlorine pulp and is acid free, in conformance with international standards for paper permanency.